Andrés Cano, Dennis Meier, Morgan Trassin (Eds.)

Multiferroics

Also of interest

Intelligent Materials and Structures
Abramovich, 2021
ISBN 978-3-11-072669-5, e-ISBN 978-3-11-072670-1

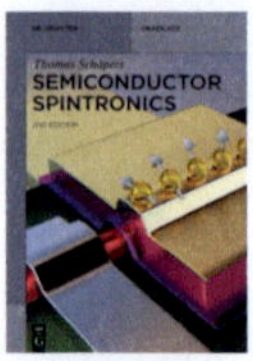

Semiconductor Spintronics
Schäpers, 2021
ISBN 978-3-11-063887-5, e-ISBN 978-3-11-063900-1

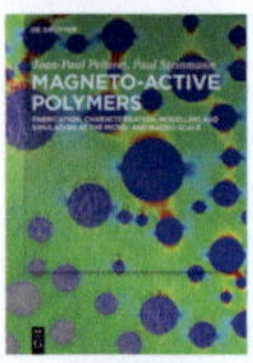

Magneto-Active Polymers.
Fabrication, Characterisation, Modelling and Simulation at the Micro- and Macro-Scale
Pelteret, Steinmann, 2020
ISBN 978-3-11-041951-1, e-ISBN 978-3-11-041857-6

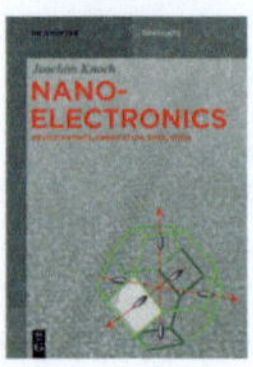

Nanoelectronics.
Device Physics, Fabrication, Simulation
Knoch, 2021
ISBN 978-3-11-057421-0, e-ISBN 978-3-11-057550-7

Multiferroics

Fundamentals and Applications

Edited by
Andrés Cano, Dennis Meier, Morgan Trassin

DE GRUYTER

Editors
Prof. Dr. Andrés Cano
CNRS
Université Grenobles Alpes
Institut Néel
38042 Grenoble
France
andres.cano@cnrs.fr

Prof. Dr. Dennis Meier
Norwegian University of Science and Technology
Department of Materials Science and Engineering
Sem Sælandsvei 12
7034 Trondheim
Norway
dennis.meier@ntnu.no

Dr. Morgan Trassin
ETH Zürich
Department of Materials
Vladimir-Prelog-Weg 1-5/10
8093 Zurich
Switzerland
morgan.trassin@mat.ethz.ch

ISBN 978-3-11-058097-6
e-ISBN (PDF) 978-3-11-058213-0
e-ISBN (EPUB) 978-3-11-058104-1

Library of Congress Control Number: 2021935414

Bibliographic information published by the Deutsche Nationalbibliothek
The Deutsche Nationalbibliothek lists this publication in the Deutsche Nationalbibliografie; detailed bibliographic data are available on the Internet at http://dnb.dnb.de.

Cover image: gonin/Gettyimages
Typesetting: Integra Software Services Pvt. Ltd.
Printing and binding: CPI books GmbH, Leck

www.degruyter.com

Preface

Progress in modern technology is driven by the discovery of new material systems and physical interactions that improve the efficiency and functionality of devices. Many exciting developments have occurred in the past two decades with the potential to revolutionize the design of devices for future information and communication technologies, promoting reduced feature size, high storage density, and low-energy consumption.

One particularly fascinating class of materials that offers advanced functional properties is multiferroics. Multiferroics simultaneously exhibit two or more ferroic orders, such as ferroelectricity, (anti-)ferromagnetism and ferroelasticity, enabling unusual interactions between spin, charge, and lattice degrees of freedom. Although the rich physics associated with the interplay between coexisting ferroic orders has fascinated scientists since the first half of the 20th century, the field truly matured only over the last two decades. Today, a fundamental understanding of the complex microscopic interactions that give rise to multiferroicity is established and it has become clear how the new physical properties can be leveraged in devices. From a technological point of view, the emergence and coupling of electric and magnetic order in multiferroics is arguably the most appealing property as it enables the electric-field control of magnetism. This possibility represents the *holy grail* in the field. The related research has led a number of advances and ramifications in a broader context, providing a promising pathway towards low-energy control of magnetically stored information and the design of power-efficient electronic devices in general.

The goal of this book is to provide a state-of-the-art summary, serving as a comprehensive reference regarding key developments in the field of multiferroics and promising future research directions. Bringing together internationally leading researchers covering both experiment and theory, the book provides a historical perspective, a comprehensive overview about the fundamental properties of multiferroic materials, as well as insight into envisioned applications.

We hope that our book will inspire newcomers to explore the fascinating world of multiferroics and help the researchers active in the field to stay up to date regarding the recent developments, modern trends, and open challenges. We thank all authors for their commitment and excellent contributions that made this project possible. Thanks to them, and the outstanding quality of their chapters, this book gathers a remarkable expertise in the theoretical description, material design, characterization and device integration of multiferroic materials. We hope that you will enjoy reading about their perspectives on the research on multiferroics and possibly become curious to learn more about multiferroics.

Andrés Cano
Dennis Meier
Morgan Trassin

https://doi.org/10.1515/9783110582130-202

Contents

List of Contributing Authors

Manuel Bibes
CNRS, Thales
Université Paris-Saclay
Unité Mixte de Physique
Palaiseau, France
manuel.bibes@cnrs-thales.fr

Eric Bousquet
Physique Théorique des Matériaux
Q-MAT, CESAM,
Université de Liège
Sart-Tilman, Belgium
eric.bousquet@uliege.be

Nicholas C. Bristowe
School of Physical Sciences
University of Kent
Canterbury, UK
n.c.bristowe@kent.ac.uk

Marco Campanini
Electron Microscopy Center, Swiss Federal Laboratories for Materials Science and Technology, Empa, Dübendorf, Zurich, Switzerland
Marco.Campanini@empa.ch

Andrés Cano
Univ. Grenoble Alpes
CNRS, Grenoble INP
Institut Néel
Grenoble, France
andres.cano@cnrs.fr

Margarita D. Davydova
Physics, Massachusetts Institute of Technology
182 Memorial Dr
Cambridge, MA, USA
davydova@phystech.edu

Rolf Erni
Electron Microscopy Center, Swiss Federal Laboratories for Materials Science and Technology, Empa, Dübendorf, Zurich, Switzerland
Rolf.Erni@empa.ch

Donald M. Evans
Department of Materials Science and Engineering
Norwegian University of Science and Technology (NTNU)
Trondheim, Norway
donald.evans@ntnu.no

Vincent Garcia
CNRS, Thales
Université Paris-Saclay
Unité Mixte de Physique
Palaiseau, France
vincent.garcia@cnrs-thales.fr

Philippe Ghosez
Theoretical Materials Physics
Université de Liège
Sart Tilman, Belgium
Philippe.Ghosez@ulg.ac.be

Elzbieta Gradauskaite
Department of Materials
ETH Zurich
Zurich, Switzerland
elzbieta.gradauskaite@mat.ethz.ch

John Heron
Department of Materials Science and Engineering
University of Michigan
Ann Arbor, MI, USA
jtheron@umich.edu

https://doi.org/10.1515/9783110582130-204

Stephan Krohns
Experimental Physics V
University of Augsburg
Augsburg, Germany
stephan.krohns@physik.uni-augsburg.de

Takashi Kurumaji
Physics, Massachusetts Institute of Technology
Cambridge, MA, USA
kurumaji@mit.eduw

Artem M. Kuzmenko
Prokhorov General Physics Institute
Russian Academy of Sciences
Moscow, Russia
kuzmenko.am@phystech.edu

Bernd Lorenz
Texas Center for Superconductivity
University of Houston
Houston, TX, USA
belorenz50@gmail.com

Thomas Lottermoser
Department of Materials
ETH Zurich
Zurich, Switzerland
thomas.lottermoser@mat.ethz.ch

Peter Lunkenheimer
Experimental Physics V
University of Augsburg
Augsburg, Germany
peter.lunkenheimer@physik.uni-augsburg.de

Dennis Meier
Department of Materials Science and Engineering,
Norwegian University of Science and Technology (NTNU)
Trondheim, Norway
dennis.meier@ntnu.no

Peter Meisenheimer
Department of Materials Science and Engineering
University of Michigan
Ann Arbor, MI, USA

Masahito Mochizuki
Department of Applied Physics
Waseda University
Tokyo, Japan
masa_mochizuki@waseda.jp

Alexander Mukhin
Prokhorov General Physics Institute of the Russian Academy of Sciences
Moscow, Russia
mukhin@ran.gpi.ru

Marvin Müller
Department of Materials
ETH Zurich
Zurich, Switzerland
marvin.mueller@mat.ethz.ch

Andrei Pimenov
Institute of Solid State Physics, Vienna University of Technology
Vienna, Austria
pimenov@ifp.tuwien.ac.at

Marta D. Rossell
Electron Microscopy Center, Swiss Federal Laboratories for Materials Science and Technology, Empa, Dübendorf, Zurich, Switzerland
marta.rossell@empa.ch

Alexey Shuvaev
Institute of Solid State Physics, Vienna University of Technology
Vienna, Austria
alexey.shuvaev@ifp.tuwien.ac.at

David Szaller
Institute of Solid State Physics, Vienna University of Technology
Vienna, Austria
david.szaller@tuwien.ac.at

Morgan Trassin
Department of Materials
ETH Zurich
Zurich, Switzerland
morgan.trassin@mat.ethz.ch

Julien Varignon
Theoretical Materials Physics
Université de Liège
Sart Tilman, Belgium
julien.varignon@gmail.com

Anatoly K. Zvezdin
Prokhorov General Physics Institute of the
Russian Academy of Sciences
Moscow, Russia
zvezdin.ak@phystech.edu

Konstantin A. Zvezdin
Prokhorov General Physics Institute of the
Russian Academy of Sciences
Moscow, Russia
zvezdin.ka@phystech.edu

Thomas Lottermoser and Dennis Meier

1 A short history of multiferroics

Abstract: The realization that materials with coexisting magnetic and ferroelectric order open up efficient ways to control magnetism by electric fields unites scientists from different communities in the effort to explore the phenomenon of multiferroics. Following a tremendous development, the field has now gained some maturity. In this article, we give a succinct review of the history of this exciting class of materials and its evolution from “ferroelectromagnets” to “multiferroics” and beyond.

Keywords: multiferroic, ferroelectric, ferromagnetic, magnetoelectric

1.1 From ferroelectromagnets to multiferroics

Hans Schmid coined the term *multiferroic* in 1993 in Ascona [1], complementing the earlier classification of *ferroic* materials by Aizu [2] . He defined multiferroics as materials that unite two or more primary ferroic states (ferroelasticity, ferroelectricity, ferromagnetism and ferrotoroidicity [1, 3]) in the same phase (Figure 1.1). A subset of these multiferroic materials is also magnetoelectric, i.e. these materials display a coupling between their electric and magnetic properties [5]. This coupling is the main reason for the world-wide interest in multiferroics, as it enables the control of magnetic order by electric fields and vice versa. While the term *magnetoelectric* originally referred to a linear coupling between electric (magnetic) field and magnetization (electric polarization), we nowadays include all types of coupling phenomena that occur between charge and spin degrees of freedom when talking about magnetoelectric multiferroics [6]. Furthermore, going beyond just primary ferroic states, the initial concept of coexisting orders has been expanded, now also including, e.g. antiferromagnetism and multi-phase materials like laminates, solid solutions, and layered (thin film) architectures [7–12]. In this work, we will use this modern interpretation when referring to multiferroics.

Although the term multiferroics appeared in literature only around the year 2000, it is important to note that the hunt for a strong coupling of magnetic and electric degrees of freedom as basis of novel voltage-controlled low-power magnetic devices began already decades earlier [13]. The research on new types of electric and magnetic long-range order really flourished during the first half of the 20th century; two outstanding events that date back to this time are the experimental discovery of

This article has previously been published in the journal Physical Sciences Reviews. Please cite as: Lottermoser, T., Meier, M. A short history of multiferroics *Physical Sciences Reviews* [Online] 2021, 6. DOI: 10.1515/psr-2020-0032

https://doi.org/10.1515/9783110582130-001

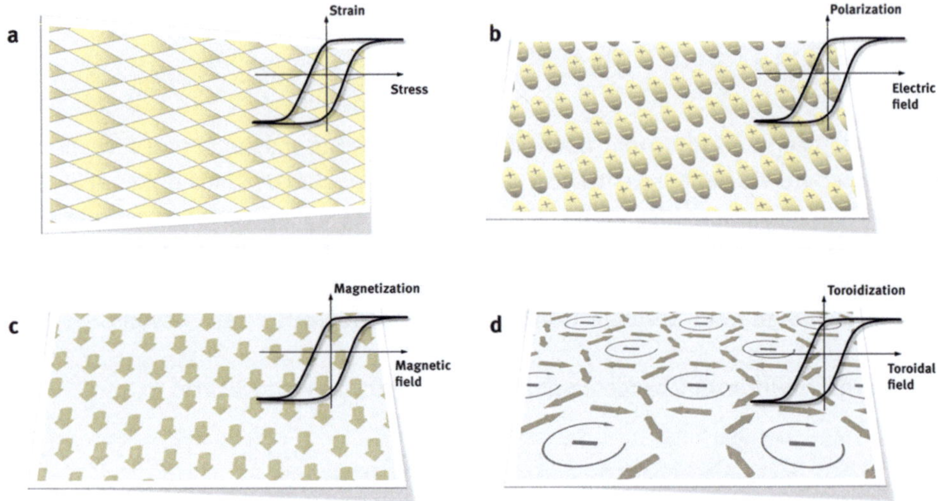

Figure 1.1: **Primary ferroic order.** Four types of ferroic order are classified as primary ferroic properties, namely ferroelasticity (a), ferroelectricity (b), ferromagnetism (c), and ferrotoroidicity (d) [1]. (a) Ferroelastics exhibit spontaneous strain, σ, which can be switched between equally stable states by application of stress. (b) Ferroelectrics develop a spontaneous electric polarisation, *P*, which switches under application of an electric field. (c) In ferromagnets, the spontaneous alignment of spins results in a macroscopic magnetic moment, *M*. This spontaneous magnetization can be switched by a magnetic field. (d) Ferrotoroidicity is discussed as fourth type of primary ferroic order [3, 4]. For example, ferro-toroidics may exhibit a vortex-like alignment of spins with a toroidization (*T*). The toroidal field required to switch the order is of the form *E* x *H*, where *E* and *H* are the electric and magnetic field, respectively. In the classical definition, a multiferroic material simultaneously shows two or more of these ferroic properties in the same phase. Nowadays, however, the term is used in a much broader context as discussed in the main text. Insets: A key characteristic of any primary ferroic order is its hysteretic response to the conjugated field (e.g. $P \leftrightarrow E$, $M \leftrightarrow H$). In the ideal multiferroic, *P* and *M* are coupled so that the magnetic order, *M*, can be switched by an electric field and the electric order, *P*, by a magnetic field *H*.

ferroelectricity by Valasek [14] and Néel's seminal work on antiferromagnetism [15]. The theoretical description of ferroelectricity and antiferromagnetism progressed rapidly, but measurements were challenging. Thus, from an experimental point of view, these physical phenomena were still rather new in the 1950s when first efforts to combine magnetic and ferroelectric order were pursued in the former Soviet Union. Smolenskii and Ioffe suggested to introduce magnetic ions into ferroelectric perovskites and create magnetic long-range order while retaining the ferroelectric state [16]. Their research led to the successful synthesis of single-crystals like $Pb(Fe_{0.5}Nb_{0.5})O_3$ and polycrystalline solid-solutions like $(1{-}x)Pb(Fe_{0.66}W_{0.33})O_3{-}xPb(Mg_{0.5}W_{0.5})O_3$, representing first multiferroics that were designed on purpose [17, 18]. Smolenskii and Ioffe referred to these systems as *ferroelectromagnets* (orginally: *seignettomagnets*).

Interestingly, two of the most intensely investigated present-day multiferroics, that is, $BiFeO_3$ and the hexagonal (h-) manganites ($RMnO_3$, $R = Sc$, Y, In, Dy-Lu), have already been identified in the early 1960s [19–22]. The celebrity of the first multiferroics era, however, were the boracites. In 1966, Asher *et al.* observed a colossal linear magnetoelectric effect in $Ni_3B_7O_{13}I$ that allowed for hysteretic switching of a multiferroic state by either electric or magnetic fields [23]. Such experimental findings were complemented by the development of a theoretical framework, predictions about emergent magnetoelectric phenomena, and the proposition of technological applications [24]. The latter were remarkably similar to modern multiferroic-based device paradigms. Moreover, classical theory tools, such as representation analysis and Landau theory, still play a key role for the description of multiferroics. Around 1970 Aizu developed a unifying classification of ferroics [2, 25]. This contributed significantly to the modern understanding of (multi-)ferroics and to Schmid's important definition [1].

Following the peak in research activities in the 1960s, the first multiferroics era petered out about a decade later. By that time, circa 50 multiferroic systems were known [26], none of which exhibited technologically feasible properties. This may explain why researchers eventually lost interest and moved on to other material classes. Of course, the research on multiferroics never stopped completely. One seminal discovery was made in 1978 by Newnham and coworkers, who reported that a spin spiral in Cr_2BeO_4 breaks spatial inversion symmetry and thereby induces a spontaneous electric polarization [27]. On just four pages the authors foreshadow much of the fascinating physics of magnetically driven (improper) ferroelectricity that should be recognized much later as key source for multiferroics with strong magnetoelectric interactions. Five years later, in 1983, Bar'yakhtar *et al.* presented a phenomenological model, elaborating how magnetic order can break inversion symmetry and, hence, induce an electric polarization [28].

The ball was set rolling again when Hans Schmid organized a conference on Magnetoelectric Interaction Phenomena in Crystals (MEIPIC-2) in 1993. The meeting and its fascinating proceedings identified and interrelated many of the phenomena, systems and theories surrounding the magnetoelectric effect [29]. Aspects crucial to the resurgence of multiferroics, such as techniques for imaging multiferroic domains and their interactions, new types of ferroic order and future multiferroic key materials, can all be traced back to MEIPIC-2.

In 2000, Spaldin (then Hill) revisited the original idea of Ioffe and Smolenskii and elaborated why in classical perovskites, displacive ferroelectricity and magnetic order are working against each other [30]. This work and a session at the 2000 March Meeting of the American Physical Society reached out to a broad audience and further prepared the stage for the second era of multiferroics. Encouraged by the knowledge of why previous attempts to expand the pool of multiferroics had stagnated and the interim progress in materials synthesis and characterization, researchers accepted the challenge and resumed the hunt for novel multiferroics of technological value.

First highlights were the discoveries of pronounced magnetoelectric interactions in h-$YMnO_3$ [31, 32], orthorhombic $TbMnO_3$ [33], and $TbMn_2O_5$ [34]. In the latter two materials, the interaction originates from non-centrosymmetric spin textures inducing a magnetically controllable electric polarization, analogous to the case of Cr_2BeO_4. In contrast to the earlier work on Cr_2BeO_4, however, the new findings succeeded in triggering world-wide attention, leading to concerted efforts in different communities, bridging materials science, condensed matter physics, and materials theory. It was this joint effort of a new generation of researchers which promoted multiferroics research to new realms (for a detailed review on the early history of multiferroics, we refer the reader to the extended work in, e.g. [13, 26, 35]).

1.2 The second era of multiferroics

Just as 40 years earlier, the new quest for multiferroics with strongly interacting magnetic and electric order had a strong focus on ferroelectrics. This time, however, the scope was to find novel types of polar states that permit the emergence of coexisting magnetic order – an approach that went far beyond a mere revival of existing concepts. Theoretical and experimental tools had advanced tremendously compared to the first era of multiferroics. It was now possible to understand multiferroicity at the atomic scale and, importantly, design new systems with unprecedented precision and complexity. This was essential for accomplishing the shift away from standard displacive ferroelectrics and towards materials where, e.g. the electric polarization is induced by the spin system rather than counteracting the magnetic ordering [36–38]. Foreshadowing such so-called improper ferroelectrics [39], Levanyuk and Sannikov already mentioned in 1974 that “a complicated change in the crystal or magnetic structure” can induce an electric polarization [40]. Now, scientists began to elaborate such “complicated change” in minute detail and improper ferroelectricity became a key component in the hunt for novel single-phase multiferroics with pronounced magnetoelectric coupling.

From an academic point of view, one of the main achievements associated with the second era of multiferroics was the development of a comprehensive framework that allowed to classify all known materials with respect to the mechanism that drives multiferroicity. Nowadays, we distinguish four classes of multiferroics with ferroelectricity driven by electronic lone pairs, geometry, charge ordering, and magnetism as summarized in Figure 1.2 [12].

The so-called **lone-pair mechanism** is based on the violation of inversion symmetry by valence electrons (Figure 1.2(a)). This mechanism is responsible for the room-temperature ferroelectricity in $BiFeO_3$ ($T_C = 1103\,K$) [46]. Here, two of the Bi^{3+} valence electrons do not participate in chemical sp-hybridized states and create a local dipole, leading to a macroscopic spontaneous electric polarization in the order

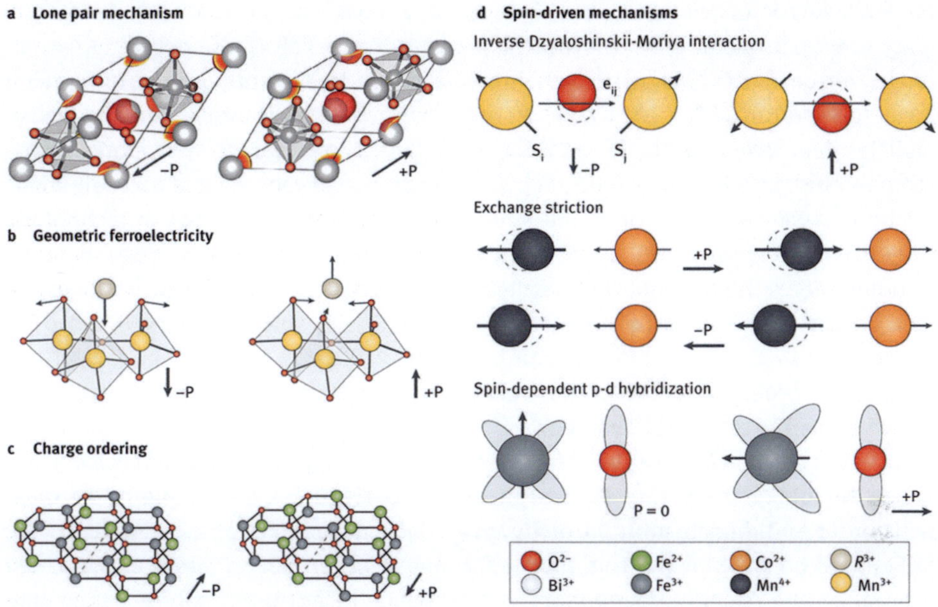

Figure 1.2: Classification of multiferroic materials. (a) Lone-pair mechanism: In $BiFeO_3$, two electrons shift away from the Bi^{3+} ion and towards the FeO_6 octahedra. This *lone pair* (illustrated by the red isosurface of the electron localization function) induces a spontaneous polarization, *P*, in the [111] direction. (b) Geometric ferroelectricity: Movements of the MnO_5 bipyramids in h-$RMnO_3$ lead to a displacement of the *R*-ions (indicated by black arrows), which leads to a ferroelectric polarization along the [001] axis [41]. (c) Ferroelectricity due to charge ordering: Alternating layers with Fe^{2+}/Fe^{3+} ratios of 2:1 and 1:2 were proposed to give rise to a spontaneous electric polarization in $LuFe_2O_3$ [42]. (d) Magnetically induced ferroelectricity: The inverse Dzyaloshinskii–Moriya interaction can lead to a canting of neighbouring magnetic spins, S_i and S_j, and thereby drive a polar displacement [43] as illustrated at the top (example: orthorhombic $TbMnO_3$ [33]). Alternatively, ferroelectricity can arise due to symmetric spin exchange as in Ca_3CoMnO_6 (middle) [44] or spin-driven modulations in chemical bonding (bottom) like in $CuFeO_2$ [45]. The Figure is reprinted with permission from Springer Nature, taken from [12]. Copyright 2016 by Springer Nature.

of 100 μC/cm^2 [47] . The polarization is the primary symmetry breaking order parameter, classifying $BiFeO_3$ as a *proper* ferroelectric. Antiferromagnetic G-type order with an additional long-range modulation and a small canted moment arises below $T_N = 643\,K$ in $BiFeO_3$ [20]. With this, $BiFeO_3$ has been holding the record for single-phase materials for more than half a century, exhibiting a large electric polarization and robust magnetoelectric coupling at room temperature. Interestingly, $BiFeO_3$ is still the only established system of its kind and all attempts to achieve further multiferroics of lone-pair type failed, including most promising candidates such as $BiMnO_3$ [48, 49].

Size effects and geometrical constraints can cause structural instabilities that lead to polar distortions and **geometric ferroelectricity** (Figure 1.2(b)). In h-$RMnO_3$, for example, a unit-cell-trimerizing lattice distortion is the driving force of the ferroelectric phase transition ($T_C \geq 1000$ K, $P \approx 6$ μC/cm^2) [41, 50–53]. As the primary order parameter is not the electric polarization, h-RMnO$_3$ are referred to as *improper* ferroelectrics. Magnetic ordering emerges independently and only at much lower temperatures ($T_N \approx 100$ K) [54]. Very similar behaviour is observed in the hexagonal ferrites, RFeO$_3$, with the additional advantage of a larger spin moment and room-temperature magnetism [55, 56]. The emergence of room-temperature multiferroicity in h-RFeO$_3$ is appealing, but strong magnetoelectric couplings at technologically relevant length scales, i.e. at the level of domains, are yet to be demonstrated. Another interesting material with geometric ferroelectricity is $BaNiF_4$, where an asymmetry of Ba^{2+} and F^- sites induces ferroelectricity [57, 58]. Although the geometric ferroelectricity in $BaNiF_4$ is too small to be of use for device applications (≈ 0.01 μC/cm^2), it is of profound interest as it involves weak magnetic order that can be reversed by switching the electric polarization.

Charge carriers can localize and form a non-centrosymmetric superlattice, leading to **ferroelectricity due to charge ordering** (Figure 1.2(c)). Long-range magnetic order and, hence, multiferroicity may arise at a separate phase transition [59]. $LuFe_2O_4$ is considered the role model for charge-order driven multiferroicity, but even after one decade of research the emergence of ferroelectricity is called into question and continues to stimulate controversial debates [42, 60]. Mixed manganites, such as $Pr_{1-x}Ca_xMnO_3$, $Y_{1-x}Ca_xMnO_3$, and $Pr(Sr_{0.1}Ca_{0.9})_2Mn_2O_7$ were discussed as well, but did not attract broader attention [59]. By all indications, charge-order driven multiferroicity is an elegant concept of strong academic interest but will not help us to design competitive functional materials.

Another way to induce acentricity in otherwise centrosymmetric structures is magnetism, which can lead to a spontaneous electric polarization and hence multiferroicity. By definition, **magnetically induced ferroelectricity** is *improper* (or *pseudoproper*) and, in many materials, originates from magnetic frustration. Non-centrosymmetric spin textures due to competing magnetic interactions evolve, for example, in Cr_2BeO_4, $TbMnO_3$, $MnWO_4$, $Ni_3V_2O_8$, $CoCr_2O_4$, and CuO [see, e.g. [9] for a review]. The non-centrosymmetric magnetic order gives rise to the so-called inverse Dzyaloshinskii-Moryia (DM) interaction [43, 61]. Enabled by relativistic spin-orbit coupling, the low magnetic symmetry is projected onto the crystal lattice and induces a small polar displacement. The orientation of the displacement is determined by the chirality of the spin system ($S_i \times S_j \neq 0$), which yields a unique one-to-one correlation between the (antiferro-)magnetic order and the electric polarization [61, 62]. Successful ongoing attempts to achieve larger polarization (> 0.1 μC/cm^2 [63]), higher ordering temperature [64–66], and viable thin film architectures [67] reflect a substantial potential for further development and correlation phenomena that are yet to be harnessed. The inverse DM interaction is the most intensively studied microscopic mechanism that leads to magnetically driven ferroelectricity but it is certainly not the only one. In the delafossites $CuFe_{1-x}Rh_xO_2$ ($0 \leq x \leq 0.15$) a spontaneous polarization of about 0.2 μC/cm^2 is induced by a screw-like spin structure ($S_i \times S_j = 0$), being driven by a complex combination of spin-orbit interaction and spin helicity [45]. Inherently larger polarization values arise in collinear magnets with non-relativistic Heisenberg-like exchange striction ($\propto S_i \cdot S_j$), such as YMn_2O_5 [68, 69], orthorhombic $HoMnO_3$ [70], and the spinel CdV_2O_4 [71]. Up to now, however, ferroelectricity due to collinear magnetism has only been observed at low temperature so that technological applications remain elusive.

In summary, comparing multiferroics with ferroelectricity driven by different mechanisms, $BiFeO_3$ and its lone-pair mechanism (Figure 1.2(a)) is still most promising when it comes to device applications. However, the pool of multiferroics with ferroelectricity due to lone-pairs has never widened since the discovery of $BiFeO_3$ in the 1960s. In contrast, the number of spin-driven multiferroics recently exploded and robust room-temperature systems with significantly improved electric and magnetic properties appear to be within reach. Thus, given the current development, it is reasonable to say that spin-driven multiferroicity may play an equally important role and possibly even dethrone the lone-pair mechanism in the future.

1.3 What's next?

The revival of multiferroics [12, 72, 73] during the second era led to a comprehensive understanding of the mechanisms that facilitate coexisting electric and magnetic order (Figure 1.2), as well as conceptually new design strategies for device architectures [67, 74, 75]. Thus, this era brought us an important step closer to multiferroic-based technology. Although the research field has truly matured over the last two decades, the race for *ideal* multiferroics is still on and scientists keep searching for the perfect material that enables low-energy electric field control of magnetism at room-temperature. In addition, research efforts that were used to focus on multiferroics are now expanding into other fields, ranging from basic cosmology-related questions [76] to novel concepts that may revolutionize information and communication technologies [77].

It is thus a perfect time to take a step back and recap what we already know about the basics of multiferroicity, available model materials, as well as opportunities for next-generation technology. For this purpose, key aspects related to the fundamentals and applications of multiferroics are reviewed in a comprehensive series of topical articles. Here, recent developments in different multiferroics are discussed, covering materials where electric and magnetic order emerge independently (type I) in Chapter 2 and 3 or jointly (type II) in Chapter 4 and Chapter 5. Other articles from the series review the characterization of multiferroics at macro- and nanoscopic length scales (Chapter 6, Chapter 7), excitations (Chapter 8, Chapter 9), and novel materials (Chapter 10) as well as domain and domain wall related phenomena (Chapter 11) and recent progress in thin films for device applications (Chapter 12).

On the one hand, the series of topical reviews is of interest for specialists to keep an overview of key discoveries within the field, despite the exploding number of publications on multiferroics. On the other hand, the comprehensive collection of articles can serve as a solid foundation for students and newcomers who are just entering the field. Because one thing is clear: the exciting journey that once started with ferroelectromagnets is by far not over yet. In fact, the third era – which will take us the beyond the classical multiferroics research as we know it from the past – is just beginning. First intriguing precursors associated with this third era of multiferroics are, for example, non-trivial topological textures such as magnetoelectric skyrmions [78, 79] and hybrid domain walls [80, 81], higher-order correlation phenomena at the level of domains [82] and emergent chemical phases at the nanoscale [83].

References

[1] Schmid H. Multi-ferroic magnetoelectrics. Ferroelectrics. 1994;162:317.
[2] Aizu K. Possible species of ferromagnetic, ferroelectric, and ferroelastic crystals. Phys Rev B. 1970;2:754–72.
[3] Schmid H. On ferrotoroidics and electrotoroidic, magnetotoroidic and piezotoroidic effects. Ferroelectrics. 2001;252:41–50.
[4] Litvin DB. Ferroic classifications extended to ferrotoroidic crystals. Acta Crystallogr Sect A. 2008;64:316–20.
[5] Dzyaloshinskii IE. On the magneto-electrical effect in antiferromagnets. Sov Phys JETP. 1960;10:628–9.
[6] Eerenstein W, Mathur ND, Scott JF. Multiferroic and magnetoelectric materials. Nature. 2006;442:759–65.
[7] Khomskii D. Classifying multiferroics: mechanisms and effects. Physics. 2009;2:20.
[8] Nan C-W, Bichurin MI, Dong S, Viehland D, Srinivasan G. Multiferroic magnetoelectric composites: historical perspective, status, and future directions. J Appl Phys. 2008;103:031101.
[9] Tokura Y, Seki S. Multiferroics with spiral spin orders. Adv Mater. 2010;22:1554–65.
[10] Barone P, Picozzi S. Mechanisms and origin of multiferroicity. Comptes Rendus Phys. 2015;16:143–52.
[11] Ma J, Hu J, Li Z, Nan C-W. Recent progress in multiferroic magnetoelectric composites: from bulk to thin films. Adv Mater. 2011;23:1062–87.
[12] Fiebig M, Lottermoser T, Meier D, Trassin M. The evolution of multiferroics. Nat Rev Mater. 2016;1:16046.
[13] Schmid H. The dice - stone 'Der Würfelstein': some personal souvenirs around the discovery of the first ferromagnetic ferroelectric. Ferroelectrics. 2012;427:1–33.
[14] Valasek J, Piezo-electric and allied phenomena in rochelle salt. Phys Rev. 1921;17:475.
[15] Néel L. Propriétés magnétiques de l'etat métallique et energie d'interaction entre atomes magnétiques. Ann Phys. 1936;11:232.
[16] Smolenskii GA, Ioffe VA Communications de Colloque International de Magnetism de Grenoble (France), 2-6 Jullet. Communication No. 1. 1958.
[17] Bokov VA, Myl'nikova IE, Smolenskii GA. Ferroelectric-antiferromagnetic crystals. Zhurnal Eksp I Teor Fiz. 1962;42:643–6.
[18] Smolenskii GA, Isupov VA, Krainik NN, Agranovskaya AI. The coexistence of the ferroelectric and ferromagnetic states. Izv Akad Nauk SSSR, Seriya Fiz. 1961;25:1333–9.

[19] Venevtsev YN, Zhdanov GS, Solov'ev SN, Bezus EV, Ivanova VV, Fedulov, SA, et al. Crystal chemical studies of substances with perovskite-type structure and special dielectric properties. Kristallografiya. 1960;5:620–6.

[20] Kiselev SV, Ozerov RP, Zhdanov GS. Detection of magnetic order in ferroelectric $BiFeO_3$ by neutron diffraction. Sov Phys Dokl. 1963;7:742.

[21] Bertaut EF, Forrat F, Fang P. Les manganites de terres rares et d'ytrium: une nouvelle classe de ferroélectriques. Comptes Rendus Acad Sci. 1963;256:1958–61.

[22] Coeuré P, Guinet F, Peuzin JC, Buisson G, Bertaut EF Ferroelectric properties of hexagonal orthomanganites of yttrium and rare earths. Proc Int Meet Ferroelectr Prag. 1966.

[23] Ascher E, Rieder H, Schmid H, Stössel H. Some properties of ferromagnetoelectric nickel-iodine boracite, $Ni_3B_7O_{13}I$. J Appl Phys. 1966;37:1404–5.

[24] Wood VE, Austin AE. Possible applications for magnetoelectric materials. Int J Magn. 1974;5:303–15.

[25] Aizu K. Possible species of 'ferroelastic' crystals and of simultaneously ferroelectric and ferroelastic crystals. J Phys Soc Japan. 1969;27:387.

[26] Smolenskii GA, Chupis IE. Ferroelectromagnets. Sov Phys Uspekhi. 1982;25:475–93.

[27] Newnham RE, Kramer JJ, Schulze WA, Cross LE. Magnetoferroelectricity in Cr_2BeO_4. J Appl Phys. 1978;49:6088–91.

[28] Bar'yakhtar VG, L'vov VA, Yablonskii DA. Inhomogeneous magnetoelectric effect. JETP Leters. 1983;37:673–5.

[29] Magnetoelectric interaction phenomena in crystals: Proceedings of the 2nd international conference (MAIPIC-2), Ascona, 1993. Gordon and Breach Science Publishers. 1994.

[30] Hill NA. Why are there so few magnetic ferroelectrics? J Phys Chem B. 2000;104:6694–709.

[31] Fiebig M, Lottermoser T, Fröhlich D, Goltsev AV, Pisarev RV. Observation of coupled magnetic and electric domains. Nature. 2002;419:818–20.

[32] Lottermoser T, Lonkai T, Amann U, Hohlwein D, Ihringer J, Fiebig M. Magnetic phase control by an electric field. Nature. 2004;430:541–5.

[33] Kimura T, Goto T, Shintani H, Ishizaka K, Arima T, Tokura Y. Magnetic control of ferroelectric polarization. Nature. 2003;426:55–8.

[34] Hur N, Park S, Sharma PA, Ahn JS, Guha S, Cheong SW. Electric polarization reversal and memory in a multiferroic material induced by magnetic fields. Nature. 2004;429:392–5.

[35] Chupis IE Ferroelectromagnets. Fifty years after discovery. arXiv 1–109. 2010.

[36] Kimura T. Spiral magnets as magnetoelectrics. Annu Rev Mater Res. 2007;37:387–413.

[37] Tokura Y, Seki S, Nagaosa N. Multiferroics of spin origin. Reports Prog Phys. 2014;77:076501.

[38] Mostovoy M. Multiferroics: a whirlwind of opportunities. Nat Mater. 2010;9:188–90.

[39] Dvořák V. Improper ferroelectrics. Ferroelectrics. 1974;7:1–9.

[40] Levanyuk AP, Sannikov DG. Improper ferroelectrics. Sov Phys Uspekhi. 1974;17:199–214.

[41] Van Aken BB, Palstra TT, Filippetti A, Spaldin NA. The origin of ferroelectricity in magnetoelectric $YMnO_3$. Nat Mater. 2004;3:164–70.

[42] De Groot J, Mueller T, Rosenberg RA, Keavney DJ, Islam Z, Kim J-W, et al. Charge order in $LuFe_2O_4$: an unlikely route to ferroelectricity. Phys Rev Lett. 2012;108:187601.

[43] Sergienko IA, Dagotto E. Role of the Dzyaloshinskii-Moriya interaction in multiferroic perovskites. Phys Rev B. 2006;73:094434.

[44] Choi YJ, Yi HT, Lee S, Huang Q, Kiryukhin V, Cheong SW. Ferroelectricity in an ising chain magnet. Phys Rev Lett. 2008;100:047601.

[45] Arima T. Ferroelectricity induced by proper-screw type magnetic order. J Phys Soc Japan. 2007;76:073702.

[46] Fedulov SA. Determination of Curie temperature for $BiFeO_3$ ferroelectric. Dokl Akad Nauk SSSR. 1961;139:1345.

[47] Wang J, et al. Epitaxial $BiFeO_3$ multiferroic thin film heterostructures. Science. 2003;299:1719–22.

[48] Belik AA. Polar and nonpolar phases of $BiMO_3$: a review. J Solid State Chem. 2012;195:32–40.

[49] Yang H, Chi ZH, Jiang JL, Feng WJ, Cao Z, Xian T, et al. Centrosymmetric crystal structure of $BiMnO_3$ studied by transmission electron microscopy and theoretical simulations. J Alloys Compd. 2008;461:1–5.

[50] Fennie CJ, Rabe K. Ferroelectric transition in $YMnO_3$ from first principles. Phys Rev B. 2005;72:100103.

[51] Lilienblum M, Lottermoser T, Manz S, Selbach SM, Cano A, Fiebig M. Ferroelectricity in the multiferroic hexagonal manganites. Nat Phys. 2015;11:1070–3.

[52] Artyukhin S, Delaney KT, Spaldin NA, Mostovoy M. Landau theory of topological defects in multiferroic hexagonal manganites. Nat Mater. 2014;13:42–9.

[53] Cano A. Hidden order in hexagonal $RMnO_3$ multiferroics (R=Dy–Lu, In, Y, and Sc). Phys Rev B. 2014;89:214107.

[54] Fiebig M, Fröhlich D, Leute S, Lottermoser T, Pavlov VV, Pisarev RV. Determination of magnetic symmetry by optical second-harmonic generation. J Magn Magn Mater. 2001; 226–230:961–2.

[55] Das H, Wysocki AL, Geng Y, Wu W, Fennie CJ. Bulk magnetoelectricity in the hexagonal manganites and ferrites. Nat Commun. 2014;5:2998.

[56] Wang W, Zhao J, Wang W, Gai Z., Balke N, Chi, M, et al. Room-temperature multiferroic hexagonal $LuFeO_3$ Films. Phys Rev Lett. 2013;110:237601.

[57] Eibschütz M, Guggenheim HJ, Wemple SH, Camlibel I, DiDomenico M. Ferroelectricity in $BaM^{2+}F_4$. Phys Lett A. 1969;29:409–10.

[58] Ederer C, Spaldin NA. Electric-field-switchable magnets: the case of $BaNiF_4$. Phys Rev B. 2006;74:020401.

[59] Van den Brink J, Khomskii DI. Multiferroicity due to charge ordering. J Phys Condens Matter. 2008;20:434217.

[60] Ruff A, Krohns S, Schrettle F, Tsurkan V, Lunkenheimer P, Loidl A. Absence of polar order in $LuFe_2O_4$. Eur Phys J B. 2012;85:290.

[61] Katsura H, Nagaosa N, Balatsky AV. Spin current and magnetoelectric effect in noncollinear magnets. Phys Rev Lett. 2005;95:057205.

[62] Mostovoy M. Ferroelectricity in spiral magnets. Phys Rev Lett. 2006;96:067601.

[63] Johnson RD, Chapon LC, Khalyavin DD, Manuel P, Radaelli PG, Martin C. Giant improper ferroelectricity in the ferroaxial magnet $CaMn_7O_{12}$. Phys Rev Lett. 2012;108:067201.

[64] Kitagawa Y, Hiraoka Y, Honda T, Ishikura T. Nakamura H, Kimura T. Low-field magnetoelectric effect at room temperature. Nat Mater. 2010;9:797–802.

[65] Rocquefelte X, Schwarz K, Blaha P, Kumar S, van den Brink J. Room-temperature spin-spiral multiferroicity in high-pressure cupric oxide. Nat Commun. 2013;4:2511.

[66] Jana R, Saha P, Pareek V, Basu A, Kapri S, Bhattacharyya, S, et al. High pressure experimental studies on CuO: indication of re-entrant multiferroicity at room temperature. Sci Rep. 2016;6:1–8.

[67] Trassin M. Low energy consumption spintronics using multiferroic heterostructures. J Phys Condens Matter. 2016;28:033001.

[68] Chapon LC, Radaelli PG, Blake GR, Park S, Cheong SW. Ferroelectricity induced by acentric spin-density waves in YMn_2O_5. Phys Rev Lett. 2006;96:097601.

[69] Radaelli PG, Chapon LC, Daoud-Aladine A, Vecchini C, Brown PJ, Chatterji, T, et al. Electric field switching of antiferromagnetic domains in YMn_2O_5: a probe of the multiferroic mechanism. Phys Rev Lett. 2008;101:067205.

[70] Picozzi S, Yamauchi K, Sanyal B, Sergienko IA, Dagotto E. Dual nature of improper ferroelectricity in a magnetoelectric multiferroic. Phys Rev Lett. 2007;99:227201.

[71] Giovannetti G, Stroppa A, Picozzi S, Baldomir D, Pardo V, Blanco-Canosa S, et al. Dielectric properties and magnetostriction of the collinear multiferroic spinel CdV_2O_4. Phys Rev B. 2011;83:060402.

[72] Spaldin NA, Cheong SW, Ramesh R. Multiferroics: past, present, and future. Phys Today. 2010;63:38–43.

[73] Spaldin NA, Ramesh R. Advances in magnetoelectric multiferroics. Nat Mater. 2019;18:203–12.

[74] Bibes M, Barthélémy A. Multiferroics: towards a magnetoelectric memory. Nat Mater. 2008;7:425–6.

[75] Multiferroics march on. Nat Mater. 2019, 18, 187–187.

[76] Spaldin NA. Multiferroics: from the cosmically large to the subatomically small. Nat Rev Mater. 2017;2:1–3.

[77] Manipatruni S, Nikonov DE, Lin CC, Gosavi TA, Liu H, Prasad B, et al. Scalable energy-efficient magnetoelectric spin–orbit logic. Nature. 2019;565:35–42.

[78] Seki S, Ishiwata S, Tokura Y. Magnetoelectric nature of skyrmions in a chiral magnetic insulator Cu_2OSeO_3. Phys Rev B Condens Matter Mater Phys. 2012;86:060403.

[79] Okamura Y, Kagawa F, Mochizuki M, Kubota M, Seki S, Ishiwata, S, et al. Microwave magnetoelectric effect via skyrmion resonance modes in a helimagnetic multiferroic. Nat Commun. 2013;4:2391.

[80] Meier D, Maringer M, Lottermoser T, Becker P, Bohatý L, Fiebig M. Observation and coupling of domains in a spin-spiral multiferroic. Phys Rev Lett. 2009;102:107202.

[81] Meier D, Leo N, Maringer M, Lottermoser T, Fiebig M, Becker P, et al. Topology and manipulation of multiferroic hybrid domains in $MnWO_4$. Phys Rev B. 2009;80:224420.

[82] Leo N, Carolus V, White JS, Kenzelmann M, Hudl M, Tolédano P, et al. Magnetoelectric inversion of domain patterns. Nature. 2018;560:466–70.

[83] Farokhipoor S, Magén C, Venkatesan S, Íñiguez J, Daumont CJM, Rubi D, et al. Artificial chemical and magnetic structure at the domain walls of an epitaxial oxide. Nature. 2014;515:379–83.

Margarita D. Davydova, Konstantin A. Zvezdin,
Alexander A. Mukhin and Anatoly K. Zvezdin

2 Spin dynamics, antiferrodistortion and magnetoelectric interaction in multiferroics. The case of $BiFeO_3$

Abstract: We present a theoretical study of the spin dynamics in perovskite-like multiferroics with homogeneous magnetic order in the presence of external magnetic and electric fields. A particular example of such material is $BiFeO_3$ in which the spin cycloid can be suppressed by application of external magnetic field, doping or by epitaxial strain. Understanding the effect of the external electric field on the spin-wave spectrum of these systems is required for devices based on spin wave interference and other innovative advances of magnonics and spintronics. Thus, we propose a model for $BiFeO_3$ in which the thermodynamic potential is expressed in terms of polarization $\boldsymbol{P}$, antiferrodistortion $\boldsymbol{\Omega}$, antiferromagnetic moment $\boldsymbol{L}$ and magnetization $\boldsymbol{M}$. Based on this model, we derive the corresponding equations of motion and demonstrate the existence of electromagnons, that is, magnons that can be excited by electric fields. These excitations are closely related to the magnetoelectric effect and the dynamics of the antiferrodistortion $\boldsymbol{\Omega}$. Specifically, the influence of the external electric field on the magnon spectra is due to reorientation of both polarization $\boldsymbol{P}$ and antiferrodistortion $\boldsymbol{\Omega}$ under the influence of the electric field and is linked to emergence of a field-induced anisotropy.

Keywords: magnetic order, antiferrodistortion, multiferroics, BFO, magnetoelectric effect, phase transitions

2.1 Introduction

In solids, the coupling between different order parameters is a highly nontrivial question. The coupling between the polarization $\boldsymbol{P}$ and magnetization $\boldsymbol{M}$ enables magnetoelectric (ME) effect which gives rise to a possibility of controlling the properties of a single material with both magnetic and electric fields. Efficient control of the magnetic order with electric field is one of the long-standing goals in the areas of spintronics and magnetic memory [1]. The magnetoelectric effect was first studied theoretically in the work [2] and then confirmed in the paper [3, 4] in Cr_2O_3 in early

This article has previously been published in the journal Physical Sciences Reviews. Please cite as: Davydova, M. D., Zvezdin, K. A., Mukhin, A. A., Zvezdin, A. K. Spin dynamics, antiferrodistortion and magnetoelectric interaction in multiferroics. The case of $BiFeO_3$ *Physical Sciences Reviews* [Online] 2020, 5. DOI: 10.1515/psr-2019-0070

https://doi.org/10.1515/9783110582130-002

1960s. Since then this effect has been discovered and intensively studied in a range of multiferroic materials (see [5] and references therein). The coexistence of both magnetic and ferroelectric order in multiferroics implies interaction between them. In such materials, aside from linear ME effect, other effects may take place, such as higher-order magnetic and electric effects, polarization switching by magnetic field [6–9] and switching of magnetization by electric field [9–11], strain- and field-induced phase transitions [9, 12, 13].

The described phenomena may occur not only in static fields, but also in dynamics [14, 15]. In this context, the new elementary excitations – electromagnons [16, 17], which unlike magnons can be excited electric field, stand out as an unusual effect at high frequencies. Their discovery in multiferroics [16] opened up new perspectives for magnonics. For example, they enable an electric guidance of spin waves, which allows to avoid a big part of the energy loss associated with using magnetic fields or spin currents. The first promising results were obtained in $BiFeO_3$, which is a perovskite-like type-I multiferroic, where a shift of the spin wave frequency constituted 30% under application of electric field with magnitude 100 kV/cm [18].

In this chapter, we will consider perovskite-like materials with ABO_3 structure, where A-ions are localized at the cubic lattice sites that are surrounded by oxygen octahedra [19]. Examples of such materials are $RFeO_3$ orthoferrites, $RMnO_3$ manganites, $BiFeO_3$, $BiMnO_3$ and $FeTiO_3$. As a particular example we will use $BiFeO_3$ (BFO), which is one of the most promising materials for spintronics and memory applications thanks to combination of magnetic and ferroelectric properties at room temperature [20–26]. Below T_C the perovskite unit cell of BFO is distorted (the symmetry group is *R*3*c*) and doubled, see Figure 2.1. The doubling occurs because oxygen octahedrons in adjacent unit cells turn clockwise and counterclockwise around the [111]-axis of the crystal. This rotation is accompanied by displacement of iron and bismuth ions and leads to distortion of octahedrons [27].

Figure 2.1: Bismuth ferrite rombohedrally distorted perovskite cell doubled by antiparallel rotation of the oxygen octahedra. Thee doubling of the unit cell of the crystal structure occurs due to the antiphase rotation of the oxygen octahedra that surround the Fe^{3+} ions (antiferrodistortion). Displacement of the oxygen and iron ions within the double cell is responsible for the spontaneous polarization.

The magnetic symmetry of BFO allows for the existence of weak ferromagnetism and a linear magnetoelectric effect [28]. However, for a long time the linear magnetoelectric effect in BFO was not found. Later it was revealed that the linear ME effect and weak ferromagnetism are suppressed because of the spin cycloid with a period of 62 nm [29, 30], lying in the plane normal to the basal plane and running along one of the second-order symmetry axes of BFO. The cycloid exists because of flexomagnetoelectric interaction with surface energy of 0.6 erg cm^{-2}, due to which the spontaneous electric polarization induces incommensurate spin modulation [30]. Because of the cycloid the average of the magnetization (canting of the Fe^{3+} sublattices leads to a weak local magnetizatione [28]) and linear ME effect are zero. Therefore, first only the quadratic ME effect had been observed, and the value of the obtained magnetoelectric susceptibility tensor elements of the order of 10^{-19} sA^{-1} had been found [31]. A lot of research concerned spin cycloid by itself and was focused on the dependence of its structure and period on temperature, magnetic anisotropy, doping and external fields [32–37]. However, under certain conditions the spin cycloid can be destroyed, and homogeneous antiferromagnetic ordering is acquired. For example, this can be acquired by applying an external magnetic field of a large magnitude [38–41], using chemical doping [42–44] and by strain [45–50] in thin films and heterostructures with $BiFeO_3$. The destruction of the spin cycloid leads to emergence of otherwise suppressed properties of the material, such as linear ME effect [51] and spontaneous magnetization [28].

We study the features of the magnon spectrum of $BiFeO_3$ in the presence of external electric and magnetic field in the case of homogeneous magnetic order. The results are applicable to thin films of BFO and doped BFO [46–48, 52]. There exists disputable relation between different results in the questions of dependence of properties of the spin waves in multiferroics on the electric field. The hint of existence of electromagnons in $BiFeO_3$ was first obtained experimentally in works [53, 54] and has been further studied later [55]. However, theoretically the situation was not clear, no electromagnons were predicted to exist in $BiFeO_3$ using atomistic molecular dynamics simulations [56], which was disproved by phenomenological theory [57] and using first-principle calculations later [58]. The questions of coupling of electric field to order parameters [59, 60] and excitations [61, 62] in BFO attracted a lot of attention as well. We confine ourselves to investigation of the two low-frequency magnon modes in the homogeneous phase, which we show to be electromagnons. We also accent our study on the mechanism of the influence of the applied static electric field on the properties of the magnon spectrum, which we connect with the reorientation of antiferrodistortion. The aim of this chapter is to provide an analytical background for understanding the mechanisms that underlie these effects. Such understanding might be required for development devices based on spin wave interference, and other advances of magnonics and spintronics (see, for example, Refs. [63, 64]).

To study the spin dynamics in BFO, we start with derivation of the Lagrangian for the system in hydrodynamic approximation. This approach allows to describe the possible low-frequency (acoustic) magnetic excitations in a perovskite-like multiferroic, BFO being an example of such. We use the Ginsburg-Landau-type approach to express the thermodynamic potential as invariant expansion in terms of the order parameters, namely polarization $\boldsymbol{P}$, antiferrodistortion $\boldsymbol{\Omega}$ (the rotational distortion of oxygen octahedra that surround the iron ions) and antiferromagnetic moment $\boldsymbol{L}$. We study the influence of the external electric field on the spin wave properties and obtain that it occurs in the form of the field-induced anisotropy in agreement with existing studies [65]. The spin-flexoelectric interaction is another way of the electric-field control of the magnon properties. This interaction was also shown to allow electric-field control of magnetization dynamics in multiferroics [66] and is of prime importance for practical purposes. In addition, it is possible to investigate the magneto-optical resonance and find out whether these modes are electromagnons.

There exists a large amount of theoretical and experimental evidence that antiferrodistortion (AFD) plays an important role in defining the magnetic properties of $BiFeO_3$ [50, 67–70]. Several recent studies helped to further improve the understanding of the mechanisms that link antiferrodistortion to other order parameters and material properties [71–75]. In particular, the reorientation of antiferrodistortion in external electric field explains the giant value of the linear magnetoelectric effect in BFO [72].

In this chapter, we show that the interactions that include antiferrodistortion lead to emergence of electric field-induced anisotropy in a multiferroic that, in turn, affects the magnon properties. The presence of the antiferrodistortion in the proposed model leads to prediction of additional effects related to spin-wave dynamics. In particular, in the electric fields of the order of MV/cm the second-order ferroelastic field-induced phase transition has been predicted [76], which may be described in terms of antiferrodistortion $\boldsymbol{\Omega}$ and polarization $\boldsymbol{P}$. It was shown that in the vicinity of this transition the magnetoelectric effect should diverge [72], and near the transition the influence of the electric field will be enhanced, which may be observed experimentally. Near the first-order ferroelectric phase transition in the electric field of the order of 10^4 V/cm reorientation of the magnetic system occurs, as it was shown for inhomogeneous phase [77]. Another transition takes place in the homogeneous phase [76], near which the ground state will change in terms of $\boldsymbol{L}_0$. This effect, together with the non-reciprocity of the obtained magnon modes (will be discussed below), leads to an additional way to tune the magnon properties with electric field. Both effects of the electric field on the spin wave properties might be of a large interest for memory technologies [78, 79].

2.2 The Lagrangian and Rayleigh dissipation function of a perovskite-like multiferroic

To obtain the dynamic equations we introduce the Lagrangian L and the Rayleigh dissipation function R. To derive them we start from the Landau-Lifshitz-Gilbert equations for a two-sublattice weak ferromagnet:

$$\dot{\boldsymbol{M}}_i = \gamma\left[\boldsymbol{M}_i, \frac{\delta\Phi}{\delta\boldsymbol{M}_i}\right] + \frac{\alpha}{M_0}\left[\boldsymbol{M}_i, \dot{\boldsymbol{M}}_i\right], \; i = 1, 2, \tag{2.1}$$

where γ is the gyromagnetic ratio of electron, α is the damping coefficient, M_0 is the absolute value of the sublattice magnetization vector and the thermodynamic potential is $\Phi \equiv \Phi(\boldsymbol{M}_1, \boldsymbol{M}_2, \boldsymbol{\Omega}, \boldsymbol{E}, \boldsymbol{H})$.

The external magnetic field is oriented $\boldsymbol{H} \parallel [111]$ in pseudocubic notation, which unravels the spin cycloid. The effect of such external magnetic field is equivalent to the strain [48], which may emerge in [111]-thin film. In the ground state the unit antiferromagnetic vector ($\boldsymbol{l} = \boldsymbol{L}/L$) $\boldsymbol{l}_0 \parallel [\bar{1}\bar{1}2]$, which is stabilized by another component of the magnetic field, parallel to $[1\bar{1}0]$. The perturbation to the ground state was also taken into account, which is caused by external electric and magnetic fields, $\boldsymbol{E}$ and $\boldsymbol{H}$ correspondingly. We introduce rhombohedral coordinate system with $OX \parallel [\bar{1}\bar{1}2]$, $OY \parallel [1\bar{1}0]$, $OZ \parallel [111]$ in pseudocubic notation. In the spherical coordinate system with OZ polar axis let the sets of angles (φ_1, θ_1) and (φ_2, θ_2) define the orientation of the two sublattice magnetizations in the assumption that length of these vectors stays constant. We obtain the Lagrangian and the Rayleigh dissipation function in terms of these variables, which will lead to the Euler-Lagrange equations:

$$L = -\frac{M_0}{\gamma}(\cos\theta_1)\frac{\partial\varphi_1}{\partial t} - \frac{M_0}{\gamma}(\cos\theta_2)\frac{\partial\varphi_2}{\partial t} - \Phi(\boldsymbol{M}_1, \boldsymbol{M}_2, \boldsymbol{\Omega}, \boldsymbol{E}, \boldsymbol{H})$$

$$R = \frac{\alpha M_0}{2\gamma}\left[\dot{\theta}_1^2 + \sin^2\theta_1^2\,\dot{\varphi}_1^2 + \dot{\theta}_2^2 + \sin^2\theta_2^2\,\dot{\varphi}_2^2\right]. \tag{2.2}$$

Next, we transfer from the description of the spin dynamics in terms of the magnetization vectors of the two sublattices $\boldsymbol{M}_1$ and $\boldsymbol{M}_2$ to dimensionless magnetization $\boldsymbol{m} = \frac{\boldsymbol{M}_1 + \boldsymbol{M}_2}{2M_0}$ and unit antiferromagnetic (Neel) vector $\boldsymbol{l} = \frac{\boldsymbol{M}_1 - \boldsymbol{M}_2}{2M_0}$.

The thermodynamic potential may be viewed as a function of the magnetization and the antiferromagnetic vector:

$$\Phi(\boldsymbol{m}, \boldsymbol{l}, \boldsymbol{P}, \boldsymbol{\Omega}, \boldsymbol{E}, \boldsymbol{H}) = \Phi_m(\boldsymbol{m}, \boldsymbol{l}, \boldsymbol{\Omega}, \boldsymbol{E}, \boldsymbol{H}) + \Phi_{fm}(\boldsymbol{l}, \boldsymbol{P}) + \Phi_{st}(\boldsymbol{P}, \boldsymbol{\Omega}, \boldsymbol{E}), \tag{2.3}$$

where Φ_m is the magnetic part of the potential, Φ_{fm} is the spin-flexoelectric part, which corresponds to inhomogeneous magnetoelectric coupling, and Φ_{st} is the ferroelectric part. The magnetic part of the thermodynamic potential, the only part which depends on the magnetization $\boldsymbol{m}$, can be written as follows:

$$\begin{aligned}\Phi_m(\boldsymbol{m}, \boldsymbol{l}, \boldsymbol{\Omega}, \boldsymbol{E}, \boldsymbol{H}) = {} & a_1 M_0^2\,\boldsymbol{m}^2 + a_2 M_0^2(\boldsymbol{m}, \boldsymbol{l})^2 + 2bM_0^2\,\boldsymbol{l}^2 + A\left[(\nabla\boldsymbol{m})^2 + (\nabla\boldsymbol{l})^2\right] \\ & - 2M_0(\boldsymbol{m}, \boldsymbol{H}_t) + E_a,\end{aligned} \tag{2.4}$$

where $a_1 = \chi_\perp/2$, a_2 and b are expansion coefficients, $\chi_\perp$ is the component of the magnetic susceptibility, which is perpendicular to the vector $\boldsymbol{l}$, A is the constant of inhomogeneous exchange, $\boldsymbol{H}_t = \boldsymbol{H} + \boldsymbol{H}_D$ is the total magnetic field, $\boldsymbol{H}_D = \frac{H_D}{\Omega_0}[\boldsymbol{\Omega}, \boldsymbol{l}]$ is the Dzyaloshinskii field and E_a is the anisotropy energy.

Let us now rewrite the unit antiferromagnetic vector in the spherical coordinate system as $\boldsymbol{l} = (\cos\varphi\sin\theta, \sin\varphi\sin\theta, \cos\theta)$ and the magnetization as $\boldsymbol{m} = (\epsilon\cos\theta\cos\varphi - \psi\sin\theta\sin\varphi, \epsilon\cos\theta\sin\varphi + \psi\sin\theta\cos\varphi, -\epsilon\sin\theta)$. The latter is parametrized using variables ε, ψ, which satisfy the relation of smallness $\epsilon, \psi \ll 1$ in the case when the canting of the magnetic sublattices is small. In this assumption the angles that determine the orientation of the sublattice magnetization vectors may be expressed as $\theta_1 = \theta + \varepsilon$, $\varphi_1 = \varphi + \psi$ and $\theta_2 = \pi - \theta + \epsilon$, $\varphi_2 = \pi + \varphi - \psi$. The Lagrangian and the Rayleigh dissipation function (2.2) acquire the following form [80]:

$$L = \frac{2M_0}{\gamma}\left(\dot{\varphi}\epsilon\sin\theta - \dot{\psi}\cos\theta\right) - \Phi(\boldsymbol{m}, \boldsymbol{l}, \boldsymbol{P}, \boldsymbol{\Omega}, \boldsymbol{E}, \boldsymbol{H})$$

$$R = \frac{\alpha M_0}{\gamma}\left[\dot{\theta}^2 + \sin^2\theta\left(\dot{\varphi}^2 + \dot{\psi}^2\right) + \dot{\epsilon}^2 + 2\sin(2\theta)\ \epsilon\ \dot{\varphi}\ \dot{\psi}\right]. \tag{2.5}$$

The Euler-Lagrange equations will be of the form $\frac{d}{dt}\left(\frac{\partial L}{\partial \dot{q}}\right) - \frac{\delta L}{\delta q} = -\frac{\partial R}{\partial \dot{q}}$, where $\frac{\delta}{\delta q} = \frac{\partial}{\partial q} - \nabla\frac{\partial}{\partial \nabla q}$ and $q = \theta, \phi, \varepsilon, \psi$. The explicit expression for them is:

$$\begin{aligned}
&-\frac{2M_0}{\gamma}\left(\dot{\varphi}\epsilon\cos\theta + \dot{\psi}\sin\theta\right) + \frac{\delta\Phi}{\delta\theta} = -\frac{\delta R}{\delta\dot{\theta}}\\
&\frac{2M_0}{\gamma}\left(\dot{\epsilon}\sin\theta + \epsilon\dot{\theta}\cos\theta\right) + \frac{\delta\Phi}{\delta\varphi} = -\frac{\delta R}{\delta\dot{\varphi}}\\
&-\frac{2M_0}{\gamma}\dot{\varphi}\sin\theta + \frac{\delta\Phi}{\delta\epsilon} = -\frac{\delta R}{\delta\dot{\epsilon}}\\
&\frac{2M_0}{\gamma}\dot{\theta}\sin\theta + \frac{\delta\Phi}{\delta\psi} = -\frac{\delta R}{\delta\dot{\psi}}.
\end{aligned} \tag{2.6}$$

This set of first-order differential equations will lead to derivation of the magnon modes. However, in some cases it is convenient to exclude the variables ε, ψ, which define the magnetization, from the thermodynamic potential as well as from the system (2.6). For this we use the identities $\frac{\delta\Phi}{\delta\epsilon} = \frac{\delta\Phi_m}{\delta\epsilon}$, $\frac{\delta\Phi}{\delta\psi} = \frac{\delta\Phi_m}{\delta\psi}$ and the expression for the magnetic part of the thermodynamic potential (2.4) to solve the last two equations from (2.6) with respect to ε, ψ (further we neglect the terms of the higher order in the small variables ε, ψ):

$$\begin{aligned}
\epsilon &= \frac{\chi_\perp}{2M_0}\left[(H_t)_x\cos\theta\cos\varphi + (H_t)_y\sin\theta\sin\varphi - (H_t)_z\sin\theta + \frac{1}{\gamma}\sin\theta\dot{\varphi}\right],\\
\psi &= \frac{\chi_\perp}{2M_0}\frac{1}{\sin\theta}\left[-(H_t)_x\sin\varphi + (H_t)_y\cos\varphi - \frac{1}{\gamma}\dot{\theta}\right].
\end{aligned} \tag{2.7}$$

This pair of equations is used to reduce the system (2.6) to a pair of the second-order differential equations:

$$\frac{\chi_\perp}{\gamma^2}\left[\ddot{\theta}+(\dot{\varphi})^2\cos\theta\sin\theta\right]+2\frac{\chi_\perp}{\gamma}\left[(H_t)_x\sin^2\theta\cos\varphi+(H_t)_y\sin^2\theta\sin\varphi\right.$$
$$\left.+(H_t)_z\sin\theta\cos\theta\right]\dot{\varphi}+\frac{\delta\Phi}{\delta\theta}=\frac{\partial R}{\partial\dot{\theta}},$$
$$\frac{\chi_\perp}{\gamma^2}\left[\sin^2\theta\ddot{\varphi}+2\dot{\theta}\dot{\varphi}\cos\theta\sin\theta\right]-2\frac{\chi_\perp}{\gamma}\left[(H_t)_x\sin^2\theta\cos\varphi+(H_t)_y\sin^2\theta\sin\varphi\right.$$
$$\left.+(H_t)_z\sin\theta\cos\theta\right]\dot{\theta}+\frac{\delta\Phi}{\delta\varphi}=\frac{\partial R}{\partial\dot{\varphi}}, \tag{2.8}$$

where the magnetic part of the thermodynamic potential depends now only on $\boldsymbol{l}$ and is:

$$\Phi_m(\boldsymbol{l},\boldsymbol{\Omega},\boldsymbol{E},\boldsymbol{H})=-\frac{\chi_\perp}{2}(\boldsymbol{H}_t)_\perp^2+A(\nabla\boldsymbol{l})^2+E_a. \tag{2.9}$$

Lastly, we view the pair of eq. (2.8) as the Euler-Lagrange equations for well-known Lagrangian and Rayleigh dissipation function [81–84] (in vector form):

$$L=\frac{\chi_\perp}{2\gamma^2}\dot{\boldsymbol{l}}^2-\frac{\chi_\perp}{\gamma}\boldsymbol{H}_t[\boldsymbol{l}\dot{\boldsymbol{l}}]-\Phi,\ R=\alpha\frac{M_0}{\gamma}\dot{\boldsymbol{l}}^2. \tag{2.10}$$

The two low-frequency magnon modes are obtained through the set of the Euler-Lagrange equations:

$$\frac{d}{dt}\frac{\partial L}{\partial\dot{\boldsymbol{l}}}=\frac{\delta L}{\delta\boldsymbol{l}}-\frac{\partial R}{\partial\dot{\boldsymbol{l}}}$$
$$\frac{\partial L}{\partial\boldsymbol{\Omega}}=0,$$
$$\frac{\partial\Phi}{\partial\boldsymbol{P}}=0. \tag{2.11}$$

An assumption that characteristic timescales in the dynamics of order parameters $\boldsymbol{P}$ and $\boldsymbol{\Omega}$ are much smaller than that of $\boldsymbol{l}$ was made; they are now viewed as the function of the angles θ and φ.

Let us now turn to a more detailed description of the thermodynamic potential (2.3), which was (with change in Φ_m according to (2.9)) used in our calculations. The spin-flexoelectric part has the form of the Lifshitz invariant:

$$\Phi_{fm}(\boldsymbol{l},\boldsymbol{P})=\beta\boldsymbol{P}(\boldsymbol{l}(\nabla\boldsymbol{l})-(\boldsymbol{l}\nabla)\boldsymbol{l}), \tag{2.12}$$

where β is the parameter of the spin-flexoelectric interaction. The ferroelectric part is:

$$\Phi_{st}(\boldsymbol{P},\boldsymbol{\Omega},\boldsymbol{E})=\Phi_{st}(\boldsymbol{P},\boldsymbol{\Omega})-\boldsymbol{PE}, \tag{2.13}$$

where the detailed invariant expansion of $\Phi_{st}(\boldsymbol{P}, \boldsymbol{\Omega})$ and its parameters may be found in Ref. [76].

The obtained Lagrangian and Rayleigh function (2.5) and (2.10) are useful expressions, which may be adapted to study spin wave properties in different materials possessing weak ferromagnetism.

2.3 The spin-wave spectrum

We consider the homogeneous phase with $\boldsymbol{l}_0 \parallel OX$ in the ground state, which can be realized by applying a stabilizing magnetic field along $[1\bar{1}0]$ axis (H_y). Also, there is another component of magnetic field H_z, parallel to [111] axis, which purpose is to unravel the spin cycloid. In the case of bulk crystal, the magnitude of the applied magnetic field must exceed critical value $H_z > H_{cr} \approx 20$ T [38]. An assumption that $H_y \ll H_z$ was made, so that we accounted only for the linear terms in the equations, describing perturbations to the initial ground state $(\boldsymbol{l}_0, \boldsymbol{\Omega}_0, \boldsymbol{P}_0)$. We consider the situation when only small external electric field compared to critical [76, 85] is present due to the same issue. Then we find the perturbation to the initial ground state $(\boldsymbol{l}_0, \boldsymbol{\Omega}_0, \boldsymbol{P}_0)$. We turn to the ferroelectric part of the potential $\Phi_{st}(\boldsymbol{P}, \boldsymbol{\Omega})$ to obtain the equilibrium values of polarization $\boldsymbol{P}_0$ and antiferrodistortion $\boldsymbol{\Omega}_0$. The equilibrium vectors $\boldsymbol{P}_0 \parallel [111]$ and $\boldsymbol{\Omega}_0 \parallel [111]$ in pseudocubic notation. We confine our consideration to the linear approximation with respect to the external electric field $\boldsymbol{E}$.

Let us specify the equations of motion (2.11) in the easy plane ground state with $\boldsymbol{l}_0 \parallel OX$. The deviations from the unperturbed ground state are expressed as $\theta = \theta_0 + \delta\theta$, $\varphi = \varphi_0 + \delta\varphi$ ($\theta_0 = \pi/2$, $\varphi_0 = 0, \pi$), $\boldsymbol{\Omega} = \boldsymbol{\Omega}_0 + \delta\boldsymbol{\Omega}$, $\boldsymbol{P} = \boldsymbol{P}_0 + \delta\boldsymbol{P}$, where $\delta\theta = \delta\theta(t, r, \boldsymbol{E}, \boldsymbol{H})$, $\delta\varphi = \delta\varphi(t, r, \boldsymbol{E}, \boldsymbol{H})$. To find variations $\delta\boldsymbol{\Omega} = \delta\boldsymbol{\Omega}(\delta\theta, \delta\varphi, \boldsymbol{E}, \boldsymbol{H})$ and $\delta\boldsymbol{P} = \delta\boldsymbol{P}(\delta\theta, \delta\varphi, \boldsymbol{E}, \boldsymbol{H})$, we solve the two last equations in (2.11). The expressions for them can be found in Appendix A.

After linearization of the equations of motion, we find the corrections to the ground state due to the presence of the static magnetic and electric fields:

$$\delta\theta^0 = \gamma^2 \frac{H_D}{\Omega_0} \eta_x^e \cos\varphi_0 \frac{H_z^0 C - H_D \omega_{20}^2}{C^2 - \omega_{10}^2 \omega_{20}^2} E_x^0$$

$$\delta\varphi^0 = \gamma^2 \frac{H_D}{\Omega_0} \eta_x^e \cos\varphi_0 \frac{H_D C - H_z^0 \omega_{10}^2}{C^2 - \omega_{10}^2 \omega_{20}^2} E_x^0. \tag{2.14}$$

Finally, the following set of equations (see full derivation in Appendix B) is obtained, which defines the magnon spectrum:

$$\begin{cases} \left(\omega_1^2(\boldsymbol{k}) - c_\omega^{(1)}\omega^2 - \omega c_x^{(1)}k_x - i\omega\dfrac{2\alpha M_0\gamma}{\chi_\perp}\right)\delta\theta + \omega_{12}^2(\boldsymbol{k})\delta\varphi = \sum(\alpha_{1i}\Delta E_i + \beta_{1i}\Delta H_i) \\ \omega_{21}^2(\boldsymbol{k})\delta\theta + \left(\omega_2^2(\boldsymbol{k}) - c_\omega^{(2)}\omega^2 - \omega c_x^{(2)}k_x - i\omega\dfrac{2\alpha M_0\gamma}{\chi_\perp}\right)\delta\varphi = \sum(\alpha_{2i}\Delta E_i + \beta_{2i}\Delta H_i). \end{cases} \quad (2.15)$$

Here and earlier superscript ‘0’ denotes the constant parts of electric or magnetic fields, ΔE_j, ΔH_j are the magnitudes of the oscillating parts of the external fields

$$\sim\exp(ikr - i\omega t)\omega_j^2(k) = \omega_{j0}^2 + d_{jx}k_x^2 + d_{jy}k_y^2 + d_{jz}k_z^2 \ (j=1,2),$$

$$\omega_{12}^2(\boldsymbol{k}) = \left(\omega_{21}^2(\boldsymbol{k})\right)^* = C + A_{yz}k_yk_z + iB_yk_y + iB_zk_z,$$

α_{ij}, β_{ij} are linear functions of frequency and wavevector, which also are affected by external static electric and magnetic fields (we omit them here due to bulkiness) and coefficients ω_{j0}, d_{ik}, $C = C(E_y^0, E_z^0, H_y^0, H_z^0)$, A_{yz}, $B_y = B_y(E_z, H_y)$ and $B_z = B_z(E_y, H_z)$, $c_{\omega,x}^{(i)}$ and $\eta_{x,y,z}^{e,(1)}$ are defined in Appendix B. The parameters ω_{j0} also depend on the external electric and magnetic field.

The mechanism of the influence of the external electric field on the magnon modes lies in the electric-field-induced anisotropy due to the variation of the antiferrodistortion in the external electric field. The reorientation of the polarization vector also plays role. However, only simultaneous consideration of both order parameters leads to an accurate description of the effect. The coefficients on the left side of eq. (2.15) do not depend on the component E_x of the electric field. On the other hand, coefficients $B_y = B_y(E_z^0, H_y^0)$ and $B_z = B_z(E_y^0, H_z^0)$ depend on the transversal component of the electric field and are relevant when the magnon wavevector forms an angle with the direction of the antiferromagnetic vector $\boldsymbol{l}_0$. The coefficients B_i arise from the spin-flexoelectric interaction, and thus the reorientation of the polarization in the external field plays a significant role in the behavior of the magnons, which propagate at an angle to $\boldsymbol{l}_0$. However, the anisotropy-like terms lead to dependence of ω_{i0} on the external electric field (see Appendix B) and this dependence is connected with the reorientation of the AFD vector.

When $\boldsymbol{k} \parallel OX$, there are two modes, which appear to be coupled due to the coefficient C, which arises when external electric and magnetic fields are applied. When $\boldsymbol{k}\perp OX$, the two modes also appear to be coupled:

$$\widetilde{\omega}_1^2(\boldsymbol{k}) = \frac{1}{2}\left(\frac{w_1^2(\boldsymbol{k})}{c_\omega^{(1)}} + \frac{w_2^2(\boldsymbol{k})}{c_\omega^{(2)}}\right) + \sqrt{\frac{|\omega_{12}^2|^2(\boldsymbol{k})}{c_\omega^{(1)}c_\omega^{(2)}} + \frac{1}{4}\left(\frac{w_1^2(\boldsymbol{k})}{c_\omega^{(1)}} - \frac{w_2^2(\boldsymbol{k})}{c_\omega^{(2)}}\right)^2}$$

$$\widetilde{\omega}_1^2(\boldsymbol{k}) = \frac{1}{2}\left(\frac{w_1^2(\boldsymbol{k})}{c_\omega^{(1)}} + \frac{w_2^2(\boldsymbol{k})}{c_\omega^{(2)}}\right) - \sqrt{\frac{|\omega_{12}^2|^2(\boldsymbol{k})}{c_\omega^{(1)}c_\omega^{(2)}} + \frac{1}{4}\left(\frac{w_1^2(\boldsymbol{k})}{c_\omega^{(1)}} - \frac{w_2^2(\boldsymbol{k})}{c_\omega^{(2)}}\right)^2}. \quad (2.16)$$

2.4 Magneto-optical resonance

Using the result (2.15) for the spin-wave spectra we can now investigate if these modes are electromagnons. We consider the case when static electric field is applied along the z-axis: $\boldsymbol{E} = (0, 0, E_z)$. This case is more simple than the general one, as for the uniform oscillations ($\boldsymbol{k} = 0$) the coupling between orthogonal oscillations is absent ($C(\boldsymbol{E} = (0, 0, E_z), \boldsymbol{H} = (0, H_y, 0)) = 0$). We also take the ground state here to be $\boldsymbol{P}_0 = (0, 0, P_0)$, $\boldsymbol{\Omega}_0 = (0, 0, \Omega_0)$ and $\boldsymbol{l}_0 = (l_0, 0, 0)$, $l_0 > 0$. We calculate the resonant contribution to the dielectric, magnetoelectric and magnetic susceptibilities from the two spin wave modes and study the influence of the external electric field on them. Here we use notations $\widetilde{\omega}_{i0}^2 = \omega_{i0}^2 / c_\omega^{(i)}$, $i = 1,\ 2$, and functions $L_{1,2}(\omega) = \widetilde{\omega}_{10,20}^2 / \widetilde{\omega}_{10,20}^2 - \omega^2$, which define the frequency dispersion of the susceptibility near the different modes.

The electric susceptibility tensor has the form $\hat{\chi}^e = (\hat{\chi}^e)_{lattice} + \Delta\hat{\chi}^e$. The spin-wave-dependent part has the following form:

$$\Delta\hat{\chi}^e = \begin{pmatrix} \Delta\chi_{xx}^e & 0 & \Delta\chi_{xz}^e \\ 0 & \Delta\chi_{yy}^e & 0 \\ -\Delta\chi_{xz}^e & 0 & \Delta\chi_{zz}^e \end{pmatrix}, \tag{2.17}$$

where the additional spin-wave contributions are:

$$\begin{aligned} \Delta\chi_{xx}^e &= -\gamma^2 \frac{H_D M_s}{\Omega_0^2} \left(\eta_x^{(1)}\right)^2 \eta_x^e \left[\frac{H_D}{\Omega_0} \eta_z^e E_z + H_D + H_y\right]^2 \frac{1}{\widetilde{\omega}_{10}^2} L_1(\omega) \\ \Delta\chi_{yy}^e &= \frac{H_D M_s}{\Omega_0^2} \left(\eta_y^e\right)^2 L_2(\omega) \\ \Delta\chi_{zz}^e &= -\frac{H_D M_s}{\Omega_0^2} (\eta_z^e)^2 L_1(\omega) \\ \Delta\chi_{xz}^e &= i\sqrt{\Delta\chi_{xx}^e \Delta\chi_{zz}^e} \frac{\omega}{\widetilde{\omega}_{10}}. \end{aligned} \tag{2.18}$$

The approximate expression for the components the magnetoelectric susceptibility (that also defines the linear magnetoelectric effect) with the spin-wave contribution:

$$\hat{\chi}^{em} = \begin{pmatrix} 0 & \chi_{xy}^{em} & 0 \\ \chi_{yx}^{em} & 0 & \chi_{yz}^{em} \\ 0 & \chi_{zy}^{em} & 0 \end{pmatrix}, \tag{2.19}$$

where the components are of the form:

$$
\begin{aligned}
\chi_{xy}^{em} &= -\gamma\frac{M_s}{\Omega_0}\eta_x^{(1)}\left[\frac{H_D}{\Omega_0}\eta_z^e E_z^0 + H_D + H_y\right]\frac{1}{\tilde{\omega}_{10}}\frac{i\omega}{\tilde{\omega}_{10}}L_1(\omega)\\
\chi_{yx}^{em} &= -\gamma\frac{M_s}{\Omega_0}\eta_y^e\left[\frac{H_D}{\Omega_0}\eta_z^e E_z^0 + H_D + H_y\right]\frac{1}{\tilde{\omega}_{20}}\frac{i\omega}{\tilde{\omega}_{20}}L_2(\omega)\\
\chi_{yz}^{em} &= -\eta_y^e\frac{M_s}{\Omega_0}L_2(\omega)\\
\chi_{zy}^{em} &= \eta_z^e\frac{M_s}{\Omega_0}L_1(\omega).
\end{aligned}
\tag{2.20}
$$

To obtain the contribution to the magnetic susceptibility, we find the approximate expression for magnetization:

$$
\boldsymbol{M} \approx \chi_\perp \boldsymbol{H}_\perp + \frac{M_s}{\Omega_0}[\boldsymbol{\Omega}, \boldsymbol{l}] - \frac{\chi_\perp}{\gamma}[\boldsymbol{l}, \dot{\boldsymbol{l}}]. \tag{2.21}
$$

After linearization, the expressions for the magnetization components become:

$$
\begin{aligned}
M_x &\approx \left(-M_s - \chi_\perp H_y^0 - \frac{M_s}{\Omega_0}\eta_z^e E_z^0\right)\delta\varphi\\
M_y &\approx M_s + \chi_\perp H_y + \frac{M_s}{\Omega_0}\eta_z^e E_z + i\omega\frac{\chi_\perp}{\gamma}\delta\theta\\
M_z &\approx \chi_\perp H_z - \frac{M_s}{\Omega_0}\eta_y^e E_y + i\omega\frac{\chi_\perp}{\gamma}\delta\varphi.
\end{aligned}
\tag{2.22}
$$

Thus, the magnetic susceptibility tensor acquires the form:

$$
\Delta\hat{\chi}^m = \begin{pmatrix} \chi_{xx}^m & 0 & \chi_{xz}^m \\ 0 & \chi_{yy}^m & 0 \\ -\chi_{xz}^m & 0 & \chi_{zz}^m \end{pmatrix}, \tag{2.23}
$$

where the approximate expressions for the components are:

$$
\begin{aligned}
\chi_{xx}^m &= \frac{\gamma^2}{\chi_\perp}\left(M_s + \chi_\perp H_y^0 + \frac{M_s}{\Omega_0}\eta_z^e E_z^0\right)^2\frac{1}{\tilde{\omega}_{20}^2}L_2(\omega)\\
\chi_{yy}^m &= \chi_\perp L_1(\omega)\\
\chi_{zz}^m &= \chi_\perp L_2(\omega)\\
\chi_{xz}^m &= i\sqrt{\chi_{xx}^m\chi_{zz}^m}\,\frac{\omega}{\tilde{\omega}_{20}}.
\end{aligned}
\tag{2.24}
$$

Thus, we have shown that the two modes are excited by both magnetic and electric ac fields and, thus, they are electromagnons. From (2.20) it is seen that contribution to the magnetoelectric tensor in the limit $\omega \to 0$ is of the order of the static magnetoelectric effect, and investigation of the magnetoelectric resonance is relevant to experiments (see, for example, the study [86] on (Eu,Y)MnO_3). In addition, due to *yz*- and *zx*-components of magnetoelectric susceptibility one can expect asymmetry in the light propagation with respect to the direction of propagation (directional anisotropy) which was studied in some multiferroics [86–91].

2.5 Influence of the electric field on the magnon spectra

In this section, we discuss the results obtained in Section 2.3. We take $\chi_\perp = 5 \times 10^{-5}$, $\beta P_0 = 0.6$ erg/cm^2, $A = 3 \times 10^{-7}$ erg/cm, $H_D = 10^5$ Oe and $K_u = 6 \times 10^5$ erg/cm^3 in our calculations. We also used notations $c_0 = \gamma\sqrt{2A/\chi_\perp}$ and $k_0 = \beta P_0/(2A)$. For the homogeneous easy-plane ground state to be stable in bulk BFO, a sufficiently high anisotropy energy or magnetic fields are required. The magnetic field is H_z = 20 T, which stabilizes the homogeneous state, and H_y = 0.2 T. However, the action of the large applied magnetic field H_z is equivalent to the action of the strain; to unravel the spin cycloid this field may be replaced by the epitaxial strain comparable to that in thin films [48], or, alternatively, a combination of much smaller magnetic field with mismatch strain can be used [92].

Our results can be as well generalized to the spin-cycloid state. In the approximation of almost free magnons the method, analogous to reducing the spectrum to the first Brillouin zone $-\pi/\Lambda \le k \le \pi/\Lambda$ (Λ is the period of the cycloid) in electron theory of metals, can be used to retrieve the sequence of magnon modes [59, 93]. Thus, our approach becomes comprehensive in terms of qualitative description of magnon spectra.

The external electric field influences the magnon spectra in a variety of ways, depending on the direction of magnon propagation, the mode polarization and the field orientation, which can be seen in Figure 2.2a and Figure 2.2b. In Ref. [94], a softening of the lower magnon mode was observed. However, we have found that in different setups the influence of the electric field may lead to a rise in magnon frequency. The most pronounced effect lies in the electric field-induced anisotropy due to the variation of the antiferrodistortion in the external electric field. We also find the linear response to electric field, some aspects of which was discussed in recent work [65].

The component of the electric field parallel to $\boldsymbol{l}_0$ does not play significant role in the magnon spectra. Both transversal components cause field-induced anisotropy, which is responsible for the frequency shift of oscillations with small wavevectors. However, for propagating magnons with $\boldsymbol{k} \perp \boldsymbol{l}_0$ we find a more prominent effect from the component, which is perpendicular to the wavevector. Also, at some magnetic-anisotropy parameters of the system (or when static external electric or magnetic field is applied) the situation in which the frequency of one of the modes is negative at certain $\widetilde{k}_0$ (which corresponds to the spin cycloid wave-vector), and

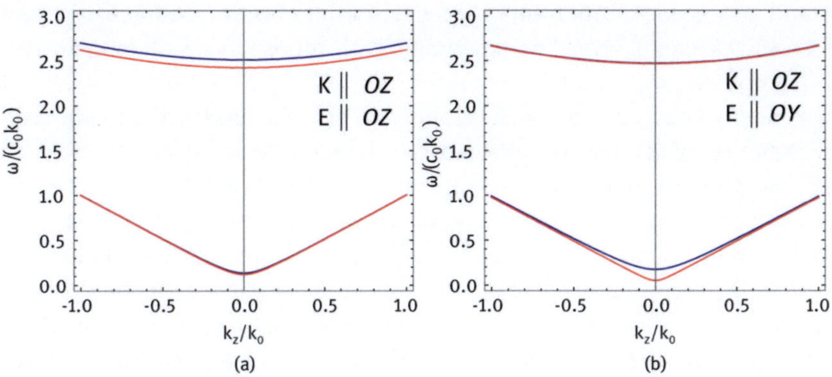

Figure 2.2: Spin-excitation spectra of the magnon modes $\tilde{\omega}_{1,2}(k_z)$ at different external electric field amplitudes.
(a) (Color online) Blue (dark gray) line: $\mathbf{E} \parallel OZ$, $E = -100$ kV/cm, red (light gray) line: $\mathbf{E} \parallel OZ$, $E = 100$ kV/cm.
(b) (Color online) Blue (dark gray) line: $\mathbf{E} \parallel OY$, $E = -100$ kV/cm, red (light gray) line: $\mathbf{E} \parallel OY$, $E = 100$ kV/cm.

the easy-plane state becomes unstable with respect to a transition to the inhomogeneous phase. However, our approach is limited to external electric fields with magnitude less than tens of MV/cm, at which such a transition occurs.

The dependence of the resonance frequency shift is illustrated in Figure 2.3. The plot illustrates the fact that the y-component of the electric field exhibits stronger influence on the lower mode (sometimes referred to as the quasi-ferromagnetic modes), whereas z-component of the electric field influences higher (quasi-antiferromagnetic) mode more. This resonance shift is caused, above all, by the electric field-induced

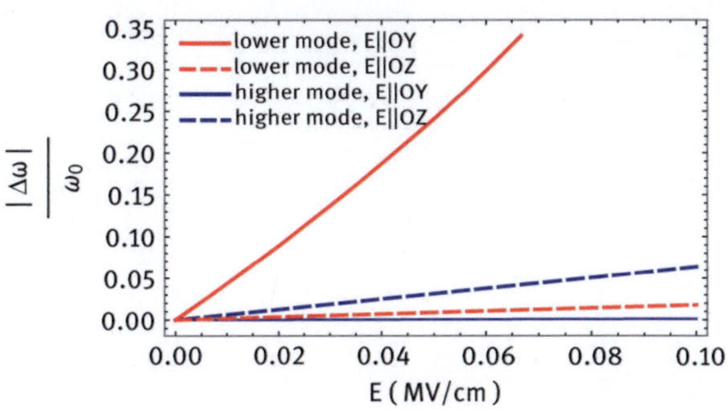

Figure 2.3: (Color online) Dependence of the absolute value of the shift of the frequency normalized tp the frequency of the mode without external electric field value at $k = 0$ of spin-excitation spectra on the value of external electric field. Red (light gray) lines correspond to lower modes, and blue (dark gray) lines correspond to higher, solid lines – $\mathbf{E} \parallel OY$, dashed lines – $\mathbf{E} \parallel OZ$.

anisotropy and may be important from the practical point of view. The frequency shift of 14% at $E \approx 180$ kV/cm, observed in experiment [94], corresponds well to the calculated shift (Figure 2.3).

As it was pointed out, our approach is limited to external electric field with magnitude less than tens of MV/cm. We estimate the electric field-induced anisotropy to be of around 30 % of the total anisotropy at this point. However, this approach can be extended to fields near critical if nonlinear changes of order parameters in high fields are taken into account. As we mentioned before, in the electric field of the order of MV/cm the second-order ferroelastic field-induced phase transition occurs. This transition may be described in terms of the antiferrodistortion $\boldsymbol{\Omega}$ as an order parameter. It was shown that in the vicinity of this transition the electric susceptibility of the antiferrodistortion diverges [72]. This susceptibility, denoted by $\eta^{e}_{x,y,z}$, influences all the parameters, defining the magnon spectrum (see (2.15) and Appendix B for more details). Thus, we expect that near the transition the influence of the electric field will be significantly enhanced.

Another effect, produced by inclusion of the antiferrodistortion into the theoretical consideration, which may be important in the light of practical applications, is connected with the non-reciprocity of the magnon spectrum with respect to the antiferromagnetic vector $\boldsymbol{l}_0$ in Figure 2.4 (see also Ref. [84]). Near the first-order ferroelectric phase transition in the fields of the order of 10 kV/cm [76, 77] this effect can be employed in a new way. The reorientation of the magnetic and electric systems in the electric field has been observed experimentally near these fields [95]. Switching of the polarization and the antiferrodistortion, which occurs at this point, will lead to switching of the orientation of the antiferromagnetic moment $\boldsymbol{l}_0$ in the ground state, which can be easily shown theoretically. However, due to dependence of the parameters in the system (2.15) (see also Appendix B) on the orientation of the antiferromagnetic moment $\boldsymbol{l}_0$, the spectrum reveals non-reciprocity

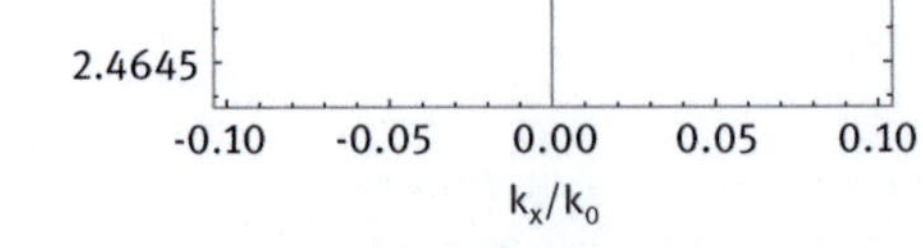

Figure 2.4: (Color online) Spin-excitation spectra of the magnon mode $\tilde{\omega}_1(k_x)$ for two possible projections of the antiferromagnetic vector $\boldsymbol{l}_0$ on the axis OX. Blue (dark gray) line corresponds to $l_0 > 0$ and red (light gray) line corresponds to $l_0 < 0$.

depending on the orientation of $\boldsymbol{l}_0$. This non-reciprocity is illustrated in Figure 2.4 in the absence of the external electric field. Thus, apart from a direct electric-field effect on the magnon spectrum, there is additional possibility for switching the properties of magnons. Apart from this, the non-reciprocity of the spin waves can be used in logic [78] and other nanoscale devices [79].

2.6 Conclusion

In summary, we have developed the theory of the spin waves in the perovskite-like multiferroic $BiFeO_3$, and revealed the role of inclusion of the antiferrodistortion in the theoretical model. We implemented the dynamic equations in the hydrodynamic approximation and consider the two low-frequency magnon modes. The mechanism of the influence of the static external electric field on the magnon modes was found to lie in the electric field-induced anisotropy. We have found that its mechanism is connected with reorientation of the antiferrodistortion in the external electric field. We also suggest that this effect must be significantly enhanced near the critical fields at which ferroelastic phase transitions occur. Apart from this, both low-frequency modes can be excited by the oscillating external electric field, i.e. they are electromagnons. The electromagnon contribution into the susceptibilities of BFO was calculated and it was found to be considerably large for the magnetoelectric susceptibility. We believe that our findings will help to build a solid theoretical foundation for further studies of the spin dynamics in BFO and to explain the existing experimental results. Understanding of the features of the spin waves in this multiferroic is required for development of promising magnonic and spintronic devices.

Acknowledgements: The work has been supported by Russian Foundation for Basic Research (grant 16-29-14005 ofi_m). AAM acknowledges the partial support of this work from the Russian Science Foundation (grant 16-12-10531).

Appendix

A Equations for variations $\delta\Omega$ and δP

We consider the case $\boldsymbol{l}_0 \parallel OX$. We find deviations $\boldsymbol{\delta\Omega}$ and $\boldsymbol{\delta P}$ from unperturbed state $(0, 0, \Omega_0)$ and $(0, 0, P_0)$ in the external electric and magnetic fields. Here and in Appendix B, the field components contain both static and oscillating parts of the external fields. The expanded form of the last two equations of the system (11):

$$\frac{\partial\Phi}{\partial\Omega_x} = -\frac{M_s}{\Omega_0}\left(H_z l_y - H_y l_z\right) - \frac{M_s H_D}{\Omega_0^2}\delta\Omega_x + \frac{M_s H_D}{\Omega_0^2} l_x(\boldsymbol{\Omega}, \boldsymbol{l}) + \frac{\partial^2\Phi}{\partial P_x \partial\Omega_x}\delta P_x + \frac{\partial^2\Phi}{\partial\Omega_x^2}\delta\Omega_x = 0 \tag{2.25}$$

$$\frac{\partial\Phi}{\partial\Omega_y} = \frac{M_s}{\Omega_0} H_z l_x - \frac{M_s H_D}{\Omega_0^2}\delta\Omega_y + \frac{M_s H_D}{\Omega_0^2} l_y(\boldsymbol{\Omega}, \boldsymbol{l}) + \frac{\partial^2\Phi}{\partial P_y \partial\Omega_y}\delta P_y + \frac{\partial^2\Phi}{\partial\Omega_y^2}\delta\Omega_y = \frac{\chi_\perp H_D}{\gamma\Omega_0}\cos\varphi_0\ \delta\dot{\varphi} \tag{2.26}$$

$$\frac{\partial\Phi}{\partial\Omega_z} = -\frac{M_s}{\Omega_0} H_y l_x - \frac{M_s H_D}{\Omega_0^2}\delta\Omega_z + \frac{M_s H_D}{\Omega_0^2} l_z(\boldsymbol{\Omega}, \boldsymbol{l}) + \frac{\partial^2\Phi}{\partial P_z \partial\Omega_z}\delta P_z + \frac{\partial^2\Phi}{\partial\Omega_z^2}\delta\Omega_z = -\frac{\chi_\perp H_D}{\gamma\Omega_0}\delta\dot{\theta} \tag{2.27}$$

$$\frac{\partial\Phi}{\partial P_x} = \beta\left(-\cos\varphi_0\ \theta_z + \varphi_y\right) + \frac{\partial^2\Phi}{\partial P_x \partial\Omega_x}\delta\Omega_x + \frac{\partial^2\Phi}{\partial P_x^2}\delta P_x - E_x = 0 \tag{2.28}$$

$$\frac{\partial\Phi}{\partial P_y} = -\beta\varphi_x + \frac{\partial^2\Phi}{\partial P_y \partial\Omega_y}\delta\Omega_y + \frac{\partial^2\Phi}{\partial P_y^2}\delta P_y - E_y = 0 \tag{2.29}$$

$$\frac{\partial\Phi}{\partial P_z} = \beta\cos\varphi_0\theta_x + \frac{\partial^2\Phi}{\partial P_z \partial\Omega_z}\delta\Omega_z + \frac{\partial^2\Phi}{\partial P_z^2}\delta P_z - E_z = 0. \tag{2.30}$$

The solution to this system of equations yields:

$$\delta\Omega_x = \eta_x^e\left(E_x - \beta\varphi_y + \beta\cos\varphi_0\theta_z\right) + \nu_1(E_z, H_y)\cos\varphi_0\delta\theta + \nu_2(E_y, H_z)\delta\varphi \tag{2.31}$$

$$\delta\Omega_y = \eta_y^e(E_y + \beta\varphi_x) - \frac{\chi_\perp H_D}{\gamma\Omega_0}\eta_y^{(1)}\cos\varphi_0\ \dot{\varphi} + \eta_y^{(1)}\frac{M_s}{\Omega_0}H_z\cos\varphi_0 + E_x\frac{H_D M_s}{\Omega_0^2}\eta_x^e\eta_y^{(1)}\delta\varphi \tag{2.32}$$

$$\delta\Omega_z = \eta_z^e(E_z - \beta\cos\varphi_0\theta_x) + \frac{\chi_\perp H_D}{\gamma\Omega_0}\eta_z^{(1)}\ \dot{\theta} - \eta_z^{(1)}\frac{M_s}{\Omega_0}H_y\cos\varphi_0 - E_x\frac{H_D M_s}{\Omega_0^2}\eta_x^e\eta_z^{(1)}\delta\theta\cos\varphi_0 \tag{2.33}$$

$$\delta P_x = \chi_x^e\left(E_x - \beta\varphi_y + \beta\cos\varphi_0\ \theta_z\right) + \tilde{\nu}_1(E_z, H_y)\delta\theta\cos\varphi_0 + \tilde{\nu}_2(E_y, H_z)\delta\varphi \tag{2.34}$$

$$\delta P_y = \chi_y^e(E_y + \beta\varphi_x) + \frac{\chi_\perp H_D}{\gamma\Omega_0}\eta_y^e\cos\varphi_0\ \dot{\varphi} - \eta_y^e\frac{M_s}{\Omega_0}H_z\cos\varphi_0 - E_x\frac{H_D M_s}{\Omega_0^2}\eta_x^e\eta_y^e\delta\varphi \tag{2.35}$$

$$\delta P_z = \chi_z^e(E_z - \beta\cos\varphi_0\theta_x) - \frac{\chi_\perp H_D}{\gamma\Omega_0}\eta_z^e\dot{\theta} + \eta_z^e\frac{M_s}{\Omega_0}H_y\cos\varphi_0 + E_x\frac{H_D M_s}{\Omega_0^2}\eta_x^e\eta_z^e\delta\theta\cos\varphi_0, \tag{2.36}$$

where $\eta_x^e = \left(\frac{\partial^2\Phi_{st}}{\partial P_x \partial \Omega_x}\right) / \left[\left(\frac{\partial^2\Phi_{st}}{\partial P_x \partial \Omega_x}\right)^2 - \left(\frac{\partial^2\Phi_{st}}{\partial P_x^2}\right)\left(\frac{\partial^2\Phi_{st}}{\partial \Omega_x^2}\right)\right]$,

$$\eta_i^e = \left(\frac{\partial^2\Phi_{st}}{\partial P_i \partial \Omega_i}\right) / \left[\left(\frac{\partial^2\Phi_{st}}{\partial P_i \partial \Omega_i}\right)^2 - \left(\frac{\partial^2\Phi_{st}}{\partial P_i^2}\right)\left(\frac{\partial^2\Phi_{st}}{\partial \Omega_i^2}\right) + \frac{H_D M_s}{\Omega_0^2}\left(\frac{\partial^2\Phi_{st}}{\partial P_i^2}\right)\right] (i = y, z),$$

$$\eta_x^{(1)} = \left(\frac{\partial^2\Phi_{st}}{\partial P_x^2}\right) / \left[\left(\frac{\partial^2\Phi_{st}}{\partial P_x \partial \Omega_x}\right)^2 - \left(\frac{\partial^2\Phi_{st}}{\partial P_x^2}\right)\left(\frac{\partial^2\Phi_{st}}{\partial \Omega_x^2}\right)\right]$$

$$\eta_i^{(1)} = \left(\frac{\partial^2\Phi_{st}}{\partial P_i^2}\right) / \left[\left(\frac{\partial^2\Phi_{st}}{\partial P_i \partial \Omega_i}\right)^2 - \left(\frac{\partial^2\Phi_{st}}{\partial P_i^2}\right)\left(\frac{\partial^2\Phi_{st}}{\partial \Omega_i^2}\right) + \frac{H_D M_s}{\Omega_0^2}\left(\frac{\partial^2\Phi_{st}}{\partial P_i^2}\right)\right] (i = y, z),$$

$$\chi_x^e = -\left(\frac{\partial^2\Phi_{st}}{\partial \Omega_x^2}\right) / \left[\left(\frac{\partial^2\Phi_{st}}{\partial P_x \partial \Omega_x}\right)^2 - \left(\frac{\partial^2\Phi_{st}}{\partial P_x^2}\right)\left(\frac{\partial^2\Phi_{st}}{\partial \Omega_x^2}\right)\right]$$

$$\chi_i^e = \left[-\left(\frac{\partial^2\Phi_{st}}{\partial \Omega_i^2}\right) + \frac{H_D M_s}{\Omega_0^2}\right] / \left[\left(\frac{\partial^2\Phi_{st}}{\partial P_i \partial \Omega_i}\right)^2 - \left(\frac{\partial^2\Phi_{st}}{\partial P_i^2}\right)\left(\frac{\partial^2\Phi_{st}}{\partial \Omega_i^2}\right) + \frac{H_D M_s}{\Omega_0^2}\left(\frac{\partial^2\Phi_{st}}{\partial P_i^2}\right)\right] (i = y, z),$$

$$\nu_1(E_z, H_y) = -\frac{M_s}{\Omega_0}\eta_x^{(1)} \frac{\frac{H_D}{\Omega_0}\eta_z^e E_z + H_D + H_y \cos\varphi_0}{1 + \frac{H_D M_s}{\Omega_0^2}\eta_z^{(1)}},$$

$$\nu_2(E_y, H_z) = \frac{M_s}{\Omega_0}\eta_x^{(1)} \frac{H_D \Omega_0 \eta_y^e E_y - H_z \cos\varphi_0}{1 + H_D M_s \Omega_0^2 \eta_y^{(1)}}$$

$$\widetilde{\nu}_1(E_z, H_y) = -\frac{\eta_x^e}{\eta_x^{(1)}}\nu_1(E_z, H_y),$$

$$\widetilde{\nu}_2(E_y, H_z) = -\frac{\eta_x^e}{\eta_x^{(1)}}\nu_2(E_y, H_z).$$

The coefficients $\eta_{x,y}^e$ correspond with the coefficient $\eta_\perp$ from article [72], and η_z^e corresponds with the coefficient $\eta_\parallel$. However, in the previous study η_x^e was equal to η_y^e, and the correction arises in the connection to the coupling between the antiferrodistortion and $\boldsymbol{l}_0$.

B Derivation of the equation for the magnon spectrum

Here we present the details of the derivation of the magnon spectrum. We start from eq. (11) in the expanded form:

$$\frac{\chi_\perp}{\gamma^2}\ddot{\theta}-\frac{\chi_\perp}{\gamma}\left(-2H_x\cos\varphi\sin^2\theta\dot{\varphi}-2H_y\sin\varphi\sin^2\theta\dot{\varphi}-H_z\sin 2\theta\dot{\varphi}\right)-\frac{\chi_\perp}{\gamma^2}\sin\theta\cos\theta\dot{\varphi}^2$$
$$=\frac{\chi_\perp}{\gamma}\left(-\dot{H}_x\sin\varphi+\dot{H}_y\cos\varphi\right)-\frac{\chi_\perp}{\gamma}\frac{H_D}{\Omega_0}\left(\dot{\Omega}_x\cos\theta\cos\varphi+\dot{\Omega}_y\cos\theta\sin\varphi-\dot{\Omega}_z\sin\theta\right)$$
$$-\frac{\chi_\perp}{\gamma}\frac{H_D}{\Omega_0}\left(\frac{\partial}{\partial\boldsymbol{\theta}}-\frac{\mathrm{d}}{\mathrm{d}t}\frac{\partial}{\partial\dot{\theta}}-\sum_{i=x,y,z}\frac{\partial}{\partial x_i}\frac{\partial}{\partial\theta_{x_i}}\right)\left([\boldsymbol{\Omega l}]\cdot[\boldsymbol{l}\dot{\boldsymbol{l}}]\right)-\alpha\frac{M_0}{\gamma}\dot{\theta}-\frac{\delta\Phi}{\delta\theta}, \tag{2.37}$$

$$\frac{\chi_\perp}{\gamma^2}\left(\sin^2\theta\ddot{\varphi}+\sin 2\theta\dot{\varphi}\dot{\theta}\right)-\frac{\chi_\perp}{\gamma}\left(2H_x\cos\varphi\sin^2\theta\,\dot{\theta}+2H_y\sin\varphi\sin^2\theta\,\dot{\theta}+H_z\sin 2\theta\,\dot{\theta}\right)$$
$$=\frac{\chi_\perp}{\gamma}\left[-\cos\theta\sin\theta(\dot{H}_x\cos\varphi-\dot{H}_y\sin\varphi)+\dot{H}_z\sin^2\theta\right]$$
$$-\frac{\chi_\perp}{\gamma}\frac{H_D}{\Omega_0}\left(\dot{\Omega}_y\sin\theta\cos\varphi-\dot{\Omega}_x\sin\theta\sin\varphi\right)$$
$$-\frac{\chi_\perp}{\gamma}\frac{H_D}{\Omega_0}\left(\frac{\partial}{\partial\boldsymbol{\varphi}}-\frac{\mathrm{d}}{\mathrm{d}t}\frac{\partial}{\partial\dot{\varphi}}-\sum_{i=x,y,z}\frac{\partial}{\partial x_i}\frac{\partial}{\partial\varphi_{x_i}}\right)\left([\boldsymbol{\Omega l}]\cdot[\boldsymbol{l}\dot{\boldsymbol{l}}]\right)-\alpha\frac{M_0}{\gamma}\sin^2\theta\dot{\varphi}-\frac{\delta\Phi}{\delta\varphi}, \tag{2.38}$$

where

$$\Phi\approx-\frac{\chi_\perp}{2}\left[\boldsymbol{H}^2-(\boldsymbol{Hl})^2\right]-\frac{M_s}{\Omega_0}(\boldsymbol{H},[\boldsymbol{\Omega},\boldsymbol{l}])-\frac{M_sH_D}{2\Omega_0^2}\left(\boldsymbol{\Omega}^2-(\boldsymbol{\Omega},\boldsymbol{l})^2\right)+E_a(\theta,\varphi)$$
$$+\,A\left[(\nabla\theta)^2+\sin^2\theta(\nabla\varphi)^2\right]+\beta\boldsymbol{P}(\boldsymbol{l}(\nabla\boldsymbol{l})-(\boldsymbol{l}\nabla)\boldsymbol{l})+\Phi_{st}(\boldsymbol{P},\boldsymbol{\Omega})-\boldsymbol{PE}, \tag{2.39}$$

and

$$\beta\mathbf{P}(l(\nabla l)-(l\nabla)l)=\beta\left[P_x\left(\sin\theta\cos\theta\sin\varphi\,\varphi_z-\cos\varphi\,\theta_z+\sin^2\theta\,\varphi_y\right)\right.$$
$$+P_y\left(-\cos\varphi\sin\theta\cos\theta\,\varphi_z-\sin\varphi\,\theta_z-\sin^2\theta\,\varphi_x\right)$$
$$\left.+P_z\left(\sin\varphi\,\theta_y+\sin\theta\cos\theta(\cos\varphi\,\varphi_y-\sin\varphi\,\varphi_x)+\cos\varphi\,\theta_x\right)\right]. \tag{2.40}$$

The second-order expansion of the thermodynamic potential has the following form (some terms, which do not contribute into the Euler-Lagrange eqs. (37) and (38) have been excluded):

$$\Phi \approx -\frac{\chi_\perp}{2}\left(H_y^2 + H_z^2\delta\theta^2 - H_yH_z\delta\theta\delta\varphi\cos\varphi_0\right)$$
$$-\frac{M_s}{\Omega_0}\left(-H_y\delta\Omega_x + (H_y(\Omega_0+\delta\Omega_z) - H_z\delta\Omega_y)\delta\theta\cos\varphi_0\right)$$
$$-\frac{M_sH_D}{2\Omega_0^2}\left[\delta\Omega_y^2 + \delta\Omega_z^2 + 2\Omega_0\cos\varphi_0\,\delta\Omega_x\delta\theta - \Omega_0^2\delta\theta^2\right] + K_u\delta\theta^2 + A\left[(\nabla\theta)^2 + (\nabla\varphi)^2\right]$$
$$+\beta\delta P_x(-\cos\varphi_0\,\theta_z + \varphi_y) + \beta\delta P_y(\cos\varphi_0\,\delta\theta\varphi_z - \cos\varphi_0\delta\varphi\theta_z - \varphi_x)$$
$$+\beta(P_0+\delta P_z)\left[\cos\varphi_0\,\delta\varphi\theta_y - \cos\varphi_0\,\delta\theta\varphi_y + \cos\varphi_0\,\theta_x\right] + \delta^2\Phi_{st}(\boldsymbol{P},\Omega,\boldsymbol{E}). \tag{2.41}$$

Substituting this expression into eqs. (37), (38) and using the result form Appendix A, the system (15) was obtained.

In the system $\omega_j^2(\boldsymbol{k}) = \omega_{j0}^2 + \sum_{i=x,y,z} d_{ji}k_i^2$ $(j=1,2)$, $\omega_{12}^2(\boldsymbol{k}) = \left(\omega_{21}^2(\boldsymbol{k})\right)^* = C + A_{yz}k_yk_z + iB_yk_y + iB_zk_z$ and the parameters are:

$$\omega_{10}^2 = \gamma^2\left(\frac{2K_u}{\chi_\perp} + H_z^2 + H_D\left(H_D + H_y\cos\varphi_0\right)\right) + \gamma^2\frac{\chi_\perp^2H_D^4}{\Omega_0^4}\left(\eta_z^{(1)}\right)^2H_y^2$$
$$-\gamma^2\frac{\chi_\perp H_D^2}{\Omega_0^2}\eta_z^{(1)}H_y\left(2H_D + H_y\cos\varphi_0\right) - \gamma^2\frac{H_DH_z}{\Omega_0}\eta_y^eE_y\cos\varphi_0$$
$$-2\gamma^2\frac{\chi_\perp H_D^3}{\Omega_0^3}\eta_z^{(1)}H_y\,\eta_z^eE_z\cos\varphi_0 + \gamma^2\frac{H_D\left(2H_D + H_y\cos\varphi_0\right)}{\Omega_0}\eta_z^eE_z + \nu_3(E_z,H_y),$$

$$c_\omega^{(1)} = 1 - \frac{\chi_\perp H_D^2}{\Omega_0^2}\eta_z^{(1)},$$

$$d_{1x} = \gamma^2\frac{2A}{\chi_\perp} - \gamma^2\frac{\chi_z^e}{\chi_\perp}\beta^2,\; d_{1y} = \frac{2A\gamma^2}{\chi_\perp},\; d_{1z} = \frac{2A\gamma^2}{\chi_\perp} - \gamma^2\frac{\chi_x^e}{\chi_\perp}\beta^2,$$

$$c_x^{(1)} = -2\gamma\frac{H_D}{\Omega_0}\beta\eta_z^e\cos\varphi_0,$$

$$A_{yz} = \gamma^2\frac{\chi_x^e}{\chi_\perp}\beta^2,$$

$$B_y = -2\frac{\gamma^2}{\chi_\perp}\beta(P_0 + \chi_z^eE_z)\cos\varphi_0 + \gamma^2\frac{H_D^2}{\Omega_0^2}\eta_x^e\beta\,\eta_z^eE_z\cos\varphi_0$$
$$+\gamma^2\frac{H_D(H_D + H_y\cos\varphi_0)}{\Omega_0}\beta\eta_x^e\cos\varphi_0 - \gamma^2\frac{H_D}{\Omega_0}\left(2\eta_z^e + \frac{\chi_\perp H_D^2}{\Omega_0^2}\eta_x^e\eta_z^{(1)}\right)\beta H_y,$$

$$B_z = \gamma^2\frac{\beta}{\chi_\perp}\left(\widetilde{\nu}_2(E_y,H_z) + 2\chi_y^eE_y - 2\frac{\chi_\perp H_D}{\Omega_0}\eta_y^eH_z\cos\varphi_0\right)\cos\varphi_0,$$

$$C = v_4(E_y, E_z, H_y, H_z),$$

$$\omega_{20}^2 = \gamma^2\left(H_y^2 + H_y H_D \cos\varphi_0\right) - \gamma^2\frac{H_D}{\Omega_0}\left(2\frac{\chi_\perp H_D^2}{\Omega_0^2}\eta_y^{(1)}H_z - H_z\right)\eta_y^e E_y\cos\varphi_0$$

$$+\ \gamma^2\frac{H_D H_y}{\Omega_0}\eta_z^e E_z\cos\varphi_0 + \gamma^2\frac{\chi_\perp^2 H_D^2 4}{\Omega_0^4}\left(\eta_y^{(1)}\right)^2$$

$$H_z^2 - \gamma^2\frac{\chi_\perp H_D^2}{\Omega_0^2}\left(\eta_z^{(1)}H_y^2 + \eta_y^{(1)}H_z^2\right) + \widetilde{v}_3(E_y, H_z),$$

$$c_\omega^{(2)} = 1 - \frac{\chi_\perp H_D^2}{\Omega_0^2}\eta_y^{(1)},$$

$$d_{2x} = \frac{2A\gamma^2}{\chi_\perp} - \gamma^2\frac{\chi_y^e}{\chi_\perp}\beta^2,\ d_{2y} = d_{1z},\ d_{2z} = d_{1y},$$

$$c_x^{(2)} = -2\gamma\frac{H_D}{\Omega_0}\beta\eta_y^e\cos\varphi_0,$$

where $v_3(E_z, H_y)$, $\widetilde{v}_3(E_y, H_z)$, $v_4(E_y, E_z, H_y, H_z)$, are the coefficients, linear in terms of E_i and H_j, $(i,j) = (z,y)||(y,z)$, which are expressed using $v_1(E_z, H_y)$, $v_2(E_y, H_z)$, $\widetilde{v}_1(E_z, H_y)$ and $\widetilde{v}_2(E_y, H_z)$ and which we omit here due to the bulkiness.

References

[1] Scott J. Data storage: multiferroic memories Nat Mater. 2007;6:256.

[2] Dzyaloshinskii IE. On the magneto-electrical effects in antiferromagnets. Sov Phys JETP. 1960;10:628.

[3] Astrov D. The magnetoelectric effect in antiferromagnetics. Sov Phys JETP. 1960;11:708.

[4] Rado GT, Folen VJ. Observation of the magnetically induced magnetoelectric effect and evidence for antiferromagnetic domains. Phys Rev Lett. 1961;7:310.

[5] Pyatakov AP, Zvezdin AK. Magnetoelectric and multiferroic media. Phys-Usp. 2012;55:557.

[6] Popov YF, Pyatakov AP, Kadomtseva AM, Vorob'ev GP, Zvezdin AK, Mukhin AA, et al. Peculiarities in the magnetic, magnetoelectric, and magnetoelastic properties of SmFe3(Bo3) 4 multiferroic. J Exp Theor Phys. 2010;111:199.

[7] Kadomtseva A, Popov YF, Vorob'ev G, Pyatakov A, Krotov S, Kamilov K. Magnetoelectric and magnetoelastic properties of rare-earth ferroborates. Low Temp Phys. 2010;36:511.

[8] Zvezdin A, Kadomtseva A, Krotov S, Pyatakov A, Popov YF, Vorob'ev G. Magnetoelectric interaction and magnetic field control of electric polarization in multiferroics. J Magn Magn Mater. 2006;300:224.

[9] Choi YJ, Zhang CL, Lee N, Cheong S-W. Cross-control of magnetization and polarization by electric and magnetic fields with competing multiferroic and weak-ferromagnetic phases. Phys Rev Lett. 2010;105:097201.

[10] Heron JT, Schlom DG, Ramesh R. Electric field control of magnetism using $BiFeO_3$-based heterostructures. Appl Phys Rev. 2014;1:021303. DOI: 10.1063/1.4870957.
[11] Matsukura F, Tokura Y, Ohno H. Control of magnetism by electric fields. Nat Nanotechnol. 2015;10:209.
[12] Golovenchits E, Sanina V. Magnetic-field-induced phase transition in $Tb_{0.95}Bi_{0.05}MnO_{3+\delta}$ multiferroic. JETP Lett. 2006;84:190.
[13] Agbelele A, Sando D, Toulouse C, Paillard C, Johnson R, Rüffer R, et al. Strain and magnetic field induced spin-structure transitions in multiferroic $BiFeO_3$. Adv Mater. 2017;29:1602327.
[14] Smirnov AI, Khlyustikov IN. Magnetoelectric effects and the stark effect in antiferromagnetic Gd_2CuO_4. Phys-Usp. 1995;38:1169.
[15] Turov EA, Nikolaev VV. New physical phenomena caused by magnetoelectric and antiferroelectric interactions in magnets. Phys-Usp. 2005;48:431.
[16] Pimenov A, Mukhin A, Ivanov VY, Travkin V, Balbashov A, Loidl A. Possible evidence for electromagnons in multiferroic manganites. Nat Phys. 2006;2:97.
[17] Pimenov A, Shuvaev A, Mukhin A, Loidl A. Electromagnons in multiferroic manganites. J Phys: Condens Matter. 2008;20:434209.
[18] Rovillain P, De Sousa R, Gallais Yt, Sacuto A, Méasson M, Colson D, et al. Electric-field control of spin waves at room temperature in multiferroic $BiFeO_3$. Nat Mater. 2010;9:975.
[19] Liu H, Yang X. A brief review on perovskite multiferroics. Ferroelectrics 2017;507:69. DOI: 10.1080/00150193.2017.1283171.
[20] Catalan G, Scott JF. Physics and applications of bismuth ferrite. Adv Mater. 2009;21:2463.
[21] Eerenstein W, Mathur ND, Scott JF. Multiferroic and magnetoelectric materials. Nature 2006;442:759.
[22] Yang SY, Martin LW, Byrnes SJ, Conry TE, Basu SR, Paran D, et al. Photovoltaic effects in $BiFeO_3$. Appl Phys Lett. 2009;95:062909.
[23] Zvezdin AK, Pyatakov AP. Phase transitions and the giant magnetoelectric effect in multiferroics. Phys-Usp 2004;47:416.
[24] Spaldin N, Ramesh R. Advances in magnetoelectric multiferroics. Nat Mater. 2019;18:203.
[25] Campanini M, Erni R, Yang C, Ramesh R, Rossell MD. Periodic giant polarization gradients in doped $BiFeO_3$ thin films. Nano Lett. 2018;18. DOI: 10.1021/acs.nanolett.7b03817.717.pMID: 29314853.
[26] Guennou M, Viret M, Kreisel J. Bismuth-based perovskites as multiferroics. C R Phys. 2015;16:182.Multiferroic materials and heterostructures / Matriaux et htrostructures multiferroques.
[27] Kubel F, Schmid H. Structure of a ferroelectric and ferroelastic monodomain crystal of the perovskite $BiFeO_3$. Acta Crystallogr Sect B: Struct Sci. 1990;46:698.
[28] Kadomtseva A, Zvezdin A, Popov Y, Pyatakov A, Vorob'ev G. Space-time parity violation and magnetoelectric interactions in antiferromagnets. JETP Lett. 2004;79:571.
[29] Sosnowska I, Neumaier TP, Steichele E. Spiral magnetic ordering in bismuth ferrite. J Phys C: Solid State Phys. 1982;15:4835.
[30] Sosnowska I, Zvezdin A. Origin of the long period magnetic ordering in $BiFeO_3$. J Magn Magn Mater. 1995;140:167.
[31] Tabares-Munoz C, Rivera JP, Bezinges A, Monnier A, Schmid H. Measurement of the quadratic magnetoelectric effect on single crystalline $BiFeO_3$. Jap J Appl Phys. 1985;24:1051.
[32] Zalesskii A, Zvezdin A, Frolov A, Bush A. 57 Fe NMR study of a spatially modulated magnetic structure in $BiFeO_3$. J Exp Theor Phys Lett. 2000;71:465.
[33] Zalesskii A, Frolov A, Khimich T, Bush A. Composition-induced transition of spin-modulated structure into a uniform antiferromagnetic state in a $Bi_{1-x}La_xFeO_3$ system studied using 57 Fe NMR. Phys Solid State. 2003;45:141.

[34] Ramazanoglu M, Ratcliff IW, Choi YJ, Lee S, Cheong S-W Kiryukhin V. Temperature-dependent properties of the magnetic order in single-crystal $BiFeO_3$. Phys Rev B. 2011;83:174434.
[35] Sosnowska I, Przeniosło R. Low-temperature evolution of the modulated magnetic structure in the ferroelectric antiferromagnet $BiFeO_3$. Phys Rev B. 2011;84:144404.
[36] Gareeva Z, Zvezdin K, Pyatakov A, Zvezdin A. Novel type of spin cycloid in epitaxial bismuth ferrite films. J Magn Magn Mater. 2019;469:593.
[37] Rusakov V, Pokatilov V, Sigov AS, Matsnev M, Gubaidulina T. Diagnostics of a spatial spin-modulated structure using nuclear magnetic resonance and Mössbauer spectroscopy. JETP Lett. 2014;100:463.
[38] Popov YF, Zvezdin AK, Vorob'ev GP, Kadomtseva AM, Murashev VA, Rakov DN. Linear magnetoelectric effect and phase transitions in bismuth ferrite $BiFeO_3$. JETP Lett. 1993;57:69.
[39] Popov YF, Kadomtseva AM, Vorobev GP, Zvezdin AK. Discovery of the linear magnetoelectric effect in magnetic ferroelectric $BiFeO_3$ in a strong magnetic field. Ferroelectrics. 1994;162:135.
[40] Popov YF, Kadomtseva AM, Krotov SS, Belov DV, Vorob'ev GP, Makhov PN, et al. Features of the magnetoelectric properties of $BiFeO_3$ in high magnetic fields. Low Temp Phys. 2001;27:478.
[41] Ruette B, Zvyagin S, Pyatakov AP, Bush A, Li JF, Belotelov VI, et al. Magnetic-field-induced phase transition in $BiFeO_3$ observed by high-field electron spin resonance: cycloidal to homogeneous spin order. Phys Rev B. 2004;69:064114.
[42] Le Bras G, Colson D, Forget A, Genand-Riondet N, Tourbot R, Bonville P. Magnetization and magnetoelectric effect in $Bi_{1-x}La_xFeO_3(0 \le x \le 0.15)$. Phys Rev B. 2009;80:134417.
[43] Gabbasova Z, Kuz'min M, Zvezdin A, Dubenko I, Murashov V, Rakov D, et al. $Bi_{1-x}R_xFeO_3$ (r=rare earth): a family of novel magnetoelectrics. Phys Lett A. 1991;158:491.
[44] Lazenka VV, Ravinski AF, Makoed II, Vanacken J, Zhang G, Moshchalkov VV. Weak ferromagnetism in La-doped $BiFeO_3$ multiferroic thin films. J Appl Phys. 2012;111:123916.
[45] Wang J, Neaton JB, Zheng H, Nagarajan V, Ogale SB, Liu B, et al. Epitaxial $BiFeO_3$ multiferroic thin film heterostructures. Science. 2003;299:1719.
[46] Bai F, Wang J, Wuttig M, Li J, Wang N, Pyatakov AP, et al. Destruction of spin cycloid in (111)c-oriented $BiFeO_3$ thin films by epitiaxial constraint: Enhanced polarization and release of latent magnetization. Appl Phys Lett. 2005;86:032511.
[47] Bea H, Bibes M, Herranz G, Zhu XH, Fusil S, Bouzehouane K, et al. Integration of multiferroic $BiFeO_3$ thin films into heterostructures for spintronics. IEEE Trans Magn. 2008;44:1941.
[48] Sando D, Agbelele A, Rahmedov D, Liu J, Rovillain P, Toulouse C, et al. Crafting the magnonic and spintronic response of $BiFeO_3$ films by epitaxial strain. Nat Mater. 2013;12:641.
[49] Lorenz M, Lazenka V, Schwinkendorf P, Bern F, Ziese M, Modarresi H, et al. Multiferroic $BaTiO_3$-$BiFeO_3$ composite thin films and multilayers: strain engineering and magnetoelectric coupling. J Phys D: Appl Phys. 2014;47:135303.
[50] Lorenz M, Wagner G, Lazenka V, Schwinkendorf P, Modarresi H, Van Bael MJ, et al. Correlation of magnetoelectric coupling in multiferroic $BaTiO_3$-$BiFeO_3$ superlattices with oxygen vacancies and antiphase octahedral rotations. Appl Phys Lett. 2015;106:012905.
[51] Wang J, Neaton JB, Zheng H, Nagarajan V, Ogale SB, Liu B, et al. Epitaxial $BiFeO_3$ multiferroic thin film heterostructures. Science. 2003;299:1719. https://science.sciencemag.org/content/299/5613/1719.full.pdf.
[52] Ke X, Zhang PP, Baek SH, Zarestky J, Tian W, Eom CB. Magnetic structure of epitaxial multiferroic $BiFeO_3$ films with engineered ferroelectric domains. Phys Rev B. 2010;82:134448.
[53] Kumar A, Murari NM, Katiyar RS. Observation of one magnon and magnon-phonon-electric dipole coupling in multiferroics bismuth ferrite thin films. Appl Phys Lett. 2008;92:152907.

[54] Cazayous M, Gallais Y, Sacuto A, de Sousa R, Lebeugle D, Colson D. Possible observation of cycloidal electromagnons in $BiFeO_3$. Phys Rev Lett. 2008;101:037601.
[55] Azeem W, Riaz S, Bukhtiar A, Hussain SS, Xu Y, Naseem S. Ferromagnetic ordering and electromagnons in microwave synthesized $BiFeO_3$ thin films. J Magn Magn Mater. 2019;475:60.
[56] Wang D, Weerasinghe J, Bellaiche L. Atomistic molecular dynamic simulations of multiferroics. Phys Rev Lett. 2012;109:067203.
[57] de Sousa R, Moore JE. Optical coupling to spin waves in the cycloidal multiferroic $BiFeO_3$. Phys Rev B. 2008;77:012406.
[58] Chang C-M, Mani B, Lisenkov S, Ponomareva I. Prediction of electromagnons in antiferromagnetic ferroelectrics from first-principles: the case of $BiFeO_3$. Ferroelectrics. 2016;494:68.
[59] Popkov AF, Kulagin NE, Soloviov SV, Gareeva ZV, Zvezdin AK. Spin wave spectra and spatially modulated structures in $BiFeO_3$. Low Temp Phys. 2014;40:58.
[60] Yang M-M, Alexe M. Light-induced reversible control of ferroelectric polarization in $BiFeO_3$. Adv Mater. 2018;30:1704908. https://onlinelibrary.wiley.com/doi/pdf/10.1002/adma.201704908.
[61] Komandin GA, Torgashev VI, Volkov AA, Porodinkov OE, Spektor IE, Bush AA. Optical properties of $BiFeO_3$ ceramics in the frequency range 0.3–30.0 THz. Phys Solid State. 2010;52:734.
[62] Livesey KL, Stamps RL. High-frequency susceptibility of a weak ferromagnet with magnetostrictive magnetoelectric coupling: using heterostructures to tailor electromagnon frequencies. Phys Rev B. 2010;81:094405.
[63] Kézsmárki I, Nagel U, Bordács S, Fishman RS, Lee JH, Yi HT. Optical diode effect at spin-wave excitations of the room-temperature multiferroic $BiFeO_3$. Phys Rev Lett. 2015;115:127203.
[64] Chumak A, Vasyuchka V, Serga A, Hillebrands B. Magnon spintronics. Nat Phys. 2015;11:453.
[65] de Sousa R, Allen M, Cazayous M. Theory of spin-orbit enhanced electric-field control of magnetism in multiferroic $BiFeO_3$. Phys Rev Lett. 2013;110:267202.
[66] Risinggård V, Kulagina I, Linder J. Electric field control of magnon-induced magnetization dynamics in multiferroics. Sci Rep. 2016;6:31800.
[67] Eliseev EA, Kalinin SV, Gu Y, Glinchuk MD, Khist V, Borisevich A, et al. Universal emergence of spatially modulated structures induced by flexoantiferrodistortive coupling in multiferroics. Phys Rev B. 2013;88:224105.
[68] Lee SS, Kim Y-M, Lee H-J, Seo O, Jeong HY, He Q, et al. Correlation between geometrically induced oxygen octahedral tilts and multiferroic behaviors in $BiFeO_3$ films. Adv Funct Mater. 2018;28:1800839. https://onlinelibrary.wiley.com/doi/pdf/10.1002/adfm.201800839.
[69] Kim Y-M, A Kumar, Hatt A , Morozovska AN, Tselev A , Biegalski MD, et al. Interplay of octahedral tilts and polar order in $BiFeO_3$ films. Adv Mater. 2013;25:2497. https://onlinelibrary.wiley.com/doi/pdf/10.1002/adma.201204584.
[70] Borisevich AY, Chang HJ, Huijben M, Oxley MP, Okamoto S, Niranjan MK, et al. Suppression of octahedral tilts and associated changes in electronic properties at epitaxial oxide heterostructure interfaces. Phys Rev Lett. 2010;105:087204.
[71] Zvezdin AK, Pyatakov AP, On the problem of coexistence of the weak ferromagnetism and the spin flexoelectricity in multiferroic bismuth ferrite. EPL (Europhys Lett). 2012;99:57003.
[72] Popkov AF, Davydova MD, Zvezdin KA, Solov'yov SV, Zvezdin AK. Origin of the giant linear magnetoelectric effect in perovskite like multiferroic $BiFeO_3$. Phys Rev B. 2016;93:094435.
[73] Xu C, Xu B, Dupé B, Bellaiche L. Magnetic interactions in $BiFeO_3$: a first-principles study. Phys Rev B. 2019;99:104420.

[74] Chen Z, Chen Z, Kuo C-Y, Tang Y, Dedon LR, Li Q, et al. Complex strain evolution of polar and magnetic order in multiferroic $BiFeO_3$ thin films. Nat Commun. 2018;9:3764.
[75] Burns SR, Sando D, Xu B, Dupé B, Russell L, Deng G, et al. Expansion of the spin cycloid in multiferroic $BiFeO_3$ thin films. NPJ Quant Mater. 2019;4:18.
[76] Kulagin N, Popkov A, Solov'ev S, Sukmanova K, Zvezdin A. Electric-field-induced structural and magnetic transformations in $BiFeO_3$ multiferroics. Phys Solid State. 2015;57:933.
[77] Popkov AF, Kulagin NE, Soloviov SV, Sukmanova KS, Gareeva ZV, Zvezdin AK. Cycloid manipulation by electric field in $BiFeO_3$ films: coupling between polarization, octahedral rotation, and antiferromagnetic order. Phys Rev B. 2015;92:140414.
[78] Jamali M, Kwon JH, Seo S-M, Lee K-J, Yang H. Spin wave nonreciprocity for logic device applications. Sci Rep. 2013;3:3160.
[79] Schneider T, Serga AA, Neumann T, Hillebrands B, Kostylev MP. Phase reciprocity of spin-wave excitation by a microstrip antenna. Phys Rev B. 2008;77:214411.
[80] Zvezdin A. Dynamics of domain walls in weak ferromagnets. Sov J Exp Theor Phys Lett. 1979;29:553.
[81] Andreev AF, Marchenko VI. Symmetry and the macroscopic dynamics of magnetic materials. Sov Phys Usp. 1980;23:21.
[82] Zvezdin AK, Mukhin AA. Lebedev Phys Inst Lett. (Kratk Soobsh Fiz.) 1981;12:10–15.
[83] Bar'yakhtar V, Ivanov B, Sukstanskii A. Nonlinear waves and the dynamics of domain walls in weak ferromagnets. JETP. 1980;51:757.
[84] Zvezdin AK, Mukhin AA. On the effect of inhomogeneous magnetoelectric (flexomagnetoelectric) interaction on the spectrum and properties of magnons in multiferroics. JETP Lett. 2009;89:328.
[85] Lisenkov S, Rahmedov D, Bellaiche L. Electric-field-induced paths in multiferroic $BiFeO_3$ from atomistic simulations. Phys Rev Lett. 2009;103:047204.
[86] Takahashi Y, Shimano R, Kaneko Y, Murakawa H, Tokura Y. Magnetoelectric resonance with electromagnons in a perovskite helimagnet. Nat Phys. 2012;8:121.
[87] Kibayashi S, Takahashi Y, Seki S, Tokura Y. Magnetochiral dichroism resonant with electromagnons in a helimagnet. Nat Commun. 2014;5:4583.
[88] Kézsmárki I, Szaller D, Bordács S, Kocsis V, Tokunaga Y, Taguchi Y, et al. One-way transparency of four-coloured spin-wave excitations in multiferroic materials. Nat Commun. 2014;5:3203.
[89] Kézsmárki I, Kida N, Murakawa H, Bordács S, Onose Y, Tokura Y. Enhanced directional dichroism of terahertz light in resonance with magnetic excitations of the multiferroic $Ba_2CoGe_2O_7$ oxide compound. Phys Rev Lett. 2011;106:057403.
[90] Takahashi Y, Yamasaki Y, Tokura Y. Terahertz magnetoelectric resonance enhanced by mutual coupling of electromagnons. Phys Rev Lett. 2013;111:037204.
[91] Kuzmenko AM, Dziom V, Shuvaev A, Pimenov A, Schiebl M, Mukhin AA, et al. Large directional optical anisotropy in multiferroic ferroborate. Phys Rev B. 2015;92:184409.
[92] Agbelele A, Sando D, Toulouse C, Paillard C, Johnson R, Rüffer R, et al. Strain and magnetic field induced spin-structure transitions in multiferroic $BiFeO_3$. Adv Mater. 2016;29:1602327.
[93] Ziman JM, Armstrong H. Principles of the theory of solids. Am J Phys. 1965;33:349.
[94] Kumar A, Scott JF, Katiyar RS. Electric control of magnon frequencies and magnetic moment of bismuth ferrite thin films at room temperature. Appl Phys Lett. 2011;99:062504.
[95] Lebeugle D, Colson D, Forget A, Viret M, Bonville P, Marucco JF, Fusil S. Room-temperature coexistence of large electric polarization and magnetic order in $BiFeO_3$ single crystals. Phys Rev B. 2007;76:024116.

Bernd Lorenz

3 Hexagonal manganites: Strong coupling of ferroelectricity and magnetic orders

Abstract: Hexagonal manganites belong to an exciting class of materials exhibiting strong interactions between a highly frustrated magnetic system, the ferroelectric polarization, and the lattice. The existence and mutual interaction of different magnetic ions (Mn and rare earth) results in complex magnetic phase diagrams and novel physical phenomena. A summary and discussion of the various properties, underlying physical mechanisms, the role of the rare earth ions, and the complex interactions in multiferroic hexagonal manganites are presented in this review.

Keywords: multiferroics, hexagonal, manganites, magnetoelectric

3.1 Introduction

Multiferroic compounds can be separated into two classes depending on whether the ferroelectricity and magnetic orders arise independently (class I) or the magnetic order breaks the inversion symmetry and induces a ferroelectric state through electronic polarization or ionic displacements (class II) [1]. This review is devoted to a comprehensive discussion of hexagonal manganites, as representatives of class I multiferroics.

The hexagonal rare earth manganites $RMnO_3$ (h-$RMnO_3$, R = rare earth or Y, Sc, In) are a typical example of class (I) multiferroics which have been studied extensively over several decades [2–5]. While the ferroelectric transition temperature, T_C, of most $RMnO_3$ is of the order of 1,000 K [6], the antiferromagnetic order sets in at T_N typically below 100 K. The first signature of a coupling between both orders had been observed in $YMnO_3$ in form of a distinct anomaly of the dielectric constant at the magnetic transition temperature T_N [7]. This observation revealed the strong interaction between the two orders even at low temperatures where the ferroelectric state is already very stable.

As a typical example, $YMnO_3$ has only one magnetic ion, Mn^{3+}, and the Mn spins order in a frustrated antiferromagnetic structure below T_N. However, other rare earth ions carry their own magnetic moment which interacts with the Mn spins. This increases the complexity of interactions between two magnetic species and the ferroelectric polarization and it results in rich phase diagrams with various phases of

This article has previously been published in the journal Physical Sciences Reviews. Please cite as: Lorenz, B. Hexagonal manganites: Strong coupling of ferroelectricity and magnetic orders *Physical Sciences Reviews* [Online] 2019, 4. DOI: 10.1515/psr-2019-0014

https://doi.org/10.1515/9783110582130-003

different magnetic symmetries and effects on the dielectric properties and the ferroelectric state. Section 3.2 provides a general introduction into the structure and physical properties of h-$RMnO_3$. In Section 3.3, we discuss the complex multiferroic phase diagram and properties of $HoMnO_3$. Section 3.4 covers the physics and multiferroic properties of h-$RMnO_3$ with nonmagnetic R-ions (R = Y, Lu, Sc, and In) whereas the remaining h-$RMnO_3$ with magnetic rare earth ions (R = Er, Tm, Yb, and Dy) are discussed in Section 3.5. A brief summary and outlook is presented in the Section 3.6.

3.2 Preliminaries: structure, ferroelectricity, and magnetism

Ferroelectricity in hexagonal manganites $RMnO_3$, with R = Y, Dy, Ho, Er, Tm, Yb, Lu, or Sc, was discovered as early as 1963 by Bertaut et al. [2]. At ambient temperature the crystal structure is described by the polar space group $P6_3cm$ (No. 185) with five- and sevenfold coordination polyhedra about the Mn and R ions, respectively [3]. Figure 3.1 shows a sketch of the structure. The origin of ferroelectricity in hexagonal $RMnO_3$ has been a matter of discussion. Two common causes of ferroelectricity in materials are (i) the "d^0-ness" where an off center displacement in perovskite ferroelectrics (like $BaTiO_3$) arises from the hybridization of empty transition metal orbitals with oxygen 2p states and the associated second-order Jahn-Teller effect [8] and (ii) the presence of a "lone pair" (s^2) set of electrons which may cause the loss of inversion symmetry through a mixing with an excited (s^1)(p^1) state [9] as, for example, realized in $BiMnO_3$ [10]. Both mechanisms do not apply to h-$RMnO_3$, so the polar lattice distortion has to be sought in other physical effects.

The structure and ferroelectricity of $YMnO_3$ have been studied experimentally and theoretically. Van Aken et al. [11] concluded from first principle calculations that the ferroelectricity in $YMnO_3$ is due to electrostatic and size effects. The polar distortion arises from the rotation of the MnO_5 bipyramids displacing the oxygen atoms from their centrosymmetric positions. The Y atoms are shifted along the c-axis forming a buckled triangular lattice in the ferroelectric state. The Y–O displacements create large local electric dipoles along the c-axis which are antiparallel (but of different magnitude) for the two inequivalent yttrium ions of the structure. Therefore, the polar state of $YMnO_3$ (and all h-$RMnO_3$) is ferrielectric.

$YMnO_3$ is centrosymmetric (space group $P6_3/mmc$, No. 195) above $T_S \simeq$1,270 K [12–16]. The transition temperature T_C into the ferroelectric $P6_3cm$ phase was reported to be about 300 K lower and the possible existence of an intermediate phase (space group $P6_3/mcm$) between T_C and T_S was suggested [13, 17]. However, the results of recent neutron scattering experiments are in favor of a direct transition from $P6_3/mmc$ to $P6_3cm$ with a tripling of the unit cell and no intermediate symmetry phase [16]. Anomalies observed by different authors near 920 K were attributed to an isosymmetric transition within the same space group, $P6_3cm$.

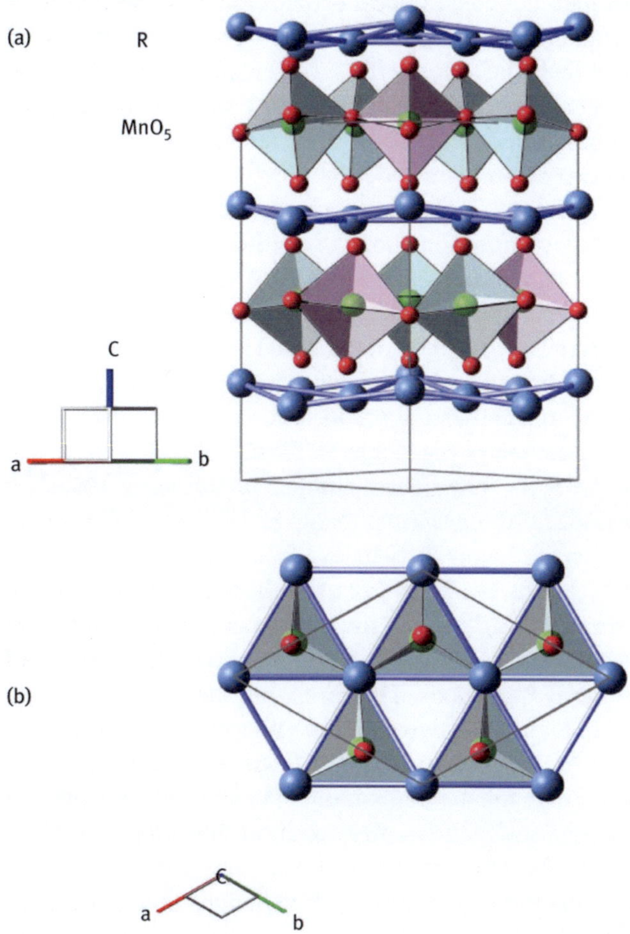

Figure 3.1: Structure of h-$RMnO_3$. (a) View along (110) and (b) view along the *c*-axis. Only one layer of MnO_5 trigonal bipyramids is shown, the next MnO_5 layer fits above or underneath the voids.

The conflicting results and interpretations have stimulated additional theoretical and experimental studies of the high-temperature phases of $YMnO_3$. The possibility of the preservation of a residual symmetry below T_S was discussed based on a Landau theory including the two components of the order parameter describing the structural distortions (tilt of the MnO_5 octahedra) which lead to the tripling of the unit cell and the coupling to the ferroelectric polarization [18]. The second transition near 920 K could then be explained as the breaking of this residual symmetry at T_C with both phases (below T_S and T_C) exhibiting a ferroelectric polarization. The existence of an intermediate high-temperature phase with $P\bar{3}c1$ symmetry was indeed verified in hexagonal $InMnO_3$ recently [19]. For $YMnO_3$, however, a more recent experimental and theoretical study provided a different explanation of the conflicting results observed so far. Second harmonic generation nonlinear optical experiments have been used to extract the ferroelectric polarization over a large temperature range up to the highest transition temperature of 1270 K [20]. The ferroelectric polarization was found to arise below 1270 K and grow continuously towards lower temperature without any anomaly near 920 K. A Monte Carlo simulation of a six state clock model, representing the hexagonal symmetry and the two-component order parameter of the trimerization distortion (including the FE polarization), led to the conclusion that both components of the order parameter and the FE polarization show a very different finite size scaling and temperature dependence of their correlation lengths. Therefore, experiments probing a relatively short length scale will detect a spurious transition at much lower temperature (~ 900 K) as compared to the true structural and FE transition temperature, 1,270 K. According to this study there exists only one structural phase transition resulting in a trimerization of the unit cell and a ferroelectric state, resolving a decades old problem of high-temperature phase transitions in $YMnO_3$.

The magnetic orders in h-$RMnO_3$ have been studied in the early 1960s by Bertaut, Koehler, and others [21, 22]. All hexagonal manganites show antiferromagnetic (AFM) order of the Mn^{3+} spins below T_N. The lattice constants and the Néel temperatures of nine h-$RMnO_3$ (R = In, Sc, Y, Dy to Lu) are shown in Table 3.1. The possible magnetic orders strongly depend on the hexagonal lattice structure, mainly the sublattice of the magnetic Mn^{3+} ions. Figure 3.2 shows a projection of the Mn sublattice along the hexagonal *c*-axis. The triangular layers of Mn^{3+} ions are stacked along *c* as shown in Figure 3.2 (two subsequent layers of Mn are distinguished by color). The R ions (not shown in Figure 3.2) are located in the open spaces between the Mn layers. Since the magnetic interactions of the Mn spins are antiferromagnetic super exchange interactions, the spins on the triangular Mn^{3+} lattice are highly frustrated and the Mn spins form a non collinear arrangement where the three spins on a triangle form an angle of 120° with one another, compatible with the hexagonal symmetry.

Table 3.1: Lattice parameters and magnetic ordering temperatures of hexagonal $RMnO_3$.

	a (nm)	*c* (nm)	T_N(K)	Ref.
$InMnO_3$	0.5891	1.148	120	[23]
$ScMnO_3$	0.5833	1.117	130	[22, 24]
$YMnO_3$	0.6148	1.144	72	[24, 25]
$DyMnO_3$	0.6182	1.145	57	[26]
$HoMnO_3$	0.6142	1.142	76	[5, 22]
$ErMnO_3$	0.6112	1.140	79–81	[5, 22]
$TmMnO_3$	0.6092	1.137	84–86	[5, 27]
$YbMnO_3$	0.6062	1.136	87–89	[5, 28]
$LuMnO_3$	0.6046	1.141	90	[24, 25]

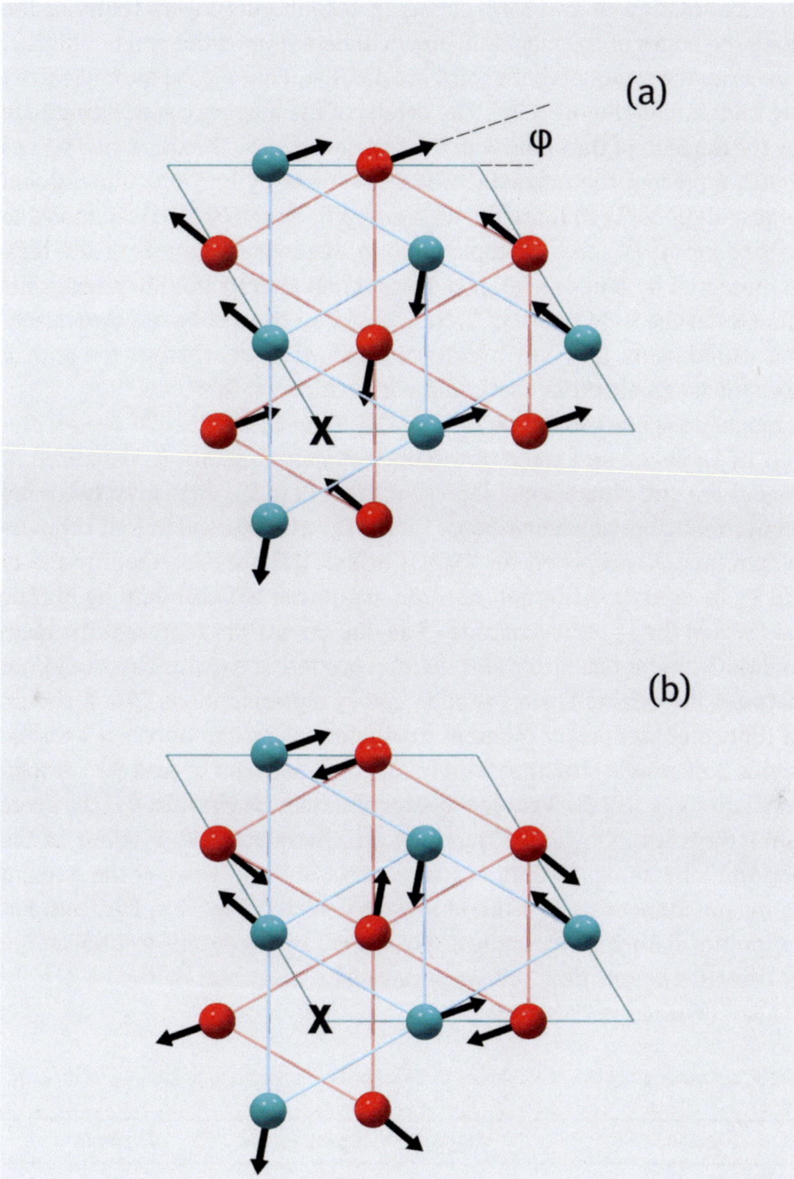

Figure 3.2: Triangular sublattices of Mn ions of h-$RMnO_3$. Two subsequent Mn layers per unit cell are distinguished by blue and red color. The magnetic unit cell (origin at *X*) and Mn spins (black arrows) are shown according to the α (a) and β (b) models. Different magnetic symmetries are defined by the angle Φ.

The relative orientation of two Mn^{3+} spins in neighboring layers (pairs in the center and along the edges of the magnetic unit cell) determine if the magnetic structure belongs to either the α model (both spins are parallel, Figure 3.2a) or the β model (both spins are antiparallel, Figure 3.2b). The details of the magnetic space group are determined by the angle Φ of the spins with the hexagonal axis. Two preferred values of Φ, 0° and 90°, represent the magnetic orders described by four one-dimensional irreducible representations, Γ_1 to Γ_4, of the little group G_k ($\mathbf{k} = 0$) which is identical to the crystal's space group $P6_3cm$. A complete group theoretical analysis of the little group G_k was presented by Muñoz et al. [29]. Bertaut and Mercier tried to identify the magnetic symmetry in the AFM phase of $YMnO_3$ based on the results of powder neutron scattering experiments [21], but a definite conclusion of whether the α or β model describes the magnetic order best could not be drawn at this time.

The four magnetic space groups represented by Γ_1 to Γ_4 irreducible representations are listed in Table 3.2 and their corresponding spin structure is visualized in Figure 3.3. Besides the one-dimensional representations Γ_1 to Γ_4, there exist two more two-dimensional irreducible representations, Γ_5 and Γ_6, as discussed in Ref. [29]. The two possible spin models proposed for $YMnO_3$ in Ref. [21] are described by the Γ_1 (β model) and Γ_3 (α model). Although neutron scattering experiments by Muñoz et al. [29] have favored the Γ_1 representation (β model, $\Phi = 90°$) to represent the magnetic order in $YMnO_3$, more recent powder neutron scattering experiments could not distinguish between the originally suggested Γ_1 and Γ_3 representations [30]. It should also be noted that combinations of different irreducible representations are possible and they describe a magnetic structure with an angle Φ between 0° and 90°. A total of four intermediate ($0° < \Phi < 90°$) magnetic structures can be visualized if the spins on the two sublattices are allowed to rotate by an arbitrary angle Φ either in the same or in opposite directions, leading to transitions within or between the α and β models. Their magnetic space groups are of lower symmetry, $P6_3$, $P\underline{6_3}$, $P3\underline{c}$, and $P3c$ [27]. The four intermediate spin structures, which may be important to understand spin rotation transitions and new phases induced by magnetic fields as, e. g. in $HoMnO_3$, are also included in Table 3.2.

Table 3.2: Magnetic symmetries of the Mn^{3+} spins in hexagonal manganites, according to [27, 29].

Irrep.	Model	Angle Φ	Space group	Example
Γ_1	β_2	90°	$P6_3cm$	$YMnO_3$
Γ_2	β_1	0°	$P6_3\underline{cm}$	$ErMnO_3$
Γ_3	α_1	0°	$P\underline{6_3}c\underline{m}$	
Γ_4	α_2	90°	$P\underline{6_3}\underline{c}m$	
$\Gamma_2 \leftrightarrow \Gamma_1$	$\beta_1 \leftrightarrow \beta_2$	$0° < \Phi < 90°$	$P6_3$	$LuMnO_3$
$\Gamma_3 \leftrightarrow \Gamma_4$	$\alpha_1 \leftrightarrow \alpha_2$	$0° < \Phi < 90°$	$P\underline{6_3}$	
$\Gamma_2 \leftrightarrow \Gamma_4$	$\beta_1 \leftrightarrow \alpha_2$	$0° < \Phi < 90°$	$P3\underline{c}$	
$\Gamma_3 \leftrightarrow \Gamma_1$	$\alpha_1 \leftrightarrow \beta_2$	$0° < \Phi < 90°$	$P3c$	

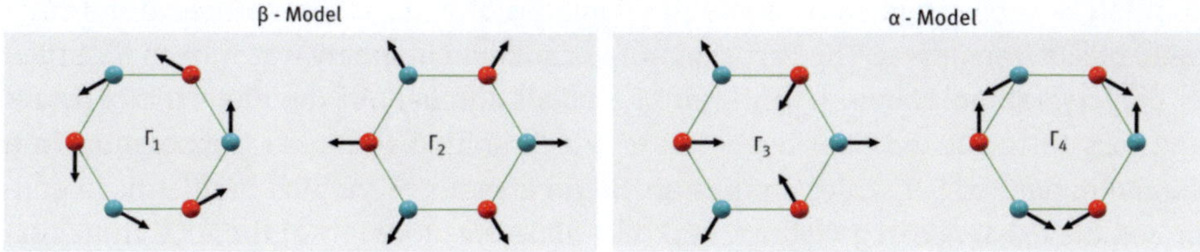

Figure 3.3: Characteristic spin structures according to the one dimensional irreducible representations of the little group G_k of $\mathbf{k} = 0$. The six Mn^{3+} ions shown represent the three ions of each layer around the origin of the magnetic unit cell, marked “X” in Figure 3.2.

The exploration of magnetic symmetries through neutron scattering experiments is based on fitting the possible magnetic structures to the scattering spectra with a limited resolution. Nonlinear optical methods can be employed in non centrosymmetric lattices [31, 32] and it was shown that the second harmonic generation (SHG) is a very sensitive technique since the second order susceptibility tensor obeys selection rules that are characteristic for different magnetic structures. Fröhlich et al. employed SHG to investigate the magnetic and ferroelectric orders in $YMnO_3$ [33] and Fiebig et al. developed the method further to extensively study the magnetic symmetry of various hexagonal manganites [34]. It was further shown that the SHG non-linear susceptibility tensor of hexagonal manganites can be used to distinguish different magnetic structures [35]. According to these studies, the symmetry of the AFM phase of $YMnO_3$ has been uniquely identified as $P\underline{6}_3c\underline{m}$ (Γ_3). Some hexagonal manganites ($ScMnO_3$, $LuMnO_3$) appear to show the coexistence of different magnetic symmetries whereas $HoMnO_3$ experiences a sudden spin rotation from $P\underline{6}_3cm$ (Γ_4) to $P\underline{6}_3c\underline{m}$ (Γ_3) within the AFM phase [34], as already proposed by Koehler et al. [22].

The overall magnetic structure of h-$RMnO_3$ is even more complex in those compounds with a magnetic rare earth ion. The rare earth moment is oriented along the c-axis due to a strong uniaxial anisotropy. Its interaction with one another and the Mn^{3+} spins results in additional phase transitions at low temperatures, as observed e. g. in $ErMnO_3$ (2.5–5 K), in $YbMnO_3$ (4 K), and in $HoMnO_3$ (5.4 K) [5, 28, 36, 37]. The complex phase diagrams of various $RMnO_3$ will be discussed in the following sections.

3.3 The complex magnetic phase diagram and magnetoelectric effects in hexagonal $HoMnO_3$

The coexistence of ferroelectric and magnetic orders in h-$RMnO_3$ raises the question if and how the two order parameters mutually interact and how different physical properties might be affected by their coupling. While the linear magnetoelectric effect is not allowed by symmetry, higher order couplings mediated through strong spin-lattice interaction could result in anomalies of dielectric quantities at the magnetic phase transitions. The first signature of such an anomaly was indeed observed in polycrystalline $YMnO_3$ [7] in form of a small anomaly of the dielectric constant and loss factor at the Néel temperature which shifted to higher temperature in a magnetic field of 5 T. Later studies on single crystals of various $RMnO_3$ have confirmed the existence of a distinct, kink-like dielectric anomaly at the AFM transition temperature [24, 38, 39], proving the existence of a strong correlation between AFM and ferroelectric orders in this class of compounds. The kink in the dielectric constant at the AFM transition could be theoretically explained as a consequence of the biquadratic magnetoelectric coupling term included in an anisotropic Heisenberg model with first and second neighbor interactions [40].

3.3.1 Dielectric properties, magnetic order, and thermodynamics of $HoMnO_3$ in the absence of magnetic fields

At zero magnetic field, $HoMnO_3$ shows the most complex magnetic phase diagram with two more distinct phase transitions below the AFM ordering temperature $T_N = 76$ K which are also accompanied by sharp anomalies in the c-axis dielectric constant $\varepsilon_c(T)$. A narrow peak of $\varepsilon_c(T)$ at $T_{SR} = 32.8$ K is followed by a step-like increase at $T_{Ho} = 5.4$ K, as shown in Figure 3.4 [41, 42]. The sharp peak of $\varepsilon_c(T)$ at T_{SR} was only observed in $HoMnO_3$ single crystals and it provides further evidence for the existence of strong magnetoelectric couplings. The symmetry of the three magnetic phases has been determined through neutron scattering [22, 37, 43, 44] and non-linear optical experiments [45, 46].

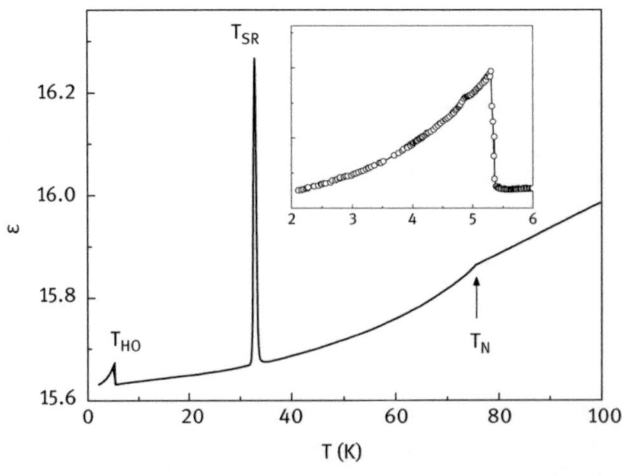

Figure 3.4: Dielectric anomalies at three magnetic phase transitions of $HoMnO_3$. The inset shows the low-T transition on an enlarged scale.

Below T_N the Mn spins order in the frustrated AFM structure with $P\underline{6_3}\underline{c}m$ symmetry (Γ_4) and at T_{SR} they rotate suddenly by 90° into a new magnetic structure described by $P\underline{6_3}c\underline{m}$ (Γ_3). Both spin structures are shown in Figure 3.3 (α-model). Below T_{SR}, the Ho^{3+} moments become partially ordered along the hexagonal c-axis [37, 47] due to their interaction with the Mn^{3+} spin system and a trigonal anisotropy term [48].

The third magnetic transition at T_{Ho} was identified as a second 90° Mn spin rotation transition to $P6_3cm$ symmetry (β model, Γ_1 in Figure 3.3) with a significant increase of both, the Mn- and Ho-sublattice magnetizations [37, 44, 47]. In the $P6_3cm$ symmetry only the Ho^{3+} moment in Wyckoff position 4b is allowed to order. This transition is most likely driven by the increasing magnetic fluctuations of the Ho^{3+} moments and their coupling to the Mn^{3+} spins.

The three sharp anomalies of ε_c at T_N, T_{SR}, and T_{Ho}, observed in high-quality single crystals of $HoMnO_3$, have been confirmed by different research groups [49–54].

The three magnetic transitions in $HoMnO_3$ are also reflected in sharp anomalies of other physical quantities [55]. The magnetization is moderately anisotropic with the in-plane magnetization slightly larger than the c-axis values. An effective magnetic moment of $\mu_{eff} = 11.43\ \mu_B$ is determined from a Curie–Weiss fit to the high temperature part of χ^{-1}. μ_{eff} is in good agreement with the value of 12.14 μ_B which is the sum of the free ion values of Mn^{3+} (5.92 μ_B) and Ho^{3+} (10.6 μ_B). Below 130 K, χ_c^{-1} deviates from the Curie–Weiss line indicating the onset of strong magnetic fluctuations between the Mn-spins. The large Ho^{3+} moments dominate the magnetic response near the AFM ordering temperature of Mn^{3+} spins, $T_N = 76$ K, and an anomaly of the magnetic susceptibility is barely detectable. However, as shown in Figure 3.5a, clear step-like changes of χ_c^{-1} at $T_{SR} = 32.8$ K and $T_{Ho} = 5.4$ K indicate the Mn spin rotation transition and the onset of AFM Ho moment order, respectively [55].

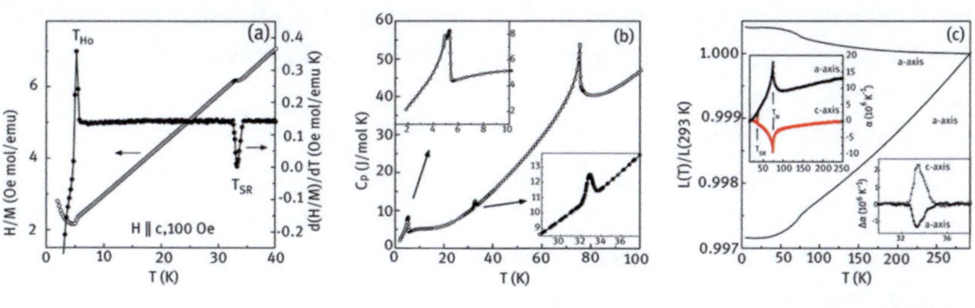

Figure 3.5: (a) c-axis inverse magnetic susceptibility [left scale] and its derivative [right scale] of $HoMnO_3$. (b) Heat capacity of $HoMnO_3$. The insets show the details at the two spin rotation transitions. (c) Relative length changes of a- and c-axes of $HoMnO_3$. Upper inset: thermal expansivities $\alpha(T)$. Lower inset: Change of $\alpha(T)$ at the first spin rotation transition.

The thermodynamic signature of all transitions is revealed by distinct anomalies of the heat capacity (Figure 3.5b) [51, 53, 55–58]. $C_p(T)$ exhibits a λ-shaped peak at T_N characteristic for a second order phase transition. The second order nature of the Néel transition is proven in a study of the critical exponents which are close to the values expected for the three dimensional Heisenberg universality class [59]. At the spin rotation transition temperature T_{SR}, the heat capacity $C_p(T)$ exhibits a sharp peak, consistent with a first order phase transition (lower inset in Figure 3.5b) [55]. The entropy change across T_{SR} is relatively small, $\Delta S = 0.04$ J/(mol K). This value was shown to be consistent with the changes of volume and magnetization across the transition and the Clausius–Clapeyron equation, valid at first order transitions, is fulfilled [60]. A sharp increase of $C_p(T)$ at T_{Ho} indicates the second Mn^{3+} spin rotation and the onset of partial Ho^{3+} moment order (upper inset in Figure 3.5b) [55].

The sharp dielectric anomalies at all three magnetic phase transitions (Figure 3.4) reveal a sizable magnetoelectric effect originating from strong spin–lattice interaction which also results in distinct anomalies of the thermal expansion coefficients at T_N, T_{SR}, and T_{Ho} [60]. There exists a remarkable anisotropy in the thermal response of the lattice constants of $HoMnO_3$. The hexagonal a-axis shrinks with decreasing temperature whereas the c-axis expands. The negative thermal expansivity of the c-axis is apparently caused by a gain in magnetic exchange energy with the increase of spin fluctuations upon cooling even above T_N. The relative length changes of a and c below room temperature are shown in Figure 3.5c (main panel) with the thermal expansivities, $\alpha_{a,c}$, displayed in the upper inset. At T_N, α_a and α_c exhibit λ-shaped anomalies, consistent with the second order nature of the Néel transition and in agreement with the specific heat anomaly (Figure 3.5b). The spin rotation transition at T_{SR} is characterized by a sudden change of the lattice constants, as shown by the sharp peaks of α_a and α_c (lower inset in Figure 3.5c). The response of a and c to the magnetic phase transitions is always opposite in sign, i.e. whenever the a-axis expands, the c-axis shrinks and vice versa. The thermal expansion measurements provided the first and unambiguous evidence for extraordinarily strong spin–lattice interaction in the hexagonal rare earth manganites.

Further signatures of strong spin–lattice coupling was derived from the enhancement of phonon frequencies below the Néel temperature of $HoMnO_3$ [61] and $LuMnO_3$ [62]. Furthermore, the thermal conductivity of $HoMnO_3$ and other h-$RMnO_3$ suddenly increases with the onset of magnetic ordering, indicating a strong dynamic coupling between acoustic phonons and low-energy spin fluctuations [49]. The atomic displacements and bond distances between different ions in the structure have been studied through high resolution neutron scattering [63] and X-ray absorption spectroscopy [64]. The results show that different displacements of ion pairs happen at the three magnetic phase transitions of $HoMnO_3$. The structural distortion at T_N is dominated by the change of in-plane distances between Mn ions, the first spin rotation transition causes a change of Mn–Mn and nearest neighbor Mn–Ho distances, and the low-temperature transition involves a structural distortion of all ions, including Ho–Ho pairs. The sudden change of the elastic moduli at T_N and T_{SR}, studied with ultrasonic techniques [65], are further indications for the importance of the spin-lattice coupling in hexagonal manganites.

The observed dielectric anomalies at all three magnetic transitions and the strong spin-lattice coupling raise the question of how much the ferroelectric state and its electrical polarization, $P_c(T)$, are affected at the various phase transitions. Pyroelectric measurements can determine any polarization change over a chosen temperature range. For $HoMnO_3$, such measurements have been conducted [54] and the results of $\Delta P_c(T) = P_c(T) - P_c(100K)$ are shown in Figure 3.6. A minute kink of $P_c(T)$ near T_N and a sharp drop at T_{SR} are followed by an increase at T_{Ho}, demonstrating the effect different magnetic orders have on the ferroelectric state. All three anomalies of P_c correlate well with the anomalies of ε_c (shown in Figure 3.5a).

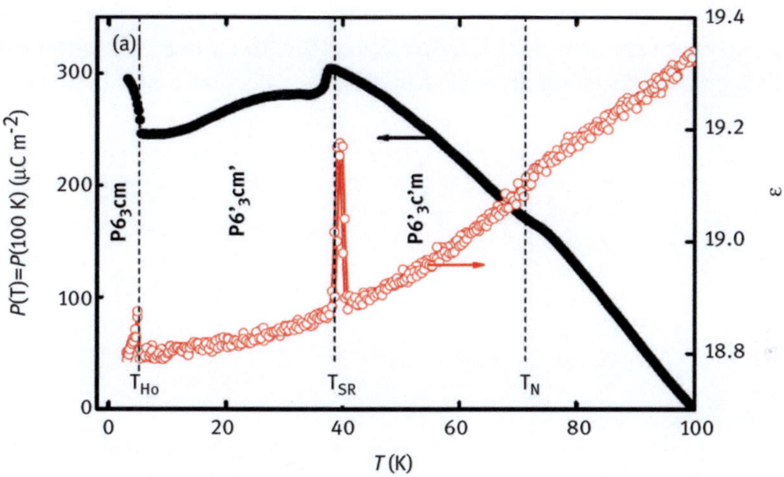

Figure 3.6: Change of ferroelectric polarization of $HoMnO_3$ below 100 K. Reprinted with permission from N. Hur et al., Phys. Rev. B **79**, 134120 (2009). Copyright (2009) by the American Physical Society.

3.3.2 The magnetic phase diagram of $HoMnO_3$

The sharp anomalies of ε_c at the three (zero field) magnetic transitions (Figure 3.4) can be used to derive the magnetic phase diagram at high fields. Previous nonlinear optical studies of $HoMnO_3$ hint for a complex phase diagram in fields oriented along the *c*-axis [27, 45]. Four different magnetic structures according to the 1D irreducible representations (Γ_1 to Γ_4, Table 3.2) had been proposed. A new high-field phase of $P6_3\underline{cm}$ symmetry was detected above 20 kOe and below 10 K. In the following we use dielectric measurements to investigate the detailed phase diagram in fields along the *a*- and *c*-axes.

The Néel temperature T_N, defined by the kink in ε_c, is barely affected by an external field up to 140 kOe. However, T_{SR} and T_{Ho} are quite sensitive to an external field depending on the field orientation. Figure 3.7 shows the dielectric constant $\varepsilon_c(T)$ for different magnetic fields H_a in the hexagonal plane near the two lower magnetic transitions. The sharp peak at T_{SR} (Figure 3.7a) broadens with increasing H_a and the peak position initially shifts to lower temperature, passes through a minimum, and increases at higher fields. The step-like increase of $\varepsilon_c(T)$ at T_{Ho} also broadens and shifts to lower temperature, as shown in Figure 3.7b. The resulting low-T phase diagram is shown in Figure 3.10a. The a-axis magnetic field suppresses the low-temperature phase (denoted “3” in Figure 3.10a) with the assigned $P6_3cm$ (Γ_1) symmetry but it stabilizes the intermediate phase “2” with $P\underline{6_3}c\underline{m}$ (Γ_3) symmetry.

The in-plane magnetic field apparently does not change the magnetic symmetries of the three zero-field phases, but it has a major effect on the phase boundaries. This is very different if the field is applied along the hexagonal c-axis. In this case the phase diagram becomes very complex at higher fields H_c with up to 6 different magnetic phases distinguished by sharp anomalies of dielectric, magnetic, and thermodynamic properties.

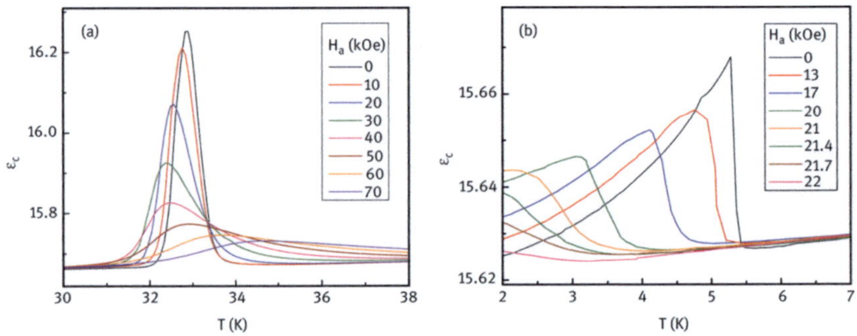

Figure 3.7: Temperature dependence of the dielectric constant $\varepsilon_c(T)$ of $HoMnO_3$ at different magnetic fields oriented along the a-axis. (a) Peak of ε_c near T_{SR}. The maximum decreases with increasing H_a. (b) Step of ε_c near T_{Ho}, moving to lower T with increasing H_a.

In the absence of magnetic fields, the c-axis dielectric constant, $\varepsilon_c(T)$, shows two sharp peaks at T_{SR} and T_{Ho}, respectively. With increasing H_c, however, both peaks evolve into plateau-like enhancements above the base line of $\varepsilon_c(T)$, both plateaus extend their temperature range, and eventually merge above 35 kOe, as shown in Figure 3.8a [41]. The plateau of $\varepsilon_c(T)$ disappears completely at higher fields, $H_c > 40$ kOe. Plotting the phase boundaries T_1 and T_2 defined by the increase and decrease of $\varepsilon_c(T)$ it becomes obvious that the sharp transition at $T_{SR}(H = 0)$ splits into two transitions in magnetic fields H_c, defining a new phase between T_1 and T_2. This phase, denoted “4” in Figure 3.10b, is characterized by a higher susceptibility with respect to external electric or magnetic fields. $\varepsilon_c(T)$ in this phase follows the dashed line in

Figure 3.8a and it reaches a peak at zero field. The enhancement of the dielectric constant in phase "4" is also demonstrated in isothermal field scans of $\varepsilon_c(H_c)$ shown in Figure 3.8b.

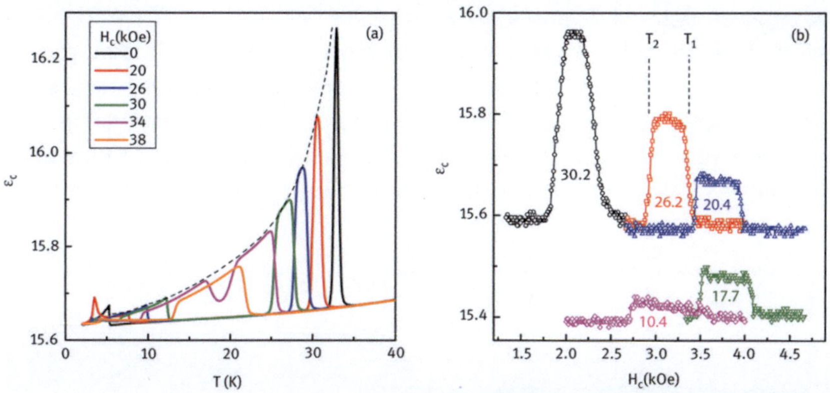

Figure 3.8: (a) Temperature dependence of $\varepsilon_c(T)$ of $HoMnO_3$ at different magnetic fields oriented along the c-axis. The dashed line represents the increase of $\varepsilon(T)$ in the intermediate $P\underline{6}_3$ phase. (b) Field dependence of ε_c at selected temperatures (T is shown underneath each curve). The two sets of data are vertically offset for better clarity. The two transition temperatures labeled T_1 and T_2 are shown for the 26 K data.

The magnetic structure in phase "4" can be derived from the magnetic symmetries of the two bordering phases "1" ($P\underline{6}_3\underline{c}m$) and "2" ($P\underline{6}_3c\underline{m}$) which are distinguished by the angle Φ of one of the three manganese spins with the hexagonal a-axis. In phases "1" and "2" Φ is locked in the symmetrical positions of 90° and 0°, respectively. It is therefore conceivable that in phase "4" Φ assumes intermediate values and changes gradually from 90° to 0° with decreasing temperature or magnetic field, resulting in a magnetic space group described by $P\underline{6}_3$. The suggested space group for phase "4" is also consistent with results of optical experiments [34]. The intermediate angle Φ in phase "4" also explains the enhanced dielectric response (through the magnetoelectric coupling) since Φ is not locked in symmetric positions as in the neighboring phases "1" and "2".

The magnetic susceptibility, measured along the c-axis, is another bulk property showing characteristic changes in the different magnetic phases of $HoMnO_3$. The derivative of the inverse susceptibility, $d(H_c/M_c)/dT$, suits best to visualize the changes in crossing the phase boundaries between phases "1", "2", and "4". Figure 3.9 displays $d(H_c/M_c)/dT$ at different magnetic fields between 100 Oe and 45 kOe. At low fields, $d(H_c/M_c)/dT$ is constant and $\chi_c(T)$ is strictly linear between T_{Ho} and T_N with a distinct kink-like anomaly at T_{SR} which results in a sharp downward peak of $d(H_c/M_c)/dT$. However, in the $P\underline{6}_3$ phase the derivative shows a finite slope which is typical for the lower symmetry phase "4". Curve C in Figure 3.9 shows a typical example of the magnetic changes at a fixed field of 33 kOe. The sequence of phases upon decreasing temperature (see phase diagram, Figure 3.10b) is $P\underline{6}_3\underline{c}m \rightarrow P\underline{6}_3 \rightarrow P\underline{6}_3c\underline{m} \rightarrow P\underline{6}_3 \rightarrow P\underline{6}_3\underline{c}m$, i.e. the system passes twice through the intermediate $P\underline{6}_3$ phase. The corresponding critical temperatures T_1 and T_2 are shown in Figure 3.9. Between pairs of T_1 and T_2 (i. e. in phase "4") the derivative increases with constant slope, as noted above. At slightly higher fields, 35 kOe and 38 kOe, the magnetic system does not pass into the $P\underline{6}_3c\underline{m}$ phase and the sequence is $P\underline{6}_3\underline{c}m \rightarrow P\underline{6}_3 \rightarrow P\underline{6}_3\underline{c}m$, as indicated by the continuous and linear increase of $d(H_c/M_c)/dT$ between the two phase transitions. Above 40 kOe no transition could be observed above 5 K, in agreement with the phase diagram of Figure 3.10b.

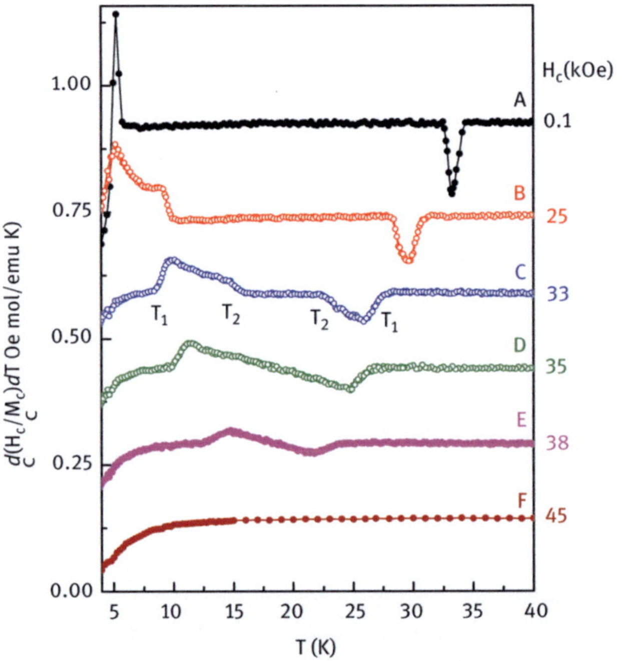

Figure 3.9: Derivative of the inverse c-axis magnetic susceptibility, $d(H_c/M_c)/dT$, at different magnetic fields. The curves are vertically offset for better clarity.

So far, the dielectric, magnetic, and thermodynamic studies have revealed up to three different magnetic phases at low fields and at temperatures above 6 K. However, at lower temperatures and higher magnetic fields the magnetic phase diagram of $HoMnO_3$ is far more complex than originally predicted [45]. A detailed analysis of the temperature and field dependence of the dielectric constant reveals a series of distinct anomalies which not only extend the phase boundaries T_1 and T_2 towards lower

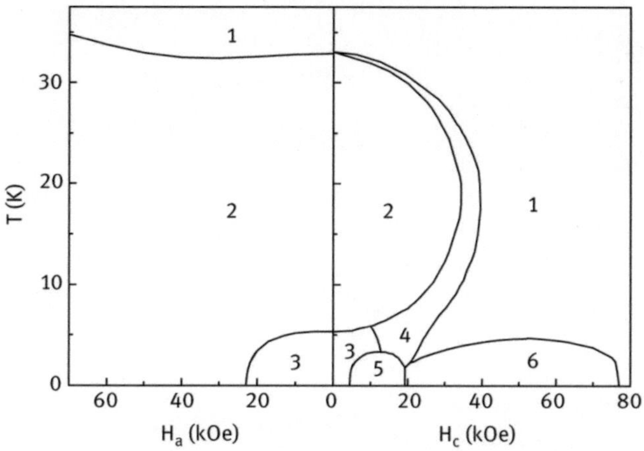

Figure 3.10: Magnetic phase diagram of $HoMnO_3$ below the Néel temperature. The phase boundaries are constructed from dielectric, magnetic, heat capacity, and thermal expansion data. The assignment of the different magnetic symmetries was made based on the results from second harmonic generation optical experiments [27, 45, 46] and neutron scattering measurements [37, 44, 63, 66]: "1" – $P\underline{6_3}\underline{c}m$, "2" – $P\underline{6_3}c\underline{m}$, "3" – $P6_3cm$, "4" – $P\underline{6_3}$.

temperatures, but also prove the existence of additional transitions and new phases [42]. The low-T phase diagram of $HoMnO_3$ will be discussed in the next section.

3.3.3 Low-temperature phases of $HoMnO_3$

Similar to the results at higher temperatures, the dielectric constant $\varepsilon_c(T, H_c)$ exhibits clear anomalies at the various phase transitions and provides the most detailed information about the phase diagram of $HoMnO_3$ at low temperatures. The low-T data for $\varepsilon_c(T)$ (at constant fields H_c) and $\varepsilon_c(H_c)$ (at constant temperature) are shown in Figure 3.11 and Figure 3.12, respectively. The various phase boundaries, denoted as T_1–T_5 in Figure 3.13, are all defined by characteristic anomalies of ε_c:

T_1 is defined by a sharp drop of ε_c in crossing from phase "4" to phase "1" (see Figure 3.8 and Figure 3.12a). A similar sharp drop of ε_c defines T_2, the transition from phase "4" to phases "2" and "3" (see Figure 3.8, Figure 3.11a,b,c, and Figure 3.12a). T_1 and T_2 merge at the zero field spin rotation transition, T_{SR}. Below about 5 K, there are two dome shaped low-T phase boundaries, labeled T_3 and T_4 in Figure 3.11 – Figure 3.13, which define the two phases "LT1" and "LT2", respectively. Both low-T high-field phases are characterized by a reduced dielectric response, ε_c drops sharply in crossing into both phases (see Figure 3.11a,b,d and Figure 3.12). Phase "LT1" extends between 6 and 20 kOe with a maximum temperature of 3.3 K (at about 13 kOe). The second low-T phase, "LT2", is stable between 20 kOe and 78 kOe

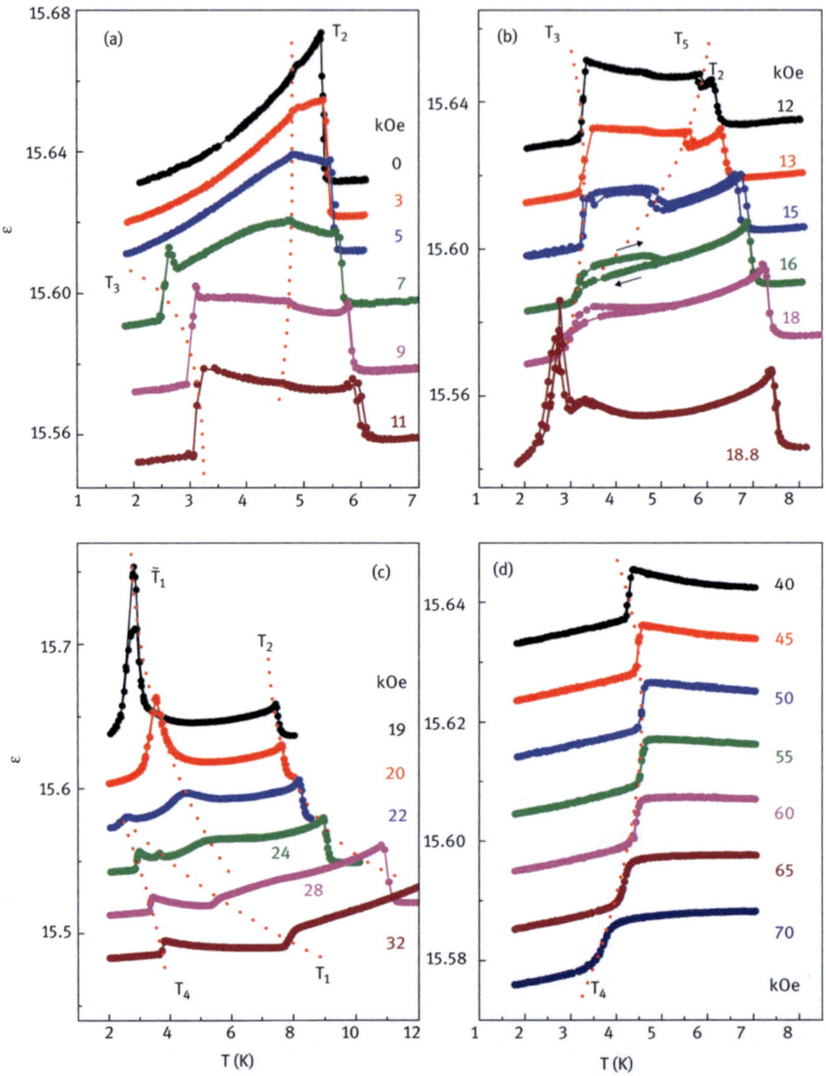

Figure 3.11: Dielectric constant $\varepsilon_c(T, H_c)$ of $HoMnO_3$. Four different field ranges are shown in (a)–(d). Different curves are vertically offset for better clarity. Anomalies of ε indicating phase boundaries are highlighted by dotted lines and labeled with temperature symbols $T_1, \ldots, T_5$.

and extends up to 4.5 K (at about 55 kOe). T_3 and T_4 merge at 20 kOe. Another phase boundary, labeled T_5, is defined by a pronounced hysteresis in thermal and field scans of ε_c, as obvious from data shown in Figure 3.11b and Figure 3.12a,b,d. The large hysteresis reveals the strong first order character of this transition. The phase boundary T_5 connects T_2 and T_3 and separates the phases "3" and "4". The low-T phase diagram, as derived from dielectric measurements, is shown in Figure 3.13. The

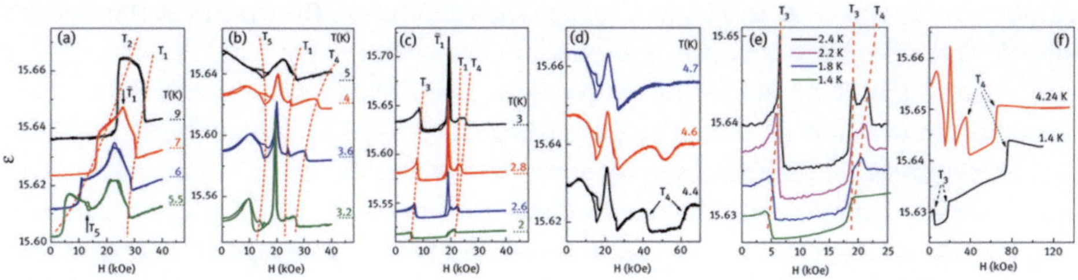

Figure 3.12: Isothermal field scans of $\varepsilon_c(H_c)$. Different curves are vertically offset. Phase boundaries T_1 to T_5 and the ε_c-peak at $\widetilde{T}_1$ are marked by dashed lines. (a)–(c) $\varepsilon_c(H_c)$ between 9 K and 2 K up to 40 kOe. (d) Emergence of the LT2 phase. (e) Valley-like anomaly of ε_c in the LT1 phase, merging of T_3 and T_4 phase boundaries. (f) $\varepsilon_c(H_c)$ data extending to 130 kOe.

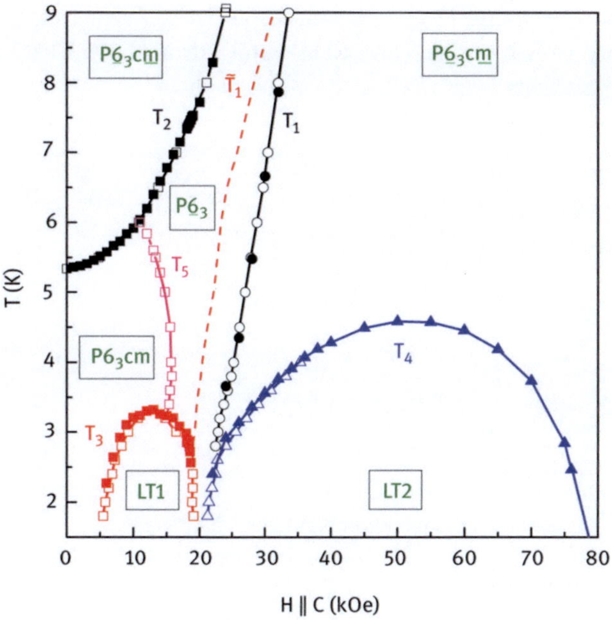

Figure 3.13: Low temperature phase diagram of $HoMnO_3$. Phase boundaries T_1 to T_5 are derived from distinct anomalies of the dielectric constant, the magnetic susceptibility, the specific heat, and magnetostriction data. The dashed line labeled $\widetilde{T}_1$ denotes the location of the peak of the dielectric and magnetic susceptibilities.

magnetic symmetries of some phases, determined by second harmonic optical studies [27, 45, 46] as well as neutron scattering experiments [37, 44, 63, 66], are indicated in the figure. The magnetic structures of the two low-temperature phases “LT1” and “LT2” have yet to be explored.

The most interesting region in the phase diagram is an area near 2 K and 20 kOe where three phase boundaries, T_1, T_3, and T_4 all come together. The dielectric constant exhibits a sharp peak in this region, indicating a softness of the system to the external electric field. This peak is clearly seen in $\varepsilon_c(T)$ at 19 kOe (Figure 3.11c) and in $\varepsilon_c(H_c)$ at 3 K (Figure 3.12b). With increasing temperature, the peak of $\varepsilon_c(H_c)$ broadens and shifts to higher temperature, as shown by the dashed line labeled $\widetilde{T}_1$ in Figure 3.13.

The softness of the dielectric response in the $P\underline{6}_3$ phase has to be related to a soft magnetic structure and the magnetoelectric coupling. The differential magnetic response to an external field is best represented by the ac magnetic susceptibility, $\chi_{ac}(H_c)$, as function of a dc bias field along the hexagonal c-axis. Figure 3.14a,b shows the results in two different temperature ranges, 5.7 K $\leq T \leq$ 25 K (a) and 1.8 K $\leq T \leq$ 5 K (b).

In the high-temperature range (Figure 3.14a), the real part $\chi'_{ac}(H_c)$ shows a significant enhancement in the $P\underline{6}_3$ phase, similar to the dielectric constant. It is conceivable that the softness originates from the fact that the angle Φ of the Mn^{3+} spins is not locked in at 0° or 90° as in the $P\underline{6}_3c\underline{m}$ and $P\underline{6}_3\underline{c}m$ phases, respectively. At lower temperature, $\chi'_{ac}(H_c)$ shows a sharp peak between 5 and 10 kOe and a drop above 20 kOe (Figure 3.14b), indicating the transitions across T_3 and T_4, respectively.

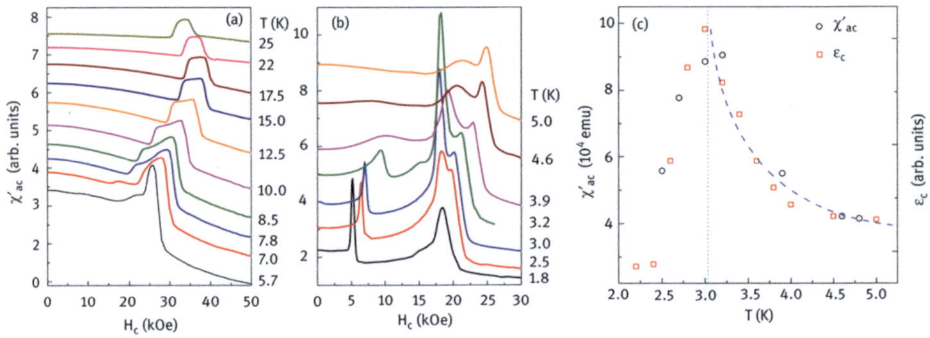

Figure 3.14: (a), (b) Isothermal field scans of the ac magnetic susceptibility at different temperatures. Different curves are vertically offset for better clarity. (c) T-dependence of the peak heights of ε_c and χ_{ac} along the line defined by $\widetilde{T}_1$.

The strongest peak of $\chi'_{ac}(H_c)$ appears at 3.2 K and near 20 kOe (Figure 3.14b) and it is consistent with the peak of ε_c (see Figure 3.12b) at the same point in the phase diagram. Plotting both peak heights on a relative scale in Figure 3.14c it becomes obvious that the magnetic and dielectric data scale very well with one another and the increase of magnetic and dielectric susceptibilities upon decreasing temperature suggest that the system is headed towards a critical point. The singular increase of χ'_{ac} and ε_c is interrupted by the T_3 phase boundary, indicated by the vertical dotted line in Figure 3.14c. One may speculate about the possible existence of a quantum critical point which is hidden in the emerging LT1 phase. The linear extrapolation of the peak position to zero temperature would locate this critical point near $H_c \simeq 15$ kOe. The observed peak of $\chi'_{ac}(H_c)$ was also confirmed by Choi et al. and attributed to a critical endpoint [67].

Hur et al. measured the change of the ferroelectric polarization P_c as function of temperature and magnetic fields H_c across the phase diagram [54]. Their dielectric data are consistent with those first reported by Yen et al. [42] (shown in Figure 3.11 and Figure 3.12). P_c shows dramatic changes at the various phase boundaries with a significant enhancement in crossing T_2 and T_4 with increasing H_c.

The complex phase diagram of $HoMnO_3$ shown in Figure 3.10 and Figure 3.13 was quantitatively confirmed by other types of measurements such as microwave studies of single crystals [68] and ultrasonic investigations of the elastic moduli [65]. Neutron scattering experiments have revealed characteristic changes of the peak intensities measured along different scattering geometries [44, 69] in agreement with the phase boundaries T_1–T_5. While most quantities used to derive the phase diagram are response functions to external stimuli, the thermodynamic signature of various phase transitions is reflected in anomalies of the heat capacity $C_p(T, H_c)$. Detailed studies reveal sharp changes and peaks of C_p at different phase boundaries [55].

In the absence of magnetic fields all three phase transitions are reflected in sharp peaks of C_p, as shown in Figure 3.5b. The following discussion focuses on the low-temperature region of the phase diagram with the large number of phase boundaries. Figure 3.15a shows C_p data below 7 K and in magnetic fields up to 20 kOe. The sharp increase of $C_p(H_c = 0)$ at T_{Ho} moves to higher temperature and follows exactly the phase boundary T_2, as noted by the dashed line on the left-hand side of Figure 3.15. Most remarkable is a narrow peak of C_p across T_3 separating the "LT1" phase. The peak rises in magnitude when the field increases to 12 kOe at $T_3 = 3.33$ K and it traces back to lower temperature with further increasing field, eventually disappearing at the upper limiting field of about 20 kOe. This smooth and continuous development and shift of the C_p-peak is consistent with the dielectric and magnetic data and it defines unambiguously the dome shaped "LT1" phase and its T_3 boundary. At magnetic fields above 20 kOe, the heat capacity develops another peak at low temperature, indicating the transition into the "LT2" phase across T_4, as shown in Figure 3.15b. The heat capacity data thus confirm the major phase transitions in $HoMnO_3$ in the magnetic field H_c oriented along the c-axis.

3.3.4 Evidence for strong spin–lattice coupling in $HoMnO_3$ through thermal expansion and magnetostriction measurements

The origin of the coupling between magnetic and dielectric properties has been sought in the strong interaction of the magnetic moments of Mn and Ho ions with the lattice. This has been verified at zero magnetic fields through thermal expansion measurements [60] showing strong anomalies of a- and c-axes at the three magnetic transitions (see Figure 3.5c). At the Néel temperature, the thermal expansivities show

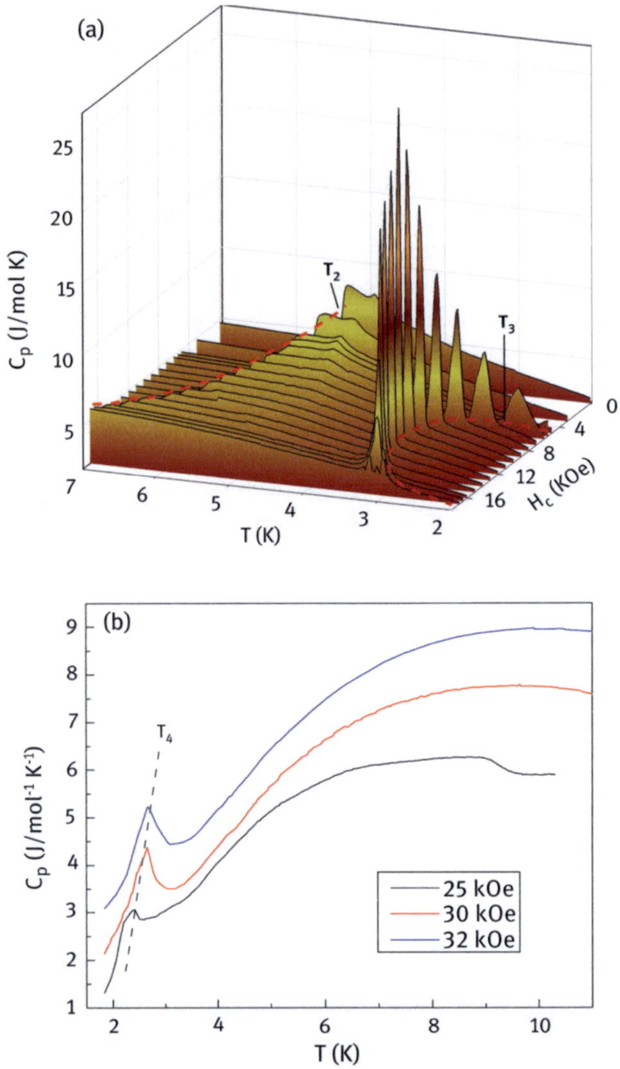

Figure 3.15: Heat capacity $C_p(T, H)$ of $HoMnO_3$. The two most prominent phase boundaries T_2 and T_3 are shown as dashed lines.

a λ-shaped peak with opposite sign for the a- and c-axes, consistent with the second order nature of the magnetic transition. This critical behavior is also reflected in heat capacity data as shown in Figure 3.5b and discussed in detail by Oleaga et al. [59]. At the first spin rotation temperature, T_{SR}, a and c exhibit a step-like change with a volume change of $\Delta V/V = 0.6 \cdot 10^{-6}$. The volume change at T_{SR}, the change of magnetization, and the entropy change at this transition were found to be consistent with the Clausius–Clapeyron equation valid at first order phase transitions [60].

It is interesting to note that the hexagonal c-axis of $HoMnO_3$ shows a negative thermal expansivity over a large temperature range. A similar negative expansion of c has been observed in the related compound $YMnO_3$ below room temperature [30, 70, 71] and even at higher temperature up to 1270 K, the transition from the centrosymmetric $P6_3/mmc$ structure to the polar $P6_3cm$ phase [14–16, 25]. It appears that the negative c-axis expansivity is a property of the ferroelectric phase.

In the low-temperature region of the phase diagram, magnetostriction measurements (Figure 3.16) display sharp anomalies at all magnetic phase boundaries identified in Figure 3.13. Above 5 K, the c-axis length increases with the field H_c and experiences a subtle change of slope at the two transitions defined by T_2 and T_1, as indicated by the dashed lines in Figure 3.16a. Below 5 K, the c-axis length reveals sharp anomalies as function of the magnetic field, as shown in Figure 3.16b. Step-like increases of c are obvious at the main phase boundaries T_1, T_3, and T_4 with increasing H_c, as indicated by the labeled dashed lines in Figure 3.16b. As an example, the inset to Figure 3.16b displays the derivative, dc/dH_c, at 3 K. At this particular temperature, the system is driven by the external field H_c twice across the phase boundary T_3, passes the critical line $\widetilde{T}_1$, and finally crosses the boundary T_1/T_4. The narrow peaks of the magnetostrictive coefficient clearly signal the various transitions and provide evidence for a strong coupling of the magnetic orders to the lattice.

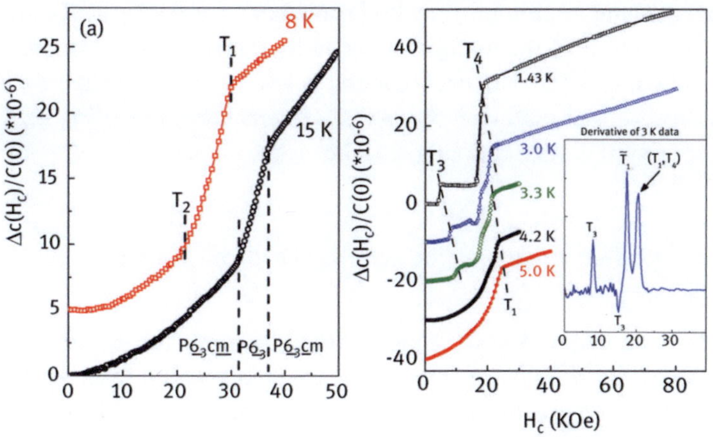

Figure 3.16: Magnetostriction of the c-axis of $HoMnO_3$ in longitudinal magnetic fields, H_c. Note that different curves are vertically offset for better clarity. (a) High-temperature range. The vertical dashed lines mark the two phase transitions from $P\underline{6}_3c\underline{m}$ to $P\underline{6}_3$ and to $P\underline{6}_3\underline{c}m$. (b) Low-temperature range. The inset to (b) shows the derivative at 3 K which clearly reveals 4 successive phase transitions.

While dilatometric measurements do reveal macroscopic length changes of, e. g. the *c*-axis with an extraordinary resolution, the microscopic displacements of single ions within the unit cell can only be investigated using other methods. e. g X-ray or neutron scattering experiments. High-resolution neutron diffraction studies of several h-$RMnO_3$ have suggested a connection between the magnetic structures (listed in Table 3.2) and the location of the Mn^{3+} ion [63]. In the $P6_3cm$ crystalline structure, the relative position of the Mn^{3+} is defined by one free parameter x_{Mn} which depends on the rare earth ion, temperature, and the magnetic order. x_{Mn} in $HoMnO_3$ increases from 0.325 above T_N to 0.335 below T_{SR}. Since the parameter x_{Mn} determines the super exchange interactions between the Mn^{3+} spins along the *c*-axis as well as within the *ab*-planes, it was proposed that the different magnetic orders realized in $HoMnO_3$ and in other hexagonal manganites (e. g. $ScMnO_3$, $YbMnO_3$) are correlated with x_{Mn} [63].

X-ray absorption spectroscopy was employed to study the local structure of $HoMnO_3$ [64] and it was shown that the bond distances between different Mn- and Ho-ions experience sudden changes at the three zero-field magnetic phase transitions. Based on a model calculation using the local spin density approximation the role of different magnetic exchange interactions (Mn–Mn, Ho–Mn, Ho–Ho) in driving the three observed magnetic phase transitions was determined. The important role of the Ho–Mn magnetic interactions was derived from an investigation of the elastic moduli revealing the magnetoelastic coupling in $HoMnO_3$ [65]. An elastic softening was observed over a wide temperature range, with pronounced anomalies at T_{SR} and T_{Ho}, and it was attributed to spin fluctuations induced by the Ho–Mn interactions. Raman studies have shown that the magnetic ordering below T_N causes a significant hardening of phonon modes that modulate the Mn–Mn interactions, providing additional evidence for strong spin–lattice coupling in $HoMnO_3$ [61].

3.3.5 Order of the Ho^{3+} magnetic moment and coupling to the Mn^{3+} spin structures in $HoMnO_3$

The Ho^{3+} moments interact with the Mn^{3+} spins resulting in a gradual polarization below T_{SR}. At lower temperature, the direct superexchange interaction between the Ho^{3+} moments results in another transition involving a second rotation of the frustrated Mn^{3+} spin system. The Ho^{3+} ions occupy two crystallographically inequivalent sites in the hexagonal structure: Ho(1) and Ho(2) corresponding to the 2a and 4b positions, respectively. The primitive cell includes six formula units $HoMnO_3$ with two Ho (1) and four Ho(2). The possible orders of the Ho-moments which are compatible with the four one dimensional representations of the magnetic space group, Γ_1 to Γ_4 (see Figure 3.3 and Table 3.2), have been analyzed by Muñoz et al. [66] and they are shown in Figure 3.17. It is worth noting that only Ho(2) moments are allowed to order in Γ_1 and Γ_4 magnetic symmetries.

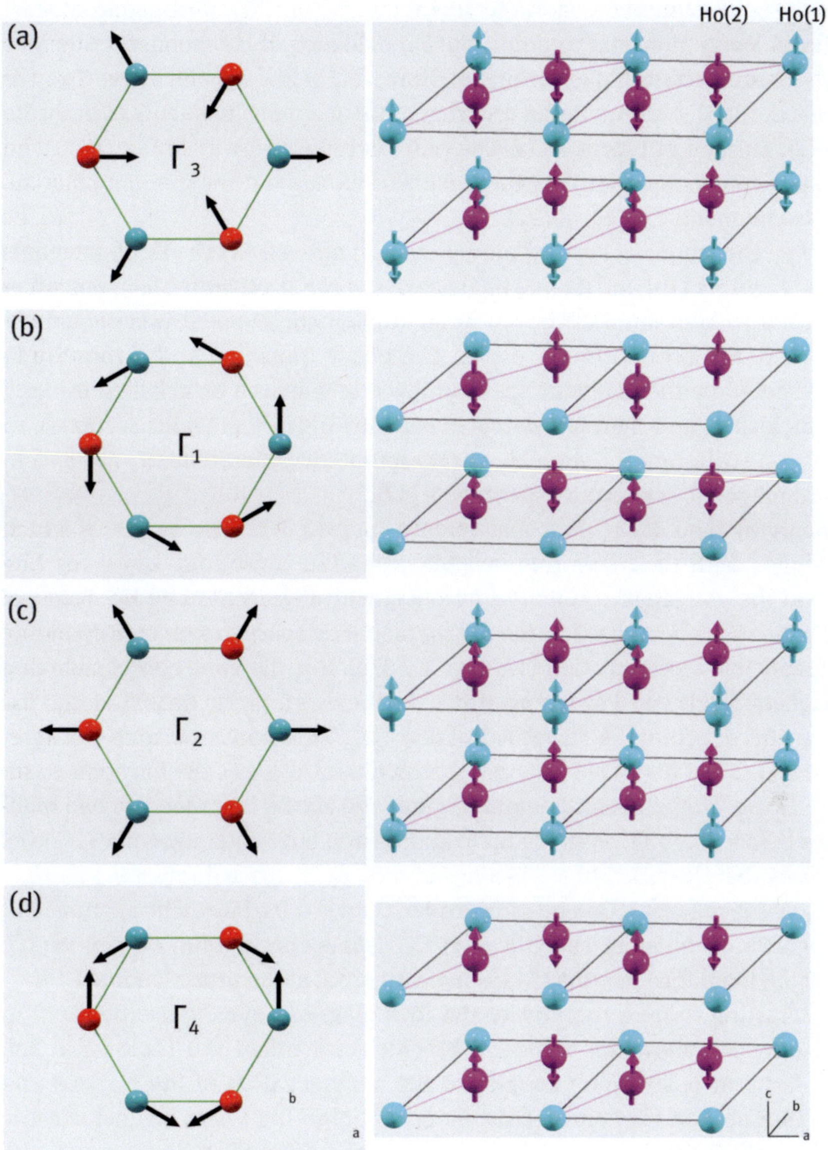

Figure 3.17: Schematic order of Mn spins (left) and Ho moments (right) in different phases of $HoMnO_3$. (a) $P\underline{6_3}c\underline{m}$, $T_{Ho} < T < T_{SR}$ in $H_c = 0$, (b) $P6_3cm$, $T < T_{Ho}$ in $H_c = 0$, (c) $P6_3\underline{cm}$ (suspected high-field symmetry, the relative alignment of the FM sublattices of Ho 2a and 4b can also be antiferromagnetic resulting in a ferrimagnetic state, possibly at lower H_c), and (d) $P\underline{6_3}\underline{c}m$, for $T_{SR} < T < T_N$ in $H_c = 0$. Note that (d) allows for a magnetic order of the Ho^{3+} moments on 4b sites, but it was not observed experimentally. Ho ions and their moments in two neighboring unit cells are shown.

Above the spin rotation transition temperature T_{SR}, the $P\underline{6_3}\underline{c}m$ magnetic symmetry (Γ_4) allows Ho(2) moments to order, but no evidence of Ho moment order was found from neutron scattering experiments. However, below T_{SR} and above T_{Ho}, the $P\underline{6_3}c\underline{m}$ space group (Γ_3) allows for an ordering of the magnetic moments of both, Ho (1) and Ho(2), as seen in Figure 3.17a. The order is induced by the interaction with the Mn^{3+} spin system and various experimental data support the gradual polarization of the Ho moments [37, 43, 47, 72].

Below T_{Ho}, the magnetic space group $P6_3cm$ (Γ_1) only allows the Ho(2) moments to order (see Figure 3.17b) and a large increase of the Ho(2) sublattice magnetization together with a sudden jump of the Mn ordered magnetic moment was indeed observed at the transition into the Γ_1 phase. The phase transition at T_{Ho} into the Γ_1 phase represented by the magnetic space group $P6_3cm$ is well established through magnetic, dielectric, and thermodynamic measurements (see previous sections), as well as neutron scattering [37, 44, 63], second harmonic generation [27, 72], and X-ray resonant magnetic scattering experiments [47].

The magnetic field along the *c*-axis mainly couples to the Ho-moments which are also aligned with the *c*-axis (the magnetocrystalline anisotropy keeps the Mn-spins close to the *ab* plane). H_c favors a ferromagnetic alignment of all Ho magnetic moments, which is only realized in the $P6_3\underline{cm}$ magnetic space group, corresponding to the Γ_2 irreducible representation (see Figure 3.17c). It is therefore conceivable that at high magnetic fields and low temperatures there exists a phase transition into the $P6_3\underline{cm}$ magnetic structure, as suggested earlier [27]. This transition from the zero-field $P6_3cm$ (Γ_1) to the high-field $P6_3\underline{cm}$ (Γ_2) structure also forces the Mn spins to rotate by 90°. However, the phase diagram (Figure 3.13) shows that there are two high-field phases below 3K, "LT1" is stable up to 20 kOe and "LT2" is stable above 20 kOe. If one assumes that the "LT2" phase belongs to the $P6_3\underline{cm}$ space group, the nature of the "LT1" phase is not clear. One possible explanation is a ferrimagnetic alignment of the Ho moments on different sublattices in LT1 (same space group $P6_3\underline{cm}$) which, with increasing field, flips into the FM Ho-moment order at the transition into LT2.

It is interesting to note that out of the four magnetic symmetries depicted in Figure 3.17 only one allows for a linear magnetoelectric effect (see Table 3.1 in Ref. [73]). The electrical polarization couples to the magnetization of the Ho ions and may result in a gain of magnetoelectric energy through the linear magnetoelectric coupling. An externally applied electric field could polarize the Ho-magnetic moment and induce a ferromagnetic order with a sizable macroscopic magnetization. Experimental evidence for an electric field-induced macroscopic magnetization in $HoMnO_3$ was found by Lottermoser et al. [73]. The possible bulk and domain boundary (linear) magnetoelectric effects have also been studied for the broader family of h-$RMnO_3$ compounds (R=Sc, Y, In, Ho, Er, Tm, Yb, Lu) [46].

Other groups have searched for the electric-field induced magnetic moment in $HoMnO_3$ and the symmetry allowed linear magnetoelectric effect at low temperatures and high fields. In field-dependent polarization measurements below 3 K and magnetic fields above 30 kOe, Hur et al. could not detect a linear magnetoelectric effect within their experimental resolution [54]. Using X-ray resonant magnetic scattering, Nandi et al. have elucidated the role of Ho^{3+} ions and studied the magnetic order of the Ho moments [47]. The results confirm the onset of Ho moment order at T_{SR} and a change of the order at T_{Ho}. However, measurements in high electric fields were found to be identical to the results of zero field experiments and the suggested change of the Ho moment order in electric fields could not be confirmed. Small angle neutron scattering experiments conducted on single crystals of $HoMnO_3$ in magnetic and electric fields have indicated that the ferromagnetic moment in electric fields possibly arises from uncompensated spins in the antiferromagnetic domain walls rather then from bulk magnetism of the Ho moments [74]. This issue is still not decided and more studies have to be conducted.

3.4 Hexagonal manganites $RMnO_3$ with nonmagnetic R-ions

As is obvious from the case of $HoMnO_3$, the presence of the rare earth magnetic moments induces additional complexity into the frustrated magnetic orders and the magnetoelectric coupling. Therefore, h-$RMnO_3$ compounds with nonmagnetic R-ions have been studied to investigate the magnetoelectric coupling between the ferroelectric state and the Mn-spin system. Those R-ions forming the hexagonal manganite structure are Y, Lu, Sc, and In.

3.4.1 Multiferroic $YMnO_3$

$YMnO_3$ is structurally very similar to $HoMnO_3$ because of the almost equal size of the Y^{3+} and Ho^{3+} ions, with nearly identical lattice parameters and Néel temperatures (see Table 3.1). $YMnO_3$ with the much simpler magnetic system (only Mn^{3+} spins) has been studied in detail and compared to $HoMnO_3$ and other h-$RMnO_3$. The high-temperature structures and phase transitions in $YMnO_3$ [14–17] are still under discussion but most studies confirm the structural transition to the $P6_3cm$ phase with the onset of ferroelectricity near 1,270 K [20]. Ferroelectricity in $YMnO_3$ was discovered in 1963 [2] and the coexistence of the ferroelectric state with antiferromagnetism at low temperatures was reported in 1964 [6]. The coupling between ferroelectric and magnetic orders was first found in $YMnO_3$ in form of a small anomaly of the dielectric constant at the Néel temperature [7].

The ferroelectricity in the $P6_3cm$ phase originates from a buckling of the MnO_5 polyhedra and the displacements of the O^{2-} and Y^{3+} ions into an off center position along the c-axis [11], as shown in Figure 3.18. The ferroelectric polarization arises mainly from the fact that the two Y^{3+} ions in different Wyckoff positions are shifted in opposite directions along the c-axis, creating local dipolar moments of different sign which are not completely compensated. The net dipolar momentum, in combination with the displacements of the oxygen ions, generate the macroscopic polarization. The origin of the structural instability was attributed to a zone boundary phonon instability resulting in an improper ferroelectric state [75, 76]. However, polarization dependent X-ray absorption measurements have suggested the existence of strong anisotropic hybridization of the Y 4*d*–O 2*p* bonds which creates the off-center displacements of Y^{3+} and the resulting large anomalies in the Born effective charges as the possible cause of the ferroelectric instability [77].

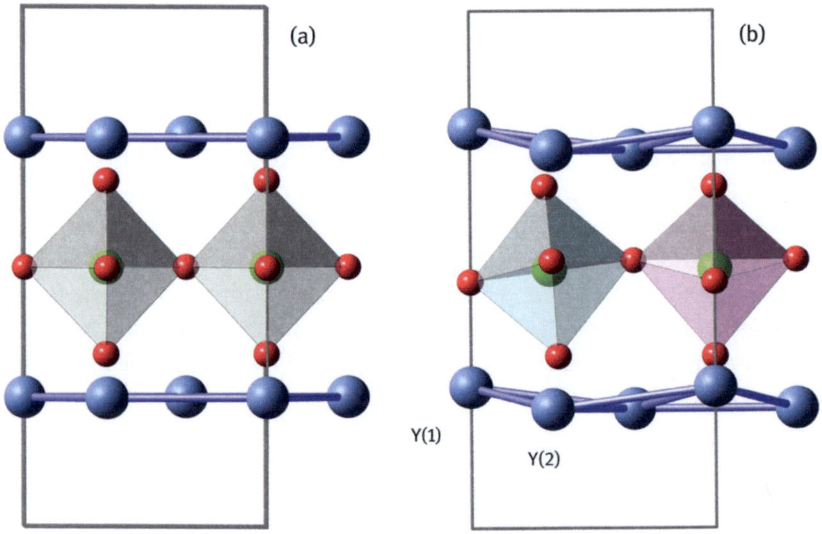

Figure 3.18: (a) Undistorted structure of $YMnO_3$ with ions in centrosymmetric positions. (b) Distorted structure in the ferroelectric state.

The magnetic order in $YMnO_3$ below the Néel temperature $T_N = 72$ K has been studied through magnetization, neutron scattering, and second harmonic generation optical experiments. The *T*-dependent magnetic susceptibility of a floating zone grown single crystal of $YMnO_3$ shows a sharp kink at T_N (Figure 3.19a) which is consistent with earlier reports for single crystals [78–80]. This kink is less clear in powder or polycrystalline samples, possibly due to sample quality or grain size effects [7, 24, 29, 81–84].

The magnetic symmetry of $YMnO_3$ below T_N was a matter of discussion for many years. Based on neutron scattering experiments, two possible magnetic structures have been suggested, $P\underline{6}_3c\underline{m}$ (Γ_3, α-model) or $P6_3cm$ (Γ_1, β-model), according to Table 3.2 [21]. Subsequent studies did favor of the $P\underline{6}_3c\underline{m}$ magnetic symmetry [85]. The difficulty to distinguish between the two magnetic symmetries using neutron scattering experiments was also found in more recent studies [30, 86]. Other data, however, appear to support the $P6_3cm$ (Γ_1) symmetry [29, 87, 88]. An even lower magnetic symmetry, $P\underline{6}_3$ with a spin tilt angle of $\Phi = 11°$ was proposed from neutron diffraction and polarimetric measurements [43].

The uncertainty in the exact determination of the magnetic symmetry through neutron scattering experiments lies in the fact that the neutron spectra have to be fitted to different models of the magnetic structure. It turns out that the above mentioned models (Γ_1 and Γ_3) produce very similar diffraction spectra. Therefore, alternative methods which are more sensitive to the magnetic symmetry, have to be employed. For $YMnO_3$, the results of second harmonic generation optical experiments suggest the $P\underline{6}_3c\underline{m}$ (Γ_3) magnetic symmetry, the same symmetry as in $HoMnO_3$ below the spin rotation transition [33, 34].

One particular feature of the interacting Mn^{3+} spin system is the quasi two-dimensional (2D) character with the spins confined to the ab-plane, the geometric frustration, and the weak interaction between adjacent planes. As a consequence, signatures of diffuse scattering as well as unconventional short range spin fluctuations have been reported above T_N as well as in the ordered phase [81, 82, 89–91]. The suppression of the thermal conductivity between T_N and room temperature is consistent with the presence of strong spin fluctuations in $YMnO_3$ [49]. It was further suggested that the presence of a residual magnetic contribution to the heat capacity, C_p, is possibly due to a magnetic glassy state that coexists with the ordered antiferromagnetism in some $RMnO_3$ (R = Y, Lu, Sc) [24, 78]. However, the deviation of $C_p(T)$ from the Debye law at low temperatures could also be explained by an additional Einstein contribution since the critical scaling of $C_p(T)$ near T_N fits very well to a 3D Heisenberg model ruling out a 2D or chiral model for the magnetic system [59, 92]. It appears that the true nature of the magnetic order and the spin fluctuations below and above T_N has yet to be revealed.

Magnetoelectric effects at the Néel temperature were first detected in form of subtle anomalies of the dielectric constant in polycrystalline samples of $YMnO_3$, $LuMnO_3$, and $ScMnO_3$ [7, 24]. Using single crystals of $YMnO_3$ and $LuMnO_3$, Katsufuji et al. [78] reported a sudden drop of the dielectric constant ε_{ab}, providing a much clearer picture of the magnetic order effects on dielectric quantities. The dielectric anomalies in $YMnO_3$ neat T_N suggest a strong spin-lattice interaction which is not complicated by the presence of a rare earth moment. Subtle anomalies of the lattice parameters at T_N, although nearly within the resolution of the data, have been reported from neutron scattering studies [30]. High resolution thermal expansion measurements provide a clearer picture of the lattice response to the magnetic transition, as shown in Figure 3.19b,c. Over a wide temperature range, the c-axis exhibits a negative expansion coefficient (similar to $HoMnO_3$) whereas the a-axis shrinks with decreasing temperature, in agreement with data from neutron scattering experiments [29]. At T_N, the a-axis experiences an enhanced contraction whereas the c-axis shows an increased expansion (Figure 3.19b), consistent with the results of scattering experiments [71, 93]. The λ-shaped anomalies of the thermal expansivities (Figure 3.19c) reflect the second order nature of the magnetic transition and the presence of critical spin fluctuations near T_N.

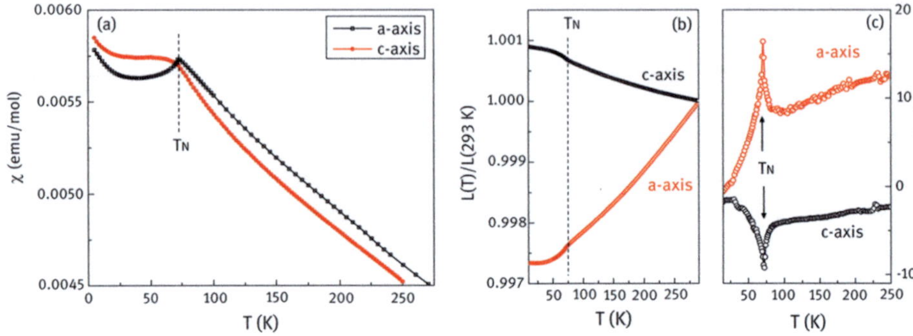

Figure 3.19: Magnetic susceptibility (a) and thermal expansion (b,c) of a floating zone grown single crystal of $YMnO_3$ along *a*- and *c*-axes. The Néel temperature is marked by the dashed line.

The ionic displacements in the unit cell of $YMnO_3$ and $LuMnO_3$ have been studied through high resolution X-ray diffraction and neutron scattering experiments [70]. At the magnetic transition temperature T_N, all atoms in the unit cell show giant displacements, two orders of magnitude larger than in typical magnetic materials, explaining the internal magnetoelectric effect, i.e. the coupling between the ferroelectric polarization and the antiferromagnetic order. The displacement of the Mn^{3+} ions from their ideal symmetric position ($x = 1/3$) results in the coupling to the electric dipole moments [93]. The Mn^{3+} displacements have an effect on the magnetic exchange coupling parameters and, possibly, determine the magnetic orders at low temperature [63]. For example, the deviation from the $x = 1/3$ position for the Mn^{3+} is of opposite sign in $YMnO_3$ and $LuMnO_3$, resulting in different antiferromagnetic structures below T_N. Furthermore, external magnetic fields have a significant influence on the atomic positions in the unit cell as revealed in high-field neutron scattering experiments [84].

Further evidence of strong magnetoelastic effects in $YMnO_3$ and other hexagonal $RMnO_3$ was found in studies of other properties as, for example, the elastic moduli [94]. Inelastic neutron scattering experiments showed the signature of strong coupling between magnons and phonons which appear to hybridize in one mixed elementary excitation [88], similar to the electromagnons studied in multiferroic orthorhombic manganites [95, 96]. Raman and infrared absorption investigations of the phonon spectrum of $YMnO_3$ and $LuMnO_3$ have found a kink at T_N and an abnormal hardening in the magnetically ordered phase, indicating strong spin–phonon coupling [97–99]. Similar phonon hardening was also found in compounds with magnetic rare earth ions, $HoMnO_3$ and $ErMnO_3$ [61, 100].

3.4.2 Multiferroic properties of $LuMnO_3$ and $ScMnO_3$

Similar to Y^{3+}, Lu^{3+} and Sc^{3+} have no magnetic moment, but both ions are significantly smaller resulting in smaller lattice parameters of $LuMnO_3$ and $ScMnO_3$ (see Table 3.1) and stronger magnetic exchange interactions. Therefore, the Néel temperatures systematically increase from 72 K ($YMnO_3$) to 90 K ($LuMnO_3$) and to 130 K ($ScMnO_3$). The in-plane magnetic anisotropy decreases and is quite small for $ScMnO_3$. The lower anisotropy makes the two magnetic structures with locked spin angles $\Phi = 0°$ and $\Phi = 90°$ less favorable (see Table 3.2).

The magnetic phase transition in $LuMnO_3$ at $T_N = 90$ K is accompanied by distinct anomalies of the bulk magnetic susceptibility and the heat capacity [24, 78, 101]. High-quality single crystals reveal a clear peak of the magnetic susceptibility at T_N, as expected for an antiferromagnetic phase transition [78, 101, 102]. The magnetic contribution to the heat capacity of $LuMnO_3$ and $ScMnO_3$ was extracted and the change of magnetic entropy below T_N amounts to the expected value for spin 2 (the spin value of Mn^{3+}) [24].

Bulk magnetic measurements of polycrystalline samples of $ScMnO_3$ have revealed irreversible behavior (difference between field cooled and zero-field cooled data) and a magnetic hysteresis loop at low temperatures, which has been interpreted as evidence for a ferromagnetic moment [24, 29, 103, 104]. This is in contrast to the possible magnetic structures ($P\underline{6}_3c\underline{m}$ or $P\underline{6}_3$) derived from neutron scattering [104] and SHG experiments [105] which do not allow for a ferromagnetic moment. The source of the irreversibility in the bulk magnetization is most likely a consequence of the polycrystalline nature of the studied samples and the associated grain boundary contributions to the magnetization. This effect was carefully investigated in $HoMnO_3$ by comparing magnetization measurements of single crystals, showing no irreversibility, and polycrystalline ceramic samples which reveal clear signatures of irreversible magnetic properties [106].

Above room temperature, the lattice parameters a and c show a significantly different temperature dependence, as compared to $YMnO_3$. Whereas the a-axis has a similar positive thermal expansivity, the c-axis is nearly temperature independent in $LuMnO_3$ and it has a (normal) positive thermal expansivity in $ScMnO_3$ (in stark contrast to the negative expansivity in $YMnO_3$). The c-axis length scales with the buckling angle of the MnO_5 bipyramids and the decrease of this angle with increasing temperature results in the contraction of the c-axis in $YMnO_3$ [25]. The buckling angle in $ScMnO_3$ increases with temperature resulting in the normal expansion of the c-axis.

At low temperatures the c-axis of $LuMnO_3$ experiences a sizable increase with decreasing temperature below 200 K, signaling the onset of magnetic fluctuations [70, 101, 107]. This increase and the related anomalies in $YMnO_3$ and $HoMnO_3$ show that the buckling of the MnO_5 units, resulting in the ferroelectric distortion, are not the only mechanism affecting the thermal expansion behavior of the c-axis. The displacements of the Mn^{3+} ions from their $x = 1/3$ ideal position in the magnetically ordered phase have an additional effect on the *c*-axis. This displacement is opposite in $YMnO_3$ (x increases below T_N) and $LuMnO_3$ (x decreases below T_N) [70, 107]. It should be noted that the deviation from $x = 1/3$ determines the sign of the magnetic coupling between Mn^{3+} spins in adjacent planes and, therefore, has a profound effect on the magnetic structure of different $RMnO_3$ [63].

There is no consensus with regard to the magnetic symmetries in the ordered phase of $LuMnO_3$. Various proposals, derived from neutron scattering experiments, include the $P\underline{6}_3$ symmetry (α-model) with an intermediate Mn^{3+}-spin angle of $\Phi = 55°$ [22], the $P\underline{6}_3\underline{c}m$ symmetry (α model, $\Phi = 90°$, Γ_4) [25, 108], and the $P6_3\underline{cm}$ symmetry (β model, $\Phi = 0°$, Γ_2) [25, 86]. Second harmonic generation optical experiments suggested a spin reorientation from $P\underline{6}_3\underline{c}m$ (α model, $\Phi = 90°$, Γ_4) to $P\underline{6}_3c\underline{m}$ (α model, $\Phi = 0°$, Γ_3) magnetic symmetry with an intermediate phase ($P\underline{6}_3$) coexisting with the two other phases below 60 K [34]. The spatial coexistence of phases with different magnetic symmetry can make the correct determination of the Mn^{3+} spin order below T_N through neutron scattering experiments very difficult.

Unlike in $LuMnO_3$, a spin reorientation in $ScMnO_3$ was clearly detected in neutron scattering experiments [29, 63, 104], starting between 60 and 80 K. However, the details of the magnetic symmetries over the whole temperature range below T_N are still disputed. Bieringer et al. [104] assigned the $P\underline{6}_3$ magnetic symmetry with a spin angle of $\Phi = 80°$ (close to $P\underline{6}_3\underline{c}m$, Γ_4) to the high-temperature phase and suggested a smooth spin rotation to a low-temperature angle of $\Phi = 15°$ (closer to $P\underline{6}_3c\underline{m}$, Γ_3). However, Muñoz et al. described the high-temperature magnetic order between 130 K and 75 K by the $P6_3\underline{cm}$ symmetry (β model, $\Phi = 0°$, Γ_2) with a continuously increasing angle Φ from 17° at 75 K to 54° at 1.8 K (magnetic symmetry $P6_3$). The SHG optical experiments support the $P\underline{6}_3\underline{c}m$ symmetry (α model, $\Phi = 90°$, Γ_4) below T_N coexisting with $P\underline{6}_3c\underline{m}$ (α model, $\Phi = 0°$, Γ_3) below 60 K, and all three phases (including $P\underline{6}_3$) coexisting at the lowest temperatures within different regions of the same crystal [27, 34, 109].

The proposed spin reorientations in $LuMnO_3$ and $ScMnO_3$ are realized in a magnetic system of exclusively Mn^{3+} spins with no rare earth moment involved. This fact makes it unlikely that the spin reorientation in the sister compound $HoMnO_3$ may be driven by the interaction of the rare earth moment with the Mn-spins. The change in the in-plane magnetic anisotropy with decreasing temperature has to be considered as the origin of those phenomena. The smaller magnetic anisotropy of $ScMnO_3$ makes this compound more susceptible to spin reorientations and magnetic structures with an intermediate spin angle Φ. $ScMnO_3$ is magnetically "soft" with respect to an in-plane spin rotation which is also reflected in the photoinduced control of the average Mn^{3+} spin angle at low temperature [105].

3.4.3 $InMnO_3$: Ferroelectric or not?

The structure of $InMnO_3$, first synthesized in 1992 [110], was determined to be hexagonal (space group $P6_3cm$) [111], similar to other $RMnO_3$. The onset of magnetic order at $T_N = 120$ K has been verified through bulk magnetic and neutron scattering experiments, with additional anomalies at 40K and 15 K indicating possible spin reorientation transitions [111–113]. However, the latter transitions below T_N have been questioned and attributed to possible impurity phases [114]. The magnetic phase transition at 120 K is supported by heat capacity measurements showing a distinct peak at T_N [114, 115]. Similar to other hexagonal manganites, the low-frequency dielectric constant exhibits a kink at T_N, indicative of strong spin–lattice interactions.

The lattice parameters of $InMnO_3$, as determined from X-ray and neutron scattering experiments [23, 111, 114, 116], deviate significantly from other hexagonal $RMnO_3$. While a and c of nearly all hexagonal $RMnO_3$ increase linearly with the Shannon ionic radius of the R^{3+} (Figure 3.20), the c-axis of $InMnO_3$ is significantly larger and the a-axis is slightly smaller than the linear extrapolation (shown as dashed lines in Figure 3.20) [111]. The large c-axis length implies a reduced magnetic interplane exchange coupling. This effect is further enhanced by the almost ideal position of the Mn^{3+} ion in the unit cell ($x_{Mn} \approx 1/3$) which renders the two superexchange pathways

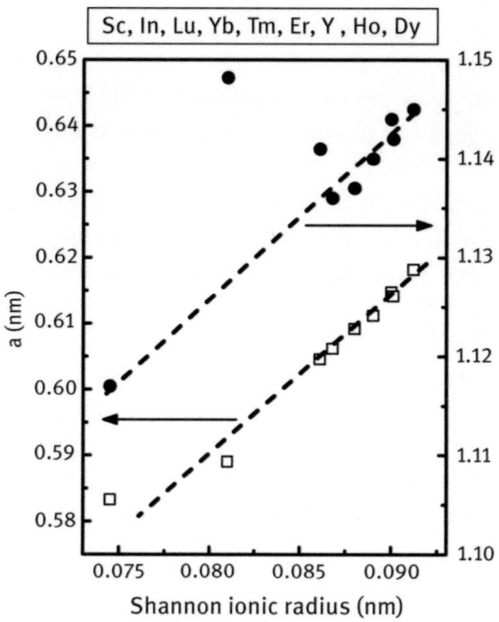

Figure 3.20: Lattice parameters a (left scale, open squares) and c (right scale, filled circles) of hexagonal $RMnO_3$ as function of the R-ionic radius, increasing from Sc to Dy. The dashed lines indicate the approximately linear dependence, except for $InMnO_3$.

between Mn^{3+} ions in adjacent planes nearly identical, increasing the magnetic frustration between spins of neighboring planes.

The magnetic symmetry of the ordered Mn^{3+} spins was determined to be either Γ_2 or Γ_4 (Figure 3.3), with the spins of two subsequent unit cells along the *c*-axis reversed by 180°, resulting in a doubling of the unit cell along the *c*-axis [111, 116]. Strong diffuse scattering of neutrons observed above T_N are consistent with the existence of 2D short range correlations of Mn^{3+} spins at higher temperatures, the highest degree of magnetic frustration in the $RMnO_3$ system, and the stabilization of 3D long range order through weak interplane interactions at much lower temperatures. Pseudodipolar magnetic interactions have been proposed to account for the magnetic long range order along the *c*-axis with the modulation vector $\vec{q} = (0, 0, 1/2)$ [116].

The existence of ferroelectric order in $InMnO_3$, associated with the noncentrosymmetric space group $P6_3cm$, was a matter of discussion for several decades. Abrahams [117] estimated the ferroelectric transition temperature as $T_c \approx 540$ K, using the known structural data [111]. Measurements of the $P-E$ hysteresis loops of polycrystalline samples indicate a ferroelectric transition at 465 K [113]. In a theoretical study, the ferroelectricity in $InMnO_3$ was attributed to an intra-atomic $4d_{z^2}$–$5p_z$ orbital mixing of In and a covalent bonding ($4d_{z^2}$(In)-$2p_z$(O)) along the *c*-axis [118].

On the contrary, there have been numerous reports arguing that $InMnO_3$ is not ferroelectric. Belik et al. [114, 115] have not found any changes of the dielectric permittivity, the number of phonon excitations, or other quantities sensitive to structural changes at high temperatures up to 900 K. The relaxor-like property of the $P-E$ hysteresis loop was reproduced and attributed to extrinsic, non-ferroelectric effects like high-loss dielectric behavior caused by defect-induced conductivity. The absence of ferroelectricity is incompatible with the polar space group $P6_3cm$ used to fit numerous X-ray diffraction and neutron scattering data. However, an alternative centrosymmetric structure, described by the trigonal space group $P\bar{3}c1$ (No. 165), was shown to fit X-ray spectra almost equally well. A comprehensive study of $InMnO_3$ employing X-ray diffraction, piezoelectric force microscopy, and optical SHG measurements combined with density functional calculations came to the conclusion that the lattice symmetry is centrosymmetric, $P\bar{3}c1$ [119]. No evidence for ferroelectricity or ferroelectric domains was found in SHG and piezoelectric force measurements.

It is worth to keep a closer look at the two proposed lattice structures of $InMnO_3$, which are displayed and compared in Figure 3.21 [119]. The two structures, $P6_3cm$ and $P\bar{3}c1$, are distinguished by the way the MnO_5 bipyramids are tilted from their ideal position and the displacements of the In^{3+} ions. In $P6_3cm$ symmetry, the In^{3+} ions on crystallographic sites 2a and 4b are shifted in opposite directions along the *c*-axis, resulting in an uncompensated dipolar moment and a macroscopic polarization (see left column in Figure 3.21a). In the centrosymmetric $P\bar{3}c1$ structure, however, one In^{3+} ion remains in the high-symmetry centrosymmetric 2b position whereas the ions in 4a position are displaced equally in opposite directions, maintaining an inversion center (right column of Figure 3.21a). Since the structure has an inversion center, no macroscopic polarization is allowed. The opposite displacements of In^{3+} rather result in an antiferroelectric state, as shown in the right column of Figure 3.21a. The corresponding Mn^{3+} spin orders are shown in Figure 3.21b. Band structure calculations for $InMnO_3$ revealed the important role of the degree of covalency of the R–O chemical bond [119]. Lower covalency favors the $P6_3cm$ structure over the $P\bar{3}c1$ structure. However, for $InMnO_3$, the energy difference between both structures was found to be small.

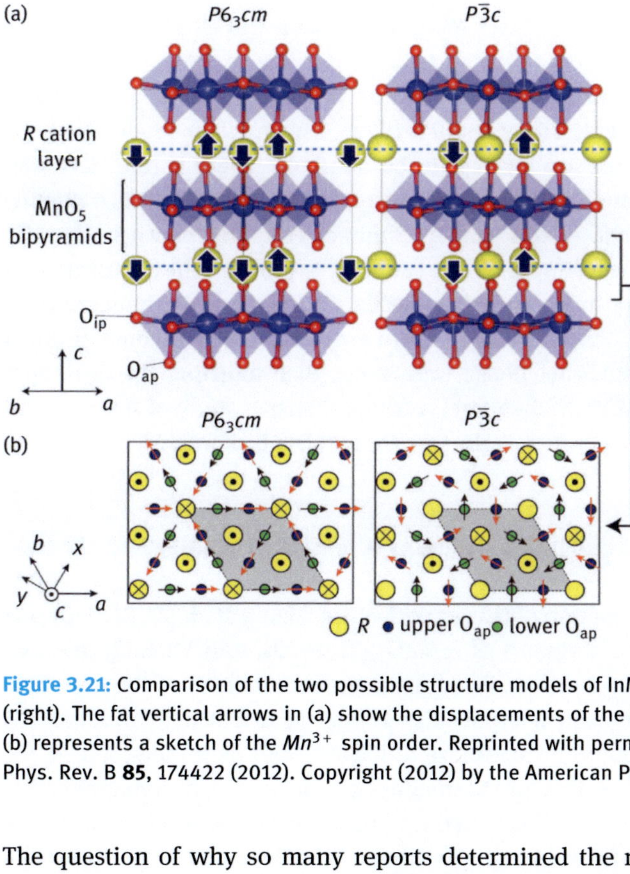

Figure 3.21: Comparison of the two possible structure models of $InMnO_3$, $P6_3cm$ (left) and $P\bar{3}c1$ (right). The fat vertical arrows in (a) show the displacements of the In^{3+} from their ideal positions. (b) represents a sketch of the Mn^{3+} spin order. Reprinted with permission from Y. Kumagai et al., Phys. Rev. B **85**, 174422 (2012). Copyright (2012) by the American Physical Society.

The question of why so many reports determined the noncentrosymmetric $P6_3cm$ structure for $InMnO_3$ but others favor the centrosymmetric $P\bar{3}c1$ structure was recently addressed in a careful study of sample synthesis and preparation conditions [19]. It was shown that the $P6_3cm$ structure is the ground state of $InMnO_3$ but that the $P\bar{3}c1$ structure exists at high temperatures and can be quenched to low temperature. Polycrystalline samples of $InMnO_3$, cooled slowly after synthesis, showed the ferroelectric P–E hysteresis loops (at 100 K) and the characteristic multiferroic domain structures in dark-field scanning transmission electron microscopy measurements. This study is apparently consistent with the small energy difference between the ground states of both structures and it solves the conundrum arising from different published works.

The true value of the ferroelectric polarization is difficult to extract from polycrystalline samples. Single crystals of $InMnO_3$ have been grown by Yu et al. [23] and their structure analysis is consistent with the $P6_3cm$ symmetry. Polarization measurements of small single crystals yielded a room temperature value of $P_c = 4.4 \mu C/cm^2$ which is comparable with P_c values of $YMnO_3$.

3.5 Hexagonal manganites with magnetic rare earth ions

The magnetic moment of the rare earth ion R^{3+} has an effect on the Mn^{3+} spin system although the crystalline anisotropy restricts the R^{3+} magnetic moment along the c-axis whereas the Mn^{3+}-spins are located in the ab-plane. There is no direct coupling between the two magnetic moments, however, the strong spin-lattice interaction results in magnetoelastic effects and an indirect coupling between the R^{3+} moments and Mn^{3+} spins. An example of this coupling is the gradual polarization of the Ho^{3+} moments below T_{SR} in $HoMnO_3$ (see Section 3.3.5). At lower temperatures, the superexchange interactions between the rare earth moments become stronger and frequently result in additional phase transitions, also affecting the Mn^{3+} spin order (as e. g. at T_{Ho} in $HoMnO_3$). Therefore, a comprehensive study of the multiferroic phase diagrams of $RMnO_3$ (with R = Er, Tm, Yb, and Dy) is of interest.

3.5.1 Multiferroic properties and phase diagrams of $ErMnO_3$, $TmMnO_3$, and $YbMnO_3$

Based on SHG and Faraday rotation optical experiments, Fiebig et al. [27, 36, 120] proposed the multiferroic phase diagrams of $ErMnO_3$, $TmMnO_3$, and $YbMnO_3$ in c-axis magnetic fields. According to those studies, the magnetic symmetry of the high-field phase of all three compounds was determined as $P6_3\underline{cm}$ (Γ_2 representation, Figure 3.3 and Figure 3.17c). The stability of this phase at high H_c can be understood by the energy gain through the coupling of the magnetic field to the ferromagnetically aligned rare earth moments [36]. The low-field phase below the Néel temperature has the $P\underline{6_3c}m$ symmetry and it extends to about 40 kOe in $ErMnO_3$, 35 kOe in $TmMnO_3$, and 45 kOe in $YbMnO_3$ and than traces back to low temperatures and low fields with a typical 'nose' shape of the phase boundaries [27].

In contrast to the results of Ref. [27], bulk measurements of the magnetic and dielectric quantities have shown that the $P\underline{6_3c}m$ phase extends to significantly higher magnetic fields, even above 140 kOe in the case of $YbMnO_3$ [5]. Bulk measurements such as dc or ac magnetization and heat capacity can be utilized to detect the thermodynamic nature of the phase transitions. As shown for $HoMnO_3$ in Sections 3.3.2 and 3.3.3, the dielectric constant ε_c is extraordinarily sensitive to magnetic phase transitions because of the strong spin lattice coupling in all hexagonal manganites. The measurement of $\varepsilon_c(T, H)$, supplemented by other physical quantities, can therefore be used to study the high-field phase diagrams of h-$RMnO_3$ compounds.

$ErMnO_3$

In zero magnetic field, $ErMnO_3$ passes through two phase transitions at $T_N = 80$ K and $T_{Er} = 2.4$ K, as revealed by two distinct peaks of the heat capacity (Figure 3.22a) [5, 121]. The magnetic susceptibility, $\chi_c(T)$, is dominated by the Er^{3+} magnetic moment but it experiences distinct slope changes at T_N and T_{Er}, as shown in Figure 3.22b. Although Sugie et al. [39] found a magnetic hysteresis loop at low temperatures indicative of a possible ferro- or ferrimagnetic moment, no such hysteresis effects could be detected in high-quality single crystals [5, 120], as shown in Figure 3.23a. The dielectric constant, shown in Figure 3.22c, has a kink at T_N, similar to $HoMnO_3$ [5, 39]. The sharpness of the Néel transition is verified by the narrow peak of $d\varepsilon_c/dT$ (inset in Figure 3.22c).

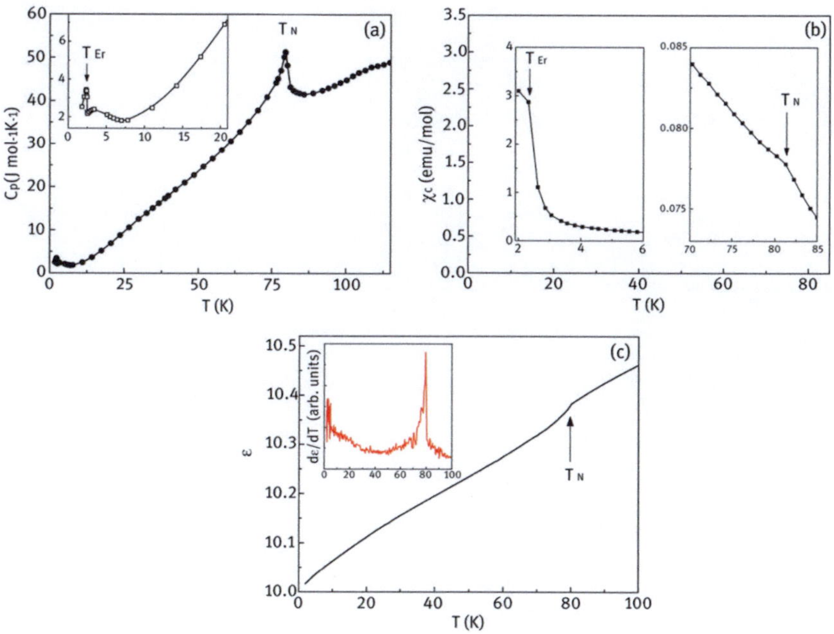

Figure 3.22: (a) Heat capacity at zero magnetic field, (b) magnetization at $H_c = 200$ Oe, and (c) dielectric constant of multiferroic $ErMnO_3$. The two transitions at T_N and T_{Er} are marked by vertical arrows. The inset to (c) shows the derivative, $d\varepsilon/dT$.

Magnetization measurements up to $H_c = 50$ kOe show two metamagnetic transitions at low temperatures (Figure 3.23a). The first transition, a sudden change of slope happens at a field of 450 Oe (1.85 K) and the critical field decreases to 270 Oe at 2.3 K. The extrapolation of this phase boundary to $H_c = 0$ coincides with the low-T peak of the heat capacity (Figure 3.22a) at $T_{Er} = 2.4$ K. The second steep increase of $M(H_c)$ at about 8 kOe (1.85 K) signals another magnetic phase transition (Figure 3.22a). The critical field of this transition increases quickly with temperature and is already above the maximum field of 50 kOe at 14 K. The derived phase boundary is named $\widetilde{T}_N$ in the following discussion. Temperature dependent magnetic susceptibility data, shown in Figure 3.23b on a logarithmic scale, exhibit a subtle slope change at $\widetilde{T}_N$ resulting in a clear minimum of the derivative, dM/dT, shown in the inset of

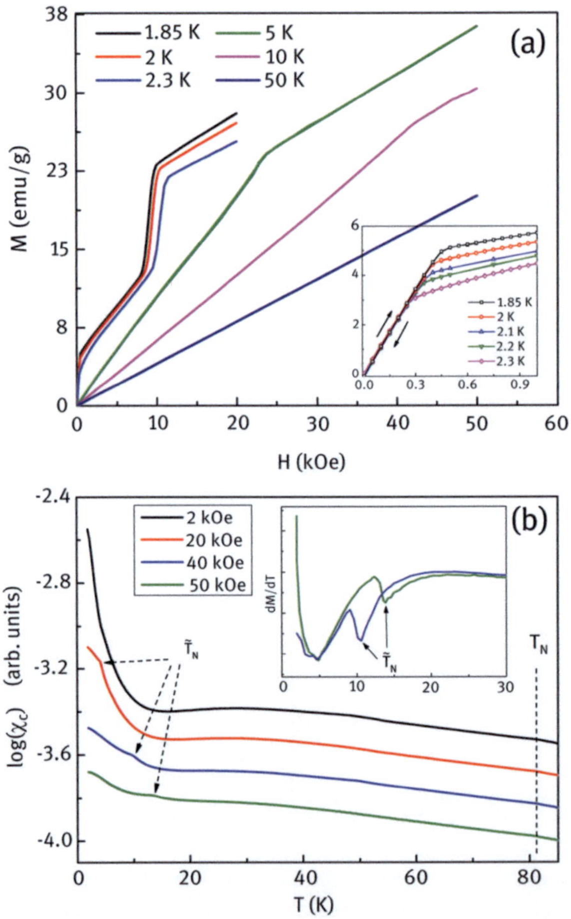

Figure 3.23: c-axis magnetization of $ErMnO_3$ up to $H_c = 50$ kOe. (a) M vs. H at various temperatures and (b) dc susceptibility on a logarithmic scale at various fields. The different curves in (b) coincide above about 50 K, but are vertically offset. Inset to (b): Derivative dM/dT with sharp minima at $\widetilde{T}_N$ and another minimum near 4–5 K.

Figure 3.23b. The Néel temperature T_N, indicated by the dashed vertical line, decreases only very little in fields up to 50 kOe.

Dielectric measurements up to $H_c = 140$ kOe complete the magnetic phase diagram of $ErMnO_3$. Below 20 K and in fields up to 70 kOe, $\varepsilon_c(H_c)$ shows a sharp kink (Figure 3.24a) which coincides with the phase boundary $\widetilde{T}_N$ as defined by the magnetic data above. The T-dependence of $\varepsilon_c$in Figure 3.24b reveals two kink-like anomalies, which represent the Néel temperature T_N, and the lower anomaly at $\widetilde{T}_N$. Whereas T_N slightly decreases with the field, $\widetilde{T}_N$ increases and merges with T_N at about 130 kOe and 65 K. At very low temperatures (see inset in Figure 3.24b), $\widetilde{T}_N$ quickly decreases with decreasing field and is extrapolated to 7 kOe at $T = 0$.

The resulting phase diagram is shown in Figure 3.24c. The phase boundaries are defined by anomalies of the dc magnetization, ac magnetic susceptibility, dielectric constant, and heat capacity, as indicated in the figure. The magnetic order below T_N is remarkably stable with respect to H_c, extending up to 130 kOe in the intermediate temperature range. This is in clear contrast to earlier reports [36] but in excellent agreement with recent studies up to 40 kOe [120]. At low temperatures and low fields the phase diagram of $ErMnO_3$ shows some interesting details, displayed in the inset to Figure 3.24c. The phase transition temperature T_{Er} quickly decreases from 2.4 K and extrapolates to about 500 Oe at zero temperature. $\widetilde{T}_N$ rises from 7 kOe and there is a clear gap between T_{Er} and $\widetilde{T}_N$ at $T = 0$ K. Another possible phase boundary, defined by an inflection point of the dc magnetic susceptibility and shown as dotted line connecting T_{Er} and $\widetilde{T}_N$ was proposed by Meier et al. [120]. The minimum of the derivative (shown in the inset to Figure 3.23b) clearly defines this inflection point near 4 K. However, since the inflection point of the dc magnetization is also observed at higher magnetic fields (see 40 and 50 kOe data of dM/dT in the inset of Figure 3.23b) and the dotted line extends into the high-field phase, it is not clear whether this anomaly identifies an additional phase transition. It

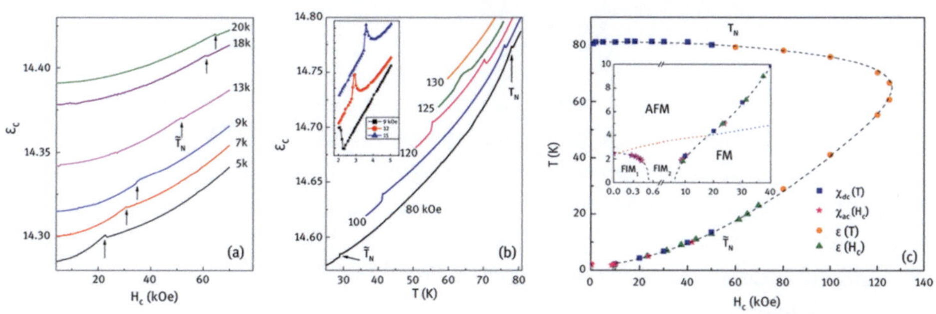

Figure 3.24: (a) Isothermal dielectric constant $\varepsilon_c(H_c)$ of $ErMnO_3$ at different temperatures below 20 K. The arrows mark the anomaly at the $\widetilde{T}_N$ phase boundary. (b) $\varepsilon_c(T)$ of $ErMnO_3$ at high fields up to 130 kOe. The two transitions at T_N and $\widetilde{T}_N$ are shown for the 80 kOe data. The different curves in (b) are vertically offset for better clarity. The inset to (b) shows the $\varepsilon_c(T)$ anomaly near $\widetilde{T}_N$ at low magnetic fields. (c) Magnetic phase diagram of $ErMnO_3$.

should also be noted that the dielectric constant $\varepsilon(T)$ measured at different magnetic fields between 3 and 40 kOe does not show any anomaly at the temperature defined by the inflection point of $\chi_c(T)$.

Different magnetic symmetries below T_N have been proposed from neutron scattering experiments, including $P\underline{6}_3$ (α model) with a spin angle to the hexagonal a-axis ϕ=70° [22] and two possible solutions fitting the spectra equally well, $P\underline{6_3}\underline{c}m$ (α model, Γ_4) and $P6_3\underline{cm}$ (β model, Γ_2) [122]. Second harmonic generation optical studies are in favor of the $P\underline{6_3}\underline{c}m$ symmetry at all temperatures between T_N and $\widetilde{T}_N$ [27, 34].

The magnetic structures at low temperatures and high c-axis magnetic fields have been addressed in a comprehensive investigation of $ErMnO_3$ employing bulk magnetic measurements, neutron scattering data, and second harmonic generation optical studies [120]. According to this work, the $P\underline{6_3}\underline{c}m$ (Γ_4) magnetic symmetry below T_N changes at T_{Er} through a Mn^{3+} spin rotation to $P6_3\underline{cm}$ (Γ_2). The latter transition is apparently driven by the strong interaction between the Er^{3+} moments on crystallographic 2a sites, which are not allowed to order in Γ_4 symmetry. According to Meier et al. [120], the phase below T_{Er} (labeled FIM_1 in Figure 3.24c and Ref. [120]) is characterized by a ferrimagnetic order of the two magnetic sublattices of Er ions on 2a and 4b sites in a multidomain state. At high magnetic fields, beyond the $\widetilde{T}_N$ phase boundary, the magnetic moments of the two Er sublattices become aligned with the field. The intermediate phase (labeled FIM_2), stable below the dotted line in Figure 3.24c, was proposed to be the single domain state of the FIM_1 phase.

The various microscopic interactions leading to the complex phase diagram of $ErMnO_3$ include the magnetic exchange between Er^{3+} moments and manganese spins, as shown in optical studies of the magnon spectra and crystal field transitions [123], as well as the strong spin lattice coupling revealed in the phonon hardening below T_N [100]. The latter does explain the sharp anomalies of the dielectric constant across different phase boundaries.

$TmMnO_3$

$TmMnO_3$ was less intensively studied in the past. The structural transition from nonpolar space group to polar $P6_3cm$ symmetry was found at 1600 K whereas at about 800 K a gradual transition from an antiferroelectric to a ferroelectric state was suggested [124]. Magnetic order of Mn^{3+} spins sets in at $T_N = 84$ K with magnetic symmetry described by $P\underline{6_3}\underline{c}m$ (Γ_4) [27, 34]. The Tm^{3+} moments on 4b sites become gradually polarized below T_N, as revealed by ^{169}Tm Mössbauer experiments [125]. The Néel temperature $T_N = 84$ K is consistent with magnetization and dielectric measurements [5, 126, 127]. In the absence of magnetic fields, the dielectric constant shows a clear kink at T_N, but no further anomaly at lower temperatures (Figure 3.25a). Unlike $HoMnO_3$, $ErMnO_3$, and $YbMnO_3$, in $TmMnO_3$ there is no indication of an additional phase transition at low temperatures associated with a possible Tm^{3+} moment order on 2a sites.

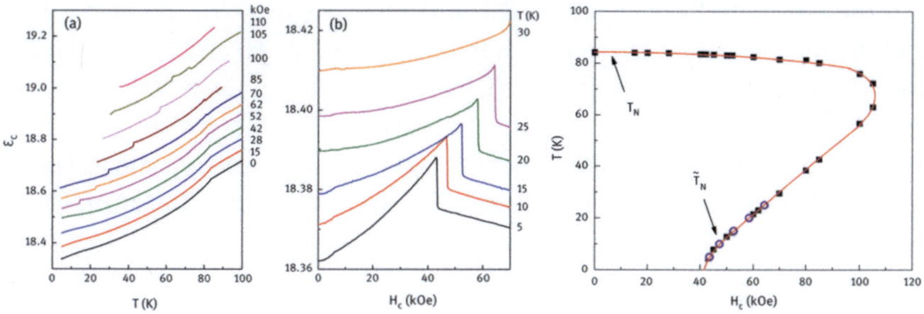

Figure 3.25: Temperature and field dependence of the dielectric constant of $TmMnO_3$. (a) $\varepsilon(T)$ at different fields. (b) Isothermal data $\varepsilon(H_c)$. (c) Phase diagram of $TmMnO_3$ derived from anomalies of $\varepsilon_c(T, H_c)$ shown in (a), (b).

With increasing magnetic field H_c, the kink of $\varepsilon_c(T)$ at T_N shifts slightly to lower temperature and a second step of ε_c develops at low T starting at 42 kOe, similar to the dielectric constant of $ErMnO_3$ (Figure 3.25a). The critical temperature, $\widetilde{T}_N$, increases with H_c and merges with T_N just below 110 kOe and about 65 K. At low temperatures, $\varepsilon(T, H_c)$ exhibits a sharp step as seen in Figure 3.25b. A sharp dip of the low temperature thermal conductivity was observed with increasing field H_c near 40 kOe [128], coinciding with the $\widetilde{T}_N$ phase boundary. Furthermore, the field dependence of the c-axis magnetization shows a distinct change of slope at $\widetilde{T}_N$ [128]. The phase diagram, derived from the dielectric data is shown in Figure 3.25c. Similar to $ErMnO_3$, the stability of the magnetic phase below T_N extends to much higher fields as earlier proposed [27]. So far, $TmMnO_3$ has the simplest phase diagram among all hexagonal $RMnO_3$ with magnetic R-ions. Only one magnetically ordered phase with $P\underline{6_3}\underline{c}m$ structure is enclosed by the T_N-$\widetilde{T}_N$ phase boundary.

$YbMnO_3$

$YbMnO_3$ shows two magnetic transitions at $T_N = 90$ K and at $T_{Yb} = 4$ K [5, 39]. Both transitions are clearly verified in sharp peaks of the heat capacity, shown in Figure 3.26a. An additional metamagnetic transition was found in field dependent isothermal magnetization data above 30 kOe and at low temperatures (Figure 3.26b). The critical temperature of this transition is labeled $\widetilde{T}_N$ in the phase diagram (Figure 3.27c). Both transitions at T_{Yb} and $\widetilde{T}_N$ are also defined by distinct increases of $\chi_c(T)$ in Figure 3.26c. The inset to Figure 3.26b shows the sharp increase of $\chi_c(T)$ at low field (100 Oe) indicating a possible ferrimagnetic order of the ytterbium sublattices. T_{Yb} decreases quickly with increasing H_c and this phase vanishes above 800 Oe, as shown in the inset to Figure 3.27c. The low temperature phase diagram of $YbMnO_3$ is quite similar to that of $ErMnO_3$ discussed above. In particular, the steep increase of the isothermal magnetization at small fields is typical for the FIM_1 phase discussed in Ref. [120] for $ErMnO_3$. The step-like increase of $M(H_c)$ above 30 kOe could be understood as the transition to the complete ferromagnetic (FM) alignment of the Yb^{3+} moments. The saturation of the magnetization at the lowest temperatures (see 40 and 50 kOe data in Figure 3.26c) is consistent with the proposed ferromagnetic high-field state. This FM state is most likely a single domain state since no significant difference between field-cooled and zero field-cooled data was observed. Some of the results for the temperature and field dependence of the magnetization discussed above have been confirmed in recent studies [28, 57, 129].

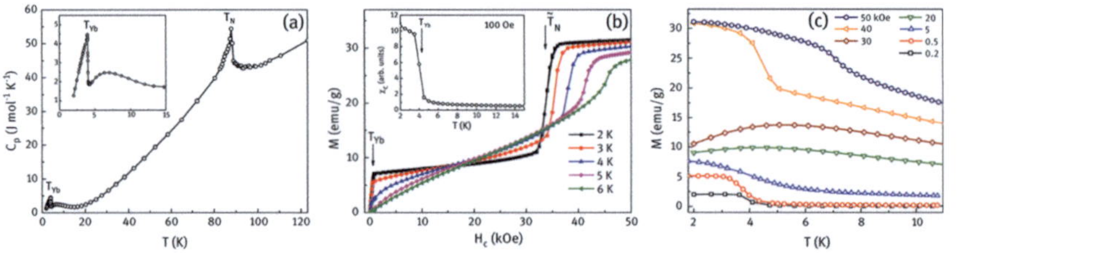

Figure 3.26: (a) Heat capacity of $YbMnO_3$ in zero magnetic field. (b) Isothermal magnetization of $YbMnO_3$ as function of *c*-axis fields. The two vertical arrows indicate the two phase transitions. Inset: Temperature dependence of the susceptibility at 100 Oe. The sharp increase at $T_{Yb} = 4$ K is due to the ferrimagnetic order of the Yb^{3+} moments on 2a and 4b crystallographic sites.
(c) Temperature dependence of the magnetization in fields up to 50 kOe. The low- and high-field transitions are well defined by sharp increases of $M(T)$.

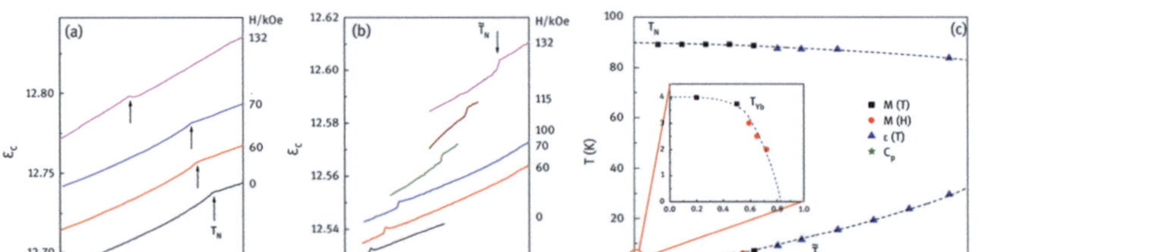

Figure 3.27: (a) and (b) Dielectric constant of $YbMnO_3$ at magnetic fields up to 132 kOe. The kink-like anomalies define T_N in (a) and $\widetilde{T}_N$ in (b). (c) Magnetic phase diagram of $YbMnO_3$. The inset shows the low-field phase boundary on an enlarged scale.

The phase boundaries T_N and $\widetilde{T}_N$ are determined up to 140 kOe by following the characteristic kink-like anomalies of the dielectric constant $\varepsilon_c(T)$, similar to the analysis for $ErMnO_3$ and $TmMnO_3$. Data shown in Figure 3.27a and b suggest that the phase enclosed by T_N and $\widetilde{T}_N$ extends beyond the maximum field of 140 kOe. The complete phase diagram of $YbMnO_3$ derived from thermodynamic, magnetic, and dielectric measurements is summarized in Figure 3.27c. The phase assignment could be similar to $ErMnO_3$. Below T_N, SHG optical measurements suggest the $P\underline{6_3}\underline{c}m$ magnetic symmetry. This symmetry allows for a gradual polarization of Yb moments on 4b sites. The low-temperature transition almost certainly does involve the order of all Yb^{3+} moments with a ferrimagnetic arrangement of Yb moments on 2a and 4b crystallographic sites. This requires a rotation of the Mn^{3+} spins by 90° and a change to $P6_3\underline{cm}$ magnetic symmetry. At higher magnetic fields, the flop of the Yb^{3+} moments at 2a positions results in the alignment of all *f* moments with the external field at $\widetilde{T}_N$.

The physical picture of the magnetic orders of Mn^{3+} spins and Yb^{3+} moments on 4b and 2a sites was first proposed by Fabrèges et al. [28] based on a comprehensive study combining magnetization measurements, ^{170}Yb Mössbauer, and neutron scattering experiments. The ordered magnetic moments of Mn, Yb(4b), and Yb(2a) could be separately determined and fit to a mean field model. According to the model, the Yb(4b) moment order below T_N is driven by the molecular field generated by the Mn spins and the Yb(2a) order below T_{Yb} results from the Yb-Yb exchange interactions. The order of Mn spins and rare earth moments has been studied independently by Salama et al., arriving at similar conclusions [130, 131].

3.5.2 Metastable hexagonal $DyMnO_3$

The stable structure of $DyMnO_3$ at ambient conditions is orthorhombic but the metastable hexagonal structure can be synthesized using special chemical techniques [26, 132]. h-$DyMnO_3$ is ferroelectric, similar to all other hexagonal manganites. The lattice parameters *a* and *c* are the largest in the series h-$RMnO_3$ (Table 3.1) because of the large Dy^{3+} ion, resulting in a reduction of the magnetic exchange couplings and the lowest Néel temperature, $T_N = 57$ K [133]. The magnetic properties of $DyMnO_3$, first studied by Ivanov et al. [132], are dominated by a sharp peak in the ac magnetic susceptibility at $T_{Dy} \approx 6$ K and a remanent magnetization below this temperature, indicating a possibly ferrimagnetic state at low *T*. T_{Dy} was found to increase in external magnetic fields oriented along the *c*-axis. Subsequent studies of the magnetic properties are in qualitative agreement with the original data [47, 133], however, the values for T_N and T_{Dy} scatter among different investigations, ranging between $T_N = 57$ K, $T_{Dy} = 3$ K [133] and $T_N = 68$ K, $T_{Dy} \approx 8$ K [47]. The differences may be due to small variations in the chemical composition of the samples used by different groups.

The magnetic order of the Dy^{3+} moments was studied using X-ray resonant magnetic scattering [47]. Between T_N and T_{Dy} the $P\underline{6_3}c\underline{m}$ (Γ_3) symmetry was assigned to the Dy^{3+} moment system. Below T_{Dy}, the $P6_3\underline{cm}$ (Γ_2) magnetic symmetry which allows for a ferrimagnetic moment was proposed. However, X-ray resonant magnetic scattering experiments only determine the magnetic symmetry of the ordered Dy sublattices, but no information about the magnetic structure of the Mn^{3+} spins is provided.

Neutron scattering and second harmonic generation optical measurements have been employed to study the Mn-spin and Dy moment orders of $DyMnO_3$ [134]. Interestingly, it was found that both subsystems order in different magnetic symmetries between T_N and T_{Dy}. Whereas the $P\underline{6}_3c\underline{m}$ (Γ_3) symmetry of the Dy moments was confirmed, the Mn spin order was described by $P\underline{6}_3\underline{c}m$ (Γ_4). Therefore, h-$DyMnO_3$ is a unique hexagonal manganite which allows for two different magnetic symmetries in a certain temperature range, apparently due to the weak *3d-4f* coupling between the order parameters describing the Mn^{3+} spin order and the Dy^{3+} moments.

Below T_{Dy} and in *c*-axis magnetic fields the Mn^{3+} spins and the Dy^{3+} moments were found to follow the same symmetry, $P6_3\underline{cm}$ (Γ_2). This particular magnetic order is stabilized by the magnetic field and the coupling to the ferromagnetic moment along the *c*-axis which is allowed only in this symmetry, similar to other h-$RMnO_3$.

3.6 Summary and outlook

Hexagonal manganites $RMnO_3$ exhibit ferroelectric order with high transition temperatures, frustrated magnetism at lower temperatures, and magnetoelectric effects mediated by strong spin lattice interactions. They represent an interesting class of type (I) multiferroics, the series of compounds includes all h-$RMnO_3$ with rare earth elements (R) from Dy to Lu as well as the non rare earth compounds $InMnO_3$, $ScMnO_3$, and $YMnO_3$. The hexagonal lattice structure results in highly frustrated magnetic subsystems of Mn spins and rare earth moments and a wealth of physical phenomena, phase transitions between different magnetic orders, spin reorientations, and magnetoelectric as well as magnetoelastic effects.

$YMnO_3$ and $HoMnO_3$ have been studied most extensively. $YMnO_3$ is of particular interest because of the missing rare earth magnetic moment which simplifies the magnetic subsystem and opens the possibility to investigate the magnetoelectric interaction between the ferroelectricity and the frustrated Mn spin orders. Nevertheless, new phenomena have been discovered in the ferroelectric and magnetic domain walls of $YMnO_3$ [135, 136].

The most complex phase diagram of all $RMnO_3$ was revealed for R=Ho in an external magnetic field oriented along the *c*-axis. A multitude of different phases are clearly identified through anomalous properties of magnetic, thermodynamic, and dielectric quantities. While the magnetic symmetries and structures of various phases have been determined through neutron scattering and non linear optical measurements, the precise magnetic symmetries of some high-field phases have yet to be explored. The magnetic and ferroelectric orders in $HoMnO_3$ are strongly correlated due to exchange interactions between Mn spins, Ho moments, and Mn–Ho exchange, the coupling of magnetic moments to the lattice and the ferroelectric order parameter. The frustration of the magnetic exchange pathways, which can be controlled by the size and properties of the R ion, makes the study of this system quite challenging but also interesting. Bulk magnetic, dielectric, heat capacity, thermal expansion, and other measurements (including neutron scattering and optical studies) have made it possible to unravel the complex phase diagram of $HoMnO_3$ and to identify the various phase boundaries, thus contributing to a more complete understanding of the complex interactions of highly frustrated magnetic systems and the origin of the magnetoelectric effect.

Other hexagonal manganites exhibit interesting properties and phenomena. $ScMnO_3$ undergoes a Mn spin reorientation transition well below its Néel temperature due to the weak in-plane magnetic anisotropy of the Mn spins. $ErMnO_3$, $YbMnO_3$, and $DyMnO_3$, similar to $HoMnO_3$, show low-temperature phases which are determined by a ferrimagnetic order of their respective f moments on the two inequivalent rare earth sublattices, which flip into a ferromagnetic alignment at high magnetic fields. The strong $f-f$ exchange interaction at low temperatures is assumed to be the cause of the f-moment alignment. The exchange with the manganese spins forces those to rotate by 90° to be compatible with the same magnetic symmetry. However, hexagonal $DyMnO_3$ was found to adopt different magnetic symmetries below T_N for the Mn spins and Dy moments, respectively, which was not observed in other $RMnO_3$. $InMnO_3$ crystallizes in two different lattice structures, depending on the synthesis conditions. The thermodynamically stable structure is ferroelectric whereas a centrosymmetric (non-ferroelectric) structure was obtained after fast cooling from high temperatures.

Hexagonal manganites, $RMnO_3$, are a class of multiferroic compounds showing a wealth of interesting properties and novel physical phenomena. Their complex behavior in external magnetic fields is a consequence of the simultaneous presence of geometric frustration, the coupling and mutual interaction of different magnetic ions, the important role of magnetic anisotropy, and the strong interaction of magnetic moments with the lattice as well as the ferroelectric order parameter. This interesting family of compounds is unique and a perfect subject for further studies. The results will provide a fundamental insight into other highly frustrated magnetic systems including more complex multiferroic materials.

While this review was focused on bulk properties of h-$RMnO_3$, recent interest has moved to the study of multiferroic domain boundaries and their potential functionalization as microelectronic devices. For example, in $YMnO_3$ it was shown that ferroelectric and antiferromagnetic domain walls (DW) are not independent of each other but a ferroelectric wall always includes an AFM wall [135]. This "clamping" effect was attributed to either the existence of the Dzyaloshinski–Moriya interaction in the domain boundaries [136] or the interaction of the AFM order parameter with the lattice strain in the DW of $YMnO_3$ [137]. Further studies of numerous hexagonal $RMnO_3$ have revealed the existence of six different multiferroic domains characterized by the direction of the ferroelectric polarization and the orientation of the AFM order parameter. The topology of those DWs separating FE domains with opposite polarizations was investigated and it turned out that six walls always merge in a vortex [138–144]. The existence of those domain vortices was explained based on the lattice geometry, the order parameter symmetry, and first principles calculations [145–148].

The electrical properties of multiferroic DWs, in contrast to the bulk properties of hexagonal $RMnO_3$, have attracted recent attention because of possible technological implications. Depending on the preparation of the bulk materials, the relative orientation of the FE polarization along the DWs, and the existence of excess oxygen or vacancies in the bulk, the DWs can be either more insulating or more conducting than the bulk [138, 144, 149, 150]. It was also demonstrated that the electrical properties of FE domain boundaries can be tuned in electric fields [151, 152] and may show a diode-like behavior depending on the relative orientation of the FE polarization across the DW [153, 154].

For technological implementations of ferroelectric domain wall properties it is necessary to minimize the amount of energy or external fields needed to manipulate their electrical properties. This can be achieved by reducing the overall size of the DWs in growing thin films of multiferroic h-$RMnO_3$. Epitaxial films of $YMnO_3$, $ErMnO_3$, and $HoMnO_3$ have been successfully deposited on various substrate and their ferroelectric and magnetic properties have been investigated [155–158]. Growing thin films of h-$RMnO_3$ on selected substrates allows further to study the effects of lattice mismatch induced strain on the multiferroic properties as well as the synthesis and study of new hexagonal manganites which otherwise form in the perovskite structure, e. g. $DyMnO_3$, $GdMnO_3$, and $TbMnO_3$ [159–162]. In the latter compound, an enhanced conductance was found in a neutral FE domain wall, demonstrating that domain wall conduction in hexagonal manganites is not limited to charged DWs [162]. From a technological point of view, there appears to be significant potential and interest in future investigations of domain wall electrical properties of h-$RMnO_3$ thin films.

Acknowledgements: This study would not have been possible without the dedicated work of two former graduate students, Clarina dela Cruz (now at Oak Ridge National Laboratory) and Fei Yen (now at Harbin Institute of Technology, Shenzhen). The majority of contributions to this review was supported in part by the US Air Force Office of Scientific Research, the T.L.L. Temple Foundation, the J. J. and R. Moores Endowment, and the State of Texas through the Texas Center for Superconductivity at the University of Houston.

References

[1] Khomskii DI. Coupled electricity and magnetism in solids: multiferroics and beyond. Preprint, arXiv:Cond-mat/1510.05174v2., 2015.
[2] Bertaut F, Forrat F, Fang P. Compt Rend Acad Sci. 1963;256:1958.
[3] Yakel HL, Koehler WC, Bertaut EF, Forrat EF. On the crystal structure of the manganese (iii) trioxides of the heavy lanthanides and yttrium. Acta Cryst. 1963;16:957.
[4] Fiebig M, Lottermoser T, Kneip MK, Bayer M. Correlations between magnetic and electrical orderings in multiferroic manganites. J Appl Phys. 2006;99:08E302.
[5] Yen F, dela Cruz C, Lorenz B, Galstyan E, Sun YY, Gospodinov M, et al. Magnetic phase diagrams of hexagonal $ErMnO_3$, $YbMnO_3$, $TmMnO_3$, and $HoMnO_3$. J Mater Res. 2007;22:2163.
[6] Smolenskii GA, Bokov VA. Coexistence of magnetic and electric ordering in crystals. J Appl Phys. 1964;35:915.
[7] Huang ZJ, Cao Y, Sun YY, Xue YY, Chu CW. Coupling between the ferroelectric and antiferromagnetic orders in $YMnO_3$. Phys Rev B. 1997;56:2623.
[8] Halasyamani PS, Poeppelmeier KR. Noncentrosymmetric oxides. Chem Mater. 1998;10:2753.
[9] Atanasov M, Reinen D. Density functional studies on the lone pair effect of the trivalent group (V) elements: I. Electronic structure, vibronic coupling, and chemical criteria for the occurrence of lone pair distortions in AX_3 molecules (A = N to Bi; X = H, and F to I). J Phys Chem A. 2001;105:5450.
[10] Seshadri R, Hill NA. Visualizing the role of bi 6s "lone pairs" in the off-center distortion in ferromagnetic $BiMnO_3$. Chem Mater. 2001;13:2892.
[11] van Aken BB, Palstra TTM, Filippetti A, Spaldin NA. The origin of ferroelectricity in magnetoelectric $YMnO_3$. Nat Mater. 2004;3:164.
[12] Lukaszewicz K, Karut-Kalicińska J. X-ray investigations of the crystal structure and phase transitions of $YMnO_3$. Ferroelectrics. 1974;7:81.
[13] Nénert G, Ren Y, Stokes H, Palstra TTM. Symmetry changes at the ferroelectric transition in the multiferroic $YMnO_3$. Preprint, arXiv: cond-mat/0504546.
[14] Nénert G, Pollet M, Marinel S, Blake GR, Meetsma A, Palstra TTM. Experimental evidence for an intermediate phase in the multiferroic $YMnO_3$. J Phys: Condens Matter. 2007;19:466212.
[15] Jeong IK, Hur N, Proffen T. High-temperature structural evolution of hexagonal multiferroic $YMnO_3$ and $YbMnO_3$. J Appl Cryst. 2007;40:730.
[16] Gibbs AS, Knight KS, Lightfoot P. High-temperature phase transitions of hexagonal $YMnO_3$. Phys Rev B. 2011;83:094111.
[17] Abrahams SC. Atomic displacements at and order of all phase transitions in multiferroic $YMnO_3$ and $BaTiO_3$. Acta Cryst B. 2009;65:450.
[18] Cano A. Hidden order in hexagonal $RMnO_3$ multiferroics (R=Dy-Lu, In, Y, and Sc). Phys Rev B. 2014;89:214107.
[19] Huang F-T, Wang X, Oh YS, Kurushima K, Mori S, Horibe Y, et al. Delicate balance between ferroelectricity and antiferroelectricity in hexagonal $InMnO_3$. Phys Rev B. 2013;87:184109.
[20] Lilienblum M, Lottermoser T, Manz S, Selbach SM, Cano A, Fiebig M. Ferroelectricity in the multiferroic hexagonal manganites. Nat Phys. 2015;11:1070.
[21] Bertaut EF, Mercier M. Structure magnetique de $MnYO_3$. Phys Lett. 1964;5:27.
[22] Koehler WC. Yakel H, Wollan EO, Cable JW. A note on the magnetic structures of rare earth manganese oxides. Phys Lett. 1964;9:93.
[23] Yu T, Gao P, Wu T, Tyson TA. Ferroelectricity in single crystal $InMnO_3$. Appl Phys Lett. 2013;102:172901.

[24] Tomuta DG, Ramakrishnan S, Nieuwenhuys GJ, Mydosh JA. The magnetic susceptibility, specific heat and dielectric constant of hexagonal $YMnO_3$, $LuMnO_3$ and $ScMnO_3$. J Phys.: Condens Matter. 2001;13:4543.
[25] Katsufuji T, Masaki M, Machida A, Moritomo M, Kato K, Nishibori E, et al. Crystal structure and magnetic properties of hexagonal $RMnO_3$ (R=Y, Lu, and Sc) and the effect of doping. Phys Rev B. 2002;66:134434.
[26] Kamegashira N, Satoh H, Ashizuka S. Synthesis and crystal structure of hexagonal $DyMnO_3$. Mater Sci Forum. 2004;449–52:1045.
[27] Fiebig M, Lottermoser T, Pisarev RV. Spin-rotation phenomena and magnetic phase diagrams of hexagonal $RMnO_3$. J Appl Phys. 2003;93:8194.
[28] Fabrèges X, Mirebeau I, Bonville P, Petit S, Lebras-Jasmin G, Forget A, et al. Magnetic order in $YbMnO_3$ studied by neutron diffraction and Mössbauer spectroscopy. Phys Rev B. 2008;78:214422.
[29] Muñoz A, Alonso JA, Martinez-Lope MJ, Casais MT, Martinez JL, Fernandez-Diaz MT. Magnetic structure of hexagonal $RMnO_3$ (R=Y, Sc): Thermal evolution from neutron powder diffraction data. Phys Rev B. 2000;62:9498.
[30] Park J, Kong U, Pirogov A, Choi SI, Park JG, Vjoi YN, et al. Neutron-diffraction studies of $YMnO_3$. Appl Phys A. 2002;74:S796.
[31] Reif J, Rau C, Matthias E. Influence of magnetism on second harmonic generation. Phys Rev Lett. 1993;71:1931.
[32] Pavlov VV, Pisarev RV, Kirilyuk A, Rasing T. Observation of a translational nonlinear magneto-optical effect in thin magnetic garnet films. Phys Rev Lett. 1997;78:2004.
[33] Fröhlich D, Leute S, Pavlov VV, Pisarev RV. Nonlinear optical spectroscopy of the two-order-parameter compound $YMnO_3$. Phys Rev Lett. 1998;81:3239.
[34] Fiebig M, Fröhlich D, Kohn K, Leute S, Lottermoser T, Pavlov VV, et al. Determination of the magnetic symmetry of hexagonal manganites by second harmonic generation. Phys Rev Lett. 2000;84:5620.
[35] Iizuka-Sakano T, Hanamura E, Tanabe Y. Second-harmonic-generation spectra of the hexagonal manganites $RMnO_3$. J Phys: Condens Matter. 2001;13:3031.
[36] Fiebig M, Degenhardt C, Pisarev RV. Interaction of frustrated magnetic sublattices in $ErMnO_3$. Phys Rev Lett. 2002;88:027203.
[37] Lonkai T, Hohlwein D, Ihringer J, Prandl W. The magnetic structures of $YMnO_3$ and $HoMnO_3$. Appl Phys A. 2002;74:S843.
[38] Iwata N, Kohn K. Dielectric anomalies at magnetic transitions of hexagonal rare earth manganese oxides $RMnO_3$. J Phys Soc Jpn. 1998;67:3318.
[39] Sugie H, Iwata N, Kohn K. Magnetic ordering of rare earth ions and magnetic–electric interaction of hexagonal $RMnO_3$ (R=Ho, Er, Yb or Lu). J Phys Soc Jpn. 2002;71:1558.
[40] Zhong CG, Fang JH. The coupling effect between ferroelectric and frustrated antiferromagnetic ordering in hexagonal ferroelectromagnet. Solid State Commun. 2003;128:449.
[41] Lorenz B, Litvinchuk AP, Gospodinov MM, Chu CW. Field-induced reentrant novel phase and a ferroelectric-magnetic order coupling in $HoMnO_3$. Phys Rev Lett. 2004;92:087204.
[42] Yen F, dela Cruz CR, Lorenz B, Sun YY, Wang YQ, Gospodinov MM, et al. Low temperature dielectric anomalies in $HoMnO_3$: the complex phase diagram. Phys Rev B. 2005;71:180407(R).
[43] Brown PJ, Chatterji T. Neutron diffraction and polarimetric study of the magnetic and crystal structures of $HoMnO_3$ and $YMnO_3$. J Phys: Condens Matter. 2006;18:10085.
[44] Vajk OP, Kenzelmann M, Lynn JW, Kim SB, Cheong SW. Magnetic order and spin dynamics in ferroelectric $HoMnO_3$. Phys Rev Lett. 2005;94:087601.

[45] Fiebig M, Degenhardt C, Pisarev RV. Magnetic phase diagram of $HoMnO_3$. J Appl Phys. 2002;91:8867.
[46] Fiebig M, Lottermoser T, Lonkai T, Goltsev AV, Pisarev RV. Magnetoeletric effects in multiferroic manganites. J Mag Mag Mater. 2005;290–291:883.
[47] Nandi S, Kreyssig A, Tan L, Kim JW, Yan JQ, Lang JC, et al. Nature of Ho magnetism in multiferroic $HoMnO_3$. Phys Rev Lett. 2008;100:217201.
[48] Condran SG, Plumer ML. A model of magnetic order in hexagonal $HoMnO_3$. J Phys: Condens Matter. 2010;22:162201.
[49] Sharma PA, Ahn JS, Hur N, Park S, Kim SB, Lee S, et al. Thermal conductivity of geometrically frustrated, ferroelectric $YMnO_3$: extraordinary spin–phonon interactions. Phys Rev Lett. 2004;93:177202.
[50] Zhou HD, Denyszyn JC, Goodenough JB. Effect of Ga doping on the multiferroic properties of $RMn_{1-x}Ga_xO_3$ (R=Ho, Y). Phys Rev B. 2005;72:224401.
[51] Zhou HD, Lu J, Vasic R, Vogt BW, Janik JA, Brooks JS, et al. Relief of frustration through spin disorder in multiferroic $Ho_{1-x}Y_xMnO_3$. Phys Rev B. 2007;75:132406.
[52] Vasić R, Zhou HD, Jobiliong E, Wiebe CR, Brooks JS. Probing multiferroicity and spin–spin interactions via dielectric measurements on Y-doped $HoMnO_3$ in high magnetic fields. Phys Rev B. 2007;75:014436.
[53] Zhou HD, Vasić R, Lu J, Brooks JS, Wiebe CR. The effect of Er doping on the multiferroics of $Ho_{1-x}Er_xMnO_3$. J Phys: Condens Matter. 2008;20:035211.
[54] Hur N, Jeong IK, Hundley MF, Kim SB, Cheong SW. Giant magnetoelectric effect in multiferroic $HoMnO_3$ with a high ferroelectric transition temperature. Phys Rev B. 2009;79:134120.
[55] Lorenz B, Yen F, Gospodinov MM, Chu CW. Field-induced phases in $HoMnO_3$ at low temperatures. Phys Rev B. 2005;71:014438.
[56] Zhou HD, Janik JA, Vogt BW, Jo YJ, Balicas L, Case MJ, et al. Specific heat of geometrically frustrated and multiferroic $RMn_{1-x}Ga_xO_3$ (R=Ho, Y). Phys Rev B. 2006;74:094426.
[57] Midya A, Das SN, Mandal P, Pandya S, Ganesan V. Anisotropic magnetic properties and giant magnetocaloric effect in antiferromagnetic $RMnO_3$ crystals (R=Dy, Tb, Ho, Yb). Phys Rev B. 2011;84:235127.
[58] Liu P, Wang XL, Cheng ZX, Du Y, Kimura H. Structural, dielectric, antiferromagnetic, and thermal properties of the frustrated hexagonal $Ho_{1-x}Er_xMnO_3$ manganites. Phys Rev B. 2011;83:144404.
[59] Oleaga A, Salazar A, Prabhakaran D, Cheng JG, Zhou JS. Critical begavior of the paramagnetic to antiferromagnetic transition in orthorhombic and hexagonal phases of $RMnO_3$ (R=Sm, Tb, Dy, Ho, Er, Tm, Yb, Lu, Y). Phys Rev B. 2012;85:184425.
[60] dela Cruz CR, Yen F, Lorenz B, Wang YQ, Sun YY, Gospodinov MM, et al. Strong spin–lattice coupling in multiferroic $HoMnO_3$: thermal expansion anomalies and pressure effect. Phys Rev B. 2005;71:060407(R).
[61] Litvinchuk AP, Iliev MN, Popov VN, Gospodinov MM. Raman and infrared-active phonons in hexagonal $HoMnO_3$ single crystals: magnetic ordering effects. J Phys: Condens Matter. 2004;16:809.
[62] Souchkov AB, Simpson JR, Quijada M, Ishibashi H, Hur N, Ahn JS, et al. Exchange interaction effects on the optical properties of $LuMnO_3$. Phys Rev Lett. 2003;91:027203.
[63] Fabréges X, Petit S, Mirebeau I, Pailhés S, Pinsard L, Forget A, et al. Spin–lattice coupling, frustration, and magnetic order in multiferroic $RMnO_3$. Phys Rev Lett. 2009;103:067204.
[64] Tyson TA, Wu T, Ahn KH, Kim SB, Cheong SW. Local spin-coupled distortions in multiferroic hexagonal $HoMnO_3$. Phys Rev B. 2010;81:054101.
[65] Poirier M, Lemyre JC, Lahaie PO, Pinsard-Gaudart L, Revcolevschi A. Enhanced magnetoelastic coupling in hexagonal multiferroic $HoMnO_3$. Phys Rev B. 2011;83:054418.

[66] Muñoz A, Alonso JA, Martinez-Lope MJ, Casais MT, Martinez JL, Fernandez-Diaz MT. Evolution of the magnetic structure of hexagonal $HoMnO_3$ from neutron powder diffraction data. Chem Mater. 2001;13:1497.
[67] Choi YJ, Lee N, Sharma PA, Kim SB, Vajk OP, Lynn JW, et al. Giant magnetic fluctuations at the critical endpoint in insulating $HoMnO_3$. Phys Rev Lett. 2013;110:157202.
[68] Lemyre JC, Poirier M. Microwave investigation of the phase diagram of hexagonal multiferroic $HoMnO_3$. Phys Rev B. 2009;79:094423.
[69] Vajk OP, Kenzelmann M, Lynn JW, Kim SB, Cheong SW. Neutron-scattering studies of magnetism in multiferroic $HoMnO_3$. J Appl Phys. 2006;99:08E301.
[70] Lee S, Pirogov A, Kang M, Jang KH, Yonemura M, Kamiyama T, et al. Giant magneto-elastic coupling in multiferroic hexagonal manganites. Nature (London), 2008;451:805.
[71] Chatterji T, Ouladdiaf B, Henry PF, Bhattacharya D. Magnetoelastic effects in multiferroic $YMnO_3$. J Phys: Condens Matter. 2012;24:336003.
[72] Lottermoser T, Fiebig M. Magnetoelectric behavior of domain walls in multiferroic $HoMnO_3$. Phys Rev B. 2004;70:220407(R).
[73] Lottermoser T, Lonkai T, Amann U, Hohlwein D, Ihringer J, Fiebig M. Magnetic phase control by an electric field. Nature (London). 2004;430:541.
[74] Ueland BG, Lynn JW, Laver M, Choi YJ, Cheong SW. Origin of electric-field-induced magnetization in multiferroic $HoMnO_3$. Phys Rev Lett. 2010;104:147204.
[75] Fennie CJ, Rabe KM. Ferroelectric transition in $YMnO_3$ from first principles. Phys Rev B. 2005;72:100103.
[76] Stengel M, Fennie CJ, Ghosez P. Electrical properties of improper ferroelectrics from first principles. Phys Rev B. 2012;86:094112.
[77] Cho DY, Kim JY, Park BG, Rho KJ, Park JH, Noh HJ, et al. Ferroelectricity driven by Y d^0-ness with rehybridization in $YMnO_3$. Phys Rev Lett. 2007;98:217601.
[78] Katsufuji T, Mori S, Masaki M, Moritomo Y, Yamamoto N, Takagi H. Dielectric and magnetic anomalies and spin frustration in hexagonal $RMnO_3$ (R=Y, Yb, and Lu). Phys Rev B. 2001;64:104419.
[79] Aken BBV, Bos JWG, de Groot RA, Palstra TTM. Asymmetry of electron and hole doping in $YMnO_3$. Phys Rev B. 2001;63:125127.
[80] Aikawa Y, Katsufuji T, Arima T, Kato K. Effect of Mn trimerization on the magnetic and dielectric properties of hexagonal $YMnO_3$. Phys Rev B. 2005;71:184418.
[81] Lonkai T, Tomuta DG, Hoffmann JU, Schneider R, Hohlwein D, Ihringer J. Magnetic two-dimensional short-range order in hexagonal manganites. J Appl Phys. 2003;93:8191.
[82] Park J, Park JG, Jeon GS, Choi HY, Lee C, Jo W, et al. Magnetic ordering and spin-liquid state of $YMnO_3$. Phys Rev B. 2003;68:104426.
[83] Dixit A, Smith AE, Subramanian MA, Lawes G. Suppression of multiferroic order in hexagonal $YMn_{1-x}In_xO_3$. Solid State Commun. 2010;150:746.
[84] Singh AK, Patnaik S, Kaushik SD, Siruguri V. Dominance of magnetoelastic coupling in multiferroic hexagonal $YMnO_3$. Phys Rev B. 2010;81:184406.
[85] Bertaut EF, Pauthenet R, Mercier M. Sur des proprietes magnetiques du manganite d'ytterbium. Phys Lett. 1965;18:13.
[86] Kozlenko DP, Kichanov SE, Lee S, Park JG, Savenko BN. Pressure-induced spin fluctuations and spin reorientation in hexagonal manganites. J Phys: Condens Matter. 2007;19:156228.
[87] Janoschek M, Roessli B, Keller L, Gvasaliya SN, Conder K, Pomjakushina E. Reduction of the ordered magnetic moment in $YMnO_3$ with hydrostatic pressure. J Phys: Condens Matter. 2005;17:L425.
[88] Petit S, Moussa F, Hennion M, Pailhés S, Pinsard-Gaudart L, Ivanov A. Spin phonon coupling in hexagonal multiferroic $YMnO_3$. Phys Rev Lett. 2007;99:266604.

[89] Sato TJ, Lee SH, Katsufuji T, Masaki M, Park S, Copley JRD, Takagi H. Unconventional spin fluctuations in the hexagonal antiferromagnet $YMnO_3$. Phys Rev B. 2003;68:014432.
[90] Chatterji T, Ghosh S, Singh A, Regnault LP, Rheinstädter M. Spin dynamics of $YMnO_3$ studied via inelastic neutron scattering and the anisotropic hubbard model. Phys Rev B. 2007;76:144406.
[91] Demmel F, Chatterji T. Persistent spin waves above the Néel temperature in $YMnO_3$. Phys Rev B. 2007;76:212402.
[92] Tachibana M, Yamazaki J, Kawaij H, Atake T. Heat capacity and critical behavior of hexagonal $YMnO_3$. Phys Rev B. 2005;72:064434.
[93] Lee S, Pirogov A, Han JH, Park JG, Hoshikawa A, Kamiyama T. Direct observation of a coupling between spin, lattice and electric dipole moment in multiferroic $YMnO_3$. Phys Rev B. 2005;71:180413(R).
[94] Poirier M, Laliberté F, Pinsard-Gaudart L, Revcolevschi A. Magnetoelastic coupling in hexagonal multiferroic $YMnO_3$ using ultrasound measurements. Phys Rev B. 2007;76:174426.
[95] Pimenov A, Mukhin AA, Ivanov VY, Travkin VD, Balbashov AM, Loidl A. Possible evidence for electromagnons in multiferroic manganites. Nat Phys 2006;2:97.
[96] Senff D, Link P, Hradil K, Hiess A, Regnault LP, Sidis Y, et al. Magnetic excitations in multiferroic $TbMnO_3$: evidence for a hybridized soft mode. Phys Rev Lett. 2007;98:137206.
[97] Fukumura H, Matsui S, Harima H, Kisoda K, Takahashi T, Yoshimura T, et al. Raman scattering on multiferroic $YMnO_3$. J Phys: Condens Matter. 2007;19:365239.
[98] Zaghrioui M, Phuoc VT, Souza RA, Gervais M. Polarized reflectivity and lattice dynamics calculation of multiferroic $YMnO_3$. Phys Rev B. 2008;78:184305.
[99] Vermette J, Jandl S, Mukhin AA, Ivanov VY, Balbashov A, Gospodinov MM, et al. Raman study of the antiferromagnetic phase transitions in hexagonal $YMnO_3$ and $LuMnO_3$. J Phys: Condens Matter. 2010;22:356002.
[100] Vermette J, Jandl S, Gospodinov MM. Raman study of spin-phonon coupling in $ErMnO_3$. J Phys Condens Matter. 2008;20:425219.
[101] Lewtas HJ, Boothroyd AT, Rotter M, Prabhakaran D, Müller H, Le MD, et al. Magnetic excitations in multiferroic $LuMnO_3$ studied by inelastic neutron scattering. Phys Rev B. 2010;82:184420.
[102] Skumryev V, Kuz'min MD, Gospodinov M, Fontcuberta J. Anisotropic paramagnetic response of hexagonal $RMnO_3$. Phys Rev B. 2009;79:212414.
[103] Xu HW, Iwasaki J, Shimizu T, Satoh H, Kamegashira N. Structure, magnetic susceptibility and heat capacity of $ScMnO_3$. J Alloys Compd. 1995;221:274.
[104] Bieringer M, Greedan JE. Magnetic structure and spin reorientation transition in $ScMnO_3$. J Solid State Chem. 1999;143:132.
[105] Fiebig M, Fröhlich D, Lottermoser T, Pisarev RV. Photoinduced instability of the magnetic structure of hexagonal $ScMnO_3$. Phys Rev B. 2002;65:224421.
[106] Galstyan E, Lorenz B, Nartyrosyan KS, Yen F, Sun YY, Gospodinov MM, et al. Magnetic hysteretic phenomena in multiferroic $HoMnO_3$ single crystals and polycrystals with nano- and micrometer particle size. J Phys Condens Matter. 2008;20:325241.
[107] Aken BBV, Palstra TTM. Influence of magnetic on ferroelectric ordering in $LuMnO_3$. Phys Rev B. 2004;69:134113.
[108] Bieringer M, Greedan JE, Wills AS. Investigation of magnetic structure evolution in the substitutional solid solution $Sc_xLu_{1-x}MnO_3$. Appl Phys A. 2002;74:S601.
[109] Fiebig M, Fröhlich D, Lottermoser T, Kohn K. Spin-angle topography of hexagonal manganites by magnetic second-harmonic generation. Appl Phys Lett. 2000;77:4401.
[110] Giaquinta DM, zur Loye HC. $InMnO_3$: a new transition metal oxide with an unusual ABO_3 structure. J Am Chem Soc. 1992;114:10952.

[111] Greedan JE, Bieringer M, Britten JF, Giaquinta DM, zur Loye HC. Synthesis, crystal structure, and unusual magnetic properties of $InMnO_3$. J Solid State Chem. 1995;116:118.

[112] Vajenine GV, Hoffmann R, zur Loye HC. The electronic structures and magnetic properties of one-dimensional ABO_6 chains in Sr_3ABO_6 (A=Co, Ni; B=Pt, Ir) and two-dimensional MO_3 sheets in $InMO_3$ (M-Fe, Mn). Chem Phys. 1996;204:469.

[113] Serrao CR, Krupanidhi SB, Bhattacharjee J, Waghmare UV, Kundu AK, Rao CNR. $InMnO_3$: a biferroic. J Appl Phys. 2006;100:076104.

[114] Belik AA, Kamba S, Savinov M, Nuzhnyy D, Tachibana M, Takayama-Muromachi E, et al. Magnetic and dielectric properties of hexagonal $InMnO_3$. Phys Rev B. 2009;79:054411.

[115] Rusakov DA, Belik AA, Kamba S, Savinov M, Nuzhnyy D, Kolodiazhnyi T, et al. Structural evolution and properties of solid solutions of hexagonal $InMnO_3$ and $InGaO_3$. Inorg Chem. 2011;50:3559.

[116] Fabrèges X, Mirebeau I, Petit S, Bonville P, Belik AA. Frustration-driven magnetic order in hexagonal $InMnO_3$. Phys Rev B. 2011;84:054455.

[117] Abrahams SC. Ferroelectricity and structure in the $YMnO_3$ family. Acta Cryst B. 2001;57:485.

[118] Oak MA, Lee JH, Jang HM, Goh JS, Choi HJ, Scott JF. *4d-5p* orbital mixing and asymmetric in *4d*-o *2p* hybridization in $InMnO_3$: a new bonding mechanism for hexagonal ferroelectricity. Phys Rev Lett. 2011;106:047601.

[119] Kumagai Y, Belik AA, Lilienblum M, Leo N, Fiebig M, Spaldin NA. Observation of persistent centrosymmetricity in the hexagonal manganite family. Phys Rev B. 2012;85:174422.

[120] Meier D, Ryll H, Kiefer K, Klemke B, Hoffmann JU, Ramesh R, et al. Mutual induction of magnetic 3d and 4f order in multiferroic hexagonal $ErMnO_3$. Phys Rev B. 2012;86:184415.

[121] Sekhar MC, Lee S, Choi G, Lee C, Park JG. Doping effects of hexagonal manganites $Er_{1-x}Y_xMnO_3$ with triangular spin structure. Phys Rev B. 2005;72:014402.

[122] Park J, Kong U, Choi SI, Park JG, Lee C, Jo W. Magnetic structure studies of $ErMnO_3$. Appl Phys A. 2002;74:S802.

[123] Standard EC, Stanislavchuk T, Sirenko AA, Lee N, Cheong SW. Magnons and crystal-field transitions in hexagonal $RMnO_3$ (R=Er, Tm, Yb, Lu) single crystals. Phys Rev B. 2012;85:144422.

[124] Massa NE, del Campo L, Meneses DDS, Echegut P, Martinez-Lope MJ, Alonso JA. Phonons and hybrid modes in the high and low temperature far infrared dynamics of hexagonal $TmMnO_3$. J Phys: Condens Matter. 2014;26:275901.

[125] Salama HA, Stewart GA. Exchange-induced Tm magnetism in multiferroic h-$TmMnO_3$. J Phys: Condens Matter. 2009;21:386001.

[126] Zhou JS, Goodenough JB, Gallardo-Amores JM, Morán E, Alario-Franco MA, Caudillo R. Hexagonal versus perovskite phase of manganite $RMnO_3$ (R=Y, Ho, Er, Tm, Yb, Lu). Phys Rev B. 2006;74:014422.

[127] Uusi-Esko K, Malm J, Imamura N, Yamauchi H, Karppinen M. Chracterization of $RMnO_3$ (R=Sc, Y, Dy-Lu): high-pressure synthesized metastable perovskites and their hexagonal precursor phases. Mat Chem Phys. 2008;112:1029.

[128] Wang XM, Zhao ZY, Fan C, Liu XG, Li QJ, Zhang FB, et al. Low-temperature heat transport, specific heat, and magnetic properties of the hexagonal $TmMnO_3$ single crystals. Phys Rev B. 2012;86:174413.

[129] Adem U, Mostovoy M, Bellido N, Nugroho AA, Simon C, Palstra TTM. Scaling behavior of the magnetocapacitance of $YbMnO_3$. J Phys: Condens Matter. 2009;21:496002.

[130] Salama HA, Stewart GA, Ryan DH, Elouneg-Jamroz M, Edge AVJ. A Mössbauer sepctroscopy investigation of h-$YbMnO_3$. J Phys: Condens Matter. 2008;20:255213.

[131] Salama HA, Voyer CJ, Ryan DH, Stewart GA. Magnetic order of the rare earth sublattice in h-$YbMnO_3$. J Appl Phys. 2009;105:07E110.

[132] Ivanov VY, Mukhin AA, Prokhorov AS, Balbashov AM, Iskhakova LD. Magnetic properties and phase transitions in hexagonal $DyMnO_3$ single crystals. Phys Solid State. 2006;48:1726.
[133] Harikrishnan S, Rößler S, Kumar CMN, Bhat HL, Rößler UK, Wirth S, et al. Phase transitions and rare-earth magnetism in hexagonal and orthorhombic $DyMnO_3$ single crystals. J Phys: Condens Matter, 2009;21:096002.
[134] Wehrenfennig C, Meier D, Lottermoser T, Lonkai T, Hoffmann JU, Aliouane N, et al. Incompatible magnetic order in multiferroic hexagonal $DyMnO_3$. Phys Rev B. 2010;82:100414(R).
[135] Fiebig M, Lottermoser T, Fröhlich D, Goltsev AV, Pisarev RV. Observation of coupled magnetic and electric domains. Nature (London). 2002;419:818.
[136] Hanamura E, Hagita K, Tanabe Y. Clamping of ferroelectric and antiferromagnetic order parameters of $YMnO_3$. J Phys Condens Matter. 2003;15:L103.
[137] Goltsev AV, Pisarev RV, Lottermoser T, Fiebig M. Structure and interaction of antiferromagnetic domain walls in hexagonal $YMnO_3$. Phys Rev Lett. 2003;90:177204.
[138] Choi T, Horibe Y, Yi HT, Choi YJ, Wu W, Cheong SW. Insulating interlocked ferroelectric and structural antiphase domain walls in multiferroic $YMnO_3$. Nat Mater. 2010;9:253.
[139] Meier D, Leo N, Jungk T, Soergel E, Becker P, Bohatý L, et al. Translation domains in multiferroics. Preprint, arXiv: cond-mat/1008.3290.
[140] Chae SC, Lee N, Horibe Y, Tanimura M, Mori S, Gao B, et al. Direct observation of the proliferation of ferroelectric loop domains and vortex-antivortex pairs. Phys Rev Lett. 2012;108:167603.
[141] Geng Y, Lee N, Choi YJ, Cheong SW, Wu W. Collective magnetism at multiferroic domains walls. Nano Lett. 2012;12:6055.
[142] Chae SC, Horibe Y, Jeong DY, Lee N, Iida K, Tanimura M, et al. Evolution of the domain topology in a ferroelectric. Phys Rev Lett. 2013;110:167601.
[143] Geng Y, Das H, Wysocki AL, Wang X, Cheong SW, Mostovoy M, et al. Direct visualization of magnetoelectric domains. Nat Mater. 2014;13:163.
[144] Schaab J, Krug IP, Nickel F, Gottlob DM, Doganay H, Cano A, et al. Imaging and characterization of conducting ferroelectric domain walls by photoemission electron microscopy. Appl Phys Lett. 2014;104:232904.
[145] Kumagai Y, Spaldin NA. Structural domain walls in polar hexagonal manganites. Nat Commun. 2013;4:1540.
[146] Mettout B, Toledano P, Lilienblum M, Fiebig M. Combinatorial model for the ferroelectric domain-network formation in hexagonal manganites. Phys Rev B. 2014;89:024103.
[147] Meier QN, Lilienblum M, Griffin SM, Conder K, Pomjakushina E, Yan Z, et al. Global formation of topological defects in the multiferroic hexagonal manganites. Phys Rev X. 2017;7:041014.
[148] Holtz ME, Shapovalov K, Mundy JA, Chang CS, Yan Z, Bourret E, et al. Topological defects in hexagonal manganites: Inner structure and emergent electrostatics. Nano Lett. 2017;17:5883.
[149] Du Y, Wang XL, Chen DP, Dou SX, Cheng ZX, Higgins M, et al. Domain wall conductivity in oxygen deficient multiferroic $YMnO_3$ single crystals. Appl Phys Lett. 2011;99:252107.
[150] Ruff E, Krohns S, Lilienblum M, Meier D, Fiebig M, Lunkenheimer P, et al. Conductivity contrast and tunneling charge transport in the vortexlike ferroelectric domain patterns of multiferroic hexagonal $YMnO_3$. Phys Rev Lett. 2017;118:036803.
[151] Meier D, Seidel J, Cano A, Delaney K, Kumagai Y, Mostovoy M, et al. Anisotropic conductance at improper ferroelectric domain walls. Nat Mater. 2012;11:284.
[152] Mundy JA, Schaab J, Kumagai Y, Cano A, Stengel M, Krug IP, et al. Functional electronic inversion layers at ferroelectric domain walls. Nat Mater. 2017;16:622.
[153] Wu W, Horibe Y, Lee N, Cheong SW, Guest JR. Conduction of topologically protected charged ferroelectric domain walls. Phys Rev Lett. 2012;108:077203.

[154] Schaab J, Skjaervo SH, Krohns S, Dai X, Holtz ME, Cano A, et al. Electrical half-wave rectification at ferroelectric domain walls. Nat Nano. 2018:13:1028.
[155] Marti X, Sanchez F, Hrabovsky D, Fabrega L, Ruyter A, Fontcuberta J, et al. Exchange biasing and electric polarization with $YMnO_3$. Appl Phys Lett. 2006;89:032510.
[156] Jang SY, Lee D, Lee JH, Noh TW, Jo Y, Jung MH, et al. Oxygen vacancy induced re-entrant spin glass behavior in multiferroic $ErMnO_3$ thin films. Appl Phys Lett. 2008;93:162507.
[157] Singh AK, Snure M, Tiwari A, Patnaik S. Effect of epitaxial strain on the magneto-electric coupling of $YMnO_3$ thin films. J Appl Phys. 2009;106:014109.
[158] Lee D, Kim HS, Jang SY, Joh KW, Noh TW, Yu J, et al. Double polarization hysteresis loop induced by the domain pinning by defect dipoles in $HoMnO_3$ epitaxial thin films. Phys Rev B. 2010;81:012101.
[159] Lee JH, Murugavel P, Lee D, Noh TW, Jo Y, Jung MH, et al. Multiferroic properties of epitaxially stabilized hexagonal $DyMnO_3$ thin films. Appl Phys Lett. 2007;90:012903.
[160] Choi WS, Moon SJ, Seo SSA, Lee D, Lee H, Murugavel P, et al. Optical spectroscopic investigation on the coupling of electronic and magnetic structure in multiferroic hexagonal $RMnO_3$ (R=Gd, Tb, Dy, and Ho) thin films. Phys Rev B. 2008;78:054440.
[161] Choi WS, Kim DG, Seo SSA, Moon SJ, Lee D, Lee JH, et al. Electronic structures of hexagonal $RMnO_3$ (R=Gd, Tb, Dy, and Ho) thin films: Optical spectroscopy and first-principles calculations. Phys Rev B. 2008;77:045137.
[162] Kim DJ, Connell JG, Seo SSA, Gruverman A. Domain wall conductivity in semiconducting hexagonal ferroelectric $TbMnO_3$ thin films. Nanotechnology. 2016;27:155705.

Takashi Kurumaji

4 Spiral spin structures and skyrmions in multiferroics

Abstract: In this article, we focus on (1) type-II multiferroics driven by spiral spin orderings and (2) magnetoelectric couplings in multiferroic skyrmion-hosting materials. We present both phenomenological understanding and microscopic mechanisms for spiral spin state, which is one of the essential starting points for type-II multiferroics and magnetic skyrmions. Two distinct mechanisms of spiral spin states (frustration and Dzyaloshinskii–Moriya [DM] interaction) are discussed in the context of the lattice symmetry. We also discuss the spin-induced ferroelectricity on the basis of the symmetry and microscopic atomic configurations. We compare two well-known microscopic models: the generalized inverse DM mechanism and the metal-ligand *d-p* hybridization mechanism. As a test for these models, we summarize the multiferroic properties of a family of triangular-lattice antiferromagnets. We also give a brief review of the magnetic skyrmions. Three types of known skyrmion-hosting materials with multiferroicity are discussed from the view point of crystal structure, magnetism, and origins of the magnetoelectric couplings. For exploration of new skyrmion-hosting materials, we also discuss the theoretical models for stabilizing skyrmions by magnetic frustration in centrosymmetric system. Several basic ideas for material design are given, which are successfully demonstrated by the recent experimental evidences for the skyrmion formation in centrosymmetric frustrated magnets.

Keywords: multiferroics, skyrmion, geometrical frustration, Dzyaloshinskii-Moriya interaction, triangular-lattice magnets

4.1 Introduction

In this review, we focus on type-II multiferroics associated with spiral spin orders. A spiral spin order is a type of modulated magnetic structure, in which the orientation of spins on an array of magnetic ions gradually rotate site by site so as the trajectory of the spin vectors forms a spiral. Versatile forms of spiral show up in many relatively simple crystal lattices, which have been attracted much attention parallel to the advancement of the neutron scattering techniques [1–10]. The spiral order is

This article has previously been published in the journal Physical Sciences Reviews. Please cite as: Rzelewska-Piekut, M., Kurumaji, T. Spiral spin structures and skyrmions in multiferroics *Physical Sciences Reviews* [Online] 2020, 5. DOI: 10.1515/psr-2019-0016

https://doi.org/10.1515/9783110582130-004

closely related with the spatial inversion symmetry breaking of the system with the emergence of left-handed or right-handed spin spiral. Indeed, this feature enables the strong coupling between electricity and magnetism in solids, as experimentally pioneered by the discovery of a giant electric polarization change by applied magnetic field in $TbMnO_3$ [11]. Coexistence of spiral spin order and ferroelectricity provides one of the simplest realizations of type-II multiferroics, where the magnetic transition simultaneously induces ferroelectricity [12]. The main interest of the first half of this review is how the spiral spin states are characterized and how the ferroelectricity emerges from them.

As an extension of the spiral spin state, we also review the magnetic skyrmions. Skyrmion is a topologically stable field configuration, originally proposed in the theory of nucleon by Skyrme [13]. This idea was later introduced into magnetic systems and a spin vortex-like soliton state is considered as a magnetic skyrmion [14]. Recently stable triangular-lattice states of nanometric magnetic skyrmions were discovered in chiral helimagnetic alloys MnSi [15], and $Fe_{1-x}Co_xSi$ [16], and also in a multiferroic Cu_2OSeO_3 [17]. Up to date, three multiferroic materials are identified to host skyrmions with observation of magnetoelectric responses. The origin of skyrmion and multiferroicity in these systems are described in the latter half of this review.

This article is organized as follows: Section 4.2 describes the origin of spiral spin order in light of both phenomenology and microscopic spin model. In Section 4.3, we discuss the ferroelectricity induced by the spiral spin structure and the underlying microscopic mechanisms. In Section 4.4, we review a family of triangular-lattice multiferroics to illustrate how the above principles are implemented in real materials. In Section 4.5, we introduce the magnetic skyrmion and the role of lattice symmetry for the skyrmion formation. In Section 4.6, we review the known skyrmion-hosting multiferroics up to date focusing on crystal structure, phase diagram, and magnetoelectric effect. In Section 4.7, we discuss the emergence of skyrmions in centrosymmetric systems and suggest relevant theoretical models and material design principles. Section 4.8 summarizes this article.

As a matter of fact, ferroelectrics with spiral spin orders consist only one of the branches of type-II multiferroics, and other types of spin-induced ferroelectricity are not covered by this review. Comprehensive description for wider class of multiferroics and different magnetoelectric coupling mechanisms can be found in other review articles [18–24]. Thin film growth of multiferroic materials is another important branch for activating magnetoelectric property in application, which is lacked in this text due to the limitation of pages. One of the breakthroughs in this research field is epitaxial growth of room-temperature multiferroic $BiFeO_3$ [25]. Recent development is reviewed in Ref. [26]. As for skyrmion, we do not cover topics on metallic materials, thin films, and heterostructures, which are otherwise reviewed in Refs [27, 28].

4.2 Spin spiral orders and microscopic origins

Two essential features for a spiral spin order are the direction of the spin modulation vector (**q**), and the orientation of the spin-spiral plane, on which the spins rotate. Figure 4.1 illustrates typical spiral spin structures along an array of magnetic ions. A flat spin-spiral plane perpendicular (Figure 4.1(a)) and parallel (Figure 4.1(b)) to the **q** vectors characterize the proper-screw and cycloidal structure, respectively. In contrast to the antiferromagnetic nature of them, a conical structure (Figure 4.1(c)) possesses a homogeneous ($\boldsymbol{q}=0$) net component of the magnetic moment in addition to a **q**-modulation. Multiferroics with this type of spin structure, e. g. $CoCr_2O_4$, hexaferrites, and Mn_2GeO_4 [29–31], are of particular importance due to the coexistence of ferroelectricity and ferromagnetism. A list of various types of spiral magnets can be found in, for example, Refs [8, 20, 21, 32, 33].

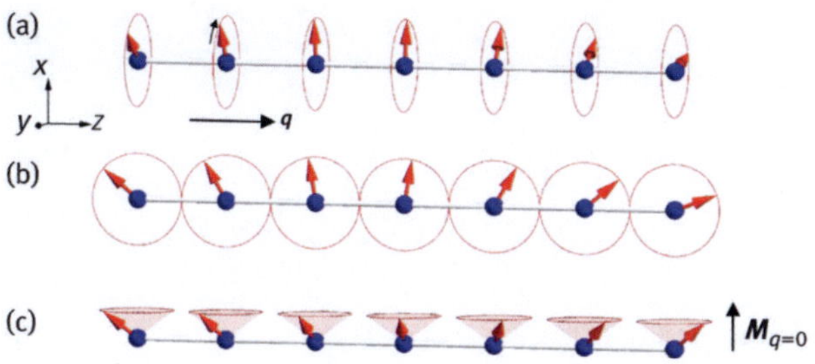

Figure 4.1: Schematic illustrations of typical spiral spin structures on a one-dimensional array of magnetic sites: (a) proper-screw, (b) cycloidal, and (c) (transverse) conical spiral structures. The vector **q** along the **z** axis represents a spin modulation direction. Spin structure in (c) has a homogeneous magnetization $\boldsymbol{M}_{q=0}$ perpendicular to the **q** vector.

A modulated magnetic order parameter $\boldsymbol{M}(\boldsymbol{r})$ is expressed as follows

$$\boldsymbol{M}(\boldsymbol{r})=\boldsymbol{M}_{\boldsymbol{q}}e^{\mathrm{i}\boldsymbol{q}\cdot\boldsymbol{r}}+\boldsymbol{M}_{\boldsymbol{q}}^{*}e^{-\mathrm{i}\boldsymbol{q}\cdot\boldsymbol{r}}, \tag{4.1}$$

where $\boldsymbol{r}$ is the spatial variable, $\boldsymbol{M}_{\boldsymbol{q}}$ is a complex vector defining the spin-spiral plane as

$$\boldsymbol{M}_{\boldsymbol{q}}=\frac{M_0}{2}(\boldsymbol{u}+\mathrm{i}\eta\boldsymbol{v})e^{\mathrm{i}\varphi}. \tag{4.2}$$

M_0 is the amplitude of the magnetic moment, $\boldsymbol{u}$ and $\boldsymbol{v}$ are independent (unit) vectors for the spin-spiral plane, $\eta(=\pm1)$ is the helicity (handedness) of the spiral, and φ is the phase of the modulation at $\boldsymbol{r}=0$. The helicity correlates with the spatial inversion symmetry of the lattice as manifested from the fact that the inversion operation, $\boldsymbol{r}\to-\boldsymbol{r}$, on eq. (4.1) reverses the helicity of the spiral. Therefore, if the original crystal has the inversion symmetry, the free energy for the spiral spin state contains the degeneracy for the helicity. We note that noncentrosymmetric magnets, in which the host lattice lacks the inversion center, constitute another group of spiral-spin materials because they forbid the degeneracy of the spin helicity. Nevertheless, in the latter class several example type-II multiferroics have been found [17, 34], which include Cu_2OSeO_3, a skyrmion-hosting multiferroic. In these cases, the orientation of the $\boldsymbol{q}$ vector plays a key role for the multiferroicity.

Origins of those two different types of the spiral-spin state can be well understood through a phenomenological approach exploiting the magnetic free energy. The magnetic phase transition to a spatially modulated magnetic state can be described by the free energy functional $F_0[\boldsymbol{M}(\boldsymbol{r})] = \int \mathrm{d}r^3 \left[\frac{1}{2}A_0\boldsymbol{M}^2 + \frac{1}{4}B\boldsymbol{M}^4\right]$ with an additional term V_{ih} for the spatial derivative of $\boldsymbol{M}$, $\partial_i\boldsymbol{M}$. For brevity, we discuss the spiral spin order with $\boldsymbol{q}||\boldsymbol{z}$, and $\boldsymbol{u} = \boldsymbol{x}$, $\boldsymbol{v} = \boldsymbol{y}$ in eq. (4.2). Up to the fourth-order of $\partial_i\boldsymbol{M}$, the forms of V_{ih} in the cases of centrosymmetric and noncentrosymmetric lattices, respectively, are given as follows [32]

$$V_{\mathrm{ih}}[\boldsymbol{M}(\boldsymbol{r})]$$

I. centrosymmetric case:

$$= \int \mathrm{d}r^3 \left[\frac{1}{2}\alpha\left(\partial_z\boldsymbol{M}_\perp\right)^2 + \frac{1}{4}\beta\left(\partial_z^2\boldsymbol{M}_\perp\right)^2\right] \tag{4.3}$$

II. noncentrosymmetric case:

$$= \int \mathrm{d}r^3 \left[\frac{1}{2}\alpha\left(\partial_z\boldsymbol{M}_\perp\right)^2 + \mathrm{i}\sigma\left(M_x\partial_z M_y - M_y\partial_z M_x\right)\right], \tag{4.4}$$

where $\boldsymbol{M}_\perp = (M_x, M_y)$. Case (I) was introduced in Ref. [35] and studied in detail by Ref. [36]. The second term in case (II) is so-called Lifshitz invariant, allowed by symmetry in a noncentrosymmetric lattice due to the relativistic effect of electrons [37]. The form of the Lifshitz invariant is predetermined by the type of crystallographic asymmetry, which directly connects to the type of skyrmion that is realized in the system as discussed later in Sections 4.5 and 4.6.

Case (I): With $\alpha > 0$ the ferromagnetic unmodulated state is stabilized. Here we focus on the case with $\alpha < 0$ that is enabled by the presence of the competing magnetic interactions. In the latter case, V_{ih} has local minimums in terms of q suggesting an instability towards a modulated state. At low temperature, V_{ih} is supposed to be relevant in the free energy to induce a phase transition to the modulated state. Equation (4.3) can be minimized by the characteristic magnetic modulation $q_{\min} = \sqrt{-\alpha/\beta}$. Given that V_{ih} is composed solely of even order terms of the spatial derivative, the spin spiral allows the degeneracy for $q_{\min}$, and $-q_{\min}$, i. e. for the spin helicity.

Case (II): even in the case for the ferromagnetic interaction for the first term in V_{ih}, ($\alpha > 0$), the free energy possesses the instability towards the spiral state. The characteristic modulation is determined by the ratio of σ to α as $|q_{\min}| = |\sigma|/\alpha$, and the helicity reflects the sign of σ, which is further connected to the chirality or the orientation of polarity of the basal lattice. Experimentally, a fixed helicity of the spiral spin state in chiral magnets were demonstrated in Refs [38–44].

The above two types of spiral spin state can be reproduced by microscopic spin Hamiltonians for one-dimensional spin chains as follows:
I. centrosymmetric case:

$$H = -J_1 \sum_{\langle i,j \rangle} \boldsymbol{S}_i \cdot \boldsymbol{S}_j - J_2 \sum_{\langle\langle i,j \rangle\rangle} \boldsymbol{S}_i \cdot \boldsymbol{S}_j \tag{4.5}$$

II. noncentrosymmetric case:

$$H = \sum_{\langle i,j \rangle} \left(-J_1 \boldsymbol{S}_i \cdot \boldsymbol{S}_j + \boldsymbol{D} \cdot \left[\boldsymbol{S}_i \times \boldsymbol{S}_j \right] \right). \tag{4.6}$$

$\langle , \rangle$ and $\langle\langle , \rangle\rangle$ represent the nearest neighbor and the next nearest neighbor spin pair, respectively. The first model shows the instability towards a spiral spin order (with spin-helicity degeneracy) in the case J_1 and J_2 are competing, e. g. $J_1 > 0$, $J_2 < 0$. The second model is characteristic for systems that possess the second antisymmetric exchange interaction term, Dzyaloshinskii–Moriya (DM) interaction term [45, 46]. The spin-spiral plane and the spin helicity are set by the orientation and the sign of the DM vector ($\boldsymbol{D}$). The orientation of the $\boldsymbol{D}$ is determined by the asymmetry between the *i-j* bond, which is related with the form of Lifshitz invariant in the continuous limit (see Section 4.5).

The length scale of the spin modulation is decided by $q = [\cos^{-1}(-J_1/4J_2)]/a$ for case (I), and $q = [\tan^{-1}(-D/J_1)]/a$ for case (II), where a is the lattice constant. They coincide with the above phenomenological models in the limit of $-J_1 - 4J_2 \sim +0$ and $J \gg |D|$, respectively. The DM interaction is of relativistic origin and the magnitude of D is usually hundred times smaller than the symmetric exchange interaction J_1. Therefore, in case (II), the spiral wavelength is typically on the order of $10 \sim 100$ nm. In contrast, for case (I), J_1 and J_2 can be on the same order of magnitude, which may result in modulation wavelength shorter than 10 nm. The different length scale of the spiral spin state for above two mechanisms is directly connected to the size of skyrmion as shown in Sections 4.6 and 4.7.

As the last comment, we note the extension of the model (I) to the two and three-dimensional lattice with further neighbor interactions. This is relevant in the rare-earth intermetallics [47] where Ruderman–Kittel–Kasuya–Yosida (RKKY) interactions [48–50] dominates the magnetic order. In general, $\boldsymbol{q}$ vector for the spiral modulation in the model (I) is related with local maximums of the Fourier transformation of exchange interactions, $J(\boldsymbol{q}) = \sum_i e^{i\boldsymbol{q}\cdot(\boldsymbol{r}_i - \boldsymbol{r}_j)} J_{ij}$, where i runs among all sites at $\boldsymbol{r}_i$ in the lattice. In the case of a triangular lattice, for example, local maximums of $J(\boldsymbol{q})$ are located at six positions in the $\boldsymbol{q}$-space, which are related with the rotational symmetry of the lattice. This fact is crucial for the formation of skyrmion-lattice state in a centrosymmetric system as discuss in Section 4.7.

4.3 Origin of ferroelectricity in spiral spin systems

In this section, we discuss the origin of ferroelectricity in spin systems that belong to the case (I). As pointed out above, the symmetry lowering from a nonpolar to a polar group driven by the magnetic ordering is essential for observable ferroelectric polarization in a type-II multiferroic. The link between the spiral spin order and ferroelectricity was first recognized in a study of olivine-type Cr_2BeO_4 [51], which shows ferroelectricity below the Néel temperature at $T_N = 28\,\mathrm{K}$ with a cycloidal spin order of Cr ions [52, 53]. The emergence of spin-driven ferroelectricity is phenomenologically explained in Refs [54, 55]. They discussed the symmetrically allowed form of the homogeneous component of electric polarization $P(\boldsymbol{r})(=\boldsymbol{P}_0 e^{i0\cdot\boldsymbol{r}})$ in terms of $\boldsymbol{M}(\boldsymbol{r})$ as follows:

$$P_{0i} = f_{ik\alpha\beta}\left[M_\alpha \partial_k M_\beta - M_\beta \partial_k M_\alpha\right], \tag{4.7}$$

where i, α, β, and k are Cartesian coordinates (x, y, z), and $f_{ik\alpha\beta}$ is a rank-four polar tensor antisymmetric with α and β ($f_{ik\alpha\beta} = -f_{ik\beta\alpha}$). The form of $f_{ik\alpha\beta}$ follows the symmetry of the lattice as listed in Ref. [56]. The first-order derivative on the right-hand side of eq. (4.7) suggests that the sign of helicity determines the sign of P_0. Indeed, the correlation between the spin helicity and direction of the ferroelectric polarization in multiferroics was experimentally demonstrated with the polarized neutron diffraction [57, 58], and circularly polarized x-ray diffraction [59].

Here, as an example, we consider a form of $f_{ik\alpha\beta}$ in a crystal belonging to D_{2h} group with a spiral spin order with $\boldsymbol{q} = (0, 0, q)$, i. e. for $k = z$ (Figure 4.2(a)–(c)). Two components, f_{yzyz} and f_{xzxz} are nonzero, and f_{izxy} are zero for any i. This indicates that the electric polarization emerges only along y and x directions for the spin-spiral plane in yz and zx planes, respectively. This is consistent with the ferroelectricity accompanying with the cycloidal spin order in Cr_2BeO_4. Related discussions can be found in Ref. [60].

The understanding of the microscopic mechanism of the spin-driven ferroelectricity accelerated following the discovery of the large nonlinear magnetoelectric response in a perovskite-type multiferroic $TbMnO_3$ [11]. Reference [61] considered the inverse effect of the DM interaction, which produces finite electric polarization by a shift of the oxygen ion located in the middle of two magnetic ions to minimize the DM energy of canted spin configuration (inverse DM mechanism). Reference [62] calculated the electronic contribution in a similar context, and local electric polarization was predicted in the presence of the spin-orbit interaction even without the shift of the oxygen ion. This mechanism was called spin-current mechanism. Both theories converge at the form of local $\boldsymbol{P}$ as follows:

$$\boldsymbol{P} = \boldsymbol{e}_{ij} \times \left(\boldsymbol{S}_i \times \boldsymbol{S}_j\right), \tag{4.8}$$

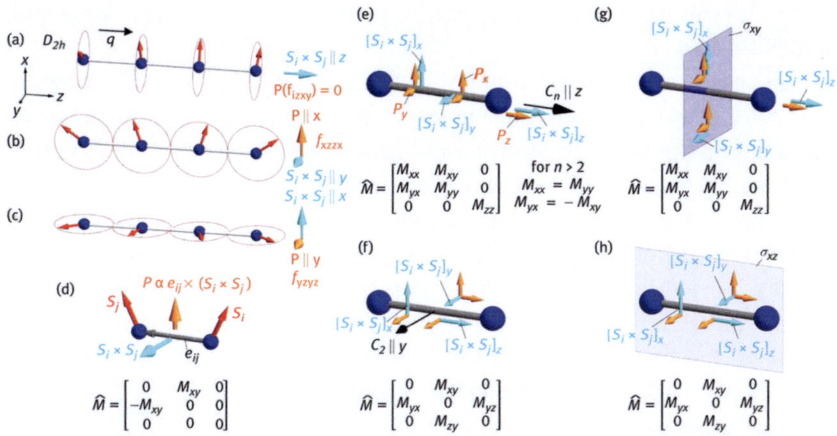

Figure 4.2: (a)-(b) Relationship between vector spin chirality $\boldsymbol{S}_i \times \boldsymbol{S}_j$ (cyan arrow) and symmetry-allowed spontaneous polarization vector $\boldsymbol{P}$ (orange arrow) in the array of magnetic ions (blue spheres) with D_{2h} symmetry. (d) Relationship between spin moments $\boldsymbol{S}_i$ and $\boldsymbol{S}_j$ (red arrows) at two magnetic ions, and $\boldsymbol{P}$ for eq. (4.8). (e)-(h) Relationship between each component of $\boldsymbol{S}_i \times \boldsymbol{S}_j$ and symmetry-allowed $\boldsymbol{P}$ (neighboring orange arrows) in a two-site cluster for the generalized inverse DM mechanism. Symmetry operation at each cluster is denoted by (e)-(f) a black arrow ($C_n||z$: n-fold rotation axis along the z axis, $C_2||y$: 2-fold rotation axis along the y axis), and (g)-(h) a blue plane (σ_{xy}: xy-mirror plane at the center of the bond, σ_{xz}: xz-mirror plane parallel to the bond). The tensor $\hat{M}$ for eq. (4.9) is also indicated.

where $\boldsymbol{e}_{ij}$ is the unit vector connecting the magnetic ions (Figure 4.2(d)). This formula successfully reproduces the selection rules in D_{2h} group, and was experimentally demonstrated in many multiferroics [18].

Further experimental studies tested the spin-current mechanism in a lot of multiferroics, and had revealed that a number of materials do not obey the selection rules in Figure 4.2(a)–(c) since these rules were deduced in a relatively high-symmetric magnetic cluster. In order to cover wider class of multiferroics, the form of the inverse DM formula in a general crystallographic environment was deduced as follows [63, 64]:

$$P_\alpha = M_{\alpha\beta}\left[\boldsymbol{S}_i \times \boldsymbol{S}_j\right]_\beta, \tag{4.9}$$

where second-rank tensor $M_{\alpha\beta}\left(\equiv \hat{M}\right)$ is determined by the symmetry for magnetic ions at site i and j. Figure 4.2(e)–(h) give the schematic illustrations of the selection rules for a certain symmetry and nonzero components of $\hat{M}$. For example, when the cluster has only the n-fold rotation symmetry around the bond, the emergence of P parallel to the bond is allowed even in the case of $\boldsymbol{S}_i \times \boldsymbol{S}_j || \boldsymbol{e}_{ij}$ (Figure 4.2(e)), which is clearly forbidden in eq. (4.8). Indeed, such violation is experimentally reported in various multiferroics such as $Cu_3Nb_2O_8$ [65], $CaMn_7O_{12}$ [66], $RbFe(MoO_4)_2$ [67],

$NaFeSi_2O_6$ [68], $BiFeO_3$ [69], $CuFeO_2$ [70], $CuCrO_2$ [71, 72], α-$NaFeO_2$ [73], MnI_2 [74], CoI_2 [75], and $NiBr_2$ [76].

The last six compounds belong to a family of triangular-lattice antiferromagnet, which is discussed in the next section. We give a brief review on the crystal structure and the relationship between electric polarization and the spiral spin structure in several cases that requires consideration of the crystallographic site symmetry (thus can be described adequately using eq. (4.9)). As a matter of fact, the origin of multiferroicity in the triangular-lattice antiferromagnets is still under debate. We also discuss the *d-p* hybridization mechanism [77], which is proposed as another origin of electric polarization in these systems. We hope the discussion in this section shed light on the origin of spin-induced polarization, and may be useful for application to more complex multiferroics or materials with spin textures like magnetic skyrmionic systems.

4.4 Multiferroic triangular-lattice antiferromagnets

A triangular-lattice magnet with antiferromagnetic interactions is a prototype of a geometrically frustrated spin system [78, 79]. Competing interaction among spins not only becomes a source of a spin-disordered states such as spin liquid states [80] but also stabilizes various types of helimagnetic order and 120°-type noncollinear order, which breaks the inversion symmetry. Multiferroics have been discovered in many systems hosting the triangular lattice such as $CuFeO_2$ [70], $CuCrO_2$ [71, 72], and α-$NaFeO_2$ [73] in delafossite-type crystal structure, transition metal dihallides MX_2 (MnI_2 [74], NiI_2, CoI_2 [75], and, $NiBr_2$ [76]), and $RbFe(MoO_4)_2$ [67]. Here, we describe several triangular-lattice multiferroics with magnetic modulation in the lattice-plane to see how the electric polarization varies with the spin structure in the same basal crystal structure. Detailed reviews on magnetism and multiferroicity can be found in Ref. [81] for delafossite-type oxides, and in Ref. [82] for MX_2 compounds.

4.4.1 Crystal structures

Figure 4.3(a)–(c) illustrate three different types of crystal structure for triangular-lattice multiferroics: delafossite type, CdI_2 type, and $CdCl_2$ type, respectively. A triangular-lattice layer is composed of edge-shared MX_6 octahedra (Figure 4.3(d)), and stacks along the *c* axis. Each layer is connected with the others through a Cu layer or by van der Waals forces. The site symmetry of a given magnetic ion belongs to D_{3d} as shown in Figure 4.3(e). Each triangular-lattice layer is a copy of the others related with a translational operation, i. e. the global space group symmetry is $P\bar{3}m1$ or $R\bar{3}m$ depending on the direction of the translation. Mirrors as well as two-fold rotation axes are not perpendicular to each other exemplify the different symmetry

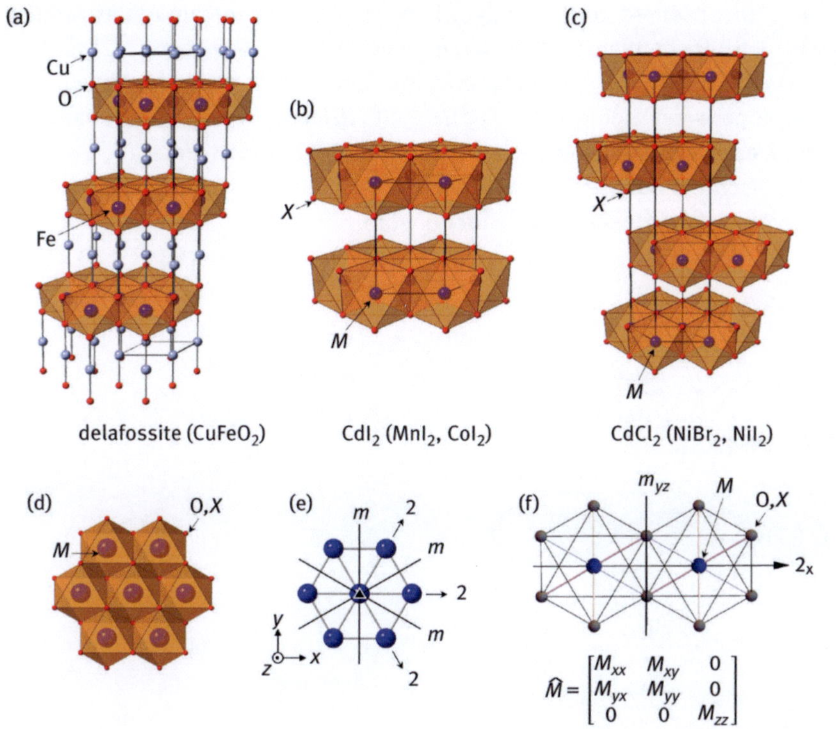

Figure 4.3: (a)-(c) Crystal structures of delafossite-type for $CuFeO_2$, CdI_2-type for MnI_2 and CoI_2, and $CdCl_2$-type for $NiBr_2$ and NiI_2. Gray sphere is for Cu. Blue sphere is for Fe or M (transition-metal). Red sphere is for oxygen or X (halogen). A hexagonal unit cell is shown by black lines.
(d) A common atomic configuration for a triangular-lattice layer in 3(a-c). (e) Symmetry-operations (m: a mirror plane perpendicular to the basal lattice; 2: 2-fold rotation axis along the bond direction) in a triangular-lattice are indicated. A black triangle indicates a 3-fold rotation axis along the c axis. (f) A local atomic configuration and symmetry operations for two neighboring magnetic ions. The tensor $\hat{M}$ is also indicated.

from D_{2h}. This is a source of the uncanceled polarization due to the generalized inverse DM mechanism.

4.4.2 Magnetism and ferroelectricity

Here, we describe three different triangular-lattice multiferroics in view of the magnetic phase diagram and relationship between P and q. We introduce $\boldsymbol{q}_{\text{in}}$ as the in-plane magnetic modulation vector.

$CuFeO_2$ (proper screw, $\boldsymbol{q}_{\text{in}}$//(110)). The magnetic ground state at zero field is a four sublattice collinear phase [83]. As shown in the H-T phase diagram (Figure 4.4(a)), application of H along the c axis induces multiple metamagnetic transitions [84, 85].

Kimura *et al.* [70] observed the ferroelectricity in the first metamagnetic phase (Figure 4.4(b)) with incommensurate magnetic modulation $\boldsymbol{q} = (q_{\text{in}}, q_{\text{in}}, 3/2)$ for $1/4 < q_{\text{in}} < 1/5$ (~ 0.72 nm) [84, 85]. The magnetic structure of this phase has been confirmed as the proper-screw spiral state (Figure 4.4(c, d)) [86]. Doping of nonmagnetic Al or Ga stabilizes this multiferroic phase to the zero-field ground state [87, 88]. The studies for the electric-field effect on the magnetic domain [89] revealed the relationship $\boldsymbol{P} || \boldsymbol{q}_{\text{in}}$, which clearly breaks the rule for eq. (4.8).

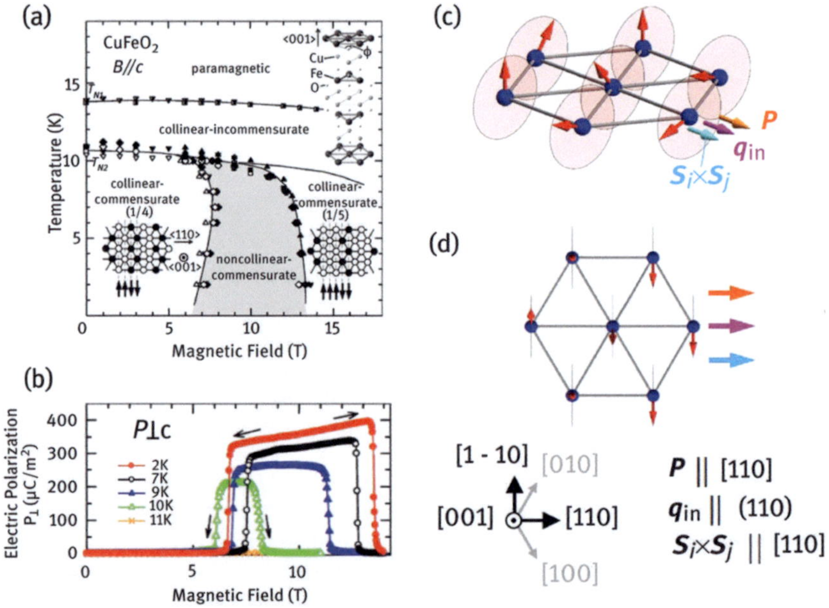

Figure 4.4: (a) Magnetic phase diagram for $CuFeO_2$ under H||[001]. (b) Magnetic field dependence of P perpendicular to the c axis. (c) Schematic illustration of the magnetic ground state. Orange (magenta) [cyan] arrow indicates $\boldsymbol{P}$ ($\boldsymbol{q}_{\text{in}}$) [$\boldsymbol{S}_i \times \boldsymbol{S}_j$]. Red circular plane indicates the spin-spiral plane. (d) Spin configuration viewed from [001] axis. Crystallographic axes are also shown.
Figures 4(a) and (b) are reproduced with permission from Ref. [70]. (c) (2006) American Physical Society.

MnI_2(proper screw, $\boldsymbol{q}_{\text{in}} || (1\bar{1}0)$). Proper-screw type helimagnetic ground state (Figure 4.5(c, d)) was found by Cable *et al.* in Ref. [90]. In zero field, there are three successive phase transitions [86] from the paramagnetic state ($T > T_{N1} = 3.95$ K) to the ground state ($T < T_{N3} = 3.45$ K). In $T_{N1} > T > T_{N2} = 3.8$ K, the spin order is the centrosymmetric incommensurate ($\boldsymbol{q} = (0.1025, 0.1025, 0.5)$, ~ 2.02 nm) collinear state with spin lying in the *ab*-plane. After the continuous rotation of $\boldsymbol{q}$ vector in the reciprocal space in $T_{N3} < T < T_{N2}$, the spiral order with $\boldsymbol{q} = (0.181, 0, 0.439)$ (~ 1.9 nm in the in-plane direction, ~ 1.56 nm in the out-of-plane direction) is stabilized [90, 91]. Due to the slanted $\boldsymbol{q}$

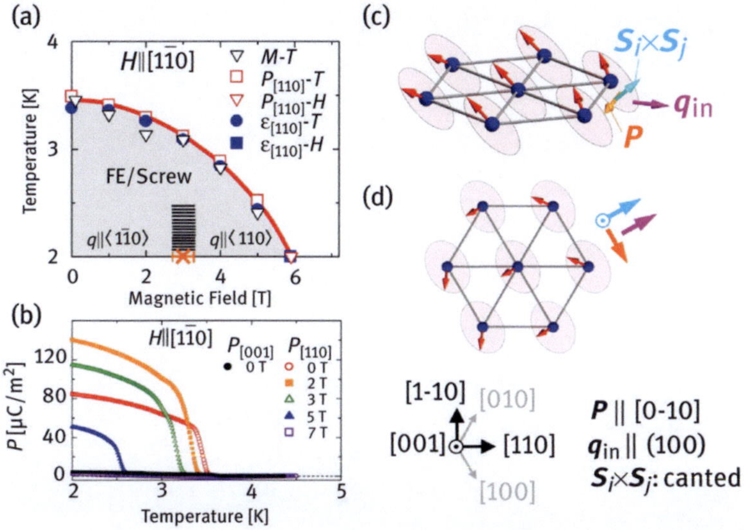

Figure 4.5: (a) Magnetic phase diagram for MnI_2 under H||[1$\bar{1}$0]. (b) Temperature dependence of P at each magnetic field. (c)–(d) Schematic illustration of the magnetic ground state. Figures 5(a) and (b) are reproduced with permission from Ref. [74]. (c) (2011) American Physical Society.

vector, the spin-spiral plane is canted from the triangular-lattice plane (Figure 4.5(c)). The emergence of ferroelectricity is observed only below T_{N3} [74] (Figure 4.5(a, b)), and the relationship between $\boldsymbol{P}$ and $\boldsymbol{q}_{\text{in}}$ is confirmed as $\boldsymbol{P} \perp \boldsymbol{q}_{\text{in}}$ (Figure 4.5(c, d)). More interestingly, detailed study on the magnetoelectric effect suggests that the application of the in-plane magnetic field for $3\ \text{T} < \mu_0 H < 6\ \text{T}$ induces the other proper-screw spin order with the relationship $\boldsymbol{P}||\boldsymbol{q}_{\text{in}}||(110)$ (Figure 4.5(a)).

First-principle calculation with the spin-orbit interaction successfully reproduces the relationship between $\boldsymbol{P}$ and $\boldsymbol{q}_{\text{in}}$ for both spiral states [64]. The complicated H-T phase diagram is explained by the frustrated exchange interactions with multiple local minimum in $J(\boldsymbol{q})$ [92], and by a spin Hamiltonian with magnetic dipole-dipole interaction competing with the hexagonal single-site anisotropy [93, 94]. The emergence of P in the zero-field ground state is consistent with the rule in eq. (4.8) due to the canted spin-spiral plane, while eq. (4.8) fails to describe the field-induced second multiferroic state. A complete explanation of P in the different $\boldsymbol{q}$-states is beyond eq. (4.8).

$NiBr_2$(cycloid, $\boldsymbol{q}_{\text{in}}$//(110)). In zero-field, the ground state of $NiBr_2$ is the long-wavelength cycloidal spin state for $\boldsymbol{q}_{\text{in}}||(110)$ (~ 6.89 nm) with spins lying in the *ab*-plane (Figure 4.6(c, d)) [95–98]. Ferroelectricity is observed only in this phase [76], with P mainly parallel to the basal lattice plane (Figure 4.6(a, b)). This is consistent with eq. (4.8), whereas a weak violation showing a component of P perpendicular to the c axis is also reported.

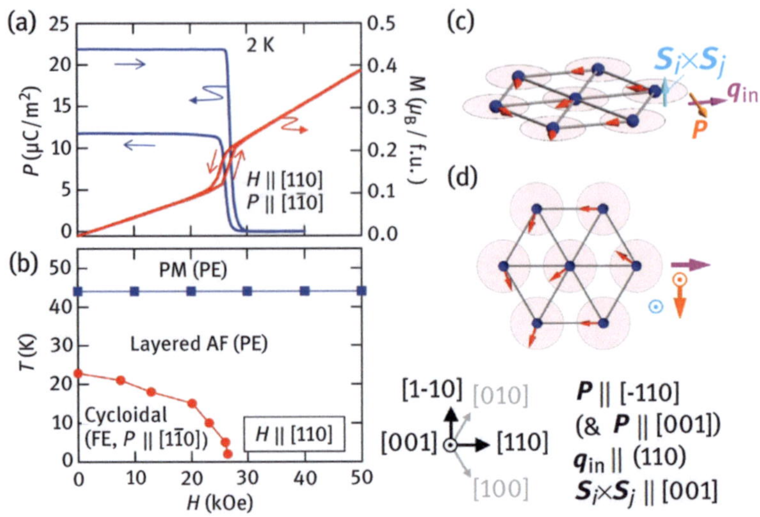

Figure 4.6: (a) Magnetic field dependence of P along $[1\bar{1}0]$ axis and M for H||[110]. (b) Magnetic phase diagram for $NiBr_2$ under H||[110]. (c)–(d) Schematic illustration of the magnetic ground state. Figures 6(a) and (b) are reproduced with permission from Ref. [76]. (c) (2011) American Physical Society.

As exemplified above, the triangular-lattice multiferroics host various types of helimagnetic structures and associated ferroelectricity in the identical local crystallographic structure. In order to explain the violation of eq. (4.8), we apply the generalized inverse DM model (eq. (4.9)) in the next subsection.

4.4.3 Generalized inverse DM mechanism in a triangular-lattice magnet

Before going into the application of eq. (4.9), we discuss the tensor $f_{ik\alpha\beta}$ for the point group of the triangular lattice (D_{3d}), which corresponds to the continuous limit of the generalized inverse DM mechanism. We show Table 4.1 for the non-zero component of $f_{ik\alpha\beta}$ for a given $\boldsymbol{q}$ vector in the triangular-lattice (xy) plane (Figure 4.3(e)) with different orientation of the spin-spiral plane. In the case of

Table 4.1: Non-zero component of $f_{ik\alpha\beta}$ and relationship with each other for $\boldsymbol{q}_{\text{in}}$ along the x and y axes and for various orientations of the spin-spiral plane in the magnet with D_{3d} symmetry.

Spiral plane	xy	yz	zx
$\boldsymbol{q}_{\text{in}}\|\|x$	$f_{yxxy}(=-f_{xyxy})$ $f_{yxxy}(=-f_{xyxy})$	$f_{xxyz}(=-f_{yyyz})$ $f_{xxyz}(=-f_{yyyz})$	$f_{yxzx}(=f_{yyyz})$ $f_{zxzx}(=-f_{zyyz})$
$\boldsymbol{q}_{\text{in}}\|\|y$	$f_{xyxy} f_{xyxy}$	$f_{yyyz}(=f_{xyzx})\ f_{zyyz}$	$f_{xyzx} f_{xyzx}$

$CuFeO_2$ ($\boldsymbol{q}_{\text{in}}||x$, spiral-plane$||yz$), observed $\boldsymbol{P}||x$ is consistent with the presence of f_{xxyz}. As for MnI_2, finite f_{xyxy} and f_{xyzx} indicate that spiral state with $\boldsymbol{q}_{\text{in}}||y$ shows $\boldsymbol{P}||x$ regardless of the orientation of the spiral plane either xy or zx. In $NiBr_2$, on the other hand, the f_{yxxy} predicts $\boldsymbol{P}||y$ in the cycloidal state for $\boldsymbol{q}_{\text{in}}||x$, but out of plane P is supposed to be zero ($f_{zxxy} = 0$), which is inconsistent with the experiment. Except for the last case, the discussion of the continuous limit reconciles the relationship between $\boldsymbol{P}$ and $\boldsymbol{q}_{\text{in}}$ on the triangular lattice.

The application of eq. (4.9) on the microscopic atomic configuration of the triangular lattice is shown in Figure 4.3(f). Presence of a mirror plane and a two-fold rotation axis decides the allowed component of $\hat{M}$, which includes additional components to eq. (4.8). Indeed, the generalized tensor component is deduced with the microscopic quantum theory in Ref. [99] and the first-principle calculations [64, 100]. Zhang *et al.* [100] applied the generalized inverse DM scheme to all of the nearest neighbor bonds on a triangular lattice and deduced the direction of the global $\boldsymbol{P}$ for various spiral spin structure as shown in Figure 4.7. As tabulated in Table 4.2, the generalized inverse DM scheme reproduces all the $\boldsymbol{P}$ directions and relationship for each component predicted by the f-tensor analysis in addition to the P_z component in the xy-cycloid for $\boldsymbol{q}_{\text{in}}||y$ as observed in $NiBr_2$ [76]. Note that the last component falls to zero in the long-

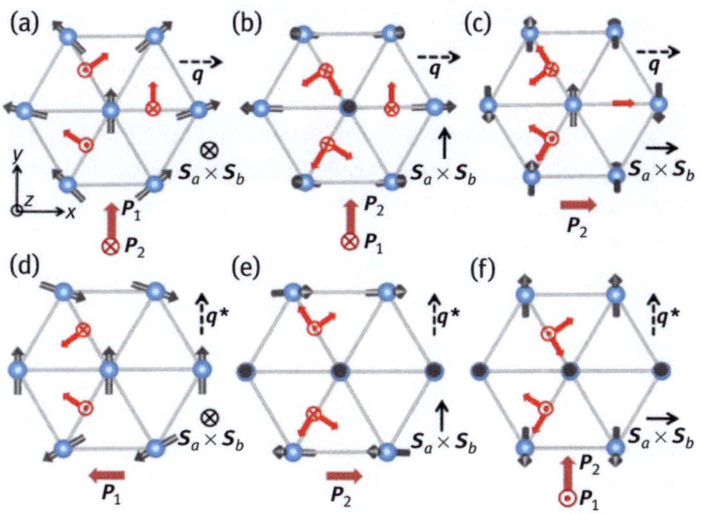

Figure 4.7: The distribution of the local electric polarization (red arrow) for each bond and resultant total electric polarization (dark red arrow) in a triangular-lattice for various spiral spin structures. Spin orders in (a)-(c) have a $\boldsymbol{q}$ vector (dashed arrow) along the x axis, and the spin-spiral planes in (a) xy, (b) zx, and (c) yz plane, respectively. In (d)-(f), spin modulates along the y axis, and the spin-spiral plane is parallel to (d) xy, (e) zx, and (f) yz plane, respectively. Each spin structure corresponds to (a) $NiBr_2$, (c) $CuFeO_2$, and (e) or (d) MnI_2 (except for the cant of the spin-spiral plane). Reproduced with permission from Ref. [100]. (c) (2017) American Physical Society.

Table 4.2: Prediction of $\boldsymbol{P}$ for various spiral structures by the generalized inverse DM mechanism (reproduced from Ref. [100]). Sign of some components and P_z in *xy*-spiral for $\boldsymbol{q}_{\text{in}}||x$ are rewritten by the author for consistency with Table 4.1.

Spiral plane	*xy*	*yz*	*xz*
$\boldsymbol{q}_{\text{in}}\|\|x$			
P_x	0	$M_{xx}\left(-\sin qa + \frac{1}{2}\sin(qa/2)\right) + \frac{3}{2}M_{yy}\sin(qa/2)$	0
P_y	$M_{yz}(\sin qa + \sin(qa/2))$	0	$M_{yy}\left(\sin qa - \frac{1}{2}\sin(qa/2)\right) - \frac{3}{2}M_{xx}\sin(qa/2)$
P_z	$-M_{zz}(\sin qa - 2\sin(qa/2))$	0	$M_{zy}(\sin qa + \sin(qa/2))$
$\boldsymbol{q}_{\text{in}}\|\|y$			
P_x	$-\sqrt{3}M_{yz}\sin(\sqrt{3}qa/2)$	0	$\frac{\sqrt{3}}{2}(M_{yy} - M_{xx})\sin(\sqrt{3}qa/2)$
P_y	0	$-\frac{\sqrt{3}}{2}(M_{yy} - M_{xx})\sin(\sqrt{3}qa/2)$	0
P_z	0	$\sqrt{3}M_{zy}\sin(\sqrt{3}qa/2)$	0

wavelength limit ($q_{\text{in}} \to 0$). Experimental observations of this component in a cycloidal spin state in such as $NiBr_2$ [76], CoI_2 [75], and α-$NaFeO_2$ [73] suggest the failure of the discussion in the continuous limit is due to the relatively short wavelength of the spin modulation of the frustration origin

4.4.4 *d-p* metal-ligand hybridization mechanism

The above discussion consistently explains the multiferroicity in a wide class of the triangular-lattice helimagnets, while here we discuss another approach giving a different scenario for the violation of eq. (4.8), i. e. the *d-p* hybridization mechanism. This approach includes an indispensable element for understanding spin-induced ferroelectricity, and also serves as a reliable scheme in the context of designing new multiferroic/skyrmion materials. The *d-p* hybridization mechanism was first pointed out in Ref. [77], where Arima discussed the modulation of the hybridization between metal (*M*) and ligand (*X*) ions depending on the orientation of the spin at

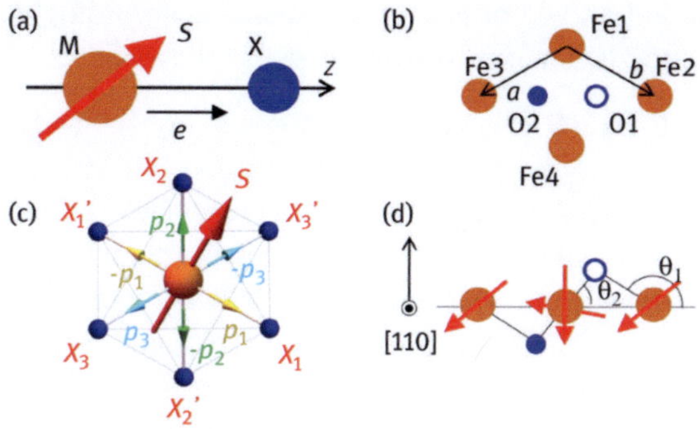

Figure 4.8: (a) M and X indicate a magnetic ion and a ligand ion, respectively. Red arrow is a spin moment vector ($\boldsymbol{S}$) on M. $\boldsymbol{e}$ is a unit vector pointing from M to X. (b) and (d) Fe_4O_2 cluster relevant to the calculation of the electric polarization due to the d-p hybridization mechanism. Orange circle indicates Fe ion, and open (closed) blue circle indicates oxygen ion above (below) the Fe-triangular-lattice plane. (c) MX_6 octahedron. Yellow, green, and cyan arrows indicate local electric dipole moments ($\boldsymbol{p}_i$) due to the d-p hybridization mechanism between respective M-X_i bonds. Figures 8(a), (b), and (d) are reproduced with permission from Ref. [77]. (c) (2007) The Physical Society of Japan.

M through the spin-orbit interaction. The spin-dependent local polarization is predicted along the M-X bond direction (Figure 4.8(a)), which is usually expressed as

$$\boldsymbol{P} = \alpha \boldsymbol{e}(\boldsymbol{e} \cdot \boldsymbol{S})^2. \tag{4.10}$$

Indeed, this formula successfully accounts for the magnetoelectric response in a non-centrosymmetric magnet $Ba_2CoGe_2O_7$ [101], and found relevance for rare-earth ions in chiral magnets $RFe_3(BO_3)_4$ (R: rare earth) [102]. Phenomenologically, the d-p hybridization mechanism provides the same formula as the magnetic anisotropy effect discussed for one of the origins of the linear magnetoelectric effect [103]. In this context, the local polarization operator $\hat{\boldsymbol{P}}$ is expressed by the quadratic form with the spin operator $\hat{\boldsymbol{S}}$ as $\hat{P}_i = \tau_{ijk}\left(\hat{S}_j\hat{S}_k + \hat{S}_k\hat{S}_j\right)$, where τ_{ijk} is the rank-three polar tensor with the same symmetry as the piezoelectric tensor for the site symmetry of the magnetic ion.

The application of the above mechanism on a centrosymmetric triangular-lattice layer for $CuFeO_2$ and MX_2 compounds immediately leads to exact *zero* polarization owing to the local inversion symmetry at M site as shown in Figure 4.8(c). It is true in general that the use of eq. (4.10) as it is cannot predict the spin-induced ferroelectricity when the magnetic ion does not have piezoelectric site-symmetry. In Ref. [77], the higher-order effect was considered for the hybridization among one X ion and neighboring three different M ions on the cluster as shown in Figure 4.8(b, d). It was hypothesized that one of the p-orbitals on the X ion hybridizes with different M simultaneously, and was assumed a competition of covalency, i. e. if the covalency between Fe2 and O1 grows, those for Fe1 and O1 or Fe4 and O1 weakens accordingly. The local polarization along O1-Fe1 bond is modified by this effect as follows:

$$P_{\overrightarrow{\mathrm{O1Fe1}}} = \alpha \cos 2(\boldsymbol{Q}\cdot\boldsymbol{r} - q - \theta_1)\cdot\{1 - \beta\cos 2(\boldsymbol{Q}\cdot\boldsymbol{r} - \theta_1) - \gamma\cos 2(\boldsymbol{Q}\cdot\boldsymbol{r} + q - \theta_1)\}, \quad (4.11)$$

where α, β, and y are constant, and θ_1 and θ_2 are geometric parameters determined by the configuration between X and M. The total sum of related formula for all bonds leads to $P_x \propto \alpha\beta \sin 2q$, and zero for all other directions, which reproduces the ferroelectricity in the proper-screw spin structure for $\boldsymbol{P} \| \boldsymbol{q}_{\mathrm{in}} \|$ [110], and importantly, explains the inversion of P_x by the inversion of the helicity (or $q \to -q$). This higher-order *d-p* hybridization effect is applicable to various spiral spin structures on a triangular lattice [104], and accounts for the emergence of ferroelectricity, for example, $\boldsymbol{P} = (P_0 \sin 2q, 0, 0)$ for $q_{\mathrm{in}} \| y$ with slanted spin-spiral plane like in MnI_2, and $\boldsymbol{P} = \left(0, P_{y0}\sin 2q + P_{y1}\sin 4q, P_{z0}(\sin 4q - 2\sin 2q)\right)$ for the cycloidal state with $q_{\mathrm{in}} \| x$ as in $NiBr_2$, where P_0, P_{y0}, P_{y1}, and P_{z0} are parameters. As a matter of fact, the prediction of the relationship between $\boldsymbol{q}_{\mathrm{in}}$ and $\boldsymbol{P}$ (but not magnitude/q-dependence) is identical with that in the generalized inverse DM mechanism.

Although the above discussion considers a higher-order effect in terms of the spin-orbit interaction than the inverse DM mechanism, further quantitative analysis may be necessary for clarifying which effect is dominant in specific compounds. Zhang *et al.* are taking an interesting approach with the first-principle calculation in Ref. [100]. They focused on the fact that $\boldsymbol{P}$ for these two mechanisms show different $\boldsymbol{q}$ dependence, and calculated the polarization in $CuFeO_2$ as a function of spin rotation angle ($\sim q_{\mathrm{in}}$) for a fictitious noncollinear order. According to their calculation, the inverse DM mechanism contributes more significantly than the *d-p* mechanism. Similar approach on the other triangular-lattice multiferroics will offer useful insights on their origin of the spin-induced polarization.

4.5 Skyrmion in noncentrosymmetric magnets

The following three sections are devoted to a discussion on the magnetic skyrmion. Typical spin configurations of a skyrmion are classified into three types as shown in Figure 4.9(a)–(c). Magnetic moment at the center of a vortex is down and continuously rotates to up when approaching outward. The magnetic moments on the two-dimensional plane can be projected to wrap a surface of a Bloch sphere once. This feature represents a non-zero topological number defined as

$$\Phi = \frac{1}{4\pi}\int \boldsymbol{n}\cdot\left(\frac{\partial \boldsymbol{n}}{\partial x}\times\frac{\partial \boldsymbol{n}}{\partial y}\right)\mathrm{d}x\mathrm{d}y = \pm 1. \quad (4.12)$$

A skyrmion is topologically protected from a trivial ferromagnetic state with $\Phi = 0$. In the beginning, the stability of this magnetic soliton on a ferromagnetic thin-film

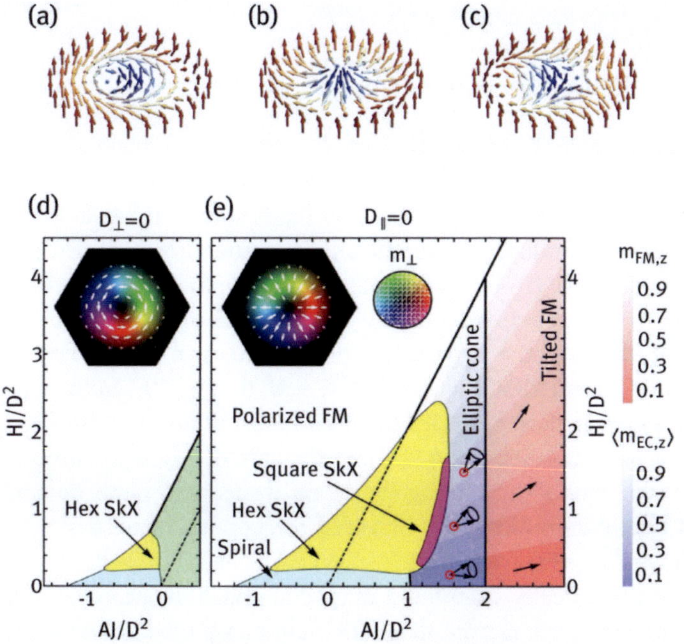

Figure 4.9: Schematic illustration of the magnetic moment configuration of (a) Bloch type, (b) Néel type, and (c) anti-vortex type skyrmion. (d)-(e) Phase diagram as a function of AJ/D^2 and HJ/D^2 for (d) $D_\perp = 0$, and (e) $D_{||} = 0$. Easy-axis(plane) anisotropy corresponds to $A < 0$ ($A > 0$). Insets: Unit cell in the hexagonal (Hex) skyrmion crystal (SkX) phase with white arrows indicating the projection of magnetization on the *xy*-plane. Thick lines denote continuous transitions, while thin lines indicate first-order transitions. Figures 9(d) and (e) are reproduced with permission from Ref. [111]. (c) (2016) American Physical Society.

has been mainly discussed [14, 105–108] until the theoretical prediction for the thermodynamic stability as a magnetic phase in Refs [109, 110].

According to Refs [109, 110], Lifshitz invariant in the free energy for a noncentrosymmetric magnet plays a key role to stabilize the skyrmion state. They considered the following form of free energy.

$$F[\boldsymbol{M}(\boldsymbol{r})] = \int \mathrm{d}r^3 \left\{ \frac{1}{2}\alpha \sum_{\mathrm{i}} (\nabla \mathrm{M_i})^2 - \frac{1}{2}\beta \mathrm{M}_z^2 - HM_z - \frac{1}{2}\boldsymbol{M} \cdot \boldsymbol{H_M} + w_{\mathrm{DM}} \right\}, \tag{4.13}$$

where the first term is the ferromagnetic exchange energy, the second term represents the uniaxial anisotropy, third term is the Zeeman energy, fourth term is the magnetostatic energy, and the last term is the Lifshitz invariant due to DM interaction. The particular type of skyrmion is determined by the form of w_{DM}, which is strongly influenced by the point-group symmetry of the lattice as follows:

$$D_n : w_{\mathrm{DM}} = D_\perp \boldsymbol{M} \cdot [\nabla_\perp \times \boldsymbol{M}] + D_z \boldsymbol{M} \cdot [\nabla_{\boldsymbol{z}} \times \boldsymbol{M}] \tag{4.14}$$

$$C_{nv}: w_{\mathrm{DM}} = D_{||}\boldsymbol{M}\cdot[(\boldsymbol{z}\times\nabla)\times\boldsymbol{M}] \tag{4.15}$$

$$D_{2d}: w_{\mathrm{DM}} = D\left[M_z\partial_y M_x - M_x\partial_y M_z + M_z\partial_x M_y - M_y\partial_x M_z\right], \tag{4.16}$$

where $\nabla_\perp = (\partial_x, \partial_y, 0)$, and $\nabla_z = (0, 0, \partial_z)$. Chiral group D_n possesses an instability for a proper-screw spin modulation along the *xy*-plane by the first term ($\propto D_\perp$) and along the *z* axis by the second term ($\propto D_z$) while only the first term is related with the formation of Bloch-type skyrmion (Figure 4.9(a)). The difference between this and the case discussed in the case (II) in Section 4.2 is that the modulation vector $\boldsymbol{q}$ for the spiral is allowed in any direction in the *xy*-plane, which enables the circularly swirling spin configuration in the two-dimensional space. As a special case, a cubic system (chiral *T* or *O* group) allows $w_{\mathrm{DM}} = D\boldsymbol{M}\cdot[\nabla\times\boldsymbol{M}]$ for $D_\perp = D_z$. In the case of polar group C_{nv}, w_{DM} tends to twist spins for the cycloidal configuration in the *xy* plane, favoring Néel-type skyrmion (Figure 4.9(b)). In the case of D_{2d}, w_{DM} can be a source of proper-screw type modulation along the *x* and *y* axis with opposite spin helicity (and cycloidal modulation along the diagonal directions), consistent with an antivortex-type spin winding (Figure 4.9(c)).

Figure 4.9(e, d) give numerically calculated magnetic phase diagrams at zero temperature in the parameter space for the magnetic field along the *c* (*z*) axis and the uniaxial anisotropy [111]. Chiral system stabilizes Bloch-type skyrmion only in a region for easy-axis anisotropy and finite magnetic field due to the competition with a conical state with $\boldsymbol{q}||\boldsymbol{H}||c$ [104]. In contrast, Néel-type skyrmion in the polar system shows wider stability even compatible with the easy-plane anisotropy. This is because w_{DM} in eq. (4.15) confines the spin modulation in the plane perpendicular to the *c* axis unlike the case for the chiral system.

Experimental observation of the skyrmion lattice state was first reported by Mühlbauer *et al.* using small-angle neutron scattering (SANS) [15], which is followed by the real-space observation of the triangular-lattice formation of skyrmions in B20 type alloys by Lorentz transmission electron microscopy by Yu *et al.* [16] (Figure 4.10(d,e)). A skyrmion triangular-lattice state is expressed by a superposition of three spiral spin modulations as follows:

$$\boldsymbol{M}(\boldsymbol{r}) = M_0\boldsymbol{z} + \sum_i \left(\boldsymbol{M}_i e^{\mathrm{i}\boldsymbol{q}_i\cdot\boldsymbol{r}} + \boldsymbol{M}_i^* e^{-\mathrm{i}\boldsymbol{q}_i\cdot\boldsymbol{r}}\right), \tag{4.17}$$

where the first term is a uniform component along the magnetic field, and $\boldsymbol{q}_i$ (i =\,1,2,3) are in the *xy* plane rotated by 120° with each other as shown in Figure 4.10(a). Figure 4.10(c) shows a six-fold SANS pattern representing a skyrmion lattice state observed in MnSi [113]. In the competition to the conical state with $\boldsymbol{q}||H$, skyrmion-lattice phase is stabilized by thermal fluctuations, and emerges in the vicinity of the transition temperature under a finite magnetic field as shown in Figure 4.10(b).

Table 4.3 shows a list of reported skyrmion-hosting materials up to date. As tabulated, Bloch-type skyrmion is observed exclusively in cubic systems, and

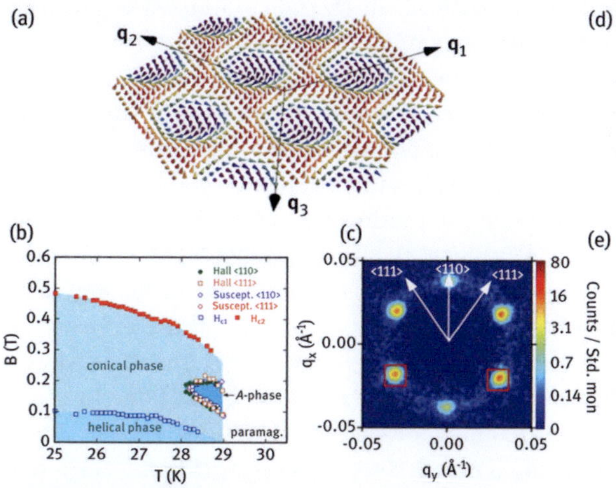

Figure 4.10: (a) Schematic illustration of triangular-lattice of Bloch-type skyrmion. $\boldsymbol{q}_i$ ($i = 1, 2, 3$) indicate magnetic modulation wave vector consisting the skyrmion lattice. (b) Magnetic phase diagram of MnSi for $B||[100]$. In the conical phase, the helix is aligned parallel to the magnetic field. (c) Typical SANS intensity in the *A* phase of MnSi. (d) The experimentally observed real-space images of the spin texture for the skyrmion crystal in $Fe_{0.5}Co_{0.5}Si$ under a weak magnetic field. White arrows represent the lateral magnetization distribution as obtained by TIE analysis of the Lorentz TEM data. (e) is a magnified view of (d). Figures 10(c) and (b) are reproduced with permission from Refs [113, 114], respectively. (c) (2009, 2011) American Physical Society. Figures 10(d) and (e) are reproduced with permission from Ref. [16]. (c) (2010) Springer Nature.

Table 4.3: Skyrmion-hosting material list. In parentheses, space group, lattice spacing for skyrmion crystal (nm), and references are indicated.

	Bloch	**Néel**	**anti-vortex**
Metal	MnSi ($P2_13$, 20 nm, [15]) FeGe ($P2_13$, 70 nm, [115]) $Fe_{1-x}Co_xSi$ ($P2_13$, 90 nm, [116]) Fe_2Mo_3N ($P4_132$, 120 nm, [117]) $Co_xZn_yMn_z$ ($P4_132$, 125 ~ 150 nm, [118, 119])	–	$Mn_{1.4}PtSn$, $Mn_{1.4}Pt_{0.9}Pd_{0.1}Sn$($I\bar{4}2m$, 150 ~ 350 nm, [120])
Insulator (semiconductor)	Cu_2OSeO_3 ($P2_13$, 60 nm, [17])	GaV_4S_8 ($R3m$, 22 nm, [121]) GaV_4Se_8 ($R3m$, 22 nm, [122, 123]) $VOSe_2O_5$ ($P4cc$, 185 nm, [124])	–

absent in the group D_n. For example, $Cr_{1/3}NbS_2$ in D_3 [125] stabilizes the proper-screw spin state for $q||c$ due to the hard-axis anisotropy [116]. Chiral ferrimagnet $LiMnFeF_6$ ($P321$), for example, is a possible candidate for Bloch-type skyrmion [126] with the detailed magnetic property yet to be explored. The number of known Néel-type skyrmion-hosting bulk compounds is less than host for Bloch-type skyrmions, despite that in polar magnets do not appear less frequently than in the chiral magnets. In the former case, searching for materials with small uniaxial anisotropy is key. For example, a family of polar ferrimagnet $M_2Mo_3O_8$ (M = Mn, Fe, Zn) [127, 128] with space group $P6_3mc$ satisfies the prerequisite for hosting Néel-type skyrmion, however detailed investigation of the magnetic phase diagram revealed the absence of the skyrmion state [129–131]. Instead, they show a collinear spin state, and exhibit metamagnetic or spin-flop transition under $H||c$ due to strong easy axis anisotropy. Another route is to look for Néel-type skyrmionic state in metallic systems. In this context, $Gd_3Ni_8Sn_4$ ($P6_3mc$) [132], $Eu_3Cu_8Sn_4$ ($P6_3mc$) [133], and $EuNiGe_3$ ($I4mm$) [134] are promising candidates. Anti-vortex type skyrmion is recently observed in metallic Heusler compounds [120], and is yet to be found in insulating systems. Multiferroic $Ba_2CuGe_2O_7$, for example, belongs to D_{2d} where a cycloidal modulation is observed in the c plane while no signature of skyrmion lattice state has been found [135, 136]. Hoffmann *et al.* provides a list of possible materials for anti-vortex type skyrmion in Ref. [137]. In the next section, we review three types of skyrmion-hosting multiferroics Cu_2OSeO_3, $GaV_4(S, Se)_8$, and $VOSe_2O_5$, and discuss their magnetoelectric responses and microscopic mechanisms.

4.6 Skyrmion-hosting multiferroics

4.6.1 Cu_2OSeO_3

Crystal structure and magnetism. Single crystals of Cu_2OSeO_3 were synthesized by chemical vapor transport reaction method [138], and were revealed to host the cubic chiral structure of $P2_13$ identical with the B20 type alloys [139]. The crystal structure is outlined by a distorted pyrochlore lattice (corner-sharing tetrahedra network) of Cu ion with an oxygen atom at the center of the tetrahedron, along with SeO_3 polyanions that decorates within the framework. Each tetrahedron is composed of three equivalent Cu sites and one that is distinct from the former (Figure 4.11(b)). Ferromagnetic-like transition at 58.8 K is observed [140] and 3-up-1-down type ferrimagnetic structure was first proposed as shown in Figure 4.11(c) [141].

Discovery of skyrmion-lattice state was reported in Ref. [17] via real-space imaging. This thermodynamic phase spans a pocket region in the H-T phase diagram as in B20 type compounds. SANS measurements [142, 143] have revealed the proper-screw spin ground state in zero field with the period of around 60 nm as well as the emergence of the six-fold pattern for skyrmion lattice state under a magnetic field.

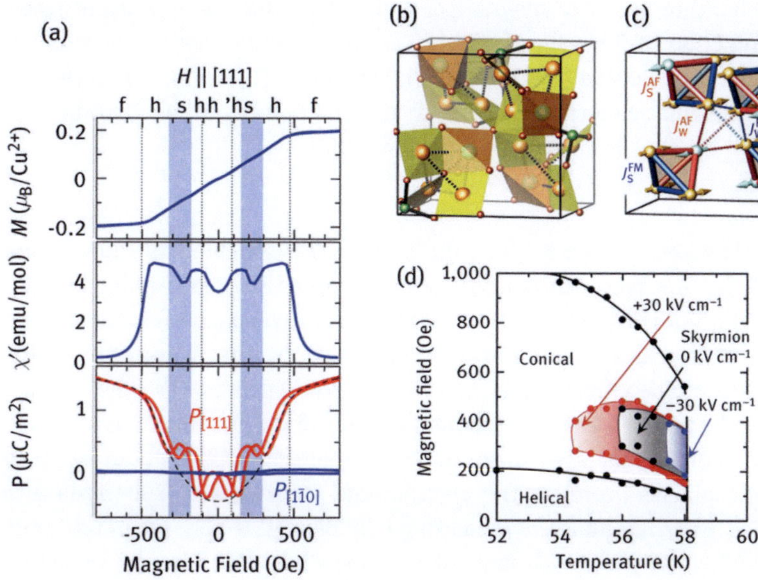

Figure 4.11: Magnetic field dependence of magnetization *M*, ac susceptibility χ', and electric polarization *P* of Cu_2OSeO_3 at 57 K for *H*||[111]. The labels s, h, h', and f denote skyrmion, helimagnetic (single *q*-domain), helimagnetic (multiple *q*-domain), and ferrimagnetic states, respectively. (b) The crystal structure of Cu_2OSeO_3. Cyan (orange) sphere is Cu(1) (Cu(2)). Red (green) sphere is O (Se). Dashed lines highlight tetrahedra, each composed of one Cu(1) and three Cu(2) atoms. (c) Schematic illustration of 3-up-1-down type ferrimagnetic structure in the distorted pyrochlore lattice of Cu^{2+} ion, a network of corner-shared tetrahedra. Red (blue) bond indicates antiferromagnetic (ferromagnetic) interaction, and the thickness represents the strength of a certain coupling. (d) The magnetic phase diagram near the transition temperature under electric fields of -30 (blue), 0 (black), and +30 kV/cm (red), determined by the magnetic susceptibility measurements. Figure 11(a) are reproduced with permission from Ref. [144]. (c) (2012) American Physical Society. Figures 11(b)-(d) are reproduced with permission from Refs [146, 152], respectively. (c) (2014, 2016) Springer Nature.

Figure 4.11(a) shows the magnetic-field dependence of electric polarization associated with the phase transitions to and from the skyrmion state, suggesting a considerable magnetoelectric coupling in this compound.

Microscopic mechanism of electric polarization. Unlike the cases described in Section 4.3, the effect of the *d-p* hybridization between Cu^{2+} and O^{2-} ions are suggested to be the leading origin of electric polarization [144]. We note that, however, the application of the *d-p* model on a single Cu^{2+} ion site predicts exactly vanishing polarization since the spin quadrupole operator $\hat{Q}_{ii}^{\alpha\beta} = \hat{S}_i^\alpha \hat{S}_i^\beta + \hat{S}_i^\beta \hat{S}_i^\alpha$ for $S = 1/2$ is isotropic with respect to the orientation of the spin. According to ab initio calculations in Refs [145, 146], the exchange interactions among Cu sites can be separated into two components: strong intra-tetrahedron coupling and relatively weak inter-tetrahedron coupling as shown in Figure 4.11(c). It is proposed that the individual tetrahedrons are viewed as a building-block of magnetism with a local magnetic moment [146–148]. The wave function of four-site cluster of Cu^{2+} is expressed not by the product state $|\uparrow\uparrow\uparrow\Downarrow>$ (where the double (single) arrow labels the Cu(1) (Cu(2)) site in the tetrahedron) but by the entangled triplet form with effective spin $S_{\mathrm{eff}} = 1$ as follows:

$$|\Psi> \equiv \left| M = 1, S_{\mathrm{eff}} = 1 > = \frac{1}{\sqrt{12}}(3|\uparrow\uparrow\uparrow\Downarrow> - |\downarrow\uparrow\uparrow\Uparrow> - |\uparrow\downarrow\uparrow\Uparrow> - |\uparrow\uparrow\downarrow\Uparrow>), \right. \quad (4.18)$$

here only the $M = 1$ wavefunction is given for brevity. This scheme has shown good agreements with the observed magnetic/optical properties such as a reduced magnetic moment [141], ESR [147], Raman [149], and far infrared spectra [150]. The electric polarization at the local tetrahedron is expressed by the expectation value $<\Psi|\hat{Q}_{ij}^{\alpha\beta}|\Psi>$ of the multi-centered quadrupole operators such as $\hat{Q}_{ij}^{\alpha\beta} = \hat{S}_i^{\alpha}\hat{S}_j^{\beta} + \hat{S}_i^{\beta}\hat{S}_j^{\alpha}$, which reproduces the magnetoelectric response in terms of the spin moment rotation.

Electric-field-control of the spin state. Different value of electric polarization in different magnetic phases promises the electric-field control of skyrmion state in this compound as experimentally demonstrated in Refs [151–153]. Okamura *et al.* observed the stabilization/destabilization of the skyrmion state through measurement of magnetic susceptibility and magnetic resonance under an electric field. This effect has been directly demonstrated in Ref. [154] using SANS technique.

4.6.2 GaV_4X_8 (X = S, Se)

The crystal structure of GaV_4X_8 at room temperature belongs to the noncentrosymmetric cubic space group $F\bar{4}3m$ [155–158] This is a lacunar spinel structure, a derivative of the spinel structure with ordered deficiency of Ga at the A site. It is alternatively viewed as a rock salt network of $(V_4S_4)^{5+}$ and $(GaS_4)^{5-}$ (Figure 4.12(a)). Four V sites form a tetrahedral cluster with an effective spin $S_{\mathrm{eff}} = 1/2$. This V cluster is Jahn-Teller active leading to the structural phase transition at low temperature by the elongation along one of the <111> directions as shown in Figure 4.12(b) [155]. The space group is lowered to a polar $R3m$ satisfying the prerequisite for the formation of Néel-type skyrmion. Sulfide (X = S) shows the structural phase transition at $T_s = 38/46$ K [150], and ferromagnetic like transition at $T_C = 13$ K [156, 159]. Nakamura *et al.* [160] reported a pocket of a metamagnetic phase around the transition temperature.

Emergence of the skyrmion state in this pocket region and cycloidal spin state in zero field are experimentally reported by Kézsmárki *et al.* [121] (Figure 4.12(c)) using SANS measurement and real-space observation by atomic force microscopy combined with the theoretical calculation. Isostructural selenide (X = Se) has $T_s = 41$ K and $T_c = 17.5$ K [122, 161]. Fujima *et al.* [122] first clarified the magnetic phase diagram in single crystals. As shown in Figure 4.12(d), the metamagnetic phase between the zero-field ground state and the forced ferromagnetic state is proposed to be the skyrmion phase, which is supported by Monte Carlo simulation with DM interaction.

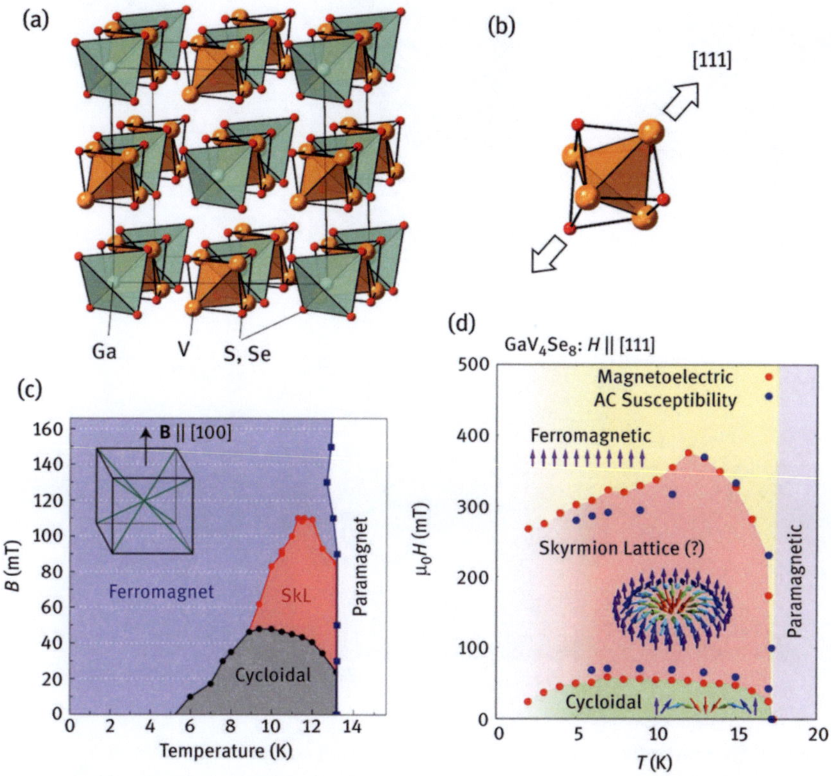

Figure 4.12: (a) Crystal structure of GaV_4X_8. (X = S, Se) (b) A V_4X_4 cluster in the low-temperature polar phase. White arrows represent the elongation along the [111] axis. (c)-(d) Magnetic phase diagram of (c) GaV_4S_8 for H||[100], and (d) GaV_4Se_8 for H||[111]. Figure 12(c) is reproduced with permission from Ref. [121]. (c) (2015) Springer Nature. Figure 12(d) is reproduced with permission from Ref. [122]. (c) (2017) American Physical Society.

A following SANS experiment confirmed the long-wavelength magnetic modulation, which is consistent with the skyrmion/cycloidal spin state [123]. Unlike in the sulfide with easy-axis anisotropy, skyrmion phase survives at the lowest temperature in the selenide due to the easy-plane anisotropy [162]. Extended stability of this skyrmion state is consistent with the theoretical calculation in Ref. [111].

We note that GaV_4X_8 belongs to the type-I multiferroics because the crystallographic transition to the polar group and the magnetic transition occurs at different temperatures. Magnetoelectric response has been observed as a change of electric polarization with a background of crystallographic pyroelectric effect [122, 163]. The d-p hybridization mechanism is regarded negligible because of the effective spin $S_{\mathrm{eff}} = 1/2$ nature at each V_4 cluster. $\boldsymbol{S}_i \cdot \boldsymbol{S}_j$ type exchange striction mechanism [163] and $\boldsymbol{e}_{ij} \times (\boldsymbol{S}_i \times \boldsymbol{S}_j)$ type inverse DM mechanism [122] are suggested as possible origins,

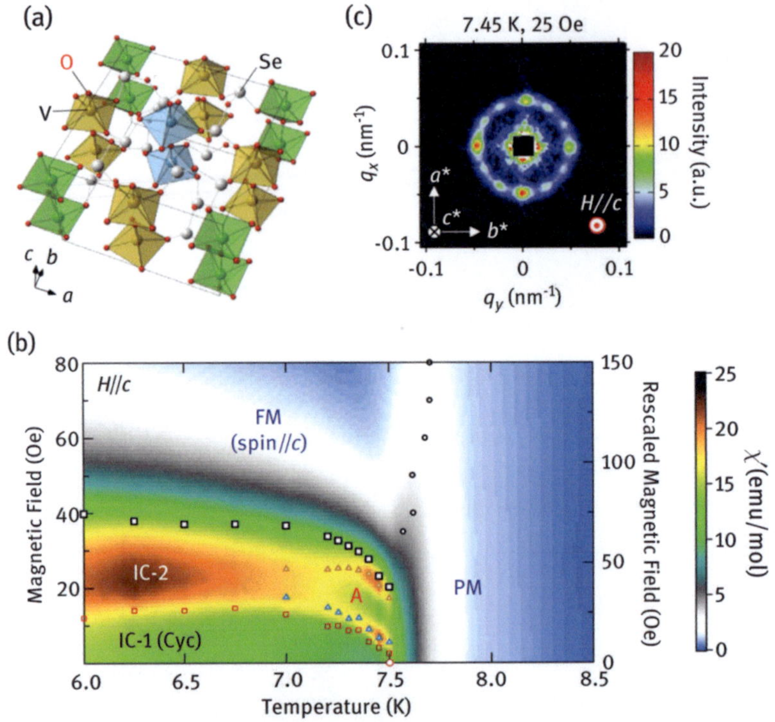

Figure 4.13: Crystal structure of $VOSe_2O_5$. Blue, green, and yellow tetragonal pyramids are inequivalent VO_5 polyhedra in a unit cell. (b) Color plot of χ' for $H||c$, and the magnetic phase diagram. Each magnetic phase is indicated as paramagnetic (PM); ferrimagnetic; cycloidal IC-1 (Cyc); incommensurately modulated magnetic order (IC-2); and skyrmion triangular-lattice state (*A*). (c) Small angle neutron scattering pattern for the *A* phase. Reproduced with permission from Ref. [124]. (c) (2017) American Physical Society.

which are further supported by the first-principle calculation [164]. No electric-field effect has been reported so far.

4.6.3 $VOSe_2O_5$

Recently, a new member joined in the insulating skyrmion-hosting materials. $VOSe_2O_5$ forms a square lattice of VO_5 pyramids (Figure 4.13(a)), which belongs to a unique tetragonal polar space group *P*4*cc* [165] compatible with existence of Néel-type skyrmions. Ferrimagnetic like transition at around 8 K was reported [166], and 3-up-1-down spin structure was suggested from the neutron diffraction and the first-principle calculation [167].

The magnetization measurement and SANS experiment using single crystals have unraveled versatile magnetic phases under $H||c$ as shown in Figure 4.13(b) [124].

Small pocket region in the vicinity of transition temperature has been identified, and twelve-fold SANS pattern due to multidomain of triangular skyrmion-lattice state has been observed (Figure 4.13(c)). Measurement of electric polarization under magnetic field detected the magnetoelectric susceptibility on the order of 0.5 ps/m [168]. Magnetoelectric response within the skyrmion state remains unexplored.

4.7 Skyrmion in a centrosymmetric magnet

In the preceding two sections, the stabilization of a skyrmion state requires DM interaction originating from the spatial inversion symmetry breaking (case (II) in Section 4.2). Indeed, skyrmion states have been observed in many forms of materials not only in bulk noncentrosymmetric crystals but also in magnetic thin-films or heterostructures without inversion symmetry at the surface/interface [28, 169, 170]. On the other hand, competing magnetic interactions may become the source of spin modulated state in a centrosymmetric system as in the case (I) in Section 4.2. Indeed, there has been many theoretical studies [171–178] on the stabilization of multiple-q state in centrosymmetric magnets, where competing interaction (e. g. RKKY) and/or geometric frustration play an important role, and possible multiple-q states (not skyrmion state) have been experimentally reported such as in $CeAl_2$ [179], Nd [180], TmS [181], $GdNi_2B_2C$ [182], $CeAuSb_2$ [183], $MnSc_2S_4$ [184], and $SrFeO_3$ [185].

Theoretical study for the stabilization of a skyrmion-lattice state in a centrosymmetric frustrated spin system was triggered by Okubo *et al.* [186], which was followed by Refs [187–191]. In an itinerant system, helicity-dependent current-induced motion of skyrmion is predicted [188, 191] while multiferroic property is predicted in insulating systems [187]. Since the symmetric exchange interaction does not restrict the type of skyrmion, energy difference among skyrmions for Bloch, Néel, and anti-vortex type is basically determined by anisotropic exchange interactions and weak dipole-dipole interaction [191]. This facilitates controlling the helicity or the topological number of skyrmion by external perturbations, providing a new degree of freedom in future skyrmion-based memory. Moreover, the size of skyrmion in centrosymmetric system is free from the limitation of the magnitude of relativistic spin-orbit interaction, which allows potentially more compact skyrmions preferable for information storage devices.

In the following subsections, we introduce the theories for the physical origin of skyrmion in centrosymmetric system, several material design principles, and experimental report of realization of the skyrmion-hosting magnets.

4.7.1 Theoretical models

There are mainly three proposed mechanisms for skyrmion stabilization in centrosymmetric systemml: (1) thermal fluctuations [186], (2) uniaxial anisotropy [187, 188], and (3) biquadratic spin interaction involving RKKY interaction in Kondo lattice [190].

(1) *Thermal fluctuations.* Okubo *et al.* [186] discussed the frustrated spin Hamiltonian on triangular lattice

$$H = -J_1 \sum_{\langle i,j \rangle} \boldsymbol{S}_i \cdot \boldsymbol{S}_j - J_{2,3} \sum_{\langle\langle i,j \rangle\rangle} \boldsymbol{S}_i \cdot \boldsymbol{S}_j - H \sum_i S_{iz}, \tag{4.19}$$

where the first term is for the nearest neighbor interaction, the following term is for the second (third) nearest neighbor interaction (Figure 4.14(a)), and the third term is

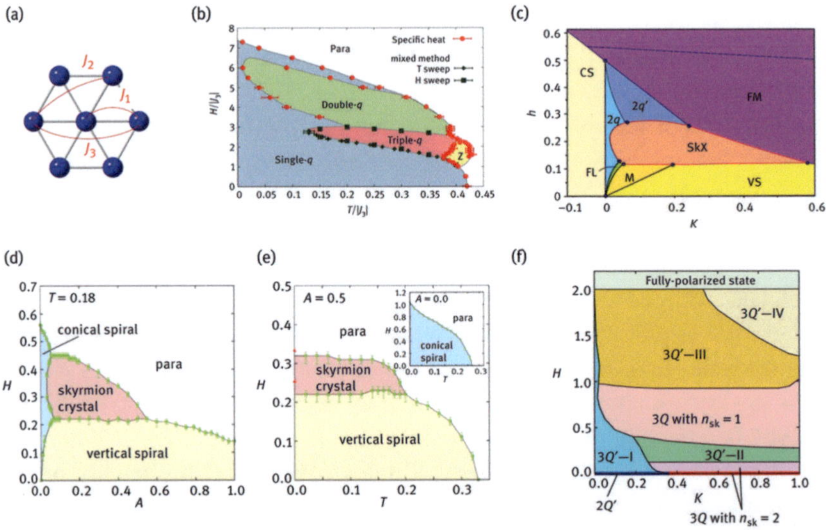

Figure 4.14: (a) First, second, and third nearest neighbor interactions, J_1, J_2, and J_3, respectively, on a triangular lattice. (b) Phase diagram of the J_1-J_3 model (eq. (4.19)) with $J_1/J_3 = -1/3$, obtained by Monte Carlo simulation. (c) Zero-temperature phase diagram of a frustrated triangular-lattice antiferromagnet with uniaxial anisotropy. Fully polarized ferromagnetic state (FM); conical spiral (CS), 2q-state, 2q'-state, vertical spiral (VS), VS plus two in-plane sinusoidal modulations (M); flop state (FL), and skyrmion crystal (SkX). $J_2 = 0.5$, magnetic anisotropy, K, and magnetic field, h, are measured in units of $J_1 = 1$. (d) A-H phase diagram with the J_1-J_3 model for $J_3/|J_1| = 0.5$ at $T = 0.18$ where A is the parameter for the uniaxial anisotropy. (e) T-H phase diagram for $A = 0.5$ (easy axis). The inset shows the phase diagram for $A = 0$. (f) Magnetic phase diagram for the model in eq. (4.20) on the triangular lattice. The magnetic phase for 3Q with $n_{sk} = 1$ corresponds to the triangular-lattice skyrmion state. Figures 14(b), (d)-(c), and (f) are reproduced with permission from Refs [186, 189, 190], respectively. (c) (2012, 2016, 2017) American Physical Society. Figure 14(c) is reproduced with permission from Ref. [187]. (c) (2015) Springer Nature.

the Zeeman energy. As discussed in Section 4.2, certain values of the exchange interactions enable the formation of a spiral state with some $\boldsymbol{q}$ vectors, and there are six degenerate directions in the lattice plane due to the rotational symmetry, which is essential for the formation of multiple-q state. Although the higher-harmonics in the modulation, which is inevitable in the multiple-q state, costs exchange interaction, effect of thermal fluctuations preferably stabilizes skyrmion state than the single-q state at finite temperature. Figure 4.14(b) shows the normalized H-T phase diagram calculated by Monte Carlo method. In addition to the single-q spiral phase as the ground state, (triple-q) skyrmion-lattice state appears at finite temperature under a magnetic field. Due to the mirror symmetry on the xy spin components, skyrmion lattice and the antivortex-type skyrmion lattice state are degenerate.
(2) *Uniaxial anisotropy.* In order to stabilize the skyrmion state at zero temperature, the effect of the uniaxial anisotropy term $-A\sum_i (S_{iz})^2$ was considered in Refs [187–189]. Phenomenologically, the exchange interaction cost due to the higher harmonics is compensated by this term [192]. Figure 4.14(c) provides the calculated magnetic phase diagram as a function of H and A, which shows the stabilization of the skyrmion state under a magnetic field by the easy-axis anisotropy. This is basically reproduced in Figure 4.14(d), and skyrmion phase appears as a metamagnetic phase in finite temperature Figure 4.14(e).

(3) *Biquadratic spin interaction.* As an alternative approach to reduce the higher harmonics cost, Hayami *et al.* [190] developed the theory for the effect of conduction electron in an itinerant magnet, and considered the higher-order term in addition to RKKY-type quadratic term. Under certain assumptions, they deduced the following spin Hamiltonian:

$$H = 2\sum_{\nu}\left[-J\boldsymbol{S}_{\boldsymbol{Q}_\nu}\cdot\boldsymbol{S}_{-\boldsymbol{Q}_\nu} + K\left(\boldsymbol{S}_{\boldsymbol{Q}_\nu}\cdot\boldsymbol{S}_{-\boldsymbol{Q}_\nu}\right)^2\right], \tag{4.20}$$

where $\boldsymbol{Q}_\nu$ ($\nu = 1, 2, 3$) is the $\boldsymbol{q}$ vector for the spiral structure determined by Lindhardt function of conduction electrons. Figure 4.14(f) shows the magnetic phase diagram with respect to K and H. Under a magnetic field, the magnetic phase with nonzero topological number ($|n_{\mathrm{sk}}| = 1$) is obtained. Furthermore, they predicted the other topological spin state in zero field in the stronger K region [193]. The relevance of the above spin Hamiltonian and an estimation of the magnitude of K in realistic materials remain elusive.

4.7.2 Material design

In order to realize the skyrmion-hosting centrosymmetric magnetic materials, we propose following design principles.

1. *Geometrically frustrated lattice.* Promising materials will have a magnetic lattice forming triangular, kagome, square, or honeycomb lattice, which frequently host magnetic frustration.
2. *Ferromagnetic interlayer coupling.* The antiferromagnetic interlayer interaction should be avoided in order to stabilize the multiple-q state with three $\boldsymbol{q}$ vectors lying in the same two-dimensional plane, i. e. q along the c axis should be zero. This excludes the most of delafossite or MX_2 type compounds from a list of candidates. It is relatively easier to realize $q_c = 0$ in rare earth intermetallics.
3. *Weak anisotropy.* $CeAuSb_2$, for example, hosts interesting multiple-q state, but the magnetic structure is collinear due to strong Ising type anisotropy of Ce^{3+} ion [183]. In order to stabilize the noncoplanar spin texture with topological number, materials with weak-anisotropic ions such as Gd^{3+}, Eu^{2+}, Mn^{2+} and Fe^{3+} may be more promising.
4. *Symmetry for multiferroicity.* In the case of designing multiferroic skyrmion material, one has to consider whether the target compound shows electric polarization in the skyrmion state. Leonov *et al.* predicted the emergence of electric polarization due to a skyrmion. They considered the case for the system belonging to the point group D_{3d}, and deduced the spin Hamiltonian for magnetoelectric coupling [187]

$$H_{\mathrm{ME}} = -\mathrm{Im}\sum_{\boldsymbol{q}}\Big[g_1 E^z S^z_{-\boldsymbol{q}}\left(q^x S^x_{\boldsymbol{q}} + q^y S^y_{\boldsymbol{q}}\right) + g_2(q^x E^y - q^y E^x) S^x_{-\boldsymbol{q}} S^y_{\boldsymbol{q}} + g_3 S^z_{-\boldsymbol{q}}\left((q^x E^x - q^y E^y) S^y_{\boldsymbol{q}} + (q^x E^y + q^y E^x) S^x_{\boldsymbol{q}}\right)\Big], \tag{4.21}$$

where g_i ($i = 1,2,3$) are coupling constants corresponding to the tensor $f_{ik\alpha\beta}$ discussed in the subsection 4.4.3. This formula predicts the emergence of polarization along the c axis in the Néel-type skyrmion state ($\propto g_1$) but not in Bloch type. The conclusion does not change in the other symmetry for O_h, C_{3i}, D_{6h}, and D_{4h}. The same discussion in the symmetry T_h, C_{6h}, and C_{4h} predict the emergence of polarization in Bloch type skyrmion state. Note that the above discussion is in the continuous limit, and an additional (but basically weak) component may possibly be allowed due to a short magnetic-modulation wavelength in a frustrated system as discussed in Section 4.4.

Very recently, we have reported a rare earth intermetallic compound Gd_2PdSi_3 as a promising candidate [194]. The crystal structure is composed of (1) a triangular lattice of (2) Gd [195], and shows the antiferromagnetic transition at around 20 K [196, 197] into a magnetic phase with (3) in-plane spin modulation for the wavelength 2.5 nm [194]. Skyrmion lattice state is detected by the topological Hall effect significantly enhanced by the squeezed skyrmion radius due to magnetic frustration. This finding is followed by another discovery of skyrmion-hosting magnet

$Gd_3Ru_4Al_{12}$, which forms breathing kagome lattice of Gd atoms [198]. These results support the relevance of the above design principles, which may lead to the future discovery of a skyrmion-hosting centrosymmetric system with multiferroicity.

4.8 Summary and outlook

In this article, we reviewed two types of origin of spiral spin structure (frustration and DM interaction) on the basis of the lattice symmetry. The frustrated spin interactions in a centrosymmetric system produce a spiral order with the degeneracy of the spin helicity, which is one of the keys of the multiferroicity in a spiral magnet. As for the microscopic mechanism of the spin-induced ferroelectricity, the generalization of the inverse DM mechanism provides a consistent description in most experimentally known cases. The contribution from the *d-p* hybridization has to be considered with caution. We also reviewed the magnetic skyrmions. The emergence of skyrmion is closely related with the noncentrosymmetry of the basal lattice. Several multiferroic materials have been discovered to host a skyrmion state, where magnetoelectric coupling has been observed. Theoretical studies have revealed the possibility of the skyrmion state in a centrosymmetric magnet due to magnetic frustration while only a few examples (metals) are experimentally reported. Searching for insulating counterpart which potentially realizes multiferroic property in the skyrmion phase is left for future work. Interesting electric-field controllability may be possible in such a system.

References

[1] Herpin A. Theorie du Magnetisme. Paris: Press Universitaires de France, 1968.

[2] Erickson RA. The antiefrromagnetic structure of manganese dioxide by neutron diffraction. Phys Rev. 1952;85:745.

[3] Herpin A, Mériel P. Étude de l'antiferromagnétisme helicoidal de $MnAu_2$ par diffraction de neutrons. J Phys (Paris). 1961;22:337.

[4] Kaplan TA. Classical spin-configuration stability in the presence of competing exchange forces. Phys Rev. 1959;116:888.

[5] Yoshimori A. A new type of antiferromagnetic structure in the rutile type crystal. J Phys Soc Jpn. 1959;14:807.

[6] Villain J. La structure des substances magnetiques. J Phys Chem Solids. 1959;11:303.

[7] Chattopadhyay T. Modulated magnetic phases in rare earth metallic systems. Science. 1994;264:226.

[8] Cox DE. Neutron-diffraction determination of magnetic structures. IEEE Trans Magn. 1972;8:161.

[9] Rossat-Mignod J. Magnetic structures. Methods Exp Phys. 1987;23:69.

[10] Koehler WC, Moon RM, Trego AL, Mackintosh AR. Antiferromagnetism in chromium alloys. I. neutron diffraction. Phys Rev. 1966;151:405.

[11] Kimura T, Goto T, Shintani H, Ishizaka K, Arima T, Tokura Y. Magnetic control of ferroelectric polarization. Nature. 2003;426:55.
[12] Khomskii D. Classifying multiferroics: Mechanisms and effects. Physics. 2009;2:20.
[13] Skyrme TH. A non-linear field theory. Proc R Soc. 1961;A260:127.
[14] Belavin AA, Polyakov AM. Metastable states of two-dimensional isotropic ferromagnets. JETP Lett. 1975;22:503.
[15] Mühlbauer S, Binz B, Jonietz F, Pfleiderer C, Rosch A, Neubauer A, et al. Skyrmion lattice in a chiral magnet. Science. 2009;323:915.
[16] Yu XZ, Onose Y, Kanazawa N, Park JH, Han JH, Matsui Y, et al. Real-space observation of a two-dimensional skyrmion crystal. Nature. 2010;465:901.
[17] Seki S, Yu XZ, Ishiwata S, Tokura Y. Observation of skyrmions in a multiferroic material. Science. 2012;336:198.
[18] Cheong S-W, Mostovoy M. Multiferroics: a magnetic twist for ferroelectricity. Nat Mater. 2007;6:13.
[19] Fiebig M. Revival of the magnetoelectric effect. J Phys D: Appl Phys. 2005;38:R123.
[20] Wang KF, Liu J-M, Ren ZF. Multiferroicity: the coupling between magnetic and polarization orders. Adv Phys. 2009;58:321.
[21] Tokura Y, Seki S, Nagaosa N. Multiferroics of spin origin. Rep Prog Phys. 2014;77:076501.
[22] Dong S, Liu J-M, Cheong S-W, Ren Z. Multiferroic materials and magnetoelectric physics: symmetry, entanglement, excitation, and topology. Adv Phys. 2015;64:519.
[23] Figbig M, Lottermoser T, Meier D, Trassin M. The evolution of multiferroics. Nat Rev Mater. 2016;1:16046.
[24] Bousquet E, Cano A. Non-collinear magnetism in multiferroic perovskites. J Phys: Cond Mater. 2016;28:123001.
[25] Wang J, Neaton JB, Zheng H, Nagarajan V, Ogale SB, Liu B, et al. Epitaxial $BiFeO_3$ multiferroic thin film heterostructures. Science. 2003;299:1719.
[26] Martin LW, Chu Y-H, Ramesh R. Advances in the growth and characterization of magnetic, ferroelectric, and multiferroic oxide thin films. Mater Sci Eng R. 2010;68:89.
[27] Nagaosa N, Tokura Y. Topological properties and dynamics of magnetic skyrmions. Nat Nanotech. 2013;8:899.
[28] Jiang W, Chen G, Liu K, Zang J, te Velthusis SG, Hoffmann A. Skyrmions in magnetic multilayers. Phys Rep. 2017;704:1.
[29] Yamasaki Y, Miyahara S, Kaneko Y, He J-P, Arima T, Tokura Y. Magnetic reversal of the ferroelectric polarization in a multiferroic spinel oxide.Phys Rev Lett. 2006;96:207204.
[30] Ishiwata S, Taguchi Y, Murakawa H, Onose Y, Tokura Y. Low-magnetic-field control of electric polarization vector in a helimagnet. Science. 2008;319:1643.
[31] White JS, Honda T, Kimura K, Kimura T, Niedermayer Ch, Zaharko O, et al. Coupling of magnetic and ferroelectric hysteresis by a multicomponent magnetic structure in Mn_2GeO_4. Phys Rev Lett. 2012;108:077204.
[32] Izyumov Y. Modulated, or long-periodic, magnetic structures of crystals. Usp Fiz Nauk. 1984;144:439.
[33] Kishine J-I, Ovchinnikov AS. Theory of monoaxial chiral helimagnet. Solid State Phys. 2015;66:1.
[34] Murakawa H, Onose Y, Tokura Y. Electric-field switching of a magnetic propagation vector in a helimagnet. Phys Rev Lett. 2009;103:147201.
[35] Hornreich RM, Luban M, Shtrikman S. Critical behavior at the onset of $\vec{k}$-space instability on the λ line. Phys Rev Lett. 1975;35:1678.

[36] Michelson A. Phase diagrams near the Lifshiftz point. II. Systems with cylindrical, hexagonal, and rhombohedral symmetry having an easy plane of magnetization. Phys Rev B. 1977;16:585.
[37] Dyzaloshinskii IE. Theory of helicoidal structures in antiferromagnets. I. nonmetals. Sov Phys JETP. 1965;19:960.
[38] Tanaka M, Takayoshi H, Ishida M, Endoh Y. Crystal chirality and helicity of the helical spin density wave in MnSi. I. Convergent-beam electron diffraction. J Phys Soc Jpn. 1985;54:2970.
[39] Ishida M, Endoh Y, Mitsuda S, Ishikawa Y, Tanaka M. Crystal chirality and helicity of the helical spin density wave in MnSi. II. polarized neutron diffraction. J Phys Soc Jpn. 1985;54:2975.
[40] Grigoriev SV, Chernyshov D, Dyadkin VA, Dmitriev V, Maleyev SV, Moskvin EV, et al. Crystal handedness and spin helix chirality in Fe1-*x*Co*x*Si. Phys Rev Lett. 2009;102:037204.
[41] Shibata K, Yu XZ, Hara T, Morikawa D, Kanazawa N, Kimoto K, et al. Towards control of the size and helicity of skyrmions in helimagnetic alloys by spin–orbit coupling. Nat Nanotech. 2013;8:723.
[42] Dyadkin V, Prsa K, Grigoriev SV, White JS, Huang P, Ronnow HM, et al. Chirality of structure and magnetism in the magnetoelectric compound Cu_2OSeO_3.Phys Rev B. 2014;89:140409(R).
[43] Kousaka Y, Koyama T, Ohishi K, Kakurai K, Hutanu V, Ohsumi H, et al. Monochiral helimagnetism in homochiral crystals of $CsCuCl_3$. Phys Rev Mater. 2017;1:071402(R).
[44] Matsumura T, Kita Y, Kubo K, Yoshikawa Y, Michimura S, Inami T, et al. Chiral soliton lattice formation in monoaxial helimagnet Yb(Ni1-*x*Cu*x*)3Al9. J Phys Soc Jpn. 2017;86:124702.
[45] Dzyaloshinskii IE. A thermodynamic theory of “weak” ferromagnetism of antiferromagnetics. J Phys Chem Solids. 1958;4:241.
[46] Moriya T. Anisotropic superexchange interaction and weak ferromagnetism. Phys Rev. 1960;120:91.
[47] Gignoux D, Schmitt D. Frustration in rare earth intermetallic compounds. J Alloys Comp. 2001;326:143.
[48] Ruderman MA, Kittel C. Indirect exchange coupling of nuclear magnetic moments by conduction electrons. Phys Rev. 1954;96:99.
[49] Kasuya T. A theory of metallic ferro- and antiferromagnetism on Zener’s model. Prog Theor Phys. 1956;16:45.
[50] Yosida K. Magnetic properties of Cu-Mn alloys. Phys Rev. 1957;106:893.
[51] Newnham RE, Kramer JJ, Schulze WA, Cross LE. Magnetoferroelectricity in Cr_2BeO_4. J Appl Phys. 1978;49:6088.
[52] Santoro RP, Newnham RE, Am J. Magnetic properties of chromium chrysoberyl. Ceram Soc. 1964;47:491.
[53] Cox DE, Frazer BC, Newnham RE, Santoro RP. Neutron diffraction investigation of the spiral magnetic structure in Cr_2BeO_4. J Appl Phys. 1969;40:1124.
[54] Baryakhtar VG, Lvov VA, Yablonskii DA. Inhomogeneous magnetoelectric effect. Sov Phys JETP Lett. 1983;37:565.
[55] Baryakhtar VG, Stefanovskii EP, Yablonskii DA. Theory of magnetic structure and electric polarization of the Cr_2BeO_4 system. Sov Phys JETP Lett. 1985;42:258.
[56] Birss RP. Symmetry and magnetism. Amsterdamml: North-Holland Pub. Co., 1964
[57] Siratori K, Akimitsu J, Kita E, Nishi M. A new method of controlling the sense of the screw spin structure. J Phys Soc Jpn. 1980;48:1111.
[58] Yamasaki Y, Sagayama H, Goto T, Matsuura M, Hirota K, Arima T, et al. Electric control of spin helicity in a magnetic ferroelectric. Phys Rev Lett. 2007;98:147204.

[59] Sagayama H, Abe N, Taniguchi K, Arima T, Yamasaki Y, Okuyama D, et al. Observation of spin helicity using nonresonant circularly polarized x-ray diffraction analysis. J Phys Soc Jpn. 2010;79:043711.
[60] Mostovoy M. Ferroelectricity in spiral magnets. Phys Rev Lett. 2006;96:067601.
[61] Sergienko IA, Dagotto E. Role of the Dzyaloshinskii-Moriya interaction in multiferroic perovskites. Phys Rev B. 2006;73:094434.
[62] Katsura H, Nagaosa N, Balatsky AV. Spin current and magnetoelectric effect in noncollinear magnets. Phys Rev Lett. 2005;95:057205.
[63] Kaplan TA, Mahanti SD. Canted-spin-caused electric dipoles: A local symmetry theory. Phys Rev B. 2011;83:174432.
[64] Xiang HJ, Kan EJ, Zhang Y, Whangbo M-H, Gong XG. General theory for the ferroelectric polarization induced by spin-spiral order. Phys Rev Lett. 2011;107:157202.
[65] Johnson RD, Nair S, Chapon LC, Bombardi A, Vecchini C, Prabhakaran D, et al. $Cu_3Nb_2O_8$: A multiferroic with chiral coupling to the crystal structure. Phys Rev Lett. 2011;107:137205.
[66] Johnson RD, Chapon LC, Khalyavin DD, Manuel P, Radaelli PG, Martin C. Giant improper ferroelectricity in the ferroaxial magnet $CaMn_7O_{12}$. Phys Rev Lett. 2012;108:067201.
[67] Hearmon AJ, Fabrizi F, Chapon LC, Johnson RD, Prabhakaran D, Streltsov SV, et al. Electric field control of the magnetic chiralities in ferroaxial multiferroic $RbFe(MoO_4)_2$. Phys Rev Lett. 2012;108:237201.
[68] Baum M, Komarek AC, Holbein S, Fernández-Díaz MT, André G, Hiess A, et al. Magnetic structure and multiferroic coupling in pyroxene $NaFeSi_2O_6$. Phys Rev B. 2015;91:214415.
[69] Tokunaga M, Akaki M, Ito T, Miyahara S, Miyake A, Kuwahara H, et al. Magnetic control of transverse electric polarization in $BiFeO_3$. Nat Commun. 2015;6:5878.
[70] Kimura T, Lashley JC, Ramirez AP. Inversion-symmetry breaking in the noncollinear magnetic phase of the triangular-lattice antiferromagnet $CuFeO_2$. Phys Rev B. 2006;73:220401.
[71] Seki S, Onose Y, Tokura Y. Spin-driven ferroelectricity in triangular lattice antiferromagnets $ACrO_2$ (A = Cu, Ag, Li, or Na). Phys Rev Lett. 2008;101:067204.
[72] Kimura K, Nakamura H, Ohgushi K, Kimura T. Magnetoelectric control of spin-chiral ferroelectric domains in a triangular lattice antiferromagnet. Phys Rev B. 2008;78:140401(R).
[73] Terada N, Khalyavin DD, Perez-Mato JM, Manuel P, Prabhakaran D, Daoud-Aladine A, et al. Magnetic and ferroelectric orderings in multiferroic α-$NaFeO_2$. Phys Rev B. 2014;89:184421.
[74] Kurumaji T, Seki S, Ishiwata S, Murakawa H, Tokunaga Y, Kaneko Y, et al. Magnetic-field induced competition of two multiferroic orders in a triangular-lattice helimagnet MnI_2. Phys Rev Lett. 2011;106:167206.
[75] Kurumaji T, Seki S, Ishiwata S, Murakawa H, Kaneko Y, Tokura Y. Magnetoelectric responses induced by domain rearrangement and spin structural change in triangular-lattice helimagnets NiI_2 and CoI_2. Phys Rev B. 2013;87:014429.
[76] Tokunaga Y, Okuyama D, Kurumaji T, Arima T, Nakao H, Murakami Y, et al. Multiferroicity in $NiBr_2$ with long-wavelength cycloidal spin structure on a triangular lattice. Phys Rev B. 2011;84:060406(R).
[77] Arima T. Ferroelectricity induced by proper-screw type magnetic order. J Phys Soc Jpn. 2007;76:073702.
[78] Ramirez AP. Strongly geometrically frustrated magnets. Annu Rev Mater Sci. 1994;24:453.
[79] Collins MF, Petrenko OA. Triangular antiferromagnets. Can J Phys. 1997;75:605.
[80] Balents L. Spin liquids in frustrated magnets. Nature. 2010;464:199.
[81] Terada N. Spin and orbital orderings behind multiferroicity in delafossite and related compounds. J Phys: Condens Matter. 2014;26:453202.
[82] McGuire MA. Crystal and magnetic structures in layered, transition metal dihalides and trihalides. Crystals. 2017;7:121.

[83] Mitsuda S, Yoshizawa H, Yaguchi N, Mekata M. Neutron diffraction study of $CuFeO_2$. J Phys Soc Jpn. 1991;60:1885.

[84] Petrenko OA, Balakrishnan G, Lees MR, Paul DM, Hoser A. High-magnetic-field behavior of the triangular-lattice antiferromagnet $CuFeO_2$. Phys Rev B. 2000;62:8983.

[85] Mitsuda S, Mase M, Prokes K, Kitazawa H, Aruga Katori H. Field-induced magnetic phase transitions in a triangular lattice antiferromagnet $CuFeO_2$ up to 14.5 T. J Phys Soc Jpn. 2000;69:3513.

[86] Nakajima T, Mitsuda S, Kanetsuki S, Prokes K, Podlensnyak A, Kimura H, et al. Spin noncollinearlity in multiferroic phase of triangular lattice antiferromagnet $CuFe_{1-x}Al_xO_2$. J Phys Soc Jpn. 2007;76:043709.

[87] Kanetsuki S, Mitsuda S, Nakajima T, Anazawa D, Katori HA, Prokes S. Field-induced ferroelectric state in frustrated magnet $CuFe_{1-x}Al_xO_2$. J Phys: Condens Matter. 2007;19:145244.

[88] Seki S, Yamasaki Y, Shiomi Y, Iguchi S, Onose Y, Tokura Y. Impurity-doping-induced ferroelectricity in the frustrated antiferromagnet $CuFeO_2$. Phys Rev B. 2007;75:100403(R).

[89] Nakajima T, Mitsuda S, Takahashi K, Yamano M, Masuda K, Yamazaki H, et al. Comprehensive study on ferroelectricity induced by a proper-screw-type magnetic ordering in multiferroic $CuFeO_2$: Nonmagnetic impurity effect on magnetic and ferroelectric order. Phys Rev B. 2009;79:214423.

[90] Cable JW, Wilkinson MK, Wollan EO, Koehler WC. Neutron diffraction investigation of the magnetic order in MnI_2. Phys Rev. 1962;125:1860.

[91] Sato T, Kadowaki H, Iio K. Successive phase transitions in the hexagonal-layered Heisenberg antiferromagnets MnX_2 (X = Br, I). Physica B. 1995;213:224.

[92] Wu X, Cai Y, Xie Q, Weng H, Fan H, Hu J. Magnetic ordering and multiferroicity in MnI_2. Phys Rev B. 2012;86:134413.

[93] Utesov OI, Syromyatnikov AV. Cascades of phase transitions in spiral magnets caused by dipolar forces. Phys Rev B. 2017;95:214420.

[94] Utesov OI, Syromyatnikov AV. Control of multiferroic order by magnetic field in frustrated helimagnet MnI_2. Theory. J Magn Magn Mater. 2019;475:98.

[95] Day P, Dinsdale A, Krausz ER, Robbins DJ. Optical and neutron diffraction study of the magnetic phase diagram of $NiBr_2$. J Phys C. 1976;9:2482.

[96] Day P, Ziebeck KR. Incommensurate spin structure in the lowtemperature magnetic phase of $NiBr_2$. J Phys C. 1980;13:L523.

[97] Adam A, Billerey D, Terrier C, Mainard R, Pegnault LP, Rossat-Mignod J, et al. Neutron diffraction study of the commensurate and incommensurate magnetic structures of $NiBr_2$. Solid State Commun. 1980;35:1.

[98] Pollard RJ, McCann VH, Ward JB. Electronic and magnetic properties of ^{57}Fe in $NiCl_2$, $NiBr_2$, NiI_2 and CoI_2 from Mossbauer spectroscopy. J Phys C. 1982;15:6807.

[99] Miyahara S, Furukawa N. Theory of antisymmetric spin-pair-dependent electric polarization in multiferroics. Phys Rev B. 2016;93:014445.

[100] Zhang JT, Ji C, Wang JL, Xia WS, Guo BX, Lu XM, et al. Spin-induced ferroelectricity in a triangular-lattice antiferromagnet studied by magnetoelectric coupling tensors. Phys Rev B. 2017;96:235136.

[101] Murakawa H, Onose Y, Miyahara S, Furukawa N, Tokura Y. Ferroelectricity induced by spin-dependent metal-ligand hybridization in $Ba_2CoGe_2O_7$. Phys Rev Lett. 2010;105:137202.

[102] Kurumaji T, Ohgushi K, Tokura Y. Magnetoelectric responses from the respective magnetic R and Fe subsystems in the noncentrosymmetric antiferromagnets $RFe_3(BO_3)_4$ (R = Eu, Gd, and Tb). Phys Rev B. 2014;89:195126.

[103] Rado GT. Mechanism of the magnetoelectric effect in an antiferromagnet. Phys Rev Lett. 1961;6:609.
[104] Kurumaji T. Magnetoelectric responses in magnetic insulators without inversion symmetry Thesis. University of Tokyo, 2014.
[105] Dzyaloshinskii IE, Ivanov BA. Localized topological solitons in a ferromagnet. Sov Phys JETP Lett. 1979;29:592.
[106] Ivanov BA, Stephanovich VA, Zhmudskii AA. Magnetic vortices – The microscopic analogs of magnetic bubbles. J Magn Magn Mater. 1990;88:116.
[107] Abanov A, Pokrovsky VL. Skyrmion in a real magnetic film. Phys Rev B. 1998;58:889(R).
[108] Bezvershenko AV, Kolezhuk AK, Ivanov BA. Stabilization of magnetic skyrmions by RKKY interactions. Phys Rev B. 2018;97:054408.
[109] Bogdanov AN, Yablonskii DA. Thermodynamically stable "vortices" in magnetically ordered crystals. The mixed state of magnets. Sov Phys JETP. 1989;95:178.
[110] Bogdanov AN, Hubert A. Thermodynamically stable magnetic vortex states in magnetic crystals. J Magn Magn Mater. 1994;138:255.
[111] Rowland J, Banerjee S, Randeria M. Skyrmions in chiral magnets with Rashba and Dresselhaus spin-orbit coupling. Phys Rev B. 2016;93:020404(R).
[112] Wilson MN, Butenko AB, Bogdanov AN, Monchesky TL. Chiral skyrmions in cubic helimagnet films: The role of uniaxial anisotropy. Phys Rev B. 2014;89:094411.
[113] Adams T, Mühlbauer S, Pfleiderer C, Jonietz F, Bauer A, Neubauer A, et al. Long-range crystalline nature of the skyrmion lattice in MnSi. Phys Rev Lett. 2011;107:217206.
[114] Neubauer A, Pfleiderer C, Binz B, Rosch A, Ritz R, Niklowitz PG, et al. Topological Hall effect in the *A* phase of MnSi. Phys Rev Lett. 2009;102:186602.
[115] Yu XZ, Kanazawa N, Onose Y, Kimoto K, Zhang WZ, Ishiwata S, et al. Near room-temperature formation of a skyrmion crystal in thin-films of the helimagnet FeGe. Nat Mater. 2011;10:106.
[116] Miyadai T, Kikuchi K, Kondo H, Sakka S, Arai M, Ishikawa Y. Magnetic properties of $Cr_{1/3}NbS_2$. J Phys Soc Jpn. 1983;52:1394.
[117] Li W, Jin C, Che R, Wei W, Lin L, Zhang L, et al. Emergence of skyrmions from rich parent phases in the molybdenum nitrides. Phys Rev B. 2016;93:060409(R).
[118] Tokunaga Y, Yu XZ, White JS, Ronnow HM, Morikawa D, Taguchi Y, et al. A new class of chiral materials hosting magnetic skyrmions beyond room temperature. Nat Commun. 2015;6:7638.
[119] Karube K, White JS, Reynolds N, Gavilano JL, Oike H, Kikkawa A, et al. Robust metastable skyrmions and their triangular-square lattice structural transition in a high-temperature chiral magnet. Nat Mater. 2016;15:1237.
[120] Nayak AK, Kumar V, Ma T, Werner P, Pippel E, Sahoo R, et al. Magnetic antiskyrmions above room temperature in tetragonal Heusler materials. Nature. 2017;548:561.
[121] Kézsmárki I, Bordács S, Milde P, Neuber E, Eng LM, White JS, et al. Néel-type skyrmion lattice with confined orientation in the polar magnetic semiconductor GaV_4S_8. Nat Mater. 2015;14:1116.
[122] Fujima Y, Abe N, Tokunaga Y, Arima T. Thermodynamically stable skyrmion lattice at low temperatures in a bulk crystal of lacunar spinel GaV_4Se_8. Phys Rev B. 2017;95:180410(R).
[123] Bordács S, Butykai A, Szigeti BG, White JS, Cubitt R, Leonov AO, et al. Equilibrium skyrmion lattice ground state in a polar easy-plane magnet. Sci Rep. 2017;7:7584.
[124] Kurumaji T, Nakajima T, Ukleev V, Feoktystov A, Arima T, Kakurai K, et al. Néel-Type Skyrmion Lattice in the Tetragonal Polar Magnet $VOSe_2O_5$. Phys Rev Lett. 2017;119:237201.
[125] Togawa Y, Koyama T, Takayanagi K, Mori S, Kousaka Y, Akimitsu J, et al. Chiral magnetic soliton lattice on a chiral helimagnet. Phys Rev Lett. 2012;108:107202.

[126] Courbion G, Jacoboni C, de Pape R. The dimorphism of $LiMnFeF_6$: A new kind of cationic order in the structural type Na_2SiF_6. J Solid State Chem. 1982;45:127.
[127] McAlister SP, Strobel P. Magnetic order in $M_2Mo_3O_8$ single crystals (M = Mn, Fe, Co, Ni). J Magn Magn Mater. 1983;30:340.
[128] Betrand D, Kerner-Czeskleba H. Étude structurale et magnétique de molybdates d'éléments de transition. J Phys. 1975;36:379.
[129] Wang W, Pascut GL, Gao B, Tyson TA, Haule K, Kiryukhin V, et al. Unveiling hidden ferrimagnetism and giant magnetoelectricity in polar magnet $Fe_2Mo_3O_8$. Sci Rep. 2015;5:12268.
[130] Kurumaji T, Ishiwata S, Tokura Y. Doping-tunable ferrimagnetic phase with large linear magnetoelectric effect in a polar magnet $Fe_2Mo_3O_8$. Phys Rev X. 2015;5:031034.
[131] Kurumaji T, Ishiwata S, Tokura Y. Diagonal magnetoelectric susceptibility and effect of Fe doping in the polar ferrimagnet $Mn_2Mo_3O_8$. Phys Rev B. 2017;95:045142.
[132] Romaka VV, Romaka L, Romaniv I, Hlil EK, Rykavets Z, Kuzhel B. Crystallographic, magnetic and electrical characteristics of $R_3Ni_8Sn_4$ compounds (R = Y, Nd, Sm, Gd, and Tb). J Alloys Compd. 2017;701:358.
[133] Fornasini ML, Manfrinetti P, Mazzone D. Crystal structure of trieuropium octacopper tetrastannide, $Eu_3Cu_8Sn_4$. Z Kristallogr NCS. 2003;218:279.
[134] Ryan DH, Cadogan JM, Rejali R, Boyer CD. Complex incommensurate helicoidal magnetic ordering of $EuNiGe_3$. J Phys: Condens Matter. 2016;28:266001.
[135] Zheludev A, Shirane G, Sasago Y, Kiode N, Uchinokura K. Spiral phase and spin waves in the quasi-two-dimensional antiferromagnet $Ba_2CuGe_2O_7$. Phys Rev B. 1996;54:15163.
[136] Mühlbauer S, Gvassaliya S, Ressouche E, Pomjakushina E, Zheludev A. Phase diagram of the Dzyaloshinskii-Moriya helimagnet $Ba_2CuGe_2O_7$ in canted magnetic fields. Phys Rev B. 2012;86:024417.
[137] Hoffmann M, Zimmermann B, Müller GP, Schürhoff D, Kiselev NS, Melcher C, et al. Antiskyrmions stabilized at interfaces by anisotropic Dzyaloshinskii-Moriya interactions. Nat Commun. 2017;8:308.
[138] Meunier G, Bertaud M. Constantes cristallographiques de $CuSe_2O_5$, $CuSeO_3$ et $CuaSeO_4$. J Appl Cryst. 1976;9:364.
[139] Effenberger H, Pertlik F. The crystal structures of the copper(II)-oxo-selenites $Cu_2O(SeO_3)$ (cubic and monoclinic), and $Cu_4O(SeO_3)_3$ (monoclinic and triclinic). Monatsh Chem. 1986;117:887.
[140] Khon K. A new ferrimagnet Cu_2SeO_6. J Phys Soc Jpn. 1977;42:2065.
[141] Bos J-W, Colin CV, Palstra TT. Magnetoelectric coupling in the cubic ferrimagnet Cu_2OSeO_3. Phys Rev B. 2008;78:094416.
[142] Adams T, Chacon A, Wagner M, Bauer A, Brandl G, Pedersen B, et al. Long-wavelength helimagnetic order andskyrmion lattice phase in Cu_2OSeO_3. Phys Rev Lett. 2012;108:237204.
[143] Seki S, Kim J-H, Inosov DS, Georgii R, Keimer B, Ishiwata S, et al. Formation and rotation of skyrmion crystal in the chiral-lattice insulator Cu_2OSeO_3. Phys Rev B. 2012;85:220406.
[144] Seki S, Ishiwata S, Tokura Y. Magnetoelectric nature of skyrmions in a chiral magnetic insulator Cu_2OSeO_3. Phys Rev B. 2012;86:060403(R).
[145] Yang JH, Li ZL, Lu XZ, Whangbo M-H, Wei S-H, Gong XG, et al. Strong Dzyaloshinskii-Moriya interaction and origin of ferroelectricity in Cu_2OSeO_3. Phys Rev Lett. 2012;109:107203.
[146] Janson O, Rousochatzakis I, Tsirlin AA, Belesi M, Leonov AA, Rössler UK, et al. The quantum nature of skyrmions and half-skyrmions in Cu_2OSeO_3. Nat Commun. 2014;5:5376.
[147] Ozerov M, Romhanyi J, Melesi M, Berger H, Ansermet J-P, van Den Brink J, et al. Establishing the fundamental magnetic interactions in the chiral skyrmionic Mott insulator Cu_2OSeO_3 by terahertz electron spin resonance. Phys Rev Lett. 2014;113:157205.

[148] Romhanyi J, van Den Brink J, Rousochatzakis I. Entangled tetrahedron ground state and excitations of the magnetoelectric skyrmion material Cu_2OSeO_3. Phys Rev B. 2014;90:140404.R.

[149] Gnezdilov VP, Lamonova KV, Pashkevich Y, Lemmens P, Berger H, Bussy F, et al. Magnetoelectricity in the ferrimagnetic Cu_2OSeO_3: symmetry analysis and Raman scattering study. Low Temp Phys. 2010;36:550.

[150] Miller KH, Xu XS, Berger H, Knowles ES, Arenas DJ, Meisel MW, et al. Magnetodielectric coupling of infrared phonons in single-crystal Cu_2OSeO_3. Phys Rev B. 2010;82:144107.

[151] White JS, Levatic I, Omrani AA, Egetenmeyer N, Prsa K, Zivkovic I, et al. Electric field control of the skyrmion lattice in Cu_2OSeO_3. J Phys: Condens Matter. 2012;24:432201.

[152] Okamura Y, Kagawa F, Seki S, Tokura Y. Transition to and from the skyrmion lattice phase by electric fields in a magnetoelectric compound. Nat Commun. 2016;7:12669.

[153] White JS, Prsa K, Huang P, Omrani AA, Zivkovic I, Bartkowiak M, et al. Electric-field-induced skyrmion distortion and giant lattice rotation in the magnetoelectric insulator Cu_2OSeO_3. Phys Rev Lett. 2014;113:107203.

[154] White JS, Zivkovic I, Kruchkov AJ, Bartkowiak M, Magrez A, Ronnow HM. Electric-field-driven topological phase switching and skyrmion-lattice metastability in magnetoelectric Cu_2OSeO_3. Phys Rev Appl. 2018;10:014021.

[155] Pocha R, Johrendt D, Pöttgen R. Electronic and structural instabilities in GaV_4S_8 and $GaMo_4S_8$. Chem Mater. 2000;12:2882.

[156] Brasen D, Vandenberg JM, Robbins M, Willens RH, Reed WA, Sherwood RC, et al. Magnetic and crystallographic properties of spinels of the type $A_xB_2S_4$ where A = Al, Ga, and B = Mo, V, Cr. J Solid State Chem. 1975;13:298.

[157] Haeuseler H, Reil S, Elitok E. Structural investigations in the system $Ga_{0.5}V_2S_4$–$Ga_{0.5}V_2Se_4$. Int J Inorg Mater. 2001;3:409.

[158] Sahoo Y, Rstogi AK. Evidence of hopping conduction in the V_4-cluster compound GaV_4S_8. J Phys: Condens Matter. 1993;5:5953.

[159] Yadev CS, Nigam AK, Rastogi AK. Thermodynamic properties of ferromagnetic Mott-insulator GaV_4S_8. Physica B. 2008;403:1474.

[160] Nakamura H, Ikeno R, Motoyama G, Kohara T, Kajinami Y, Tabata Y. Low-field multi-step magnetization of GaV_4S_8 single crystal. J Phys: Conf Ser. 2009;145:012077.

[161] Bichler D, Johrendt D. Interplay of Magnetism and Bonding in $GaV_{4-x}Cr_xSe_8$. Chem Mater. 2011;23:3014.

[162] Ehlers D, Stasinopoulos I, Tsurkan V, Krug von Nidda H-A, Fehér T, Leonov A, et al. Skyrmion dynamics under uniaxial anisotropy. Phys Rev B. 2016;94:014406.

[163] Ruff E, Widmann S, Lunkenheimer P, Tsurkan V, Bordács S, Kézsmarki I, et al. Multiferroicity and skyrmions carrying electric polarization in GaV_4S_8. Sci Adv. 2015;10:e1500916.

[164] Zhang JT, Wang JL, Yang XQ, Xia WS, Lu XM, Zhu JS. Magnetic properties and spin-driven ferroelectricity in multiferroic skyrmion host GaV_4S_8. Phys Rev B. 2017;95:085136.

[165] Meunier PG, Bertaud M, Galy J. Cristallochimie du sélénium(+IV). I. VSe_2O_6, une structure à trois chaînes parallé1es $(VO5)_n{}^{6n-}$ indépendantes pontées par des groupements $(Se_2O)^{6+}$. Acta Crystallogr Sect B. 1974;30:2834.

[166] Trombe J-C, Gleizes A, Galy J, Renard J-P, Journaux Y, Verdaguer M. Structure and magnetic properties of vanadyl chins crystal structure of $VOSeO_3$ and comparative magnetic study of $VOSeO_3$ and $VOSe_2O_5$. Nouv J Chim. 1987;11:321.

[167] Kim S-H, Halasyamani PS, Melot BC, Seshadri R, Green MA, Sefat AS, et al. Experimental and computational investigation of the polar ferrimagnet $VOSe_2O_5$. Chem Mater. 2010;22:5047.

[168] Kurumaji T. Unpublished.

[169] Heinze S, von Bergmann K, Menzel M, Brede J, Kubetzka A, Wiesendanger R, et al. Spontaneous atomic-scale magnetic skyrmion lattice in two dimensions. Nat Phys. 2011;7:713.
[170] Romming N, Hanneken C, Menzel M, Bickel JE, Wolter B, von Bergmann K, et al. Writing and deleting single magnetic skyrmions. Science. 2013;341:636.
[171] Shiba H. Dipole-dipole interaction as a possible origin of incommensurate magnetic structures in triangular antiferromagnets. Solid State Commun. 1982;41:511.
[172] Fujita M, Nakanishi T, Machida K. Triple-**Q** vortex-lattice magnetic structures of itinerant-fermion systems on a triangular lattice. Phys Rev B. 1992;45:2190.
[173] Gekht RS. Magnetic states and phase transitions in frustrated triangular-lattice antiferromagnets. Usp Fiz Nauk. 1989;159:261.
[174] Kamiya Y, Batista C. Magnetic vortex crystals in frustrated Mott insulator. Phys Rev X. 2014;4:011023.
[175] Seabra L, Sindzingre P, Momoi T, Shannon N. Novel phases in a square-lattice frustrated ferromagnet: 1/3-magnetization plateau, helicoidal spin liquid, and vortex crystal. Phys Rev B. 2016;93:085132.
[176] Rousochatzakis I, Rössler UK, van Den Brink J, Daghofer M. Kitaev anisotropy induces mesoscopic z_2 vortex crystals in frustrated hexagonal antiferromagnets. Phys Rev B. 2016;93:104417.
[177] Yao X, Dong S. Topological triple-vortex lattice stabilized by mixed frustration in expanded honeycomb Kitaev-Heisenberg model. Sci Rep. 2016;6:26750.
[178] Wang Z, Kamiya Y, Nevidomskyy AH, Batista C. Three-dimensional crystallization of vortex strings in frustrated quantum magnets. Phys Rev Lett. 2015;115:107201.
[179] Shapiro SM, Gurewitz E, Parks RD, Kupferberg LC. Multiple-q magnetic structure in $CeAl_2$. Phys Rev Lett. 1979;43:1748.
[180] Bak P, Lebech B. "Triple-$\vec{q}$ " modulated magnetic structure and critical behavior of Neodymium. Phys Rev Lett. 1978;40:800.
[181] Lassailly Y, Vettier C, Holtzberg F, Burlet P. Evidence of 4-q magnetic structure in TmS. Physica. 1986;136B:391.
[182] Normile PS, Rotter M, Detlefs C, Jensen J, Canfield PC, Blanco JA. Magnetic ordering in $GdNi_2B_2C$ revisited by resonant x-ray scattering: Evidence for the double-q model. Phys Rev B. 2013;88:054413.
[183] Marcus GG, Kim D-J, Tutmaher JA, Rodriguez-Rivera JA, Birk JO, Niederneyer C, et al. Multi-q mesoscale magnetism in $CeAuSb_2$. Phys Rev Lett. 2018;120:097201.
[184] Gao S, Zaharko O, Tsurkan V, Su Y, White JS, Tucker GS, et al. Spiral spin-liquid and the emergence of a vortex-like state in $MnSc_2S_4$. Nat Phys. 2017;13:157.
[185] Ishiwata S, Nakajima T, Kim J-H, Inosov DS, Kanazawa N, White JS, et al. Emergent topological spin structures in a centrosymmetric cubic perovskite. arXiv: 1806.02309.
[186] Okubo T, Chung S, Kawamura H. Multiple-q states and the skyrmion lattice of the triangular-lattice Heisenberg antiferromagnet under magnetic fields. Phys Rev Lett. 2012;108:017206.
[187] Leonov AO, Mostovoy M. Multiply periodic states and isolated skyrmions in an anisotropic frustrated magnet. Nat Commun. 2015;6:8275.
[188] Lin S-Z, Hayami S. Ginzburg-Landau theory for skyrmions in inversion-symmetric magnets with competing interactions. Phys Rev. 2016;93:064430.
[189] Hayami S, Lin S-Z, Batista C. Bubble and skyrmion crystals in frustrated magnets with easy-axis anisotropy. Phys Rev B. 2016;93:184413.
[190] Hayami S, Ozawa R, Motome Y. Effective bilinear-biquadratic model for noncoplanar ordering in itinerant magnets. Phys Rev B. 2017;95:224424.

[191] Zhang X, Xia J, Zhou Y, Liu X, Zhang H, Ezawa M. Skyrmion dynamics in a frustrated ferromagnetic film and current-induced helicity locking-unlocking transition. Nat Commun. 2017;8:1717.
[192] Batista CD, Lin S-Z, Hayami S, Kamiya Y. Frustration and chiral orderings in correlated electron systems. Rep Prog Phys. 2016;79:084504.
[193] Ozawa R, Hayami S, Motome Y. Zero-field skyrmions with a high topological number in itinerant magnets. Phys Rev Lett. 2017;118:147205.
[194] Kurumaji T, Nakajima T, Hirschberger M, Kikkawa A, Yamasaki Y, Sagayama H, et al. Skyrmion lattice with a giant topological Hall effect in a frustrated triangular-lattice magnet. arXiv: 1805.10719.
[195] Kotsanidis PA, Yakinthos JK, Gamari-Seale E. Magnetic properties of the ternary rare earth silicides R_2PdSi_3 (R = Pr, Nd, Gd, Tb, Dy, Ho, Er, Tm and Y). J Magn Magn Mater. 1990;87:199.
[196] Szytula A, Hofmann M, Penc B, Slaski M, Majumdar S, Sampathkumaran EV, et al. Magnetic behaviour of R_2PdSi_3 compounds with R = Ce, Nd, Tb–Er. J Magn Magn Mater. 1999;202:365.
[197] Saha SR, Sugawara H, Matsuda TD, Sato H, Malik R, Sampathkumaran EV. Magnetic anisotropy, first-order-like metamagnetic transitions, and large negative magnetoresistance in single-crystal Gd_2PdSi_3. Phys Rev B. 1999;60:12162.
[198] Hirschberger M, Nakajima T, Gao S, Peng L, Kikkawa A, Kurumaji T, et al. Skyrmion phase and competing magnetic orders on a breathing kagome lattice. arXiv: 1812.02553.

Eric Bousquet and Andrés Cano

5 Non-collinear magnetism & multiferroicity: the perovskite case

Abstract: The most important types of non-collinear magnetic orders that are realized in simple perovskite oxides are outlined in relation to multiferroicity. These orders are classified and rationalized in terms of a mimimal spin Hamiltonian, based on which the notion of spin-driven ferroelectricity is illustrated. These concepts find direct application in reference materials such as $BiFeO_3$, $GdFeO_3$ and $TbMnO_3$ whose multiferroic properties are briefly reviewed.

Keywords: Multiferroics, perovskites, magnetism

5.1 Introduction

Perovskite oxides compose a broad class of versatile materials displaying a number of interesting properties. Many systems with the general chemical formula ABO_3 belong to this class. In their ideal cubic case, the A atom is at the cube corners while the B one is surrounded by oxygen octahedra at the body-center positions [see Figure 5.1(a)]. This structure is prone to different distortions that produce several lower-symmetry variants, notably tetragonal and orthorhombic phases [see Figure 5.1(b)–(d)]. Among these distortions, the spontaneous off-centering of the B and/or A atoms with respect to the O ones implies a polar distortion that further produces ferroelectricity as in the prototypical ferroelectrics $BaTiO_3$ and $BiFeO_3$ [see Figure 5.1(b)]. Interestingly, these materials can host magnetic atoms either at A, B or both positions.[1] The interaction of corresponding spins can be of different nature depending on the specific perovskite under consideration, which enables the emergence of a variety of long-range magnetic orders in these systems. These include non-collinear structures that coexist with, are affected by, and even promote structural changes as those involved in ferroelectricity. The perovskite setup thus provides an excellent playground for the study of multiferroicity.

1 Typically, the magnetic ABO_3 perovskites can have two types of localized spins: the spin associated to the d electrons of the transition-metal B ion and the spin coming from the f electrons of the A atom whenever this atom is a magnetic rare earth.

This article has previously been published in the journal Physical Sciences Reviews. Please cite as: Bousquet, E., Cano, A. Non-collinear magnetism & multiferroicity: the perovskite case *Physical Sciences Reviews* [Online] 2021, 6. DOI: 10.1515/psr-2019-0071

https://doi.org/10.1515/9783110582130-005

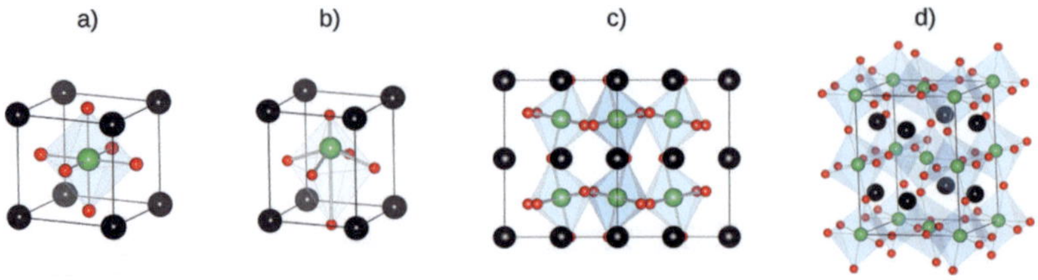

Figure 5.1: Unit cell of the ideal cubic ABO_3 perovskite (a) and its most frequent tetragonal [(b) and (c)] and orthorhombic (d) variants. A, B and O-site atoms are indicated in black, green and red respectively.

One feature that has motivated a considerable amount of research on these systems is that they generally materialize insulators that often develop antiferromagnetic (AFM) orders with weak-ferromagnetic (wFM) components. Thanks to this circumstance, the general incompatibility between ferroelectricity (which requires an insulating state) and ferromagnetism (which most often appears in metals) can potentially be bypassed. Thus, perovskites of this type have historically driven the quest of room-temperature multiferroics in their original sense – that is, as ferroelectric ferromagnets. Another interesting feature is that some of them can develop special magnetic orders that break inversion symmetry, thus further inducing ferroelectricity. This type of ferroelectricity is unconventional and has also attracted a lot of research attention since it implies, by construction, strong magnetoelectric couplings – at the qualitative level at least. These features are very appealing, especially in the field of spintronics, where they keep inspiring new device concepts.

In the following we provide an overview of the main non-collinear magnetic orders connected to multiferroicity that emerge in perovskite oxides. This overview is restricted to simple perovskites for which the ideal $Pm\bar{3}m$ cubic structure serves as a high-symmetry reference. In Section 5.2, we classify these magnetic orders and introduce a minimal theoretical framework that allows to rationalize their emergence as well as spin-driven ferroelectricity. Section 5.3 is focused on prototypical type-I multiferroics where non-collinear magnetism and ferroelectricity emerge separately. Then, in Section 5.4, we review some important perovskites materializing spin-driven ferroelectrics via both spin-canted and spin-spiral orders. Section 5.5 is devoted to the conlcusions. This article complements excellent recent reviews on multiferroic materials [1–14]. We refer the more interested reader to these publications and, very importantly, to the original references therein.

5.2 Preliminaries: a theory primer

5.2.1 Classifying spin orders in magnetic perovskites

Figure 5.2 illustrates the elementary collinear orders that are most relevant for the insulating magnetic perovskites. In the literature, these orders are customarily labeled as

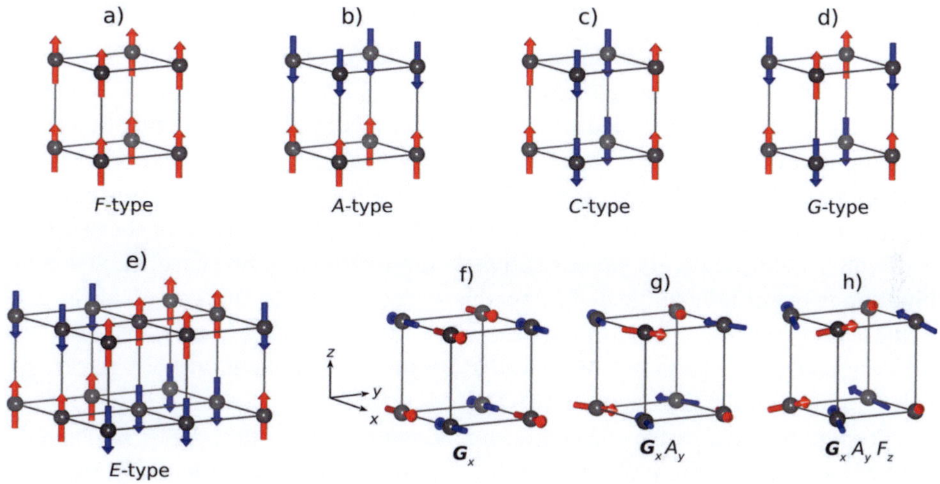

Figure 5.2: Schematic view of the *F*-type (a), *A*-type (b), *C*-type (c), *G*-type, and *E*-type (e) collinear orders that are relevant for the insulating ABO_3 perovskites. These orders refer to both *B*-site and *A*-site spins, which point along the *z* axis for the sake of illustration. Particular combinations of these orders giving rise to non-collinear configurations is illustrated in (f), (g) and (h).

- *F*-type, when all the spins point in the same direction (FM ordering),
- *A*-type, when the spins point in opposite directions in consecutive planes (AFM order of FM planes),
- *C*-type, when the spins point in opposite directions in consecutive lines (AFM order of FM lines),
- *G*-type, when nearest-neighboring spins point in opposite directions ("full" AFM ordering).

These configurations correspond to different solutions of a minimal spin Hamiltonian [see eq. (5.6) below] in the simplest case of symmetric exchange interactions between nearest neighbors. In terms of the cubic lattice with one spin per unit cell, these orders are associated to the propagation vectors $\mathbf{q} = (0, 0, 0)$, (0,0,1/2), (0,1/2,1/2), and $(1/2, 1/2, 1/2)$ respectively. Alternatively, these configurations can also be defined in terms of the relative orientation of four magnetic sublattices (one per magnetic *B* atom, for example, in the corresponding magnetic unit cell). Thus, defining the appropriate 4-spin cluster, they correspond to non-zero values of the order parameters:

$$\mathbf{F} = \mathbf{S}_1 + \mathbf{S}_2 + \mathbf{S}_3 + \mathbf{S}_4, \tag{5.1}$$

$$\mathbf{A} = \mathbf{S}_1 - \mathbf{S}_2 - \mathbf{S}_3 + \mathbf{S}_4, \tag{5.2}$$

$$\mathbf{C} = \mathbf{S}_1 + \mathbf{S}_2 - \mathbf{S}_3 - \mathbf{S}_4, \tag{5.3}$$

$$\mathbf{G} = \mathbf{S}_1 - \mathbf{S}_2 + \mathbf{S}_3 - \mathbf{S}_4. \tag{5.4}$$

Taking explicitly into account the spin direction, this gives a 4×3 basis for describing the main collinear orders in perovskites: F_x, F_y, F_z, A_x, A_y, A_z, C_x, C_y, C_z, G_x, G_y and G_z.

These elementary configurations are useful to describe some realistic orders that appear in magnetic perovskites. This is conveniently done in terms extended Bertaut's notation explained in [2], since generally one of the elementary orders dominate the overall structure. The reason is that the combination of these orders is driven by comparatively weak relativistic interactions. In any case, such a combination is eventually subjected to symmetry requirements and can yield, in particular, non-collinear structures as the ones illustrated in Figure 5.2(g) and (h). Additionally, in the presence of polar distortions, these elementary orders can develop spiral modulations as in $BiFeO_3$ [15].

From purely symmetry considerations, it can be shown that these orders yield neither, a spontaneous polarization nor a linear magnetoelectric effect in ABO_3 perovskites if they involve the B-site spins only [16–18]. However, there exists some A-site orderings compatible with the linear magnetoelectric effect [17–20]. Namely, the G_x, A_y, A_z, G_z, A_x and G_y orders. In addition, in the presence of a pre-existing B-spin order, the emergence of A-spin orders of this type can additionally induce an electric polarization, as in the ferrites discussed in Section 4.1.1 below.

Another important type of collinear order is the so-called E-type AFM order, in which spins are arranged as depicted in Figure 5.2(e). This type of collinear order breaks inversion symmetry by itself and was predicted to induce an electric polarization in [21]. This prediction was subsequently confirmed experimentally in orthorhombic $HoMnO_3$ and $YMnO_3$ [22]. In these systems, E-AFM order emerges due to specific interactions beyond the simplest exchange between nearest neighbors, notably the biquadratic interaction obtained from spin-phonon coupling [23–25].

Additionally, the orthorhombic $RMnO_3$ manganites characteristically display non-collinear spiral orders. This is due to a specific competition between isotropic exchange interactions of different nature — FM vs AFM — that yields magnetic frustration in these systems. In the simplest cycloidal case (see Figure 5.3), the spin spiral can be parametrized as

$$\mathbf{S}_i = S\cos(\mathbf{Q}\cdot\mathbf{r}_i)\widehat{\mathbf{x}} + S\sin(\mathbf{Q}\cdot\mathbf{r}_i)\widehat{\mathbf{y}}, \tag{5.5}$$

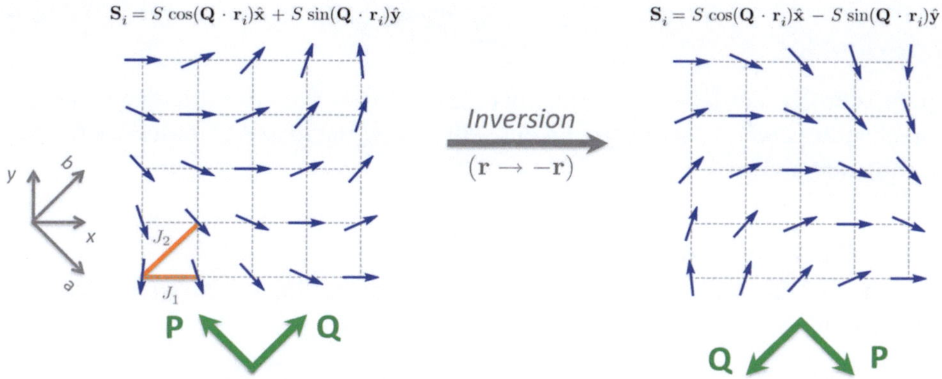

Figure 5.3: Simple spin-spiral with propagation vector Q and its transformation under inversion symmetry (P indicates the polar axis of the spiral). The blue arrows can be understood as the Mn spins in the orthogonal rare-earth manganites $RMnO_3$ for example, with crystal axes as indicated in the figure. J_1 and J_2 indicate nearest-next-nearest neighbor spins.

where the propagation vector **Q**, and hence the period of the spiral, is essentially determined by the exchange couplings. As a rule, such a period is comparable to the interatomic distances yet incommensurate with the lattice, which is an important difference with respect to the the previous orders. Similarly to the *E*-AFM case, the emergence of spiral order does not require the breaking of inversion symmetry in first place. This breaking, and the subsequent electric polarization, is, on the contrary, the consequence (rather than the cause) of the appearance of the spiral. In this sense, spiral multiferroics are frequently called improper ferroelectrics. However, if the spiral phase is preceded by a sinusoidal collinear order, a more appropriate term would be "assisted" pseudo-proper ferroelectrics from the point of view of symmetry [26, 27].

5.2.2 Generalized Heisenberg model

The emergence of the above orders can be conveniently rationalized in terms of the generalized spin Hamiltonian[2]

2 We use dimensionless spins so that the magnetic moment associated to the spin can be written as $\mu_s = g\mu_B\mathbf{S}$. Here g is the spin g-factor and μ_B the Bohr magneton (the ratio between the spin magnetic moment to its angular momentum $g\mu_B/\hbar$ is the so-called the gyromagnetic ratio). Correspondingly, the coupling to an external field $\mathbf{H}$ is given by the Zeeman coupling $-\mu_s\cdot\mathbf{H}$. Further, the total magnetization due to the spin magnetic moments is $\mathbf{M} = g\mu_B\sum_i\mathbf{S}_i$ within the so-called L-S or Russell–Saunders coupling approximation, where the sum runs over all the different spins $\mathbf{S}_i$ of the system.

$$H = -\sum_{i,j} J_{ij}^{\alpha\beta} S_i^{\alpha} S_j^{\beta}. \tag{5.6}$$

Here Latin indices label the positions of the spins in the crystal lattice while the Greek indices refer to spin components. Physically, the tensor $J_{ij}^{\alpha\beta}$ describes the following interactions.

5.2.2.1 Symmetric exchange interaction

The trace of the symmetric part yields a scalar representing the isotropic exchange interaction

$$H_{\mathrm{ex}} = -\sum_{i,j} J_{ij} \mathbf{S}_i \cdot \mathbf{S}_j, \tag{5.7}$$

where $J_{ij} = (1/3)J_{ij}^{\alpha\alpha}$. This is the primary quantum mechanical interaction between localized spins, which traces back to the overlap of the electron wavefunctions subjected to Pauli's exclusion principle (Heisenberg exchange). Whenever this overlap involves non-magnetic ligand atoms (oxygen, fluorine, etc.), the interaction is however called superexchange. This is in fact the case in the ABO_3 perovksites.

In our notation $J > 0$ favors the parallel alignment of spins and hence corresponds to a FM interaction, while $J < 0$ favors the antiparallel configuration and hence is AFM. Note that at this level the quantization axis (easy axis) is not determined yet.[3]

In most of the ABO_3 perovskites the exchange interaction between B-site spins restricts to nearest neighbors, and is AFM with a magnetization energy of the order of meV. In the important case of the manganites ($B =$ Mn), however, such an interaction is FM and much weaker.[4] This makes it necessary to take into account next-nearest-neighbor interactions, which turn out to be AFM. The competition of nearest- and next-nearest interactions in these systems yields a magnetic frustration that promotes the emergence of spiral orders as mentioned above.

5.2.2.2 Antisymmetric interactions: magnetic anisotropy

The anisotropy of the magnetic properties of a crystal is due to comparatively weak spin-orbit (relativistic) interactions. In the magnetic perovskites several anisotropic terms can be at play, and their characteristic energy can be as large as ~0.1 meV.

---

3 Depending on the crystal symmetry, the exchange interactions can be different along different directions (i. e. $J_{ij} = J(\mathbf{r}_i, \mathbf{r}_j) \neq J(|\mathbf{r}_i - \mathbf{r}_j|)$). However, they remain isotropic in the sense that they only depend on the relative orientation of the spins (but not on their orientation with respect to the lattice) unless relativistic effects are taken into account.

4 This results from the competition between t_{2g}- vs. e_g-orbital-mediated contributions to the exchange interaction between nearest neighbors.

Single-ion anisotropy

In terms of the Hamiltonian eq. (5.6), the so-called single-ion anisotropy (SIA) is associated to the traceless part of J_{ii}. This (local) interaction defines a dependence of the magnetic energy on the relative direction of the spin with respect to the crystal axes.[5] This dependence traces back to the interplay between the orbital state of the magnetic ion and the electric field produced by the surrounding charges, *i. e.* the so-called crystal field, which is further transferred to the spin via spin-orbit coupling.

In cubic perovskites, the lowest-order SIA can be written as

$$H_{sia} = \sum_i K_2 (S_{x,i}^2 S_{y,i}^2 + S_{y,i}^2 S_{z,i}^2 + S_{x,i}^2 S_{z,i}^2). \tag{5.8}$$

Thus, the spin easy-axes are along the [100], [010] and [001] directions for $K_2 > 0$, and along the [111] directions for $K_2 < 0$.

If the local environment of the spins becomes uniaxial, then the SIA transforms into

$$H_{sia} = -\sum_i K_i (\mathbf{S}_i \cdot \hat{\mathbf{n}}_i)^2, \tag{5.9}$$

where the unit vector $\hat{\mathbf{n}}_i$ represents the easy axis for the i-th spin. This axis can be different for different spins. Alternatively, this type of anisotropy can be written as

$$H_{sia} = -\sum_i \left[K_i S_{z,i}^2 + K'_i (S_{x,i}^2 - S_{y,i}^2)\right] \tag{5.10}$$

in terms of the spin components along the (local) principal axes. If $K > 0$, the anisotropy is said to be of the easy axis type while it is of the easy plane type if $K < 0$. In the latter case, the K' term determines the local direction of the spins within the xy plane.

The above already defines an interesting magnetostructural coupling that is relevant for some magnetic perovskites. In fact, if a cubic perovskite undergoes an antiferrodistortive (AFD) distortion, then the resulting SIA eq. (5.10) will depend on the relative orientation between the spins and the axial vector associated to such a distortion. In $BiFeO_3$, for example, the spins tend to be parallel to the AFD axial vector (easy-axis SIA) [28]. Further, in the presence of a polar distortion, another SIA contributions can be generally expected. In $BiFeO_3$ in particular, the polar distortion tends to orient the spins along the polar axis, which produces an extra easy-axis SIA that is in competition with the previous one [28]. In cases like this, the

5 In the relativistic case, there appears another anisotropic correction to the exchange interaction in the form of a traceless symmetric part that is non local. This is characterized by five different parameters and, when it comes to its symmetry in spin space, it can be seen as a short-range version of the dipole-dipole interaction. This correction, however, seems to play a comparatively unimportant role in the multiferroic perovskites.

effective SIA eventually depends on the relative strength of the AFD vs ferroelectric distortions.

Dzyaloshinskii–Moriya interaction

The antisymmetric part of $J_{ij}^{\alpha\beta}$ in eq. (5.6) is associated to the so-called Dzyaloshinskii-Moriya (DM) interaction (or antisymmetric anisotropic exchange). This interaction is customarily written as

$$H_{DM} = -\sum_{i,j} \mathbf{D}_{ij} \cdot (\mathbf{S}_i \times \mathbf{S}_j). \tag{5.11}$$

In terms of the $J_{ij}^{\alpha\beta}$ tensor, the DM vector is defined as $D_{ij}^{\gamma}\epsilon^{\gamma\alpha\beta} = \frac{1}{2}\left(J_{ij}^{\alpha\beta} - J_{ji}^{\beta\alpha}\right)$. This interaction emerges from the interplay between broken inversion symmetry and spin-orbit coupling [29, 30]. The DM vector contains at most three independent parameters, and is constrained by symmetry as discussed in Moriya's original paper [30]. In the case of perovskites, where the exchange interactions are mediated by the oxygen atoms (superexchange), $\mathbf{D}_{ij} \propto \mathbf{r}_i \times \mathbf{r}_j$, where $\mathbf{r}_i$ denotes the position of the spins with respect to the oxygen atom. In any case, the DM interaction favors the perpendicular alignment of spins. Thus, it is in competition with the isotropic exchange interaction between nearest-neighboring spins and represents an important source of magnetic frustration. In fact, the DM interaction explains some of the most important non-collinear features (spin canting, weak FM, weak AFM, . . .) observed in the multiferroic perovskites as discussed below.

5.2.3 Basic mechanisms for multiferroicity

Some of the magnetic orders that emerge in the magnetic perovskites break inversion symmetry and therefore are expected to induce an electric polarization. This can be understood in terms of two basic mechanisms of magnetostriction.

On the one hand, the electric polarization can emerge via the dependence of the symmetric exchange interactions on the relative displacement of the atoms (i. e. symmetric magnetostriction). That is, via the changes in overlap of the wavefunctions that are associated to such a displacement. In the ABO_3 perovskites, these interactions are mediated by the oxygen atoms (superexchange) so that $J = J(\mathbf{r}_i, \mathbf{r}_j; \mathbf{r}_{ij}^{\circ})$, where $\mathbf{r}_{i(j)}$ represents the position of the magnetic atoms and $\mathbf{r}_{ij}^{\circ}$ of the corresponding oxygen. These positions can be expressed as $\mathbf{r} = \mathbf{r}^{(0)} + \delta\mathbf{r}$, where $\delta\mathbf{r}$ accounts for the corresponding displacement. Thus, the aforementioned dependence can formally be written as

$$J(\mathbf{r}_i, \mathbf{r}_j; \mathbf{r}_{ij}^{\circ}) = J_{ij} + \mathbf{J}_{ij}^{(i)} \cdot \delta\mathbf{r}_i + \mathbf{J}_{ij}^{(j)} \cdot \delta\mathbf{r}_j + \mathbf{J}_{ij}^{\circ} \cdot \delta\mathbf{r}_{ij}^{\circ} + \ldots \tag{5.12}$$

Here $J_{ij} = J(\mathbf{r}_i^{(0)}, \mathbf{r}_j^{(0)}; \mathbf{r}_{ij}^{\circ(0)})$ and the form of vectors $\mathbf{J}_{ij}^{\alpha}$ can be deduced from symmetry considerations. In the case of polar displacements: $\mathbf{J}_{ij}^{(i)} \cdot \delta\mathbf{r}_i + \mathbf{J}_{ij}^{(j)} \cdot \delta\mathbf{r}_j + \mathbf{J}_{ij}^{\circ} \cdot \delta\mathbf{r}_{ij}^{\circ} = \mathbf{J}_{ij}{}' \cdot \mathbf{P}$. Thus, whenever $\mathbf{J}'_{ij} \neq 0$, the order of the spins can induce an electric polarization $\mathbf{P}$ since the total energy will be minimized with

$$\mathbf{P} \propto \sum_{ij} \mathbf{J}'_{ij}(\mathbf{S}_i \cdot \mathbf{S}_j). \tag{5.13}$$

This mechanism is rather general, and in fact can be triggered by purely electronic effects. The non-collinearity of the magnetic structure is actually detrimental for this mechanism. However, it works when the system has two species of spins, which is the case of perovskites like $GdFeO_3$ or $DyFeO_3$. In the case of the orthorhombic rare-earth $R$$MnO_3$ manganites with spiral order, this mechanism can be operative due to the extra features generated by the staggered a-axis component of the DM vectors. Specifically, this mechanism is ineffective if the spiral is in the bc plane [24, 31] but becomes operative for ab-spirals [24]. In the former case, the AFM ordering along the c-axis produces the cancellation of the contributions of the different ab planes.

On the other hand, the same reasonings can be applied to the DM interaction. In general, this interaction also depends on the atomic displacements (i. e. antisymmetric magnetostriction):

$$\mathbf{D}(\mathbf{r}_i, \mathbf{r}_j) = \mathbf{D}_{ij} + \hat{D}_{ij}^{(i)} \delta\mathbf{r}_i + \hat{D}_{ij}^{(j)} \delta\mathbf{r}_j + \hat{D}_{ij}^{\circ} \delta\mathbf{r}_{ij}^{\circ} + \ldots \tag{5.14}$$

and therefore can produce a spontaneous electric polarization out of the order of the spins

$$\mathbf{P} \propto \sum_{ij} \hat{D}_{ij}{}'(\mathbf{S}_i \times \mathbf{S}_j). \tag{5.15}$$

This is the so-called inverse DM mechanism [31]. It also has a purely electronic version, in which the electric polarization can be associated to the spin current generated by the vector chirality $\mathbf{S}_i \times \mathbf{S}_j$ of non-collinear spins. In this case, it is called the spin-current mechanism [32]. More phenomenologically, this type of polarization can be seen as due to coupling terms of the type $\mathbf{P} \cdot [\mathbf{M}(\nabla \cdot \mathbf{M}) - (\mathbf{M} \cdot \nabla)\mathbf{M})]$ which, in contrast to the symmetric magnetostriction, is always allowed by symmetry [33–35]. The specific form of these couplings depends on the specific symmetry of the system under consideration [36, 37]. In case of the orthorhombic $R$$MnO_3$ manganites, in particular, the polarization induced via antisymmetric magnetostriction can be expected as $\mathbf{P} \propto \sum_{ij} \hat{\mathbf{e}}_{ij} \times (\mathbf{S}_i \times \mathbf{S}_j)$, where $\hat{\mathbf{e}}_{ij}$ is the unit vector connecting the corresponding spins.

Interestingly, this type of couplings link the spins to the electric polarization also at the dynamical level, giving rise to peculiar low-energy excitations called electromagnons (see e. g. [12, 38–42]). This generic phenomenon further leads to the notion of dynamic multiferroicity [43], and become essential to understand the specific quantum critical features that emerge in multiferroic materials [44].

5.3 Non-collinear orders in type-I multiferroic perovskites ($T_N \neq T_{FE}$)

In this section we discuss the non-collinearity that appears in "pre-existing" magnetic orders of the elementary type due to polar distortions of the perovskite or oxygen-octahedra rotations. In this case, the emergence of magnetic and structural orders can be nominally decoupled (owing to the fact that Néel and ferroelectric transition temperatures are different, $T_N \neq T_{FE}$). This scenario is now called type-I multiferroicity [45] and is realized in some magnetic perovskites whose structure is compatible with wFM.

5.3.1 $BiFeO_3$

The $R\bar{3}c$ structure implies AFD distortions of the reference cubic phase and is comparatively rare among the magnetic perovskites. However, this is precisely the initial structure of the most studied multiferroic $BiFeO_3$.

The Fe spins in $BiFeO_3$ have a intrinsic tendency towards *G*-type AFM according to their dominant exchange interactions. However, below T_{FE}~1100 K, $BiFeO_3$ displays a spontaneous electric polarization along the [111] direction that lowers the symmetry to $R3c$. This switches on a DM interaction such that the *G*-AFM order that emerges at $T_N \sim 640$ K undergoes a spiral incommensurate modulation with a rather long period of ~62 nm [15]. The propagation vector is along the [$1\bar{1}0$], [$01\bar{1}$] or [$\bar{1}01$]) directions, and the spins lie in the plane formed by this vector and the [111] direction of the electric polarization (so that the spiral is of cycloidal type, see Figure 5.4). In addition, there is a second DM interaction that produces a canting of the spins out of the spiral plane that changes sign from site to site, which further gives rise to specific features in the spin-wave excitation spectrum [46–51].

Interestingly, the nominal *G*-AFM order could also be perpendicular to the [111] direction. In that case, spins would undergo a wFM canting due to SIA proportional to arctan(D/J), with J and D being the strength of exchange and DM interactions along the [111] direction respectively [28, 53, 54]. This wFM was originally observed in $BiFeO_3$ thin films below <100 nm, and is further accompanied with a magnetoelectric response of ~10 ps/m [55]. Additionally, the application of a magnetic field can induce a magnetoelectric phase closely related to the *G*-AFM one with wFM [56]. This wFM phase, however, is suppressed whenever the spontaneous electric polarization dominates over the AFD distortions.

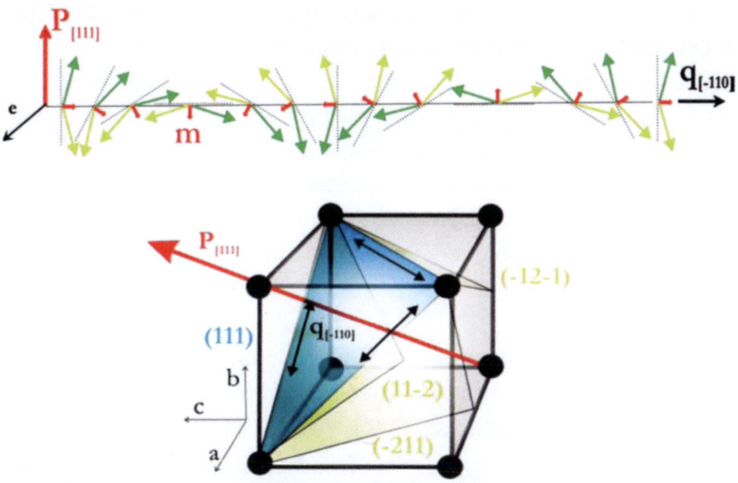

Figure 5.4: Schematic view of the cycloidal modulation of the *G*-AFM order in $BiFeO_3$ (green arrows, with a small canting indicated in red) (from [52]).

The possibility of controlling the overall magnetic order in $BiFeO_3$ using epitaxial strain was nicely demonstrated in [57]. Tensile epitaxial strain from 0% to 0.5% switches the propagation vector of the cycloid towards the [110] direction, beyond which the cycloid disappears and the wFM is released along the $[1\bar{1}0]$ direction. Compressive strain produces similar suppression of the cycloid that eventually gives rise to a wFM moment away from the [111] and $[1\bar{1}0]$ directions.

5.3.2 *Pbnm* perovskites

The *Pbnm* structure is probably the most common structure among the magnetic perovskites [58]. In this space group, all the elementary collinear orders can develop a spin canting perpendicular to the easy axis, so that wFM can be potentially realized [59, 60]. However, *Pbnm* is centrosymmetric and therefore, by definition, such a wFM is not directly compatible with a spontaneous electric polarization in this particular space-group symmetry. Thus, promoting type-I multiferroicity in these systems requires an electric-polarization engineering. Two main routes have been explored to achieve this, via epitaxial strain and via superlattices.

The first-principles investigation of $CaMnO_3$ reported in [61] illustrates the first strategy. The magnetic ground state of unstrained $CaMnO_3$ is $\mathbf{G}_x A_y F_z$, and hence contains a wFM component along the z direction. For tensile epitaxial strain, it is found a first critical strain at which there is a spin-flop to a $F_x C_y \mathbf{G}_z$ configuration so that the wFM moment flips from z to the x axis. At even higher strain, a ferroelectric transition is found due to the condensation of a polar soft mode that becomes unstable under these high-strain conditions [61, 62]. This transition, however, preserves the $F_xC_y\mathbf{G}_z$ state and hence enables a non-zero magnetoelectric effect that diverges near the structural transition. If the strain increases even further, then another magnetic transition is found from $F_xC_y\mathbf{G}_z$ to $\mathbf{A}_xG_yC_z$ so that the wFM (and the linear magnetoelectric effect) is eventually lost. The magnetoelectric coupling of strained $CaMnO_3$ has been explained in terms of a pentalinear coupling between octahedra rotations, polarization and the magnetic orders (AFM and FM), by means of which the electric-field control of the magnetization magnitude and sign can be described [63]. This mechanism should be valid for all *Pbnm* perovskites under similar conditions, which has not been explored extensively [64].

The second strategy is devised to exploit the "hybrid" improper ferroelectricity that can be engineered in $ABX_3/A'BX_3$ superlattices via uncompensated antipolar A and A' displacements [65–68]. In this way, since the wFM canting is directly linked to the out-of-phase AFD distortions (see Section 2.2.2), an electric-field switchable magnetization can in principle be obtained as well. The polarization reversal in this case mainly involves the in-phase AFD rotation with an energy barrier that can be tuned by means of epitaxial strain or chemical composition [69, 70], even if the actual mechanism can be more complex in some cases [71].

5.4 Non-collinear orders in type-II multiferroic perovskites ($T_N = T_{FE}$)

Spin-driven ferroelectricity is fundamentally different from conventional $BaTiO_3$-like ferroelecticity and defines the so-called type-II multiferroics (where $T_N = T_{FE}$) [45]. In the following, we focus on orthorhombic *Pbnm* perovskites where this special type of ferroelectricity is realized. This includes spin-canted structures with wFM components and spin-spiral AFM orders. The former require the simultaneous order of A- and B-site spins [17, 18] and happens mainly via symmetric exchange striction [72]. The latter, in contrast, is possible via B-site spins only and is mainly due to the antisymmetric magnetostriction.

5.4.1 wFM and ferroelectricity

5.4.1.1 Orthorhombic rare-earth ferrites $RFeO_3$

The series of orthorhombic rare-earth ferrites provides interesting magnetic states, including wFM phases [2]. $GdFeO_3$ and $DyFeO_3$, in addition, exemplify the phenomenon of spin-induced ferroelectricity obtained out of two magnetic species via symmetric magnetostriction.

$GdFeO_3$

In $GdFeO_3$, the Fe-spins first develop a $\mathbf{G}_x A_y F_z$ order at $T_N^{Fe} = 661\,\text{K}$ [73]. Then, at a much lower temperature ($T_N^{Gd} = 2.5\,\text{K}$), the Gd-spins also order according to a $G_x A_y$ pattern [74]. Thus, the overall order breaks both time reversal and space inversion symmetries and gives rise to multiferroicity (see Figure 5.6). The measured magnetization is about 0.37 μ_B/f.u. and the electric polarization is $\sim 0.12\,\mu\,\text{C/cm}^2$ at 2 K [74]. The latter is relatively large compared to other spin-induced ferroelectrics, as expected from the symmetric magnetostriction mechanism that operates in this case.

Figure 5.5 shows that such an electric polarization is extremely sensitive to the external magnetic field, which makes $GdFeO_3$ a strong magnetoelectric. Beyond a

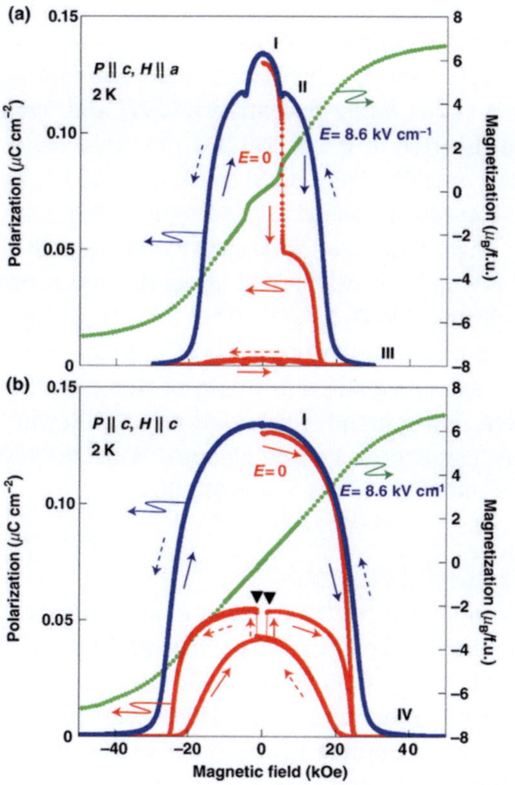

Figure 5.5: Electric polarization and magnetization measured in $GdFeO_3$ at 2 K as a function of the magnetic field (from [74]).

critical magnetic field, the polarization is completely suppressed due to a spin reorientation that changes the type of the underlying magnetic order. Changes of this type also manifest in variations in the effective internal magnetic field [75].

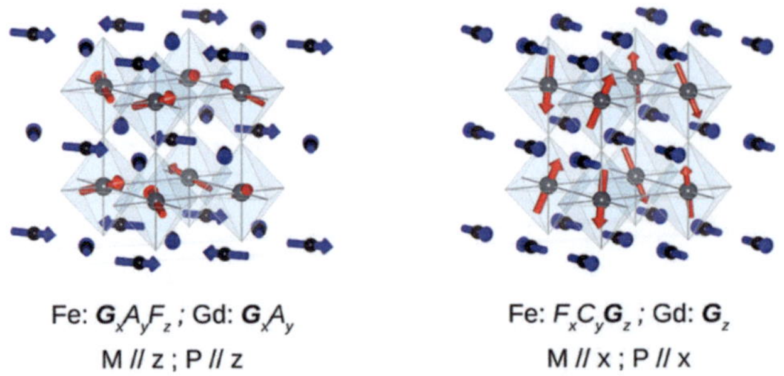

Figure 5.6: Non-collinear orders that can induce an electric polarization as in $RFeO_3$.

$DyFeO_3$

$DyFeO_3$ displays a similar but slightly more complex behavior. In this case, the initial structure of Fe-spins is $\mathbf{G}_xA_yF_z$ ($T_N^{Fe} \simeq 645$ K). This structure abruptly transforms into $A_x\mathbf{G}_yC_z$ at 35–50 K in the so-called Morin transition [76, 77] and, very recently, yet another phase has been reported at 0.5 K [78]. On the other hand, the Dy-spins develop a G_xA_y order at $T_N^{Dy} \simeq 4$K. The combined $A_x\mathbf{G}_yC_z$ and G_xA_y orders of the Fe and Dy spins do not induce an electric polarization. However, they enable a linear magnetoelectric effect that was the largest ever reported in single crystal at that time [79]. By applying a magnetic field along the c-axis the $A_x\mathbf{G}_yC_z$ structure of the Fe spins transforms into $\mathbf{G}_xA_yF_z$ at $H_c^{Fe} = 2.4$ T [80]. Then the overall order becomes equivalent to that in $GdFeO_3$ and also induces a spontaneous electric polarization [79]. The exchange-striction origin of this polarization was subsequently confirmed from first-principles calculations [81]. Beyond that, the spin-reorientation transition temperature can be changed substantially in $DyFeO_3$ thin films by means of epitaxial strain (even below 5K) [82].

5.4.1.2 Orthorhombic rare-earth chromites and nickelates

The rare-earth chromites $RCrO_3$ have magnetic properties similar to the ferrites [2]. Some systems in this series display a relaxor-like behavior in their paramagnetic phases that, however, needs to be better understood [16, 83–85]. Also, the ferroelectric polarization reported in some cases requires further clarification [83, 86, 87]. The $RNiO_3$ series, in its turn, displays intriguing metal-insulator transitions with interplay between phonons, orbital, charge and spin orders [88, 89] that can in principle be manipulated [90]. In addition, the non-collinear R-spin order in these nickelates are believed to promote interesting interplays that can potentially generate spin-induced ferroelectricity with relatively large electric polarizations [91–93]. These systems, however, remain comparatively unexplored as potential multiferroics [94–97].

5.4.2 Spiral order and ferroelectricity

5.4.2.1 Orthorhombic rare-earth manganites $RMnO_3$

The renewed interest in multiferroics was largely triggered by Kimura's seminal paper reporting on a new type of ferroelectricity of magnetic origin in $TbMnO_3$ [98]. This motivated a number of subsequent studies that revealed the family of orthorhombic rare-earth manganites $RMnO_3$ as a "model family" of spin-driven ferroelectrics (due to both, collinear and non-collinear spiral orders).

The room-temperature structure of these manganites corresponds to the *Pbnm* space group commonly observed in magnetic perovskites [see Figure 5.1(d)]. At low temperatures, Mn spins undergo different type of orderings depending on the rare-earth *R* atom (see Figure 5.7). The *R* spins tend to follow such Mn-spins and, in addition, often undergo their own additional ordering at a lower temperature. One particular feature of these systems is that the Mn nearest-neighbor exchange interactions are relatively weak and have FM character. Such a FM tendency is further in competition with next-nearest-neighbor AFM interactions. This generates magnetic frustration and, among other possible ground states, promotes the emergence of

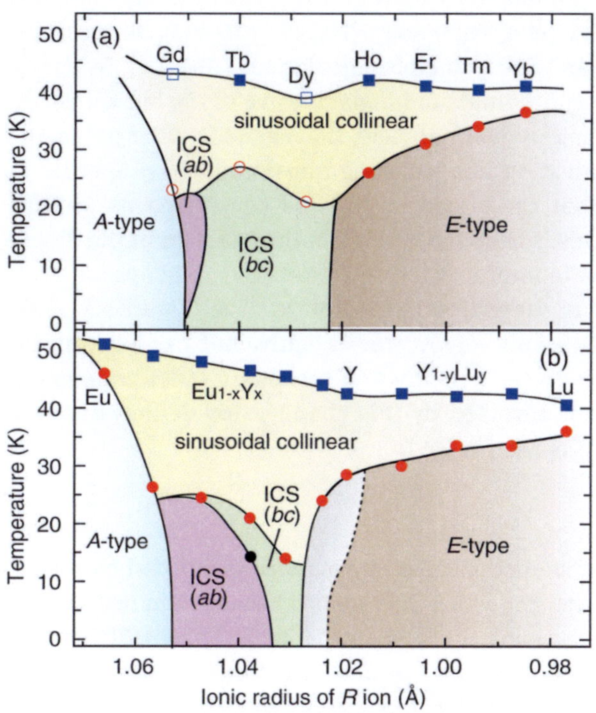

Figure 5.7: Temperature versus *R*-cation phase diagram of the orthorhombic $RMnO_3$ series, disregarding the magnetism of the *R*-cations (from Ref. [99], itself adapted from [98, 100]).

spin spirals that break inversion symmetry (see Figure 5.8). The plane of the spiral, in its turn, is determined by a subtle competition between SIA and DM interaction, and therefore depends on the specific system under consideration. When it comes to the spin-induced electric polarization, both symmetric and antisymmetric magnetostriction mechanisms can be at play in $R\mathrm{MnO_3}$.

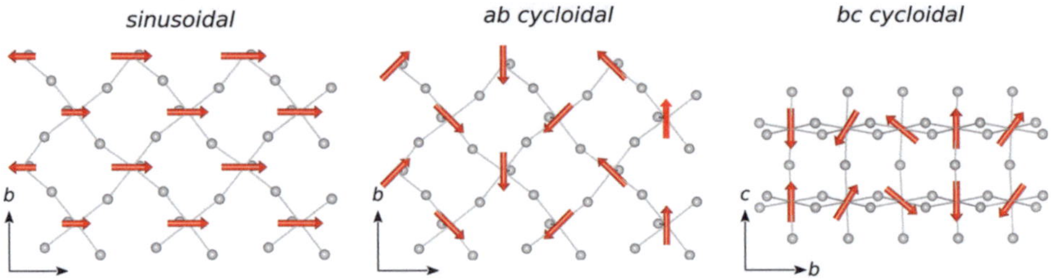

Figure 5.8: Sketch of the longitudinal spin-density-wave (collinear) and spiral (non-collinear) orders of the Mn spins observed in $R\mathrm{MnO_3}$ manganites (R = Tb, Dy, $\mathrm{Eu_{1-x}Y_x}$, and $\mathrm{Eu_{1-x}Ho_x}$).

The current understanding of the physics of these systems has emerged through a rather exhaustive experimental characterization and the use of complementary theoretical approaches such as phenomenology [26, 35, 101–103], microscopic modeling [24, 25, 31, 104–107] and first-principle calculations [108–111]. As a rule, spiral order is preceded by a longitudinal spin-density-wave of the Mn spins (see Figure 5.7). Thus, from the symmetry point of view, the magnetic order parameter associated to this magnetic transition has the same transformation properties as the electric polarization, so that these systems can be considered as pseudo-proper ferroelectrics [26, 27]. This is best appreciated in the behavior of the dielectric constant which, in contrast to improper ferroelectrics [112], displays a narrow divergence at the ferroelectric transition temperature (see Figure 5.9) [26]. Further, the reversal of the cycloidal wavevector is equivalent to changing the sign of the coupling constant between primary and secondary order parameters. Since the electric polarization is reversed by this change, this defines a sort of "ideal" magnetoelectric switching [27, 35, 109].

$\mathrm{TbMnO_3}$

$\mathrm{TbMnO_3}$ is probably of the most studied multiferroic manganite and can be considered as a reference compound. Its magnetic structure has been determined by neutron [102, 113–115] and X-ray resonant scattering experiments [116–118]. The Mn spins first order at $T_{N1}^{\mathrm{Mn}} = 42\,\mathrm{K}$ according to a collinear structure in which the spins are along the b-axis forming a longitudinal spin-density-wave in the b direction and an AFM structure along c with the wave-vector $\mathbf{q}_{\mathrm{Mn}} = (0, 0.28, 1)$. The Mn spins further develop a transverse component along the c-axis at $T_N^{\mathrm{Mn}} = 28\,\mathrm{K}$ that transforms

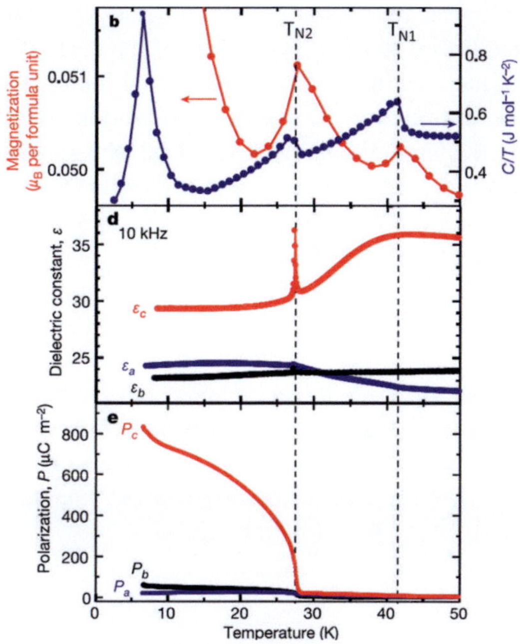

Figure 5.9: Magnetic and dielectric anomalies of $TbMnO_3$ (from Ref. [98]).

the structure into a (non-collinear) cycloidal in the *bc* plane. The main features of these orders understood in terms of magnetic frustration, SIA and DM interaction [25]. In addition, the Tb 4*f*-electron spins order at $T_N^{\mathrm{Tb}} = 7$ K also in a cycloidal fashion with $\mathbf{q}_{\mathrm{Tb}} = (0, \sim 0.42, 1)$. Mn and Tb orders can in fact be strongly coupled and affect each other.

The electric polarization in $TbMnO_3$ reaches $\approx 0.08\,\mu$ C/cm^2 [98] which is rather small compared with conventional ferroelectrics. By now, it is well established that this is mainly due to relativistic spin–orbit coupling via antisymmetric magnetostriction. Both electronic and ionic contributions to this mechanism seem to be at work, with the electronic one being extremely sensitive to the rotation of the oxygen octahedra [110]. Interestingly, there are also contributions due to symmetric magnetostriction that, however, tend to cancel each other due to alternating O shifts along the *c* direction [24, 25]. These different contributions have been probed experimentally by means of polarized X-ray scattering experiments with femtometer resolution, revealing the dominant character of the ionic contributions [119].

Due to its magnetic origin, the electric polarization can be manipulated by an external magnetic field.

Specifically, by applying the field along either *a* or *b* directions the spiral becomes commensurate with $\mathbf{q} = (0, 1/4, 1)$ and the polarization flops from $\parallel c$ to $\parallel a$. This is due to the reorientation of the plane of the spiral from *bc* to *ab* [120, 121].

Phenomenologically, this flop is reproduced by means of high-order anisotropy terms [35]. More microscopically, it was linked to the modification of the competition between between SIA and DM interactions induced by the external field [99]. The magnetic field along *c*, in contrast, suppresses the electric polarization in all crystallographic directions [122, 123]. In this case, the spiral is replaced by a simple AFM stripe order commensurate to the lattice. This can be understood as due to the reduction of the exchange energy of the spiral under the magnetic field vs. the advantage of the new state in terms of DM interaction [99]. Although the (H, T) phase diagram is reproduced qualitatively by considering the Mn spins only, the quantitative description requires accounting for the effective field generated by the Tb magnetic moments [99].

Additionally, a pressure-induced transition from the *bc* cycloidal state to the *E*-AFM state has been observed at 4–5 GPa [124, 125]. The spontaneous polarization then flops from the *c* to the *a* axis and its amplitude increases by one order of magnitude, which is in tune with the symmetric magnetostriction origin of the polarization in the collinear *E*-phase.

$DyMnO_3$

Similarly to $TbMnO_3$, $DyMnO_3$ develops an incommensurate collinear order at $T_{N1}^{Mn} = 39$ K with the wavevector $q \sim 0.8$ along the *b*-axis [123, 126]. In this case, this order is followed by the Dy-spins due to a stronger Mn-*R* coupling [113, 123, 126–128]. At $T_{N2}^{Mn} = 18$ K, the Mn order transforms into a cycloidal which is also transferred to the Dy spins. This generates an electric polarization, which compared to $TbMnO_3$ has an extra contribution due to interaction between Mn and Dy spirals [129]. The total polarization is therefore due to the combined effect of both inverse DM and exchange striction mechanisms [35, 113, 129–131]. At T_N^{Dy} = 5–6.5 K, however, the Dy–Dy interaction overcomes the Mn–Dy one and the Dy spins develop a commensurate AFM order that is decoupled from the Mn spiral. Interestingly, the ferroelectric polarization decreases from this point, which reveals a competition between Mn and Dy orders [127, 128].

$(Gd,Tb)MnO_3$

$GdMnO_3$ shows important differences compared to the previous manganites [100, 113, 126, 132–137]. This system is located near the boundary between the *A*-AFM phase and the *ab*-spiral phase. This situation makes the Mn spins to develop a longitudinal spin-density-wave with wavevector $q \sim 0.24$ along the *a*-axis at $T_N^{Mn} = 42$ K. With decreasing the temperature, the wavevector decreases and tends to the commensurate value 1/5. However, before reaching this value, the spin order transforms into the commensurate *A*-AFM ($q = 1/4$) with a wFM component at ~21 K. In this phase, $GdMnO_3$ shows very weak ferroelectricity with $P \parallel a$ that can be substantially enhanced by applying a magnetic field along the *b* axis and under pressure.

At a lower temperature, there is a Mn-induced order of the Gd spins ($T_N^{Gd} \sim$ 5–13 K) that tends to suppress ferroelectricity.

By Tb doping the Mn–O–Mn bond angles increase while the orthorhombic distortion decreases, which modifies the frustrated spin exchange interactions. This enables the control of the propagation wavevector in $Gd_{1-x}Tb_xMnO_3$, and to promote *ab* and *bc* spin spiral phases inducing ferroelectricity with $P \parallel a$ for $0.2 \leq x \leq 0.4$ and $P \parallel c$ for $0.4 \leq x$ respectively (see Figure 5.10) [100]. In $Gd_{0.7}Tb_{0.3}MnO_3$, in particular, the spiral wavevector is found to be commensurate $\mathbf{q} = (0, 1/4, 1)$ and $P \sim 300\mu C/m^2$ [137]. Since reorientation of the spiral plane from *ab* to *bc* induced by doping or by a magnetic field $\parallel b$ is accompanied with the polarization flop from *a* to *c* (see Figure 5.10), this polarization is ascribed to antisymmetric magnetostriction like in $TbMnO_3$ [137].

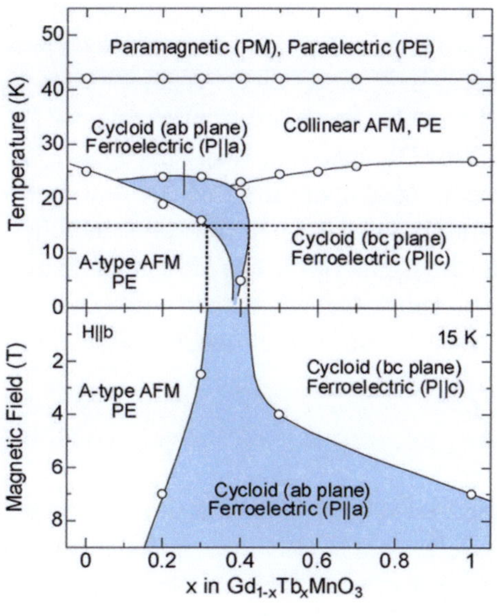

Figure 5.10: (x,T) and (x,H) phase diagram of $Gd_{1-x}Tb_xMnO_3$ (from Ref. [137], itself from [100]).

$(Eu,Y)MnO_3$ and $(Eu,Ho)MnO_3$

In principle, the $4f$ magnetic moments of the rare-earth are absent in $EuMnO_3$ (since the Eu^{3+} ion with $S = L = 3$ has a $J = 0$ non-magnetic ground state). This compound is therefore interesting to individuate the properties arising from the Mn $3d$ spins alone. This system turns out to be located near the boundary between the *A*-AFM and the *ab*-spiral phases [138–140]. Thus, like $GdMnO_3$, the Mn-spins first develop a longitudinal spin-density-wave at $T_{N1} \sim 52$ K which is followed by a transition to an *A*-AFM state with wFM canting at $T \sim 46$ K. In this case, however, no ferroelectricity has been reported and the spin-phonon couplings seems to be comparatively small [134]. Ferroelectricity can be induced by applying a strong magnetic field (>20 T) parallel to b [140, 141]. This is interpreted in terms of a field-induced commensurate order that generates an electric polarization $\parallel a$ via symmetric magnetostriction [140]. Additionally, $EuMnO_3$ has been predicted to host a peculiar insulator-metal transition associated to the transformation from A- to E-AFM order with FM stacking along c (E^*-AFM) [142].

The magnetic properties of $EuMnO_3$ can be tuned by doping with different A cations (A = La, Ca, Y, Lu, Sr, Ho) [138–140, 143–156]. Y doping is particularly interesting, since the essential multiferroic properties of $TbMnO_3$ and $DyMnO_3$ can be reproduced in this way (see Figure 5.7) [138–140, 144, 145, 147, 149, 150, 152, 154, 156–158]. Moreover, the orientation of the spiral plane (*ab* vs. *bc*) can be controlled by Y content. Further, resonant elastic x-ray scattering measurements revealed that the Eu ions actually display an induced magnetic moment (*i. e.* Van Vleck magnetism) in $Eu_{0.8}Y_{0.2}MnO_3$, with long-range AFM ordering due to the exchange coupling to Mn spins [159]. This represents the first observation of long-range order of Van Vleck moments.

In $Eu_{1-x}Ho_xMnO_3$ the magnetic ordering can also be varied from canted A-AFM to E-AFM order with intermediate spiral phases [21, 146, 160, 161]. Specifically, *ab* spiral order emerges at $x = 0.2$ which, by lowering temperature and increasing the Ho content, is reoriented to a *bc* spiral. The induced electric polarization is modified accordingly and changes from the a axis to the c axis [146]. The Ho magnetic moments, in their turn, order below 6.5 K in a spiral fashion with $\mathbf{q} = (0, 1/2, 0)$ in $HoMnO_3$ [161]. This produces a large increase in the spontaneous polarization, which reaches 2,400 μ C/m^2 [162].

R = Lu, Tm

Both $LuMnO_3$ and $TmMnO_3$ display collinear E-AFM order inducing the electric polarization via symmetric magnetostriction ($T_N^E \sim 35$ K and $P \sim$ 5,000μC/m^2 in $LuMnO_3$, and $T_N^E \sim$ 32 K and $P \sim$1,500μC/m^2 in $TmMnO_3$) [145, 163, 164]. However, resonant X-ray scattering measurements revealed that the magnetic structure of these compounds is not purely E-type, but there is an additional (alternating) spin canting along the c axis [165]. This type of canting has been predicted by theory due to the staggered SIA and DM, although along the a direction [25].

Thin films

In virtually strain-free films such as $TbMnO_3/YAlO_3(100)$, the long-range magnetic ordering has been reported to be essentially the same as in the bulk [166, 167]. In particular, the in-plane spontaneous polarization along the c axis is observed in these films down to the (6 nm) ultrathin limit. However, in contrast to the bulk, the ordering of the Tb magnetic moments gradually suppresses this polarization, which suggests a subtle modification of the Tb-Mn coupling. The same situation has been reported for $DyMnO_3$/Nb-SrTiO3(001) thin films [130]. In this case, however, the spontaneous polarization is enhanced by 800% compared to the bulk.

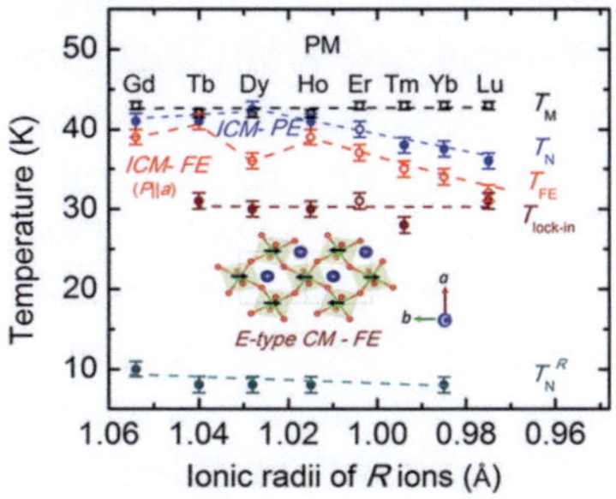

Figure 5.11: Generic phase diagram of the $RMnO_3$ series in strained thin films, showing the *R*-ion dependence of the magnetic (T_M, T_N, T_N^R and $T_{lock\text{-}in}$) and ferroelectric (T_{FE}) transition temperatures. By lowering the temperature, the non-collinear spiral phase is succeeded by the *E*-AFM phase at the lock-in temperature $T_{lock\text{-}in}$ in these films (from [168]).

In these spiral-based thin-film multiferroics, as in conventional ferroelectrics, multi-domain states can be expected to emerge due to the minimization of the corresponding depolarizing-field energy. This phenomenon, however, has specific features in these multiferroics that were first addressed in [26]. Specifically, the standard Kittel-law regime, in which the size of the domains d scales as the squared-root of the film thickness l [$d \sim (\ell l)^{1/2}$], is predicted to undergo a crossover towards an unconventional non-Kittel-law regime in which the system tends to the single-domain state as if the magnetic order would induce no ferroelectricity [$d \sim \ell(\ell/l)$].

In $GdMnO_3/SrTiO_3(001)$ wFM has been reported below 105 K, and ferroelectricity below 75 K with $P \sim 4{,}900$ C/m^2 along a [169]. The latter has been ascribed to the presence of *ab* spiral spin ordering. In $YMnO_3/SrTiO_3(110)$ the polarization is also found along the c axis [170]. In this case, the polarization decreases by applying a magnetic field in the *ab* plane and can be switched to the a axis by a magnetic field along c. These and complementary observations suggest that the magnetic order in this case can be tuned from *E*-AFM to *bc* ($\rightarrow$*ab*) spiral order and then *A*-AFM by epitaxial strain [171]. In fact, first-principle calculations subsequently predicted the emergence of *E*-AFM order in $TbMnO_3$ under sufficiently large strain (4.5 %) [172]. This prediction is in tune with the generic emergence of *E*-AFM order across the $RMnO_3$ series observed experimentally in strained films [168, 173] (see Figure 5.11).

5.5 Conclusions

We have outlined some important interconnections between non-collinear magnetism and multiferroicity, as exemplified in simple perovskite oxides. These fundamental features provide a basis for further understanding and controlling the interesting properties of these systems, and can naturally inspire the design of new multiferroic materials.

Whenever multiferroicity reduces to the superposition of two formally independent phenomena, the distortion of the crystal lattice still may have a qualitative impact on the magnetic properties. This happens via specific magnetoelectric responses, as illustrated in $BiFeO_3$. Here, the polar distortion results into the cycloidal modulation of the antiferromagnetic order of the Fe spins while epitaxial strain favors the emergence of a weak ferromagnetic component. Conversely, in spin-driven ferroelectrics, such as $GdFeO_3$ or $TbMnO_3$, changes in the magnetic structure result in qualitative changes in the underlying lattice and charge distribution (at both static and dynamical levels). In this respect, a new promising perspective to play with multiferroicity and magnetoelectricity is by means of the ultra-fast laser techniques where the magnetic properties can be tuned via resonant phonons for example [174–182].

These concepts illustrate the fundamentals of multiferroic materials, which keep attracting a considerable research attention in perovskite oxides and beyond (e. g. in van der Waals heterostructures [183, 184]).

References

[1] Arima T-h. Spin-Driven Ferroelectricity and Magneto-Electric Effects in Frustrated Magnetic Systems. J Phys Soc Japan. 2011;80:052001.

[2] Bousquet E. Cano A. Non-collinear magnetism in multiferroic perovskites. J Phys: Condens Matter. 2016;28:123001.

[3] Davydova MD, Zvezdin KA, Mukhin AA, Zvezdin AK. Spin dynamics, antiferrodistortion and magnetoelectric interaction in multiferroics. The case of $BiFeO_3$. Phys Sci Rev. 2020:20190070.

[4] Dong S, Liu J-M. Recent progress of multiferroic perovskite manganites. Mod Phys Lett B. 2012;26:30004.

[5] Evans DM, Garcia V, Meier D, Bibes M. Domains and domain walls in multiferroics. Phys Sci Rev. 2020;5:20190067.

[6] Fiebig M, Lottermoser T, Meier D, Trassin M. The evolution of multiferroics. Nat Rev Mater. 2016;1:16046.

[7] Fontcuberta J. Multiferroic $RMnO_3$ thin films. C R Phys. 2015;16:204–26.

[8] Krohns S, Lunkenheimer P. Ferroelectric polarization in multiferroics. Phys Sci Rev. 2019;4:20190015.

[9] Kurumaji T. Spiral spin structures and skyrmionsin multiferroics. Phys Sci Rev. 2019;5:20190016.

[10] Lottermoser T, Meier D. A short history of multiferroics. Phys Sci Rev. 2020, 000010151520200032, eISSN 2365-659X, ISSN 2365-6581, DOI: https://doi.org/10.1515/psr-2020-0032.

[11] Pyatakov AP, Zvezdin AK. Magnetoelectric and multiferroic media. Phys Uspekhi. 2012;55:557–81.

[12] Szaller D, Shuvaev A, Mukhin AA, Kuzmenko AM, Pimenov A. Controlling of light with electromagnons. Phys Sci Rev. 2019;5:20190055.

[13] Tokura Y, Seki S, Nagaosa N. Multiferroics of spin origin. Rep Prog Phys. 2014;77:076501.
[14] Varignon J, Bristowe NC, Bousquet E, Ghosez P. Magneto-electric multiferroics: designing new materials from first-principles calculations. Phys Sci Rev. 2019;5:20190069.
[15] Sosnowska I, Neumaier TP, Steichele E. Spiral magnetic ordering in bismuth ferrite. J Phys C: Solid State Phys. 1982;15:4835–4846.
[16] Johnson RD, Terada N, Radaelli PG. Comment on "spin-canting-induced improper ferroelectricity and spontaneous magnetization reversal in $SmFeO_3$". Phys Rev Lett. 2012;108:219701.
[17] Yamaguchi T, Tsushima K. Magnetic symmetry of rare-earth orthochromites and orthoferrites. Phys Rev B. 1973;8:5187–98.
[18] Zvezdin A, Mukhin A. Magnetoelectric interactions and phase transitions in a new class of multiferroics with improper electric polarization. JETP Lett. 2008;88:505–10.
[19] Holmes LM, Van Uitert LG. Magnetoelectric effect and metamagnetic transitions in dy alo_3. Phys Rev B. 1972;5:147–153.
[20] Velleaud G, Mercier M. A new magnetoelectric compound: $DyCoO_3$. Solid State Commun. 1975;17:237–9.
[21] Sergienko IA, S, en C, Dagotto E. Ferroelectricity in the magnetic e-phase of orthorhombic perovskites. Phys Rev Lett. 2006;97:227204.
[22] Lorenz B, Wang Y-Q, Chu C-W. Ferroelectricity in perovskite $HoMno_3$ and $YMno_3$. Phys Rev B. 2007;76:104405.
[23] Kaplan TA. Frustrated classical Heisenberg model in one dimension with nearest-neighbor biquadratic exchange: exact solution for the groundstate phase diagram. Phys Rev B. 2009;80:012407.
[24] Mochizuki M, Furukawa N, Nagaosa N. Spin model of magnetostrictions in multiferroic Mn perovskites. Phys Rev Lett. 2010;105:037205.
[25] Mochizuki M, Furukawa N, Nagaosa N. Theory of spin-phonon coupling in multiferroic manganese perovskites $RMnO_3$. Phys Rev B. 2011;84:144409.
[26] Cano A, Levanyuk AP. Pseudoproper ferroelectricity in thin films. Phys Rev B. 2010;81:172105.
[27] Tolédano P. Pseudo-proper ferroelectricity and magnetoelectric effects in $TbMnO_3$. Phys Rev B. 2009;79:094416.
[28] Weingart C, Spaldin N, Bousquet E. Noncollinear magnetism and singleion anisotropy in multiferroic perovskites. Phys Rev B. 2012;86:094413.
[29] Dzyaloshinsky I. A thermodynamic theory of "weak" ferromagnetism of antiferromagnetics. J Phys Chem Solids. 1958;4:241–55.
[30] Moriya T. Anisotropic superexchange interaction and weak ferromagnetism. Phys Rev. 1960;120:91–8.
[31] Sergienko IA, Dagotto E. Role of the Dzyaloshinskii–Moriya interaction in multiferroic perovskites. Phys Rev B. 2006;73:094434.
[32] Katsura H, Nagaosa N, Balatsky AV. Spin current and magnetoelectric effect in noncollinear magnets. Phys Rev Lett. 2005;95:057205.
[33] Bar'yakhtar VG, L'Vov VA, Yablonskiˇi DA. Inhomogeneous magnetoelectric effect. Sov J Exp Theor Phys Lett. 1983;37:673.
[34] Cano A. Kats EI. Electromagnon excitations in modulated multiferroics. Phys Rev B. 2008;78:012104.
[35] Mostovoy M. Ferroelectricity in spiral magnets. Phys Rev Lett. 2006;96:067601.
[36] Harris AB. Landau analysis of the symmetry of the magnetic structure and magnetoelectric interaction in multiferroics. Phys Rev B. 2007;76:054447.
[37] Kaplan TA, Mahanti SD. Canted-spin-caused electric dipoles: a local symmetry theory. Phys Rev B. 2011;83:174432.

[38] Cano A. Theory of electromagnon resonances in the optical response of spiral magnets. Phys Rev B. 2009;80:180416.
[39] Cazayous M, Gallais Y, Sacuto A, de Sousa R, Lebeugle D, Colson D. Possible observation of cycloidal electromagnons in $BiFeO_3$. Phys Rev Lett. 2008;101:037601.
[40] Katsura H, Balatsky AV, Nagaosa N. Dynamical magnetoelectric coupling in helical magnets. Phys Rev Lett. 2007;98:027203.
[41] Pimenov A, Mukhin AA, Ivanov VY, Travkin VD, Balbashov AM, Loidl A. Possible evidence for electromagnons in multiferroic manganites. Nat Phys. 2006;2:97–100.
[42] Valdés Aguilar R, Mostovoy M, Sushkov AB, Zhang CL, Choi YJ, Cheong S-W, et al. Origin of electromagnon excitations in multiferroic $rMnO_3$. Phys Rev Lett. 2009;102:047203.
[43] Juraschek DM, Fechner M, Balatsky AV, Spaldin NA. Dynamical multiferroicity. Phys Rev Mater. 2017a;1:014401.
[44] Narayan A, Cano A, Balatsky AV, Spaldin NA. Multiferroic quantum criticality. Nat Mater. 2019;18:223–8.
[45] Khomskii D. Classifying multiferroics: mechanisms and effects. Phys Online J. 2009;2:20.
[46] Fishman RS. Field dependence of the spin state and spectroscopic modes of multiferroic $BiFeO_3$. Phys Rev B. 2013;87:224419.
[47] Lebeugle D, Colson D, Forget A, Viret M, Bataille AM, Gukasov A. Electric-field-induced spin flop in $BiFeO_3$ single crystals at room temperature. Phys Rev Lett. 2008;100:227602.
[48] Nagel U, Fishman RS, Katuwal T, Engelkamp H, Talbayev D, Yi HT, et al. Terahertz spectroscopy of spin waves in multiferroic $BiFeO_3$ in high magnetic fields. Phys Rev Lett. 2013;110:257201.
[49] Rahmedov D, Wang D, Íñniguez J, Bellaiche L. Magnetic cycloid of $BiFeO_3$ from atomistic simulations. Phys Rev Lett. 2012;109: 037207.
[50] Ramazanoglu M, Laver M, Ratcliff W, Watson SM, Chen WC, Jackson A, et al. Local weak ferromagnetism in single-crystalline ferroelectric $BiFeO_3$. Phys Rev Lett. 2011;107:207206.
[51] Zvezdin AK, Pyatakov AP. On the problem of coexistence of the weak ferromagnetism and the spin flexoelectricity in multiferroic bismuth ferrite. EPL (Europhys Lett) 2012;99:57003–.
[52] Moubah R, Elzo M, El Moussaoui S, Colson D, Jaouen N, Belkhou R, et al. Direct imaging of both ferroelectric and antiferromagnetic domains in multiferroic $BiFeO_3$ single crystal using x-ray photoemission electron microscopy. Appl Phys Lett. 2012;100:042406.
[53] Moskvin AS, Bostrem IG. Special features of the exchange interactions in orthoferrite–orthochromites. Fiz Tverd Tela (Leningrad) 1977;19:2616–26.
[54] Moskvin AS, Sinitsyn EV. Antisymmetric exchange and four-sublattice model of orthoferrites. Fiz Tverd Tela. 1975;17:2495–7.
[55] Catalan G, Scott JF. Physics and applications of bismuth ferrite. Adv Mater. 2009;21:2463–85.
[56] Ruette B, Zvyagin S, Pyatakov AP, Bush A, Li JF, Belotelov VI, et al. Magnetic-field-induced phase transition in $BiFeO_3$ observed by high-field electron spin resonance: cycloidal to homogeneous spin order. Phys Rev B. 2004;69:064114.
[57] Sando DE. Crafting the magnonic and spintronic response of $BiFeO_3$ films by epitaxial strain. Nat Mat. 2013;12:641.
[58] Chen P, Grisolia MN, Zhao HJ, González-Vázquez OE, Bellaiche L, Bibes M, et al. Energetics of oxygen-octahedra rotations in perovskite oxides from first principles. Phys Rev B. 2018;97:024113.
[59] Bertaut EF. Symetrie cristalline et faible ferromagnetisme. J Phys Radium. 1961;22:839–41.
[60] Bertaut EF. Magnetism, vol. 3. Academic Press Inc, 1963.
[61] Bousquet E, Spaldin NA, Induced magnetoelectric response in pnma perovskites. Phys Rev Lett. 2011;107:197603.

[62] Günter T, Bousquet E, David A, Boullay P, Ghosez P, Prellier W, et al. Incipient ferroelectricity in 2.3% tensile-strained $CaMnO_3$ films. Phys Rev B. 2012;85:214120.

[63] Zhao HJ, Grisolia MN, Yang Y, Íñiguez J, Bibes M, Chen XM, et al. Magnetoelectric effects via pentalinear interactions. Phys Rev B. 2015;92:235133.

[64] Varignon J, Bristowe N, Bousquet E, Ghosez P. Magneto-electric multiferroics: Designing new materials from first-principles calculations. Phys Sci Rev. 2020;5 (2), 2365–659X, DOI: https://doi.org/10.1515/psr-2019-0069.

[65] Benedek NA, Fennie CJ. Hybrid improper ferroelectricity: a mechanism for controllable polarization-magnetization coupling. Phys Rev Lett. 2011;106:107204.

[66] Benedek NA, Mulder AT, Fennie CJ. Polar octahedral rotations: a path to new multifunctional materials. J Solid State Chem. 2012;195:11–20. Polar Inorganic Materials: Design Strategies and Functional Properties.

[67] Bristowe NC, Varignon J, Fontaine D, Bousquet E, Ghosez P. Ferromagnetism induced by entangled charge and orbital orderings in ferroelectric titanate perovskites. Nat Commun. 2015;6:6677.

[68] Song N, Rondinelli JM, Kim BG. Noncentrosymmetric structural transitions in ultrashort ferroelectric $aGaO_3$~$A'GaO_3$ superlattices. Phys Rev B. 2015;91:134104.

[69] Rondinelli JM, Fennie CJ. Octahedral rotation-induced ferroelectricity in cation ordered perovskites. Adv. Mater. 2012;24: 1961–8.

[70] Zanolli Z, Wojde l JC, Íñiguez J, Ghosez P. Electric control of the magnetization in $BiFeO_3$/$LaFeO_3$ superlattices. Phys Rev B. 2013;88:060102.

[71] Xu B, Wang D, Zhao HJ, Íñiguez J, Chen XM, Bellaiche L. Hybrid improper ferroelectricity in multiferroic superlattices: finite-temperature properties and electric-field-driven switching of polarization and magnetization. Adv Funct Mater 2015;25:3626–33.

[72] Zhao HJ, Bellaiche L, Chen XM, Íñiguez J. Improper electric polarization in simple perovskite oxides with two magnetic sublattices. Nat Commun. 2017;8:1–11.

[73] Treves D. Studies on orthoferrites at the Weizmann Institute of Science. J Appl Phys. 1965;36:1033–9.

[74] Tokunaga Y, Furukawa N, Sakai H, Taguchi Y, Arima T-H, Tokura Y. Composite domain walls in a multiferroic perovskite ferrite. Nat Mater. 2009;8:558–62.

[75] Panchwanee A, Upadhyay SK, Lalla NP, Sathe VG, Gupta A, Reddy VR. Low-temperature Raman, high magnetic field ^{57}Fe Mössbauer, and X-ray diffraction study of magnetodielectric coupling in polycrystalline $GdFeO_3$. Phys Rev B. 2019;99:064433.

[76] Gorodetsky G, Sharon B, Shtrikman S. Magnetic properties of an antiferromagnetic orthoferrite. J Appl Phys. 1968;39:1371–2.

[77] Morin FJ. Magnetic susceptibility of αFe_2O_3 and αFe_2O_3 with added titanium. Phys Rev. 1950;78:819–20.

[78] Zhao HJ,WRen, Y Yang,Íñiguez J, Chen XM, Bellaiche L., Near room-temperature multiferroic materials with tunable ferromagnetic and electrical properties, Nat Commun. 2014;5:4021.

[79] Tokunaga Y, Iguchi S, Arima T, Tokura Y. Magnetic-field-induced ferroelectric state in $DyFeO_3$. Phys Rev Lett. 2008;101:097205.

[80] Johnson C, Prelorendjo L, Thomas M. Field induced spin reorientation in orthoferrites $DyFeO_3$, $HoFeO_3$ and $ErFeO_3$. J Magn Magn Mater. 1980;15–18:557–8.

[81] Stroppa A, Marsman M, Kresse G, Picozzi S. The multiferroic phase of $DyFeO_3$: an ab initio study. New J Phys. 2010;12:093026.

[82] Khim T-Y, Eom MJ, Kim JS, Park B-G, Kim J-Y, Park J-H. Strain control spin reorientation transition in $DyFeO_3$/$SrTiO_3$ epitaxial film. Appl Phys Lett. 2011;99:072501.

[83] Rajeswaran B, Khomskii DI, Zvezdin AK, Rao CNR, Sundaresan A. Fieldinduced polar order at the néel temperature of chromium in rare-earth orthochromites: interplay of rare-earth and CR magnetism. Phys Rev B. 2012;86:214409.

[84] Sahu JR, Serrao CR, Ray N, Waghmare UV, Rao CN. Rare earth chromites: a new family of multiferroics. J Mater Chem. 2007;17:42–4.

[85] Serrao CR, Kundu AK, Krupanidhi SB,Waghmare UV, Rao CN. Biferroic $YCro_3$. Phys Rev B. 2005;72:220101.

[86] Ghosh A, Dey K, Chakraborty M, Majumdar S, Giri S. Polar octahedral rotations, cation displacement and ferroelectricity in multiferroic Sm- CrO_3. EPL (Europhys Lett) 2014;107:47012.

[87] Saha R, Sundaresan A, Rao CN. Novel features of multiferroic and magnetoelectric ferrites and chromites exhibiting magnetically driven ferroelectricity. Mater. Horiz. 2014;1:20–31.

[88] Mercy A, Bieder J, Iniguez J, Ghosez P. Structurally triggered metalinsulator transition in rare-earth nickelates. Nat Commun 2017;8:1677.

[89] Zhang Y, Schmitt MM, Mercy A, Wang J, Ghosez P. From charge- to orbital-ordered metal-insulator transition in alkaline-earth ferrites. Phys Rev B. 2018;98:081108.

[90] Hepting M, Minola M, Frano A, Cristiani G, Logvenov G, Schierle E, et al. Tunable charge and spin order in $PrNiO_3$ thin films and superlattices. Phys. Rev. Lett. 2014;113:227206.

[91] Catalan G. Progress in perovskite nickelate research. Phase Transitions. 2008;81:729–49.

[92] Giovannetti G, Kumar S, Khomskii D, Picozzi S, van den Brink J. Multiferroicity in rare-earth nickelates $rNiO_3$. Phys. Rev. Lett. 2009;103:156401.

[93] Kumar D, Rajeev KP, Alonso JA, Martínez-Lope MJ. Spin-canted magnetism and decoupling of charge and spin ordering in $NdNiO_3$. Phys Rev B. 2013;88:014410.

[94] da Silveira Lacerda LH, de Lazaro SR. Improvement of multiferroic property and change of magnetic ordering in new $ANiO_3$ (A=Ti, Ge, Zr, Sn, Hf and Pb). Comput Mater Sci. 2018;153:228–34.

[95] Ding L, Manuel P, Khalyavin DD, Orlandi F, Kumagai Y, Oba F, et al. Unusual magnetic structure of the high-pressure synthesized perovskites $ACrO_3$ (A=Sc, In, Tl). Phys Rev B. 2017;95:054432.

[96] Wang S,Wu X,Wang T, Zhang J, Zhang C, Yuan L, et al. Mild hydrothermal crystallization of heavy rare-earth chromite $ReCrO_3$ (Re= Er, Tm, Yb, Lu) perovskites and magnetic properties. Inorg Chem. 2019;58:2315– 29.

[97] Zákutná D, Vlček J, Fitl P, Nemkovski K, Honecker D, Niiòanský D, et al. Noncollinear magnetism in nanosized cobalt chromite. Phys Rev B. 2018;98:064407.

[98] Kimura T, Goto T, Shintani H, Ishizaka K, Arima T, Tokura Y. Magnetic control of ferroelectric polarization. Nature (London) 2003;426:55.

[99] Mochizuki M, Furukawa N. Theory of magnetic switching of ferroelectricity in spiral magnets. Phys Rev Lett. 2010;105:187601.

[100] Goto T, Yamasaki Y, Watanabe H, Kimura T, Tokura Y. Anticorrelation between ferromagnetism and ferroelectricity in perovskite manganites. Phys Rev B. 2005;72:220403.

[101] Harris AB. Ferroelectricity induced by incommensurate magnetism (invited). J Appl Phys. 99, 08E303 (2006).

[102] Kenzelmann M, Harris AB, Jonas S, Broholm C, Schefer J, Kim SB, Zhang CL, Cheong SW, Vajk OP, Lynn JW. Magnetic inversion symmetry breaking and ferroelectricity in $TbMnO_3$. Phys Rev Lett. 2005;95:087206.

[103] Ribeiro JL, Vieira LG. Landau model for the phase diagrams of the orthorhombic rare-earth manganites $rMnO_3$ (r = Eu, Gd, Tb, Dy, Ho). Phys Rev B. 2010;82:064410.

[104] Furukawa N, Mochizuki M. Roles of bond alternation in magnetic phase diagram of $RMnO_3$. J Phys Soc J. 2010;79:033708.
[105] Hu CD. Relationship between ferroelectricity and Dzyaloshinskii–Moriya interaction in multiferroics and the effect of bond-bending. Phys Rev B. 2008;77:174418.
[106] Mochizuki M, Furukawa N. Mechanism of lattice-distortion-induced electric-polarization flop in the multiferroic perovskite manganites. J Phys Soc Japan 2009a;78:053704.
[107] Mochizuki M, Furukawa N. Microscopic model and phase diagrams of the multiferroic perovskite manganites. Phys Rev B. 2009b;80:134416.
[108] Fedorova NS, Ederer C, Spaldin NA, Scaramucci A. Biquadratic and ring exchange interactions in orthorhombic perovskite manganites. Phys Rev B. 2015;91:165122.
[109] Malashevich A, Vanderbilt D. First principles study of improper ferroelectricity in $TbMnO_3$. Phys Rev Lett. 2008;101:037210.
[110] Malashevich A, Vanderbilt D. Dependence of electronic polarization on octahedral rotations in $TbMnO_3$ from first principles. Phys Rev B. 2009;80:224407.
[111] Xiang HJ, Wei S-H, Whangbo M-H, Da Silva JL. Spin–orbit coupling and ion displacements in multiferroic $TbMnO_3$. Phys Rev Lett. 2008;101:037209.
[112] Levanyuk AP, Sannikov DG. Reviews of topical problems: improper ferroelectrics. Sov Phys Uspekhi. 1974;17:199–214.
[113] Aliouane N, Prokhnenko O, Feyerherm R, Mostovoy M, Strempfer J, Habicht K, Rule KC, et al. Magnetic order and ferroelectricity in $RmnO_3$ multiferroic manganites: coupling between R- and MN-spins. J Phys: Condens Matter 2008;20:434215.
[114] Kajimoto R, Yoshizawa H, Shintani H, Kimura T, Tokura Y. Magnetic structure of $TbMnO_3$ by neutron diffraction. Phys Rev B. 2004;70:012401.
[115] Quezel S, Tcheou F, Rossat-Mignod J, Quezel G, Roudaut E. Magnetic structure of the perovskite-like compound $TbMnO_3$. Phys B+C 1977;86:916–8.
[116] Jang H, Lee J-S, Ko K-T, Noh W-S, Koo TY, Kim J-Y, et al. Coupled magnetic cycloids in multiferroic $TbMnO_3$ and $Eu_{3\sim4}Y_{1\sim4}MnO_3$. Phys Rev Lett. 2011;106:047203.
[117] Mannix D, McMorrow DF, Ewings RA, Boothroyd AT, Prabhakaran D, Joly Y, et al. X-ray scattering study of the order parameters in multiferroic $TbMno_3$. Phys Rev B. 2007;76:184420.
[118] Wilkins SB, Forrest TR, Beale TA, Bland SR, Walker HC, Mannix D, et al. Nature of the magnetic order and origin of induced ferroelectricity in $TbMnO_3$. Phys Rev Lett. 2009;103:207602.
[119] Walker HC, Fabrizi F, Paolasini L, de Bergevin F, Herrero-Martin J, Boothroyd AT, et al. Femtoscale magnetically induced lattice distortions in multiferroic $TbMnO_3$. Science 2011;333:1273–6.
[120] Aliouane N, Argyriou DN, Strempfer J, Zegkinoglou I, Landsgesell S, Zimmermann MV. Field-induced linear magnetoelastic coupling in multiferroic $tbmno_3$. Phys Rev B 2006;73:020102.
[121] Aliouane N, Schmalzl K, Senff D, Maljuk A, Prokeš K, Braden M. Flop of electric polarization driven by the flop of the mn spin cycloid in multiferroic $TbMnO_3$. Phys Rev Lett. 2009;102:207205.
[122] Argyriou DN, Aliouane N, Strempfer J, Zegkinoglou I, Bohnenbuck B, Habicht K, Zimmermann MV. Melting of incommensurate-ferroelectric phase with magnetic field in multiferroic $TbMnO_3$. Phys Rev B. 2007;75:020101.
[123] Kimura T, Lawes G, Goto T, Tokura Y, Ramirez AP. Magnetoelectric phase diagrams of orthorhombic $rMnO_3$ (r = Gd, Tb, and Dy). Phys Rev B. 2005;71:224425.
[124] Aoyama T, Yamauchi K, Iyama A, Picozzi S, Shimizu T, Kimura K. Giant spin-driven ferroelectric polarization in $TbMnO_3$ under high pressure. Nat Commun. 2014;5:4927.

[125] Aupiais I, Mochizuki M, Sakata H, Grasset R, Gallais Y, Sacuto A, et al. Colossal electromagnon excitation in the non-cycloidal phase of $TbMnO_3$ under pressure. NPJ Q Mater. 2018;3:60.
[126] Goto T, Kimura T, Lawes G, Ramirez AP, Tokura Y. Ferroelectricity and giant magnetocapacitance in perovskite rare-earth manganites. Phys Rev Lett. 2004;92:257201.
[127] Feyerherm R, Dudzik E, Aliouane N, Argyriou DN. Commensurate dy magnetic ordering associated with incommensurate lattice distortion in multiferroic $DyMnO_3$. Phys Rev B. 2006;73:180401.
[128] Prokhnenko O, Feyerherm R, Dudzik E, Landsgesell S, Aliouane N, Chapon LC, et al. Enhanced ferroelectric polarization by induced dy spin order in multiferroic $DyMnO_3$. Phys Rev Lett. 2007;98:057206.
[129] Feyerherm R, Dudzik E, Prokhnenko O, Argyriou DN. Rare earth magnetism and ferroelectricity in $rMnO_3$. J Phys: Conf Ser. 2010;200:012032.
[130] Lu C, Dong S, Xia Z, Luo H, Yan Z, Wang H, et al. Polarization enhancement and ferroelectric switching enabled by interacting magnetic structures in $DyMnO_3$ thin films. Sci Rep. 2013;3:3374.
[131] Zhang N, Dong S, Liu J-M. Ferroelectricity generated by spin-orbit and spin-lattice couplings in multiferroic $DyMnO_3$. Front Phys. 2012;7:408–17.
[132] Aoyama T, Iyama A, Shimizu K, Kimura T. Multiferroicity in orthorhombic $rMnO_3$ (r = Dy, Tb, and Gd) under high pressure. Phys Rev B 2015;91:081107.
[133] Arima T, Goto T, Yamasaki Y, Miyasaka S, Ishii K, Tsubota M, Inami T, Murakami Y, Tokura Y. Magnetic-field-induced transition in the lattice modulation of colossal magnetoelectric $GdMnO_3$ and $TbMnO_3$ compounds. Phys Rev B. 2005;72:100102.
[134] Ferreira WS, Agostinho Moreira J, Almeida A, Chaves MR, Aráujo JP, Oliveira JB, et al. Spin–phonon coupling and magnetoelectric properties: $EuMnO_3$ versus $GdMnO_3$. Phys Rev B. 2009;79:054303.
[135] Feyerherm R, Dudzik E, Wolter AU, Valencia S, Prokhnenko O, Maljuk A, et al. Magnetic-field induced effects on the electric polarization in $rMnO_3$(r = Dy,Gd). Phys Rev B. 2009;79:134426.
[136] Skaugen A, Shukla DK, Feyerherm R, Dudzik E, Islam Z, Strempfer J. Magnetic order in $GdMnO_3$ in magnetic fields. J Phys: Conf Ser. 2014;519:012007.
[137] Yamasaki Y, Sagayama H, Abe N, Arima T, Sasai K, Matsuura M, et al. Cycloidal spin order in the a-axis polarized ferroelectric phase of orthorhombic perovskite manganite. Phys Rev Lett. 2008;101:097204.
[138] Hemberger J, Schrettle F, Pimenov A, Lunkenheimer P, Ivanov VY, Mukhin AA, et al. Multiferroic phases of $Eu_{1-x}Y_xMnO_3$. Phys Rev B. 2007;75:035118.
[139] Ivanov VY, Mukhin AA, Travkin VD, Prokhorov AS, Kadomtseva AM, Popov YF, et al. Magnetic, dielectric and magnetoelectric properties of new family of orthorhombic multiferroic $Eu_{1-x}Y_xMNO_3$ manganites. J Magn Magn Mater. 2006;300:130.
[140] Tokunaga M, Yamasaki Y, Onose Y, Mochizuki M, Furukawa N, Tokura Y. Novel multiferroic state of $Eu_{1-x}y_xMnO_3$ in high magnetic fields. Phys Rev Lett. 2009;103:187202.
[141] Kadomtseva A, Popov Y, Vorobev G, Ivanov V, Mukhin A, Balbashov A. Specific features of the magnetic field-induced orientational transition in $EuMnO_3$. J Exp Theor Phys Lett. 2005;81:590–3.
[142] Qiu R, Bousquet E, Cano A. Pressure-induced insulator–metal transition in $EuMnO_3$. J Phys: Condens Matter. 2017;29:305801.

[143] Chen CT, Lin BN, Hsu YY, Liao JD, Cheng WH, Lin CY, et al. Anomalous variation of a-type antiferromagnetic order and two-dimensional ferromagnetic fluctuation in orbital-ordered $La_{1-x}Eu_xMnO_3$. Phys Rev B. 2003;67:214424.

[144] Danjoh S, Jung J-S, Nakamura H, Wakabayashi Y, Kimura T. Anomalous induction of ferroelectric polarization by magnetization reversal in the phase-separated multiferroic manganite $Eu_{0.8}Y_{0.2}MnO_3$. Phys Rev B. 2009;80:180408.

[145] Ishiwata S, Kaneko Y, Tokunaga Y, Taguchi Y, Arima T-h, Tokura Y. Perovskite manganites hosting versatile multiferroic phases with symmetric and antisymmetric exchange strictions. Phys Rev B. 2010;81:100411.

[146] Ivanov VY, Mukhin AA, Glushkov VV, Balbashov AM. Spontaneous reorientation of the electric polarization in $Eu_{1-x}Ho_xMnO_3$ multiferroics. Sov J Exp Theor Phys Lett. 2013;97:28–33.

[147] Kalvius GM, Litterst FJ, Hartmann O, Wäppling R, Krimmel A, Mukhin AA, Balbashov AM, Loidl A. Magnetic properties of the multiferroic compounds $Eu_{1-x}Y_xMnO_3$ ($x = 0.2$ and 0.3). J Phys Conf Ser. 2014;551:012014.

[148] Ku H, Chen C, Lin B. A-type antiferromagnetic order, 2d ferromagnetic fluctuation and orbital order in stoichiometric $La_{1-x}Eu_xMnO_3$. J Magn Magn Mater. 2004;272 276:85–87. Proceedings of the International Conference on Magnetism (ICM 2003).

[149] Moreira JA, Almeida A, Ferreira WS, Chaves MR, Kundys B, Ranjith R. Polar properties of $Eu_{0.6}Y_{0.4}MnO_3$ ceramics and their magnetic field dependence. J Phys: Condens Matter 2009;21:446002.

[150] Noda K, Akaki M, Kikuchi T, Akahoshi D, Kuwahara H. Magneticfield- induced switching between ferroelectric phases in orthorhombicdistortion- controlled $RMnO_3$. J Appl Phys. 2006;99:080000.

[151] Oliveira J, Moreira JA, Almeida A, Chaves MR, da Silva JM, Oliveira JB, et al. Phase diagram of the orthorhombic, lightly lutetium doped $EuMnO_3$ magnetoelectric system. Phys Rev B. 2011;84:094414.

[152] Pimenov A, Loidl A, Mukhin AA, Travkin VD, Ivanov VY, Balbashov AM. Terahertz spectroscopy of electromagnons in $Eu_{1-x}Y_xMnO_3$. Phys Rev B. 2008;77:014438.

[153] Tadokoro Y, Shan YJ, Nakamura T, Nakamura S. Crystal structure and characterizations of perovskite oxides $(Eu_{1-x}Sr_x)MnO_3$ ($0.0 \leq x \leq 0.5$). Solid State Ionics. 1998;108:261–7.

[154] Tokunaga M, Katakura I, Yamasaki Y, Onose Y, Tokura Y. High-field study of multiferroic properties in orthorhombic $Eu_{1-x}Y_xMnO_3$. J Phys: Conf Ser. 2009;150:042212.

[155] Troyanchuk I, Samsonenko N, Szymczak H, Nabialek A. Magnetic study of the $Ca_{1-x}Eu_xMnO_3$ ($0 \leq x \leq 1$) perovskites. J Solid State Chem. 1997;131:144–9.

[156] Yamasaki Y, Miyasaka S, Goto T, Sagayama H, Arima T, Tokura Y. Ferroelectric phase transitions of 3d-spin origin in $Eu_{1-x}Y_xMnO_3$. Phys Rev B. 2007;76:184418.

[157] Issing S, Pimenov A, Ivanov VY, Mukhin AA, Geurts J. Compositiondependent spin-phonon coupling in mixed crystals of the multiferroic manganite $Eu_{1-x}Y_xMnO_3$ ($0 \leq x \leq 0.5$) studied by raman spectroscopy. Phys Rev B. 2010;81: 024304.

[158] Murakawa H, Onose Y, Kagawa F, Ishiwata S, Kaneko Y, Tokura Y. Rotation of an electric polarization vector by rotating magnetic field in cycloidal magnet $Eu_{0.55}Y_{0.45}MnO_3$. Phys Rev Lett. 2008;101:197207.

[159] Skaugen A, Schierle E, van der Laan G, Shukla DK, Walker HC, Weschke E, et al. Long-range antiferromagnetic order of formally nonmagnetic Eu^{3+} van vleck ions observed in multiferroic $Eu_{1-x}Y_xMnO_3$. Phys Rev B. 2015;91:180409.

[160] Lorenz B, Wang YQ, Sun YY, Chu CW. Large magnetodielectric effects in orthorhombic $HoMnO_3$ and $YMnO_3$. Phys Rev B. 2004;70:212412.

[161] Muñoz A, Casáis MT, Alonso JA, Martínez-Lope MJ, Martínez JL, Fernández-Díaz MT. Complex magnetism and magnetic structures of the metastable $HoMnO_3$ perovskite. Inorg Chem. 2001;40:1020–8.
[162] Feng SM, Chai YS, Zhu JL, Manivannan N, Oh YS, Wang LJ, et al. Determination of the intrinsic ferroelectric polarization in orthorhombic $HoMnO_3$. New J Phys. 2010;12:073006.
[163] Okamoto H, Imamura N, Hauback BC, Karppinen M, Yamauchi H, Fjellvåg H. Neutron powder diffraction study of crystal and magnetic structures of orthorhombic $LuMnO_3$. Solid State Commun. 2008;146:152–6.
[164] Pomjakushin VY, Kenzelmann M, Dönni A, Harris AB, Nakajima T, Mitsuda S, et al. Evidence for large electric polarization from collinear magnetism in $TmMnO_3$. New J Phys. 2009;11:043019.
[165] Garganourakis M, Bodenthin Y, de Souza RA, Scagnoli V, Dönni A, Tachibana M, et al. Magnetic and electronic orderings in orthorhombic $rMnO_3$ (r = Tm, Lu) studied by resonant soft x-ray powder diffraction. Phys Rev B. 2012;86:054425.
[166] Glavic A, Becher C, Voigt J, Schierle E, Weschke E, Fiebig M, et al. Stability of spin-driven ferroelectricity in the thin-film limit: coupling of magnetic and electric order in multiferroic $TbMnO_3$ films. Phys Rev B. 2013;88:054401.
[167] Glavic A, Voigt J, Persson J, Su Y, Schubert J, de Groot J, et al. High quality $TbMnO_3$ films deposited on $YAlO_3$. J Alloys Compd. 2011;509:5061–3.
[168] Mukherjee S, Shimamoto K, Windsor YW, Ramakrishnan M, Parchenko S, Staub U, et al. Multiferroic phase diagram of e-type $rMnO_3$ films studied by neutron and x-ray diffraction. Phys Rev B. 2018;98:174416.
[169] Li X, Lu C, Dai J, Dong S, Chen Y, Hu N, et al. Novel multiferroicity in $GdMnO_3$ thin films with self-assembled nano-twinned domains. Sci Rep. 2014;4:7019.
[170] Fina I, Fàbrega L, Martí X, Sánchez F, Fontcuberta J. Magnetic switch of polarization in epitaxial orthorhombic $YMnO_3$ thin films. Appl Phys Lett. 2010;97:232905.
[171] Jiménez-Villacorta F, Gallastegui JA, Fina I, Marti X, Fontcuberta J. Strain-driven transition from E-type to A-type magnetic order in $YMnO_3$ epitaxial films. Phys Rev B. 2012;86:024420.
[172] Hou YS, Yang JH, Gong XG, Xiang HJ. Prediction of a multiferroic state with large electric polarization in tensile-strained $TbMnO_3$. Phys Rev B. 2013;88:060406.
[173] Shimamoto K, Mukherjee S, Bingham NS, Suszka AK, Lippert T, Niedermayer C, et al. Single-axis-dependent structural and multiferroic properties of orthorhombic $rMno_3$(r = Gd–Lu). Phys Rev B. 2017;95:184105.
[174] Afanasiev D, Hortensius JR, Ivanov B, Sasani A, Bousquet E, Blanter Y, et al. Light-driven ultrafast phonomagnetism. arXiv:1912.01938v1.
[175] Disa A, Fechner M, Nova T, Liu B, Först M, Prabhakaran D, et al. Polarizing an antiferromagnet by optical engineering of the crystal field. Nat Phys. 2020;16:937.
[176] Fechner M, Sukhov A, Chotorlishvili L, Kenel C, Berakdar J, Spaldin NA. Magnetophononics: ultrafast spin control through the lattice. Phys. Rev. Mate. 2018;2:064401.
[177] Gu M, Rondinelli JM. Nonlinear phononic control and emergent magnetism in Mott insulating titanates. Phys Rev B. 2018;98:024102.
[178] Juraschek DM, Fechner M, Balatsky AV, Spaldin NA. Dynamical multiferroicity. Phys Rev Mater. 2017b;1:014401.
[179] Juraschek DM, Fechner M, Spaldin NA. Ultrafast structure switching through nonlinear phononics. Phys Rev Lett. 2017;118:054101.
[180] Khalsa, G, Benedek N. Ultrafast optically induced ferromagnetic/antiferromagnetic phase transition in $GdTiO_3$ from first principles. NPJ Quantum Mater. 2018;3:15.
[181] Nova TF, Cartella A, Cantaluppi A, Först M, Bossini D, Mikhaylovskiy RV, et al. An effective magnetic field from optically driven phonons. Nat Phys. 2017;13:132.

[182] Radaelli PG. Breaking symmetry with light: ultrafast ferroelectricity and magnetism from three-phonon coupling. Phys Rev B. 2018;97:085145.
[183] Gong C, Kim EM, Wang Y, Lee G, Zhang X. Multiferroicity in atomic van der Waals heterostructures. Nat Commun. 2019;10:2657.
[184] Liu F, You L, Seyler KL, Li X, Yu P, Lin J, et al. Room-temperature ferroelectricity in $CuInP_2S_6$ ultrathin flakes. Nat Commun. 2016;7:1–6.

Stephan Krohns and Peter Lunkenheimer

6 Ferroelectric polarization in multiferroics

Abstract: Multiferroic materials, showing ordering of both electrical and magnetic degrees of freedom, are promising candidates enabling the design of novel electronic devices. Various mechanisms ranging from geometrically or spin-driven improper ferroelectricity via lone-pairs, charge-order or -transfer support multiferroicity in single-phase or composite compounds. The search for materials showing these effects constitutes one of the most important research fields in solid-state physics during the last years, but scientific interest even traces back to the middle of the past century. Especially, a potentially strong coupling between spin and electric dipoles captured the interest to control via an electric field the magnetization or via a magnetic field the electric polarization. This would imply a promising route for novel electronics. Here, we provide a review about the dielectric and ferroelectric properties of various multiferroic systems ranging from type I multiferroics, in which magnetic and ferroelectric order develop almost independently of each other, to type II multiferroics, which exhibit strong coupling of magnetic and ferroelectric ordering. We thoroughly discuss the dielectric signatures of the ferroelectric polarization for $BiFeO_3$, Fe_3O_4, $DyMnO_3$ and an organic charge-transfer salt as well as show electric-field poling studies for the hexagonal manganites and a spin-spiral system $LiCuVO_4$.

Keywords: multiferroics, ferroelectrics, dielectric spectroscopy, polarisation measurements

6.1 Introduction

The control of the ferroelectric polarization of multiferroic materials allows innovative magnetoelectric functionalities on a macroscopic as well as on a microscopic level, which may pave the way for next-generation building blocks for future electronic products [1–6]. Therefore, the analysis of the electrical and, hence, the ferroelectric properties is of key relevance. The objective of this manuscript is to gain fundamental insights into the underlying complex coupling mechanisms as well as a target-oriented manipulation of the magnetoelectric effect or the multiferroic ordering. Figure 6.1 provides a brief overview (adapted from Refs. [3] and [4]) on the definition of "multiferroicity" as well as "ferroic properties", "magnetodielectric effect" and "magnetoelectric effect". As there are various mechanisms giving rise to multiferroicity, we review

This article has previously been published in the journal Physical Sciences Reviews. Please cite as: Krohns, S., Lunkenheimer, P. Ferroelectric polarization in multiferroics *Physical Sciences Reviews* [Online] 2019, 4. DOI: 10.1515/psr-2019-0015

https://doi.org/10.1515/9783110582130-006

dielectric and electric-field poling studies for six magnetoelectric and multiferroic materials based on five different and important underlying mechanisms: Lone-pair multiferroicity, geometrically driven improper ferroelectricity, relaxor-ferroelectricity in magnetic materials, electric-dipole-driven magnetism and strongly coupled spin-driven improper ferroelectricity.

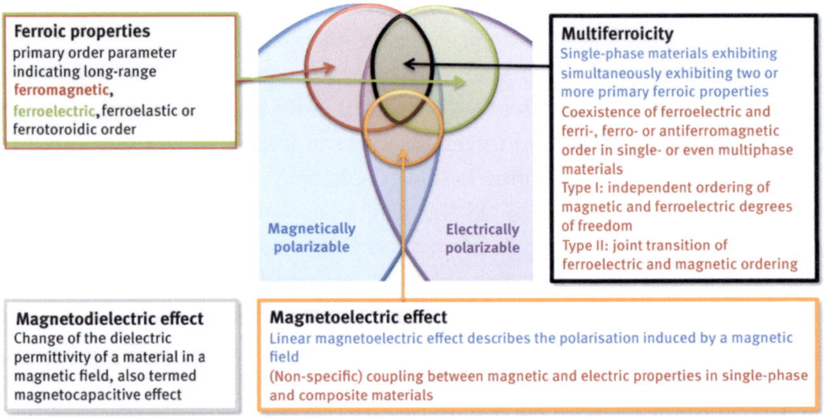

Figure 6.1: Definitions in the field of multiferroics (adapted from Ref. [4]) and the relationship between originally defined multiferroic and magnetoelectric materials (adapted from Ref. [3]). Ferroic order, especially ferromagnetism (ferroelectricity), forms a subset of magnetically (electrically) polarizable materials. In their intersection, multiferroic and magnetoelectric coupling can arise. For multiferroicity and the magnetoelectric effect the originally (blue) used and nowadays (red) commonly accepted definitions are given. The latter would lead to an enlarged intersection as e.g. antiferromagnetic order is also allowed for multiferroicity.

After an introduction on the signatures of ferroelectricity in dielectric spectroscopy and polarization measurements, we explicitly focus on the ferroelectric polarization of the following multiferroics:

Probably the most investigated single-phase multiferroic system is $BiFeO_3$ [7], which is indeed of particular interest for multiferroic applications as its multiferroicity already exists at ambient temperatures. We briefly review the dielectric and polar properties, only focusing on single crystals.

Magnetite is one of the oldest known magnetic materials. We discuss its fascinating property, namely relaxor-ferroelectricity, making Fe_3O_4 multiferroic [8]. However, the coupling of its order parameters is rather weak and occurs at temperatures below liquid nitrogen.

Electric-field poling studies of improper ferroelectrics like $YMnO_3$ and $ErMnO_3$ are rare [9–12]. Beside their multiferroic properties, these systems are in the scientific focus as they exhibit versatile ferroelectric domain and domain-wall properties [6]. Results of dielectric spectroscopy and polarization measurements combined with local-probe analysis allow manipulating the local structure via electric field and may pave the way towards their utilization in domain-wall-based electronics (as discussed, e.g. in Chapter 3).

A more exotic class of multiferroic materials is charge-transfer salts, which can exhibit improper ferroelectric ordering due to the localization of charges at organic molecules. For κ-$(BEDT\text{-}TTF)_2Cu[N(CN)_2]Cl$ (here BEDT-TTF stands for bis(ethylenedithio)-tetrathiafulvalene, often also abbreviated as ET), charge-order-driven electronic ferroelectricity was reported representing a novel type of multiferroic mechanism [13]. This allows in principle to tune magnetic ordering via an electric field, making this material class highly interesting from a fundamental point of view.

In multiferroics, the magnetic and ferroelectric order can develop almost independently from each other (i.e. type I). In contrast, type II multiferroics exhibit strong coupling of their ferroic orders [14]. Often complex magnetic ordering triggers improper ferroelectricity in these systems leading to identical ferroelectric and magnetic transition temperatures. For various compounds, like $TbMnO_3$ [15], $DyMnO_3$ [16] or $LiCuVO_4$ [17], the ferroelectric polarization can be controlled by magnetic fields. However, the reversed control is a challenge. For spin spirals (as discussed in Chapter 4), it was demonstrated that an electric field can change the helicity of these spirals from counterclockwise to clockwise rotation and vice versa [18–20]. Here, we briefly review the dielectric properties arising at the joint transition temperatures of *$DyMnO_3$* [16].

Another prime example for a strongly coupled multiferroic is the spin-spiral system *$LiCuVO_4$* [17]. We discuss its dielectric as well as its switching properties by electric fields and the impact of switching on the spin spirals.

6.2 Signatures of ferroelectricity

6.2.1 Signatures of ferroelectricity in dielectric spectroscopy

For multiferroics, due to conductivity contributions and/or extrinsic Maxwell–Wagner (MW) like relaxation phenomena arising from thin insulating barrier layers [21], it is often difficult or even impossible to unequivocally identify a polar ground state by a direct measurement of the electrical polarization. Therefore, one of the most prominent experimental methods used to identify ferroelectricity in multiferroics is dielectric spectroscopy. It allows for the detection of the significant anomalies in the temperature dependence of the real part of the dielectric permittivity (the dielectric constant ε') that are expected to occur at ferroelectric transitions. In this section, we discuss the typical experimental signatures of ferroelectricity in dielectric spectroscopy for the different classes of ferroelectric states, which in principle can also be found in multiferroics, e.g. $TbMnO_3$ (Ref. [15]) or $LiCuVO_4$ (Ref. [17]). For more details, we refer the reader to various review articles and books on ferroelectricity [22–26].

An important class of ferroelectric materials is constituted by the so-called displacive ferroelectrics. At their ferroelectric transition, a high-symmetry structure without permanent dipole moments transfers into a lower-symmetry structure with polar order. The latter arises from permanent dipole moments generated by local shifts of ions. In fact, in displacive ferroelectrics a so-called soft mode occurs, a dipolar-active transverse optical phonon that becomes soft when approaching the critical temperature. This finally results in the freezing-in of the long-wavelength ionic displacements and leads to an off-symmetry position of the involved ions that establishes long-range polar order. An important property of displacive ferroelectrics is the divergence of the static dielectric constant ε_s: In the paraelectric phase, above the ferroelectric transition at T_{FE}, the temperature dependence of ε_s can usually be well described by a Curie–Weiss law.

$$\varepsilon_s = C / (T - T_{CW}). \tag{6.1}$$

Here C is the Curie constant determined by the dipole number per volume and the dipole moment and T_{CW} denotes the Curie–Weiss temperature, depending on the average interaction strength between the dipoles. Often the Curie–Weiss temperature and the ferroelectric ordering temperature are of similar order. Figure 6.2(a) schematically indicates $\varepsilon'(T)$ for a displacive ferroelectric. Below the transition temperature, $\varepsilon'(T)$ decreases again, leading to a sharp peak at T_{FE}. The detailed shape of this peak depends on the details of the transition, especially its first- or second-order nature [25, 26]. When performing conventional dielectric spectroscopy, typically covering frequencies in the Hz–MHz range, usually no significant frequency dependence of ε' is observed for this class of ferroelectrics.

Figure 6.2(d) shows a multiferroic example of this type or ferroelectric, $LiCuVO_4$ [27]. It should be noted that this is an improper ferroelectric, where the polar order is driven by complex magnetic ordering via the inverse Dzyaloshinskii–Moriya interaction (see Section 6.3.6 for details on this material). However, this mechanism also leads to ionic displacements, which finally generates ferroelectricity. As documented in Figure 6.2(d), there is a small frequency dependence of ε' in this material, which may arise from the non-canonical origin of its polar order. However, it is much weaker than in the other two examples (Figure 6.2(e) and Figure 6.2(f)), which belong to different classes of ferroelectric as explained below.

While displacive ferroelectrics have no permanent dipole moments in their high-temperature paraelectric phase, in another class of ferroelectrics permanent dipole moments already exist above T_{FE}. In these so-called order–disorder ferroelectrics, in the paraelectric state these dipoles are statistically disordered with respect to site and time, but at the ferroelectric transition they align and polar order arises. Their temperature-dependent dielectric constant also exhibits a peak at the transition. However, in contrast to displacive ferroelectric, it also often reveals significant dispersion effects at audio and radio frequencies exhibiting the signatures of dielectric relaxation. This results from dipolar reorientations within double- or multi-well

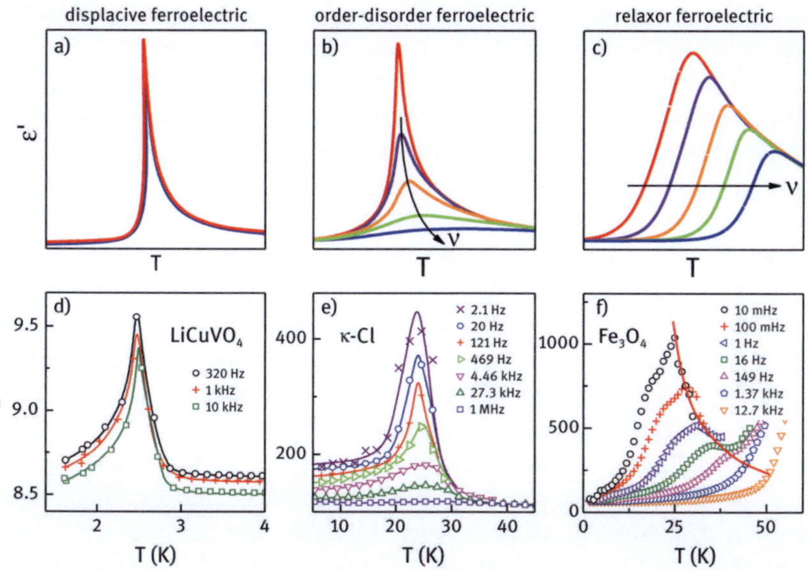

Figure 6.2: (a)–(c) Schematic plots of the temperature dependence of the dielectric constant of materials belonging to the three classes of ferroelectrics as typically detected by dielectric spectroscopy. The different lines represent $\varepsilon'(T)$ at different frequencies. Frames (d)–(f) show $\varepsilon'(\nu)$ of three multiferroics ($LiCuVO_4$ [27], κ-(BEDT-TTF)$_2$Cu[N(CN)$_2$]Cl [13] and Fe_3O_4 [8], respectively), which represent experimental examples corresponding to frames (a)–(c).

potentials, which slow down on decreasing temperature. In most cases, the detected dielectric response is almost monodispersive and approximately of Debye type, implying a single, well-defined relaxation time. In the paraelectric state, the dipolar units undergo fast reorientations between equivalent directions, while they reveal long-range order within the ferroelectric state. A schematic plot of the typical dielectric behaviour of displacive ferroelectrics is shown in Figure 6.2(b). While at low frequencies a well-defined and sharp peak arises in $\varepsilon'(T)$, it successively becomes suppressed with increasing frequency. Finally, almost no dielectric anomaly is detected at high frequencies, an effect that can occur in the kHz to MHz frequency range, depending on the material [26]. At low frequencies, only a small shift of the maxima in $\varepsilon'(T)$ is observed for different frequencies. Often the temperature dependence of the relaxation time determined from an evaluation of the frequency dependence of ε' and of the dielectric loss ε'' does not follow simple thermally activated Arrhenius behaviour. Instead, critical behaviour $\tau \propto 1/(T - T_C)^\gamma$ is frequently observed [26].

An example of a multiferroic exhibiting order–disorder ferroelectricity is provided in Figure 6.2(e). It shows $\varepsilon'(T)$ of the antiferromagnetic quasi-two-dimensional organic charge-transfer salt κ-(BEDT-TTF)$_2$Cu[N(CN)$_2$]Cl (κ-Cl) for several frequencies [13]. The strong suppression of the ferroelectric peak at high frequencies with only weak temperature shift, which is typical for order–disorder ferroelectrics (Figure 6.2(b)), is well documented. In this material, which will be discussed in more detail in Section 6.3.4, a combination of molecular dimerization and charge order leads to electronic ferroelectricity [13]. Above the charge-order transition, holes fluctuating within the dimers correspond to disordered dipoles, which finally order ferroelectrically below the transition.

The third class of polar materials treated here are the relaxor-ferroelectrics, which roughly can be regarded as ferroelectrics with a smeared-out phase transition (for reviews on these materials, see Refs. [28] and [29]). Often disorder helps promoting relaxor-ferroelectric states, which frequently is observed in solid solutions like the pseudo-cubic perovskite $PbMg_{1/3}Nb_{2/3}O_3$ [30]. Just as in the other classes of ferroelectrics, $\varepsilon'(T)$ of relaxors reveals a peak, however, with a smoothly rounded shape. Strong dispersion effects show up, which, especially in the frequency dependence of ε' and ε'', are reminiscent of the relaxation dynamics in glass-forming dipolar liquids as glycerol [31]. As indicated in the schematic plot in Figure 6.2(c), with increasing frequency the smeared-out ε'-peak shifts to higher temperatures and its amplitude becomes reduced. This results in characteristic dispersion effects below the peak temperatures. In relaxor-ferroelectrics, the observed shift of the peak temperatures in $\varepsilon'(T)$ usually signifies a Vogel–Fulcher–Tammann (VFT) law of the relaxation time $\tau(T)$ [32–34]:

$$\tau = \tau_0 \exp\left[\frac{B}{T - T_{VF}}\right] \tag{6.2}$$

Here, T_{VFT} is the divergence temperature of τ, where the dipolar dynamics finally becomes frozen, and B is an empirical constant. The temperature in eq. (2) represents the peak temperature in $\varepsilon'(T)$ and the relaxation time is deduced via $\tau = 1/(2\pi\nu)$, where ν is the frequency of the applied ac field. It should be noted that VFT behaviour is also found for the characteristic slowing down of molecular motion in glass-forming materials [31, 35]. The finding of a VFT law in relaxor-ferroelectrics usually is interpreted as being due to the glasslike freezing-in of short-range cluster-like ferroelectric order, assumed to exist in these materials [26, 28].

In most cases, the relaxation effects found in the dielectric spectra of relaxor-ferroelectrics do not show the characteristics of a simple monodispersive Debye relaxation. For the latter, the frequency dependence of the complex dielectric constant $\varepsilon^\star = \varepsilon' - i\varepsilon''$ should behave like $\varepsilon^\star = \varepsilon_\infty + (\varepsilon_s - \varepsilon_\infty)/(1 + i\omega\tau)$ with ε_∞ the high-frequency limit of ε' and $\omega = 2\pi\nu$. This leads to peaks in the dielectric-loss spectra with half widths of about 1.14 decades. However, in relaxor-ferroelectric, just like in most conventional glass-forming systems [31, 36, 37], the experimentally measured peaks are broadened. Various empirical functions are available to describe such broadened relaxation features in the dielectric spectra [38–40]. The broadening can be ascribed to a distribution of relaxation times arising from disorder [36, 37]. Usually, the occurrence of nano-scale ferroelectric clusters is assumed to explain the observed effects, in contrast to the macroscopic long-range ferroelectric domains existing in conventional ferroelectric states.

Finally, it should be noted that in relaxor-ferroelectrics, for temperatures above the dielectric-constant maximum, $\varepsilon'(T)$ often deviates from the characteristic Curie–Weiss behaviour, eq. (1), commonly found for conventional ferroelectrics. These deviations may arise due to short-range correlations between the mentioned nanoscale ferroelectric regions which are the precursors of the freezing-in of polarization fluctuations into a glasslike low-temperature state [41].

In Figure 6.2(f), we show an example of a multiferroic relaxor-ferroelectric, Fe_3O_4 [8]. It exhibits a stronger shift of the $\varepsilon'(T)$ peak with frequency than in the order–disorder system κ-Cl (Figure 6.2(e)) [13] and a broader, smeared-out peak. The ferroelectricity in this material is assumed to be driven by charge order [42], but the reason for the observed relaxor behaviour is not finally clarified. In Section 6.3.3, we provide a more detailed discussion of this material.

One should be aware that semiconducting single crystals or ceramics as well as electrically heterogeneous media often show a barrier-layer capacitance, which can be of internal or external origin [21]. Examples for internal layers are grain boundaries in polycrystalline samples. External layers are usually formed at the sample surface, e.g. by the formation of Schottky diodes when applying metallic contacts to a semiconducting material. This can lead to so-called MW relaxations, non-intrinsic effects that should not be confused with the intrinsic relaxational response found in order–disorder or relaxor-ferroelectrics. MW relaxations are caused by the presence of thin insulating barrier layers within the sample that give rise to high capacitances. The latter can lead to very high, so-called "colossal" [43], dielectric constants, which can impede the detection of an underlying ferroelectric ordering. These layers cause the artificial occurrence of a strong relaxation-like frequency-dependence of the dielectric properties, even without the presence of intrinsic relaxations in the investigated material. This can be understood when modelling the thin interface regions and the bulk sample by an appropriate equivalent circuit [21, 44]. Usually, at low frequencies the high capacitance of the layers dominates, leading to colossal dielectric constants, while at high frequencies the layer capacitors become shorted and the intrinsic properties may be detected. However, often intrinsic and non-intrinsic effect superimpose, hampering an unequivocal evaluation of the experimental data. In addition, these layers often show non-linear dielectric response (e.g. the Schottky diodes mentioned above), artificially mimicking ferroelectric hysteresis loop-like behaviour [45, 46].

Overall, great care should be taken to avoid MW effects leading to an erroneous claim of ferroelectricity or even multiferroicity. A prime example for a material showing MW relaxation due to internal and external barrier layer capacitances is $CaCu_3Ti_4O_{12}$ [47–49]. In this context, one should also mention $LuFe_2O_4$, which is often considered as a multiferroic with charge-order-driven polar order, i.e. electronic ferroelectricity [50]. However, the intrinsic nature of the detected dielectric properties of $LuFe_2O_4$ was questioned by several works, and non-intrinsic MW effects in this material seem very likely [51–53].

6.2.2 Signatures of ferroelectricity in polarization measurements

Polarization measurements are the most direct way to detect ferroelectricity in bulk materials. In the light of possible non-intrinsic contributions due to MW effects, a detailed dielectric analysis is advisable to deduce the temperature and frequency range as well as the electrical excitation fields, in which electric-field poling studies can be successfully performed. Positive-up-negative-down (PUND) measurements accompanying the hysteresis loop measurements can in addition verify the intrinsic nature of the detected polarization response [54].

Hysteresis loop measurements, for which the polarization *P* is measured in dependence of an external electric field *E*, represent the standard method for identifying ferroelectricity. This should lead to a clear hysteresis loop as schematically shown in Figure 6.3(a), which evidences the switchability of the polarization. It should be noted that, aside of the occurrence of a unique polar axis in the material, this switchability is an important characteristic of ferroelectrics. It essentially corresponds to domain nucleation, domain growth and the shifting of ferroelectric domain walls by the applied electrical field [26]. One should be aware that, at temperatures far below the ferroelectric transition, the domains often can be quite large and a possible domain-wall motion cannot by achieved by electrical fields of realistic amplitude (less than the breakdown field) [23–26, 54]. Therefore, *P*(*E*) measurements at higher temperatures, shortly below T_{FE}, often are desirable. However, at elevated temperatures Ohmic losses due to mobile charge carriers, leading to finite dc conductivity, can dominate the measured hysteresis loops, hampering the detection of switchability. Moreover, as mentioned above, non-intrinsic effects, e.g. Schottky diodes arising at the contact-sample interfaces, can lead to considerable non-linear response of the sample to an applied electrical field, which can give rise to apparent *P*(*E*) hysteresis loops, not related to ferroelectric switching [46].

An alternative to conventional *P*(*E*) measurements, partly avoiding these problems, is the so-called PUND experiments. For this experimental method, a sequence of field pulses is applied to the sample, with two positive pulses followed by two negative ones as indicated in Figure 6.3(b). For a ferroelectric state, the monitoring of the time-dependent current *I*(*t*) should reveal a response as schematically indicated in Figure 6.3(c). Ideally, a peak in the current should arise at the increasing flanks of the first and third pulses. This mirrors the switching of the polarization, involving the reorientation of the dipolar moments. This reorientation corresponds to a motion of charges and thus generates a current pulse. This pulse should, however, be absent at the second and forth pulses, which have the same polarity as the preceding ones. There the polarization was already switched into the same direction by the previous pulse and therefore no dipolar motion (and, thus, no current pulse) is expected. For non-intrinsic contributions to the polarization, the latter argument usually is not valid and *I*(*t*) looks the same for the two succeeding pulses with same polarity. In Figure 6.3(d) and Figure 3(f) we show an example, h-$ErMnO_3$ [12], for a ferroelectric hysteresis

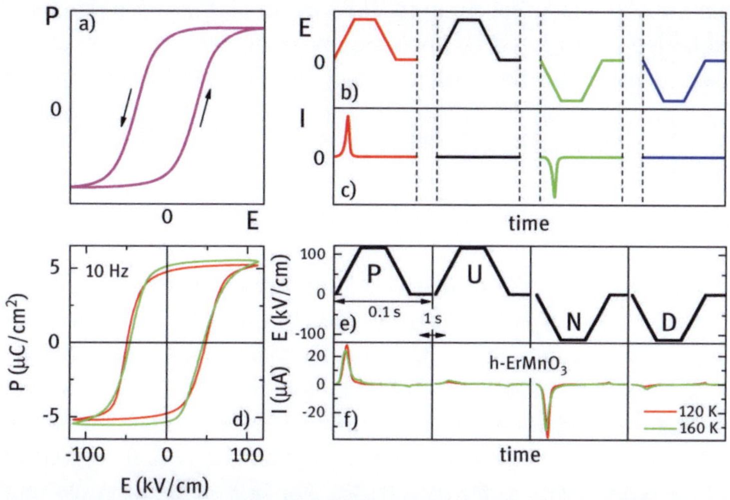

Figure 6.3: (a) Schematic plots of the *P*(*E*) hysteresis loop as expected for a ferroelectric state. (b) and (c) Schematic plot of the applied field (a) and of the polarization-induced current response (b) as expected for a PUND measurement of a ferroelectric. Frames (d)–(f) show *P*(*E*) and PUND results for h-$ErMnO_3$ [10, 12] at 120 and 160 K, which represent experimental examples corresponding to frames (a)–(c). The 160 K curve represents a measurement for which non-intrinsic effects start to contribute to the overall polarization indicated by the small peaks in the second and the forth pulses.

loop as well as the measured current of a PUND measurement (the ferroelectric properties are discussed in detail in Section 6.3.2). Both experiments are performed at 120 and 160 K. For the latter, the hysteresis loop (Figure 6.3(d)) shows an additional non-intrinsic contribution to the polarization. This non-intrinsic effect gives rise to small peaks in the current signal of the PUND measurement (Figure 6.3(f)) while the second and the forth voltage pulses are applied (Figure 6.3(e)). Therefore, this PUND method is very useful to discriminate between intrinsic and non-intrinsic polarization effects.

Finally, we want to mention that pyrocurrent measurements also can be used to check for the occurrence of ferroelectricity. They were e.g. of key importance for the first detection of the small polarizations in spin-spiral systems like $TbMnO_3$ [15]. Usually, for these experiments the sample is cooled below its ferroelectric transition temperature under an applied high electrical field, which ideally should lead to a monodomain polar state. In a subsequent heating run without external field, the pyrocurrent is monitored. When crossing the transition, the reorientation of the dipolar moments that return to their disordered states in the paraelectric phase should lead to a pyrocurrent pulse. From the integration of this time-dependent current signal, the polarization change at the transition can be determined. When applying the prepoling field in the opposite direction, the pyrocurrent also should invert its polarity. This can be regarded as a somewhat indirect check of the switchability of the polarization, prerequisite for a true ferroelectric state. It should be noted that this method is only suited for well insulating samples. Notably, in multiferroics ferroelectric ordering can often also be triggered by applying magnetic fields. In such cases, magnetocurrent measurements can be used to detect the polarization changes at these transitions.

6.3 Ferroelectric signatures in various multiferroics

In the following sections, we review the ferroelectric ordering and partially the tuneability for prominent examples of *type I* and *type II* multiferroics [14]. Ferroelectricity in type I multiferroics often originate from the lone-pair mechanism, geometrically driven effects or charge ordering. In $BiFeO_3$ [7], Bi^{3+} ions play the major role for the lone-pair mechanism featuring two outer 6s electrons not involved in chemical bonding. This "negative electron charge" gives rise to a high polarization and, in case of long-range ordered lone-pairs, leads to ferroelectricity. For this prime example of a room-temperature multiferroic system we briefly review the dielectric properties as well as the ferroelectric hysteresis loops for single crystals.

Charge-order, which is often observed in transition-metal compounds containing mixed valence ions, can result in ferroelectric ordering. Among other systems, like $TbMnO_5$ or Ca_3CoMnO_6 [55], $LuFe_2O_4$ [50] and the oldest known magnetic material, Fe_3O_4, are assumed to show this type of multiferroicity. We elucidate here the remarkable charge-order-driven relaxor-ferroelectric properties of Fe_3O_4.

The geometrically driven improper ferroelectricity observed, e.g. in the hexagonal manganite $YMnO_3$ [56] (as discussed in more detail in Chapter 3), is realized via a tilting of the MnO_5 bipyramids leading to Y displacements along the crystallographic *c*-direction at T_c far above ambient temperatures ($T_c > 900$ K) [11]. The antiferromagnetic ordering of the magnetic Mn^{3+} ions – for $ErMnO_3$ and $YMnO_3$ – occurs at temperatures around 80 K [57]. However, their complex ferroelectric domain patterns with topologically protected vortex structures and the fascinating functionalities arising at the domain walls put these multiferroics into the scientific focus for domain-wall-based nanoelectronics [5, 6, 58].

Another driving force for the revival of the research on multiferroics, especially those of type II (Chapter 4), was the discovery of magnetic ferroelectrics exhibiting strong magnetoelectric coupling about 15 years ago. In 2003, Kimura *et al.* [15] demonstrated improper ferroelectric ordering in $TbMnO_3$, which can be controlled by an external magnetic field. The revealed polarization for $TbMnO_3$ is rather low and of the order 600 $\mu C/m^2$ [15, 16]. By partially or fully replacing the Tb^{3+} ion by Dy^{3+}, the ferroelectric polarization can reach values of up to 3,000 $\mu C/m^2$. Besides these two most famous type II multiferroics, dozens of further interesting systems exist, which are discussed in detail in various reviews [14, 18, 19]. For many of them, the evolution of spin spirals due to frustrated magnetic order, spin-orbit coupling as well as exchange striction for collinear spin structures magnetically induces an improper ferroelectric polarization by breaking the inversion symmetry. This multiferroic mechanism is directly linked to some type of frustrated spin order, which makes them one of the most exciting developments in this research field [1, 4, 14]. The most discussed mechanism is the inverse Dzyaloshinskii–Moriya interaction or spin-current in spin-spiral systems [59–61]. For that mechanism the intensively investigated $DyMnO_3$ and $LiCuVO_4$ constitute prime examples.

6.3.1 Lone-pair multiferroicity in $BiFeO_3$

The functional properties, possible applications and the morphologies of multiferroic $BiFeO_3$ belonging to the rare class of materials with long-range magnetic and ferroelectric order at ambient temperatures are discussed e.g. in Ref. [62]. Here, we only briefly document the dielectric and ferroelectric properties of single crystalline $BiFeO_3$, which crystallizes (space group R3c) in a rhombohedrally distorted perovskite structure [7, 63]. The R3c symmetry allows for weak ferromagnetism as well as for spontaneous polarization along the crystallographic [111] direction. The two valence electrons of the Bi-ion, which do not participate in the chemical bonding, play a crucial role establishing the high ferroelectric polarizability leading to P values of the order of 100 $\mu C/cm^2$ [7, 64, 65]. The synthesis of large stoichiometric $BiFeO_3$ crystals turned out to be challenging. Impurities and defects can significantly impact the magnetic and dielectric properties. For the latter, impurity-induced conductivity and, thus, the formation of MW type contribution can lead to spurious effects. Moreover, the detection of intrinsic ferroelectric hysteresis loops at ambient temperatures can be hampered or even impeded due to such extrinsic contributions. Nevertheless, the overall magnetic and structural properties of $BiFeO_3$ are well characterized. The system shows polar order close to 1,100 K and antiferromagnetic ordering at about 645 K [66], which varies, most likely due to sample quality, between 595 and 650 K [67]. The displacement of Bi^{3+} and Fe^{3+} ions with respect to the ideal perovskite structure contributes to the electric polarization, which is only slightly temperature dependent [68, 69].

Figure 6.4 shows for various frequencies the temperature-dependent dielectric constant (a) and the conductivity (b) of a $BiFeO_3$ single crystal [64]. Distinct signatures of MW relaxations are indicated by the step-like increase of $\varepsilon' \approx 80$ (the intrinsic value at low temperatures) to an upper plateau value of about 10^4. The inset in (a) denotes an enlarged view of the temperature-dependent intrinsic dielectric constant as detected for high frequencies and temperatures below 300 K. Here, MW contributions do not play any role and no significant frequency dependence is observed. A dielectric analysis close to the ferroelectric transition at high temperature is hampered due to possible decomposition of $BiFeO_3$. However, e.g. Krainik *et al.* [70] tried using microwave frequencies to measure the dielectric properties up to 1,120 K. The temperature-dependent conductivity of the $BiFeO_3$ single crystal is shown in Figure 6.4(b). At low temperatures, it is mainly dominated by ac conductivity due to hopping charge transport. The inset presents the ac conductivity behaviour for 100 K in a broad frequency range pointing towards a super-linear power law, i.e. the frequency exponent exceeds 1. The origin of this behaviour still needs to be clarified; however, it is also rather common in transition metal oxides [71]. The temperature-dependent dc conductivity σ_{dc} corresponds to the regions where $\sigma'(T)$ becomes frequency independent, which leads to a merging of the curves for different frequencies. It behaves thermally activated with an activation energy of 0.39 eV. Finally, at high temperatures $\sigma'(T)$ deviates from the frequency-independent σ_{dc} because MW contributions start dominating the measured conductivity.

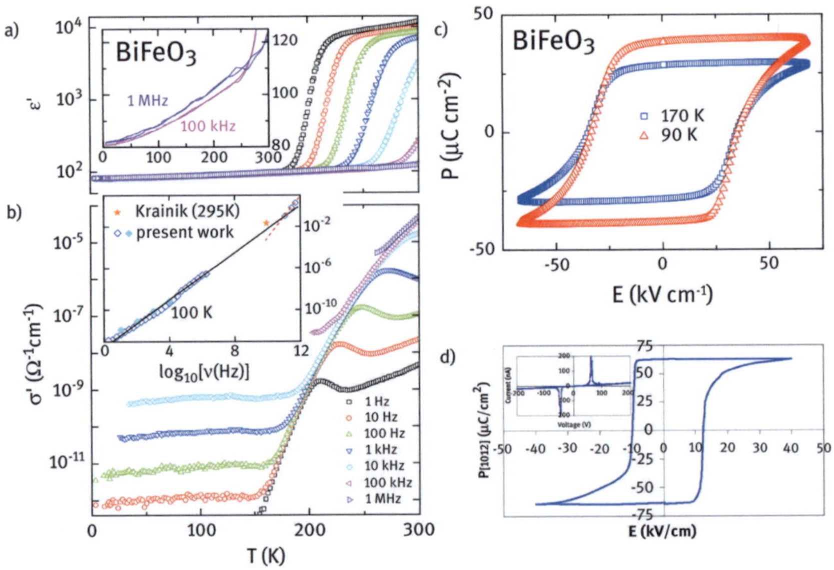

Figure 6.4: Temperature-dependent dielectric constant (a) and conductivity (b) for frequencies between 1 Hz and 1 MHz [64]. The inset in (a) denotes the temperature dependence of the dielectric constant as measured at 100 kHz and 1 MHz on an enlarged linear scale. The inset in (b) shows the frequency-dependent ac conductivity at 100 K [64, 70]. The solid line indicates power-law behaviour of the conductivity with a frequency exponent of 1.15. (c) Ferroelectric hysteresis loop measured at 6 Hz for 90 and 170 K [64]. (d) Ferroelectric hysteresis loop of a highly insulating $BiFeO_3$ single crystal at room temperature. The saturation polarization is in the order of 60 μCcm^{-2} and the coercive field is 12 kV/cm [65]. The inset shows the raw data of this polarization measurement. *[(a), (b) and (c) Reprinted with permission from Springer Nature, J. Lu et al., Eur. Phys. J. B* ***75****, 451 (2010). (d) Reprinted from D. Lebeugle, D. Colson, A. Forget, and M. Viret, Appl. Phys. Lett.* ***91****, 022907 (2007) with the permission of AIP Publishing].*

Ferroelectric hysteresis loops for that specific sample were measured at low temperatures (c.f. Figure 6.4(c)) [64]. For various temperatures below 200 K ferroelectric polarizations of the order of 30–40 $\mu C/cm^2$ and coercive fields around 20 kV/cm were detected, revealing typical hysteresis loops like in proper ferroelectrics. For another

$BiFeO_3$ bulk single crystal, exhibiting a five decades higher room temperature resistivity, Figure 6.4(d) presents a ferroelectric hysteresis determined at 300 K [65]. The values of the saturation polarization (35 μC/cm^2) and the coercive field (15 kV/cm) are rather similar and finally not too far away from the theoretically predicted values as well as from those detected in thin films [7]. Finally, conductivity matters and electric-field poling seems a feasible way of controlling the ferroelectric as well as the interesting domain properties in this rare example of a room-temperature multiferroic systems. More details on the magnetoelectric and dielectric properties are described in detail in the review by Catalan and Scott [62] as well as discussed in Chapter 2.

6.3.2 Geometrically driven improper ferroelectricity in $RMnO_3$

Geometrically driven systems, like the hexagonal manganites $RMnO_3$ (R = In, Y, Er, Dy, Ho, . . ., Lu), constitute a highly interesting class of multiferroics due to their versatile properties. Especially, the research on their complex (multi)ferroic domain structure aims at their application as functional elements in future nanoelectronics [6, 58, 72, 73]. Here, we focus on the ferroelectric properties of yttrium and erbium manganites. They display improper ferroelectricity ($T_C \approx$ 1,200–1,500 K) [74] exhibiting a cloverleaf pattern of the ferroelectric domains due to the formation of six possible domain states. The corresponding domain walls are arranged around vortex-like topological defects. At T_C, a structural transition via unit-cell tripling drives this ferroelectricity resulting in a fascinating domain pattern. An example is depicted as inset in Figure 6.5 revealing a vortex density of about 10^4/mm^2 (or 0.01/μm^2). The density of the complex vortex-domain pattern depends on how the samples were cooled from temperatures above T_C [75]. The corresponding ferroelectric domains and domain walls are relatively robust as they are attached to these topologically protected vortices [76]. Beside the effect of sample preparation on the vortex density, within these constraints electric fields should be perfectly suited for manipulating the domains and domain walls. However, electric-field poling studies of hexagonal manganites are rare [9, 11, 12]. Dielectric spectroscopy performed in a broad temperature and frequency range as well as polarization measurements enable the investigation of the macroscopic polar response to applied electric fields.

The detailed analysis of the dielectric properties allows determining the temperature- and frequency-dependent range for which e.g. single-crystalline h-$ErMnO_3$ exhibits purely intrinsic dielectric behaviour. So far, often extrinsic so-called MW polarizations arising e.g. from surface barrier layers, especially at ambient temperatures, have hampered field-poling measurements [77]. Insights into the polarization of hexagonal manganites were mainly gained from ferroelectric hysteresis loops recorded at fixed temperature and frequency, as well as pyrocurrent measurements [78–80]. However, many oxide materials [21, 47, 81] require – as already

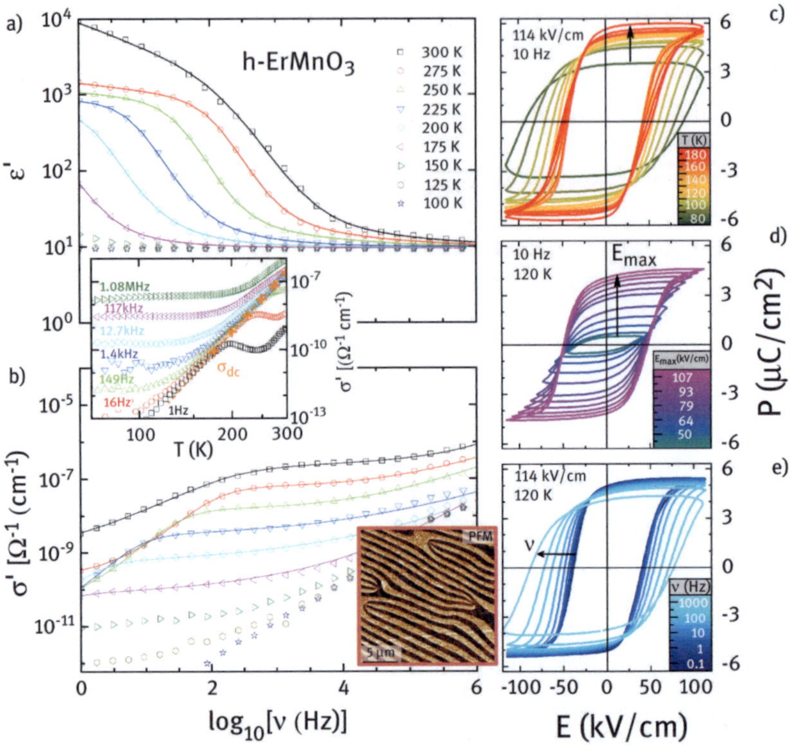

Figure 6.5: (a) Dielectric constant and (b) conductivity for various temperatures for a single crystalline h-$ErMnO_3$ sample oriented with out-of-plane polarization. The lines are fits with an equivalent circuit. The inset presents an Arrhenius representation of the conductivity featuring a thermally activated behaviour of the bulk dc conductivity, which is indicated by the pulse. (c) Temperature-, (d) electric field- and (e) frequency-dependent hysteresis-loop measurements. The lower inset denotes the polar surface pattern measured via piezo-response force microscopy [12]. *[A. Ruff et al. Frequency dependent polarisation switching in h-$ErMnO_3$, Appl. Phys. Lett.* ***112****, 182908 (2018) used in accordance with the Creative Commons Attribution (CC BY) license].*

mentioned – a thorough dielectric investigation accompanied by an equivalent-circuit analysis. This is an appropriate way for separating the intrinsic bulk dielectric properties from extrinsic effects. For example, a thorough dielectric analysis performed for $YMnO_3$ evidenced barrier-layer capacitances arising from extrinsic effects [77]. In a recent study [12], precise dielectric analyses of h-$ErMnO_3$ have determined a regime in which intrinsic ferroelectric hysteresis loops can be clearly revealed as a function of frequency, temperature and applied electric fields. Figure 6.5 shows the frequency-dependent dielectric constant ε'(a) and conductivity σ'(b) of h-$ErMnO_3$ for various temperatures. Similar to the $YMnO_3$ [77], for $T > 150$ K the dielectric constant reveals a distinct frequency-dependent step-like decrease from "colossal" [48] values of up to 10^4 to intrinsic values of about 15, which is a typical signature of a MW relaxation arising at interface layers. For low temperatures ($T < 150$ K) no step-like feature remains in ε' within the measured frequency range from 1 Hz to 1 MHz. An equivalent-circuit model is used to fit the dielectric spectra (denoted by the lines in (a) and (b)) [12, 21]. From these results the contributions of the bulk and MW-type interfaces can be separated.

For high temperatures, the conductivity $\sigma'(\nu)$ (Figure 6.5(b)) exhibits an increase from a low conductivity to a plateau, which denotes the bulk dc conductivity σ_{dc}. For higher frequencies, the bulk conductivity shows a power-law increase, which is best revealed at low temperatures. It indicates the regime of universal dielectric response, signifying hopping conductivity [82]. The temperature dependence of σ_{dc} is denoted by the plusses in the inset of (b). Its behaviour follows an Arrhenius law with an activation energy of 0.29 ± 0.01 eV, which is in excellent agreement with previously reported values [77, 83]. Consistent with the barrier-layer framework, σ_{dc} follows the left flank of the relaxation peak arising in $\sigma'(T)$ for various frequencies. Altogether, the step-like features in the dielectric properties are indeed the typical signatures of MW relaxations. The formation of a Schottky diode constituting thin barrier layers at the sample-electrode interface seems to be the dominant mechanism. However, also less conductive domain walls may play a crucial role, which would allow designing colossal-dielectric-constant materials via topological protected domain walls [77].

The saturation polarization for hexagonal manganites was calculated to be of the order of 5–6 μC/cm^2 [84]. As mentioned in Section 6.2.1, contributions from barrier-layer capacitances can superimpose intrinsic ferroelectric hysteresis loops. A proper analysis requires knowledge of the temperature and frequency range for which pure intrinsic ferroelectric switching behaviour can be detected. From the dielectric analysis of h-$ErMnO_3$, a frequency range from 0.1 Hz to 1 kHz within the temperature range from 80 to 180 K is determined for which pure intrinsic dynamic ferroelectric switching behaviour is expected to prevail. In addition, high electric activation fields E_a are required to reach saturation polarization. Figure 6.5 depicts well-defined hysteresis loops as function of temperature (c), electric excitation fields (d) and applied excitation frequency (e).

Ruff *et al.* [12] reported that measurements at temperatures around 120 K seem to be an ideal compromise of excluding non-intrinsic effects, which are confirmed by the absence of addition peaks in PUND measurements (c.f. Figure 6.6(b)), but still having achievable electric activation fields for polarization reversal. For measurements at 120 K, an excitation frequency of 10 Hz and applied fields of 114 kV/cm result in a text book-like hysteresis loop revealing good agreement with the calculated [84] saturation polarization P_s of the order of 5–6 μC/cm^2 [12]. When the temperature exceeds 150 K, the intrinsic hysteresis loop begins to be superimposed by the aforementioned barrier layer contributions, resulting in a slightly rounded saturation polarization for the highest applied electric fields (c.f. red curves in Figure 6.5(c)). To reach saturation polarization, it often requires at least an E_a exceeding two times $|E_c|$, which is of the order of 40 kV/cm (c.f. Figure 6.5(d)). In case of reduced E_a or at lower temperatures (E_c increases with decreasing temperature), only partial polarization reversal can be achieved resulting in a reduced value of P_s. In the case of h-$ErMnO_3$ it is assumed that domain growth of the predefined topological domains (see lower inset in Figure 6.5, which depicts the polar surface pattern measured by piezo-response force microscopy) [9, 12] or bubble domain formation is the underlying process for polarization reversal. One hundred and twenty kelvins is also ideally suited to investigate the dynamic frequency-dependent polarization-reversal process between 0.1 Hz and 1 kHz (Figure 6.5(e)). The coercive field strongly increases with increasing frequency, finally leading to a partial polarization switching for $\nu > 300$ Hz. Even in the case of maximum polarization, a distinct volume fraction of "opposite" domains remains, as domain walls between two topologically protected vortices are most likely not annihilated. However, these opposite domains are constricted to narrow "channel"-like domains, especially at high electric excitation fields [85]. For decreasing absolute values of the applied fields, the "channel"-like domains split up

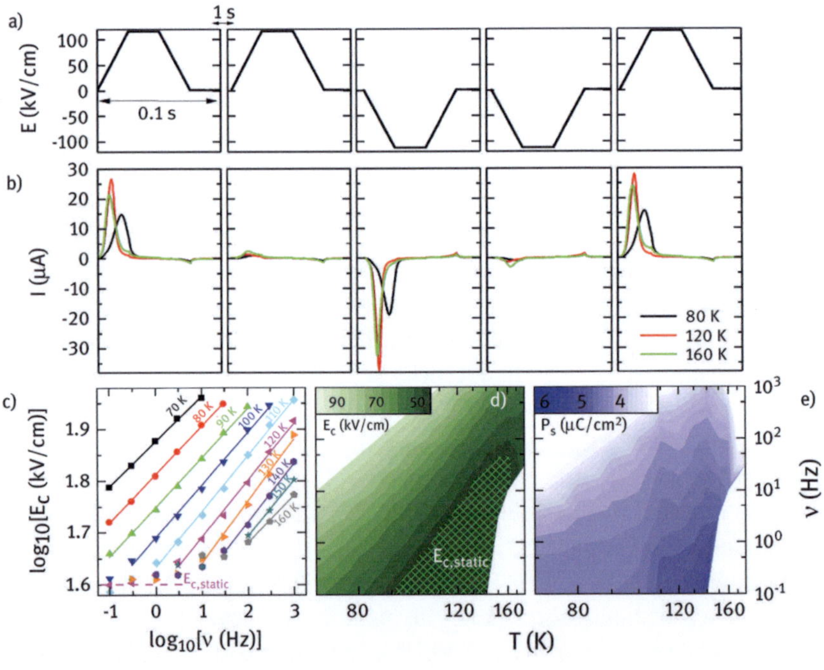

Figure 6.6: Positive-up-negative-down measurement of h-$ErMnO_3$ for three temperatures. (a) Sequence of the electrical excitation fields for the PUND measurement and (b) corresponding current signal. Peaks in frames I, III and V denote the ferroelectric switching, while the absence of peak features in II and IV (non-switching pulses) excludes artificial MW-type contributions. (c) Frequency dependence of the coercive field in double-logarithmic scale for various temperatures and excitation frequencies. Lines are linear fits following the Ishibashi–Orihara theory. Temperature- and frequency-dependent phase diagrams of (d) coercive field and (e) saturation polarization. *[A. Ruff et al. Frequency dependent polarisation switching in h-$ErMnO_3$, Appl. Phys. Lett.* ***112****, 182908 (2018) used in accordance with the Creative Commons Attribution (CC BY) license].*

again and two distinct domain walls develop. The remanent polarization denotes the balanced state of preferred domains versus the opposite "channel"-like domains.

Based on these results, special emphasis is put on the frequency-dependent polarization switching, which can be explained in terms of domain-wall movement. Interestingly, this improper ferroelectric material exhibits similar behaviour compared to proper ferroelectrics, showing pure domain-wall movement [12]. But, compared to ferroelectrics, such as $BaTiO_3$, the polarization is a secondary-order parameter. The impact of this somehow "predefined" and "fixed" domain pattern on the ferroelectric switching needs to be verified.

The hysteresis results of Figure 6.5 can be compared to a recent study by Yang *et al.* [86] who calculated the dynamic hysteresis loops based on the Ginzburg–Landau theory. The measured $P(E)$ loops ($\nu < 300$ Hz) are in perfect agreement with those calculated using the temporal evolution of the order parameters derived from the time-dependent Ginzburg–Landau equation [12, 86]. Interestingly, these loops from theory are solely derived from movements of domain walls connecting topologically protected vortices, which provides strong evidence for pure domain-wall motion as dominating process of the polarization reversal. Further interesting features are indicated in the current response of PUND measurements performed at three different temperatures (Figure 6.6(a) and Figure 6(b)). The peaks arising in response to the applied first, third and fifth electric-field pulses and the virtual absence of these peaks for pulses II and IV, indeed, strongly indicate an intrinsic ferroelectric switching (c.f. Section 6.2.2). However, the tiny but still significant humps in the non-switching pulses can be well explained in terms of the fusion process of two domain walls at sufficiently high E_a. In this case, two domain walls are so close to each other that they form the aforementioned "channel"-like domain, which itself acts as a domain-wall-like object. Releasing the applied electrical field on a very short time scale – due to the depolarization field – results in a splitting of that object back into the two distinct domain walls. This leads to a small reduction of P_s and to a tiny current response. Hence, this peak is observable at the decreasing flank of the switching pulse and on both flanks of the non-switching pulses (Figure 6.6(b)). Again, this is in perfect agreement with simulations [86].

The evolution of the temperature- and frequency-dependent hysteresis loops of h-$ErMnO_3$ is discussed in terms of pure domain-wall movement. From the Kolmogorov–Avrami–Ishibashi model, Ishibashi and Orihara [87] derived a more simplified scenario with deterministic nucleation of polar domains, which is often used to describe the switching kinetics in proper ferroelectrics. Domain-wall motion is solely responsible for the polarization reversal and it only depends on the frequency of the applied field and its waveform (usually sinusoidal). Within the scope of the Ishibashi–Orihara model, the coercive field should show a simple power-law behaviour, i.e. $E_c \propto \nu^{\beta}$. For this scenario, Figure 6.6(c) provides an analysis of the frequency-dependent logarithm of the coercive field for various temperatures while frequency- and temperature-dependent diagrams of E_c and P_s are presented in frames (d) and (e), respectively.

Linear fits yield an empirical β-exponent of about 0.103, which is of the same order as β-exponents of pure domain-wall motion in conventional ferroelectrics, like lead zirconate titanate ($\beta = 0.05$) [88] and strontium bismuth tantalate ($\beta = 0.12$) [54] as well as multiferroic domain-wall motions of $LiCuVO_4$ ($\beta = 0.08$) [89]. This constitutes another strong hint that pure domain-wall motion as proposed by Yang *et al.* [86] represents the major mechanism for polarization reversal in h-$ErMnO_3$. The phase diagrams shown in Figure 6.6(d) and Figure 6.6(e) demonstrate a thermally activated behaviour as well as the frequency-dependent kinetics of the domain-wall motion. This electric-field control of domain-wall motion may allow manipulating the real-space topological domain structure of hexagonal manganite.

6.3.3 Relaxor-ferroelectricity in magnetite

Magnetite is one of the oldest and most studied magnetic materials. Nature uses, for example, this compound in form of small crystals for the navigation of birds or magnetotactic bacteria orientating along the earth magnetic field. Altogether, the field for applications of magnetite is vast, ranging from classical applications as compass needles to magnetic recording. Besides its functionality at ambient conditions, this compound also fascinates due to its so-called Verwey transition [90, 91], which marks a metal-insulator transition at about 120 K enhancing its resistivity by at least two orders of magnitude. This is shown in Figure 6.7(a) where the temperature-dependent dc conductivity clearly signifies the Verwey transition, which emerges as a distinct jump in conductivity by about two orders of magnitude at 125 K. For temperatures below 200 K, an overall semiconducting behaviour is detected. Most likely, the charge ordering of Fe^{2+} and Fe^{3+}, arising from the mixed-valence state of iron ions in Fe_3O_4 (comprising one Fe^{2+}, two Fe^{3+} and four O^{2-} ions), plays a crucial role for the unusual properties of magnetite. Interestingly, charge-order-driven ferroelectricity was proposed to explain multiferroicity in $(PrCa)MnO_3$ [92] as well as in $LuFe_2O_4$ [50]. For the latter compound, the charge-order mechanism is controversially debated. Thorough dielectric and ferroelectric analyses [51–53] have provided strong hints that the observed "ferroelectric polarization" is of non-intrinsic origin, i.e. MW-type relaxation, representing a typical example of a transition-metal oxide with colossal dielectric constants [21]. Hence, ferromagnetic $LuFe_2O_4$ most likely does not exhibit long-range polar order, thus excluding proper multiferroicity. In contrast, it seems well justified to apply the concept of charge-order-driven multiferroicity to magnetite. This was confirmed by a thorough investigation of single-crystalline magnetite using broadband dielectric spectroscopy and frequency-dependent ferroelectric polarization measurements [8]. These results point to a regime of relaxor-like polar ordering in Fe_3O_4, arising from the continuous freezing of polar degrees of freedom and the formation of a tunnelling-dominated glasslike state at low temperatures. Within this phase, for which proper long-range ferroelectricity is excluded due to its centrosymmetric monoclinic symmetry [93], indeed distinct ferroelectric hysteresis loops are revealed (Figure 6.7(b)). Not only single-crystalline Fe_3O_4 exhibits ferroelectric properties, but also for thin films PUND measurements have evidenced the occurrence of multiferroicity at low temperatures [94]. However, based on polarization measurements alone, there may still be some doubts concerning their intrinsic origin, as in semiconducting materials Schottky-diode-induced surface layers, like in the aforementioned $LuFe_2O_4$, can lead to artificial dielectric non-linearity giving rise to hysteresis-loop-like electrical behaviour [45, 46]. Ensuring an intrinsic origin of the polar properties requires a thorough dielectric characterization in a broad temperature and frequency range.

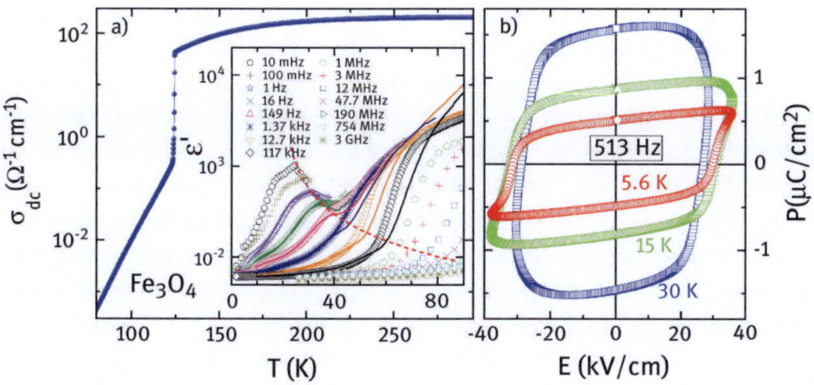

Figure 6.7: (a) Temperature-dependent dc conductivity for single-crystalline Fe_3O_4 [8]. The Verwey transition clearly shows up as jump in conductivity by two orders in magnitude. Inset: $\varepsilon'(T)$ of magnetite for various frequencies obtained with silver-paint (symbols) and sputtered gold contacts (solid lines). The dashed line, calculated assuming a Curie–Weiss law, illustrates the temperature dependence of ε_s of the main intrinsic relaxation. (b) Ferroelectric polarization P of Fe_3O_4 as a function of external electric field E at 513 Hz and three temperatures [8]. *[Reprinted with permission from F. Schrettle et al., Phys. Rev. B* ***83****, 195109 (2011). Copyright 2018 by the American Physical Society].*

The inset of Figure 6.7(a) depicts $\varepsilon'(T)$ at low temperatures (below the Verwey transition) for frequencies between 10^{-2} and 3×10^9 Hz. To reveal the influence of possible extrinsic Schottky layer-based effects, two different types of metal contacts were used [21]. Indeed a metal contact-related difference in dielectric properties is revealed, especially for low frequencies above 50 K, which is indicated by the discrepancy of the solid lines (sputtered gold contacts) and symbols (silver-paint contacts). $\varepsilon'(T)$ exhibits a distinct step-like feature from values of up to 10^3–10^4 down to its high-frequency limit of about $\varepsilon_\infty \approx 60$. For decreasing temperature, this step shifts to lower frequencies. For low temperatures, the dielectric spectra reveal at least three partially superimposing relaxation steps (c.f. 149 Hz curve, where their points of inflection are approximately located at 10, 30 and 50 K). One dielectric anomaly emerges as a broad maximum below 40 K; the corresponding static dielectric constant ε_s is strongly temperature-dependent as indicated by the dashed line. These are typical signatures of relaxor-ferroelectricity [8]. Most likely, the freezing-in of short-range clusters that are ferroelectrically ordered drives this behaviour. The mentioned strong increase of $\varepsilon_s(T)$ follows a Curie–Weiss law (dashed line in Figure 6.7(a)). At high temperatures, this relaxor-like feature is partially superimposed by extrinsic, electrode-dominated effects as evidenced by the disagreement of the $\varepsilon'(T)$ curves for different contact materials (symbols and lines in the inset of Figure 6.3). However, the relaxation feature itself is well reproduced in both measurements, clearly pointing its intrinsic nature. Moreover, even in relaxor-ferroelectrics well-developed ferroelectric hysteresis loops at low temperatures should arise. Therefore, to account for the intrinsic properties, Figure 6.7(b) shows ferroelectric hysteresis loops at sufficiently low temperatures (5.6, 15 and 30 K) and at a reasonable frequency of 513 Hz (Fe_3O_4 requires high electric excitation fields for poling and thus we have used a high-voltage booster, limiting the applicable frequency to the sub kHz-range). For 5.6 K, indeed a characteristic hysteresis loop is measured [8], which is further confirmed by the PUND results at 15 K of magnetite thin films [94]. Typical for relaxor-ferroelectrics are relatively broad loops in the nearly long-range-ordered state, i.e. at low temperatures, while more narrow hystereses should be detected in the vicinity of the transition (at higher temperatures) [28]. However, the rather strong temperature-dependent increase in conductivity of the single-crystalline Fe_3O_4 sample hampered hysteresis-loop measurements for $T > 15$ K. The intrinsic ferroelectric polarization of magnetite detected at 5.6 K is about an order of magnitude lower than that in conventional ferroelectrics.

In summary, magnetite reveals the typical signature of relaxor-ferroelectricity and a continuous slowing down of its polar dynamics [8]. Most likely the loss of inversion symmetry occurs on a local scale only due to the cluster-like short-range polar order. One can assume that the polar and charge degrees of magnetite are intimately related. Hence, the charge-order should also be of short-range type and exhibit freezing at low temperatures. However, it may not reach complete arrest. Instead, an evaluation of the relaxation dynamics of magnetite rather indicates a slowly fluctuating glasslike state at low temperatures (a “charge glass”), which is dominated by quantum-mechanical tunnelling [8]. Excess heat capacity as often found for glassy materials corroborates this scenario. A possible microscopic picture is that the suppression of long-range charge order in magnetite arises from the Fe^{2+}/Fe^{3+} *B*-site ions sitting on a strongly frustrated pyrochlore lattice. The shown results contribute to the ongoing debate of the presence of ferroelectricity in magnetite and of the discrepancy between the theoretically expected ferroelectricity and the frequent finding of a centrosymmetric structure.

6.3.4 Electric-dipole-driven magnetism in a charge-transfer salt

Charge-order-driven ferroelectricity constitutes a promising way to establish multiferroic order [42]. Charge order results from strong electronic correlations. Prime examples for such a mechanism are found in various organic charge-transfer salts having effectively ¼-filled hole bands [95, 96]. Indeed, electronic ferroelectricity was found in several members of this material class [95, 96] and even multiferroicity was detected [13, 97, 98].

An interesting recent example is the antiferromagnetic quasi-two-dimensional organic charge-transfer salt κ-$(BEDT\text{-}TTF)_2Cu[N(CN)_2]Cl$ (κ-Cl) [13]. Figure 6.8(a) shows a top view of the ac plane of κ-Cl revealing dimers of ET molecules with half-filled dimer bands [13, 99, 100]. Recent reports [101–103] demonstrate the importance of intra-dimer degrees of freedom and inter-site interactions. In a nutshell, within the ET layers adjacent molecules form dimers on which a single electron hole is delocalized above $T_{FE} \approx 25$ K. As proposed in Ref. [13], below T_{FE} the hole becomes localized, with its probability density being enlarged at one ET molecule of the dimer (c.f. Figure 6.8(a)), establishing long-range charge order (but see Ref. [104] for a contrasting view). The resulting ferroelectricity was clearly evidenced by dielectric and polarization measurements [13] as will be discussed below. Notably, κ-Cl exhibits a well-pronounced jump of the temperature-dependent conductivity close to 25 K [13, 105, 106], corroborating the occurrence of charge order at T_{FE}. At low temperatures, the hole spins constitute intralayer antiferromagnetic and interlayer ferromagnetic exchange, leading to an overall antiferromagnetic order, i.e. κ-Cl is multiferroic. As evidenced e.g. by magnetic susceptibility measurements [99, 107], at T_{FE} polar and spin order arise almost simultaneously. This was ascribed to electric-dipole-driven magnetic ordering, which is triggered by the reduction of geometrical frustration induced by the charge ordering [13]. This represents an unconventional type of multiferroic ordering where dipolar order drives spin ordering, in marked contrast to the well-known mechanism of spin-driven ferroelectricity as appearing, e.g. in $LiCuVO_4$.

The temperature-dependent dielectric constant $\varepsilon'(T)$ of κ-Cl is depicted in Figure 6.8(b) for various frequencies and the electrical field along the b direction. At around 25 K, well-pronounced peaks with values of ε' reaching up to several hundred are clearly revealed. Interestingly, the temperature of the peak maximum remains nearly constant for all frequencies while the amplitude strongly decreases with increasing frequencies. This is a typical behaviour of order–disorder-type ferroelectrics [26]. As discussed in Section 6.2.1, in this case disordered electric dipoles already exist at high temperatures and order below the phase transition at T_{FE} [25, 26], leading to an overall net polarization. For κ-Cl we assume that the electron holes at the dimers (c.f. Figure 6.8(a)) fluctuate between the two ET molecules at $T > T_{FE}$ and cooperatively lock-in at one of the two molecules for $T < T_{FE}$. Therefore, the polarization should primarily occur within the ac-plane. However, due to an inclined spatial orientation of the ET molecules [100], also a polarization along the b direction is induced.

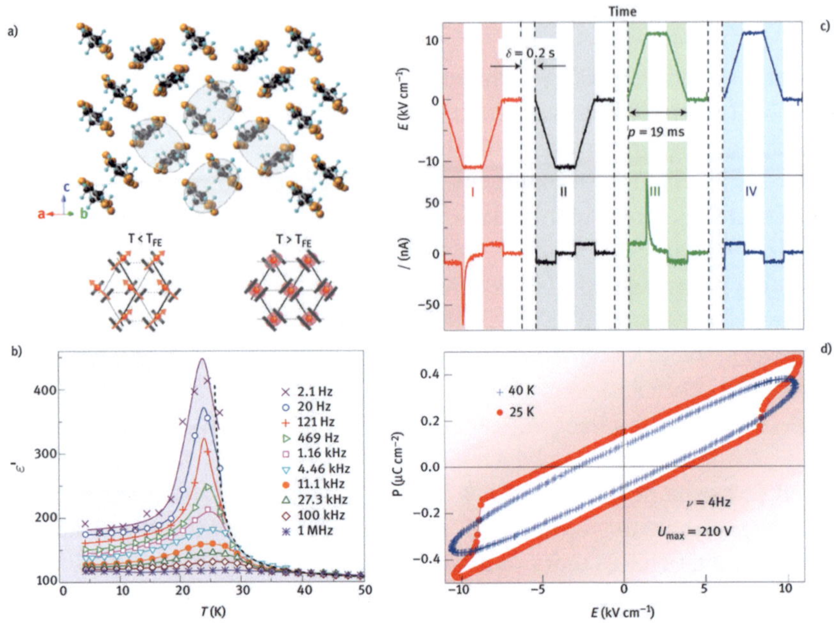

Figure 6.8: (a) Top view of the *ac* plane of κ-Cl showing the layers formed by the ET molecules [13]. In the upper part of the figure, selected dimers formed by adjacent ET molecules are highlighted by grey ellipses. The lower part of (a) schematically depicts the *ac* planes for temperatures below and above T_{FE}. Here the ET molecules are indicated by the thick grey lines. The red spheres denote the electron holes; for $T > T_{FE}$, the red shaded areas indicate their delocalization within the dimer. The orange arrows illustrate the dipolar moments proposed to arise below T_{FE} due to charge ordering. (b) Temperature dependence of the dielectric constant $\varepsilon'(T)$ for various frequencies measured along the *b* direction (solid lines are guides for the eyes) [13]. The dashed line indicates Curie–Weiss behaviour. (c) PUND measurement performed at 25 K with waiting time δ and pulse width p [13]. The upper part of the graph shows the excitation signal, the lower part the resulting time-dependent current. (d) Ferroelectric hysteresis loop measurement at temperatures below and above the ferroelectric phase transition at 4 Hz. *[Reprinted from P. Lunkenheimer et al., Nat. Mater.* ***11****, 755 (2012)].*

We found that this polarization and the corresponding peak in $\varepsilon'(T)$ are much easier to detect for the electrical field oriented along this direction. This follows from the fact that the dc charge transport along *b* in this material is much lower than within the ac planes [13], where it strongly hampers meaningful dielectric and polarization measurements [108].

It should be mentioned that the occurrence of CO in κ-Cl is still controversially discussed [104, 109, 110]. However, in the related systems κ-(BEDT-TTF)$_2$Hg(SCN)$_2$Cl and α-(BEDT-TTF)$_2$I$_3$, the occurrence of CO is undisputed. In the first system, very recently CO-driven ferroelectricity was clearly evidenced by length-change measurements and dielectric spectroscopy [111]. Moreover, in the α system, relaxor-ferroelectricity was reported, again based on similar mechanisms as discussed for κ-Cl above [112].

Dielectric spectroscopy alone usually is insufficient to provide a final proof of ferroelectricity [45, 46]. As discussed in Section 6.2.2, in addition non-linear polarization measurements should be performed, especially ferroelectric hysteresis-loop and/or PUND measurements [54]. Figure 6.8(c) shows the results of a PUND measurement applied to κ-Cl, for which four trapezoid pulses, preceded by a prepoling pulse (not shown), were applied to the sample [13]. The detected current signal reveals strong peak-like features occurring when the electric field exceeds about 10 kV/cm for the first and third pulses, signifying the switching of the polarization. Just as expected for proper ferroelectrics (c.f. Section 6.2.2), the second and forth pulses do not exhibit corresponding current features. The occurrence of ferroelectricity in κ-Cl is further confirmed by $P(E)$ measurements (Figure 6.8(d)), revealing the typical ferroelectric hysteresis loops in the ordered phase at $T < T_{FE}$. In contrast, above T_{FE} the elliptical shape of the $P(E)$ curve signals a linear polarization response with additional loss contributions from charge transport.

In summary, the reported experimental results provide strong evidence for ferroelectricity in κ-Cl, coinciding with the onset of magnetic ordering. Notably, the ferroelectric ordering in this system seems to be based on a primarily electronic mechanism, constituting a charge-order-driven ferroelectric state. Indeed, charge-transfer salts were considered as good candidates for multiferroics with a ferroelectric state driven by charge order [42]. Interestingly, the results point towards a multiferroicity mechanism where ferroelectric order drives the magnetic one, in marked contrast to the well-known spin-driven multiferroics. From a fundamental point of view, this and related materials [101, 111, 112] are highly interesting due to their exotic multiferroic properties. From a more applied perspective, the transition temperatures as well as the limited magnitudes of polarization make them unsuited for technical applications.

6.3.5 Spin-driven improper ferroelectricity in $DyMnO_3$

$DyMnO_3$ crystallizes in a perovskite structure when grown in a specific atmosphere [16]. At low temperatures ($T_N \approx 40$ K) and in zero magnetic fields the system shows a collinear sinusoidal spin structure [113], which locks in at $T_C \approx 20$ K into a non-collinear cycloidal spin spiral giving rise to improper ferroelectricity due to the inverse Dzyaloshinskii–Moriya interaction or spin-current mechanisms [59–61], with tilted spins $\mathbf{S}_i$ and $\mathbf{S}_{i+1}$ at neighbouring atomic sites i and $i + 1$. A prerequisite for ferroelectric ordering is the breaking of the inversion symmetry, which arises in a spin-spiral system due to this tilted spin configuration. The ferroelectric polarization follows a distinct symmetry relation in these spiral magnets:

$$\mathbf{P} \propto \mathbf{e} \times \mathbf{Q}, \tag{6.3}$$

where **Q** is the propagation vector of the spin spiral and $\mathbf{e} = (\mathbf{S}_i \times \mathbf{S}_{i+1})$ corresponds to the vector along the spiral axis, i.e. the normal vector of the spiral plane [59–61]. The tilting direction, i.e. clockwise or counterclockwise, of these two spins determine the direction of the normal vector. Even for a 1D element, like a spin spiral, a chirality-like vector can be defined. As a consequence of this "handedness" of the spin spiral, the polarization can be switched, if the modulation direction is fixed [19, 20, 89].

For $DyMnO_3$, the spin moments lie in the *b*-*c* plane modulating along the *b*-axis, which induce a ferroelectric order along the *c*-axis. The magnetic-field dependence of the ferroelectric ordering is a typical signature for type II multiferroics. Applying a magnetic field along different crystallographic axes causes the ferroelectric polarization vector to switch, e.g. *H* along *b* gives rise to a polarization in *a* direction for $H > 1$ *T* at 2 K. In Figure 6.9(a), the magnetic field-dependent dielectric constant, measured at 10 kHz and for various temperatures, reveals a peak feature indicating a paraelectric to ferroelectric transition [16]. Magnetocurrent measurements (c.f. Figure 6.9(b)) confirm these findings and a polarization occurs only in the presence of an applied magnetic field. For 2 K, a rather high polarization of the order of 3,000 μC/m^2 is detected in the vicinity of the ferroelectric transition. The amplitude of the polarization is temperature dependent and vanishes above the "lock-in" temperature T_C. Figure 6.9(c) shows the *P*(*E*) behaviour at 4.2 K in zero and applied magnetic fields. Indeed, in zero magnetic fields only linear electric-field-dependent contributions are detected. However, for the switched case, which is reached by applying an external magnetic field of 6 *T*, a typical ferroelectric hysteresis loop establishes, confirming the saturation polarization of the order of 3,000 μC/m^2. Within the framework of a spin-spiral mechanism, the electric-field-dependent polarization indicates that two domains, with left-handed and right-handed helicity of the spin spirals, exist, which allow $\pm P_a$ keeping the spin-spiral plane and the modulation direction constant. Thus, the volume fraction of domains with different helicities of the spin spiral can be switched by an electric field. However, it should be noted that this multiferroic effect is also discussed in terms of an exchange-striction mechanism, especially at low temperatures, due to Dy-Mn interactions [114].

6.3.6 Spin-driven improper ferroelectricity in the spin-½ chain cuprate system $LiCuVO_4$

Spin-spiral systems show the simultaneous existence and strong coupling of ferroelectric and magnetic ordering [14, 19, 59–61]. Especially, the latter allows investigating the interplay of electrical and magnetic degrees of freedom leading to complex multiferroic ordering. In this section, we focus on the multiferroic properties of a typical spin-driven ferroelectric material, the spin-½ chain cuprate $LiCuVO_4$ [17]. In general, spin-½ systems display a rich variety of exotic ground states, which often are based on competing magnetic exchange leading to frustration. One of the simplest

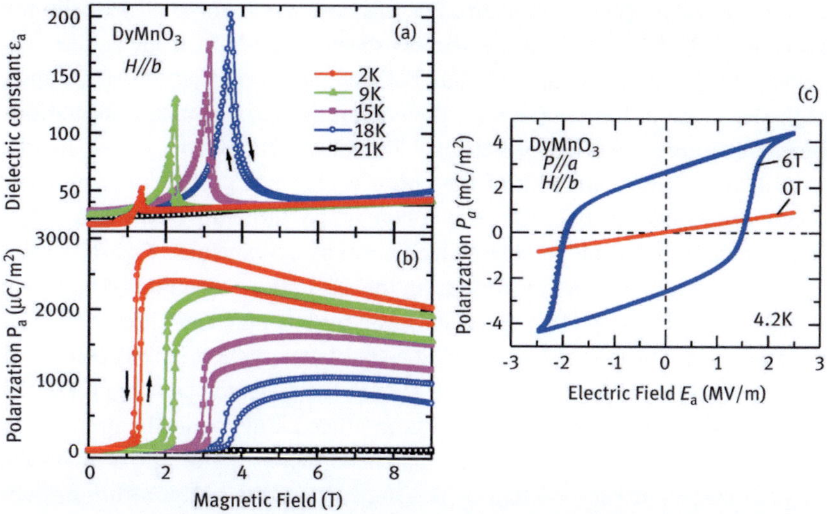

Figure 6.9: (a) Dielectric constant, (b) polarization and (c) $P(E)$ loops of a $DyMnO_3$ single crystal measured as function of magnetic field, which is applied along the b-axis. The dielectric constant and the polarization are detected along the a-direction for selected temperatures. The polarization loops are obtained at magnetic fields of zero and 6 T. *[Reprinted with permission from T. Kimura et al., Phys. Rev. B 71, 224425 (2005). Copyright 2018 by the American Physical Society].*

frustrated systems is the spin-½ chain, with competing nearest and next-nearest neighbour interactions. For $LiCuVO_4$, they lead to the formation of a spin-spiral order formed in the *ab* plane (chiral vector **e** along **c**) in the 3D ordered phase below T_N of 2.5 K with the propagation along the ***b*** direction [27]. In zero magnetic fields, ferroelectric polarization is induced along the ***a*** direction [17, 27]. Applying external magnetic fields affects the orientation of the chiral vector, which above a critical field H_1 of 2.5 T aligns along the magnetic field direction. Above H_2, which is of the order of 7.5 T, the spin spiral is transformed into a modulated collinear spin structure impeding an electrical polarization according to eq. (3). As a consequence of eq. (3) and due to the fact that the chirality vector aligns with the external magnetic field, external magnetic field between 2.5 and 7.5 T can switch the direction of the electrical polarization. Results of dielectric spectroscopy in applied fields up to 9 T on a single-crystalline sample oriented along different directions nicely demonstrate this switching behaviour of the ferroelectric polarization [17, 20, 27]. Figure 6.10(a) and Figure 10(b) show the temperature-dependent dielectric constant in various external magnetic fields for two cases: first (Figure 6.10(a)), H along the **c** direction measuring the polarization along ***a*** direction denoting the "non-switching" case as the chiral vector remains in **c** direction, when exceeding H_1. Second (Figure 6.10(b)), H is applied along ***a*** direction while detecting the dielectric properties and thus the polarization along **c**. Following eq. (3), this denotes the "switching" case. Not shown is another possibility, i.e. H along the crystallographic $\boldsymbol{b}$ direction. For this case it was revealed that the polarization vanishes at H_1 [27]. Interestingly, this is indeed a spin-helix state, as the rotation of the spin spiral is in $\boldsymbol{ac}$ plane and thus both $\mathbf{e}$ and $\mathbf{Q}$ point along $\boldsymbol{b}$. In Figure 6.10(a) a peak shows up in the temperature-dependent dielectric constant for various magnetic fields signalling the onset of improper ferroelectricity (Section 6.2.1). As discussed above, at $H > H_1$ no switching of the chiral vector is expected for H along $\mathbf{c}$. Exceeding higher magnetic fields $H > H_2$, the collinear spin structure impedes an electrical polarization and the peak in the dielectric constant vanishes. Figure 6.10(b) shows the case of switching the polarization from the $\boldsymbol{a}$ into the $\mathbf{c}$ direction by applying an external magnetic field along the $\boldsymbol{a}$ direction. Even for low magnetic fields, $H < H_1$, a small peak feature is still visible at T_N, denoting a slight experimental misalignment of the sample [20]. However, for higher field, $H_1 < H < H_2$, the dielectric peak denoting the ferroelectric polarization becomes much more pronounced. This perfectly agrees with the expectation for the switching case, where the chiral vector rotates into $\boldsymbol{a}$ direction, leading to a polarization along $\mathbf{c}$ direction. The maxima in the temperature-dependent dielectric constants, which can be easily observed as projection onto the magnetic field-dependent plane, are close to the critical fields H_1 and H_2. As the dielectric constant is a measure for the strength of the polarization, this experiment indicates that the polarization can be indeed switched by applying an external magnetic field, confirming previous magnetocurrent and magnetodielectric experiments [17, 27]. In addition, ferroelectric hysteresis-loop measurements and PUND studies are also feasible, pointing towards an electric-field poling of the orientation of the chiral vector and thus the switching of a counterclockwise to clockwise spin spiral and vice versa via electric fields [20, 89].

Figure 6.10(c) shows a ferroelectric hysteresis-loop measurement in the ordered state at 2 K (subtracted linear paraelectric background). The frequency of the electric excitation field varies from 0.1 to 300 Hz. The revealed loop resembles the typical signature of a proper ferroelectric with a rather low saturation polarization of 20 $\mu C/m^2$, which is in the order of the theoretically predicted values [115]. The shape of the hysteresis loop smears out with increasing frequency and the coercive field increases. This feature may indicate a domain-wall movement of multiferroic domains originating from clockwise and counterclockwise spin-spiral states [89]. In Figure 6.10(d), results of PUND measurements at 2 K confirm the switching of the spin-driven ferroelectric polarization: additional peak features in the current responses of the first and third pulses signal switching of the macroscopic polarization. As discussed in Section 6.2.2, the absence of such peaks in the second and forth pulses – for theses pulses the polarization was already switched by the preceding pulse – supports the intrinsic nature of ferroelectric switching in $LiCuVO_4$. However, the underlying physical mechanism of multiferroic domains is still under debate.

In summary, the temperature and magnetic field-dependent dielectric properties of the spin-driven ferroelectric $LiCuVO_4$ were investigated in detail close to the

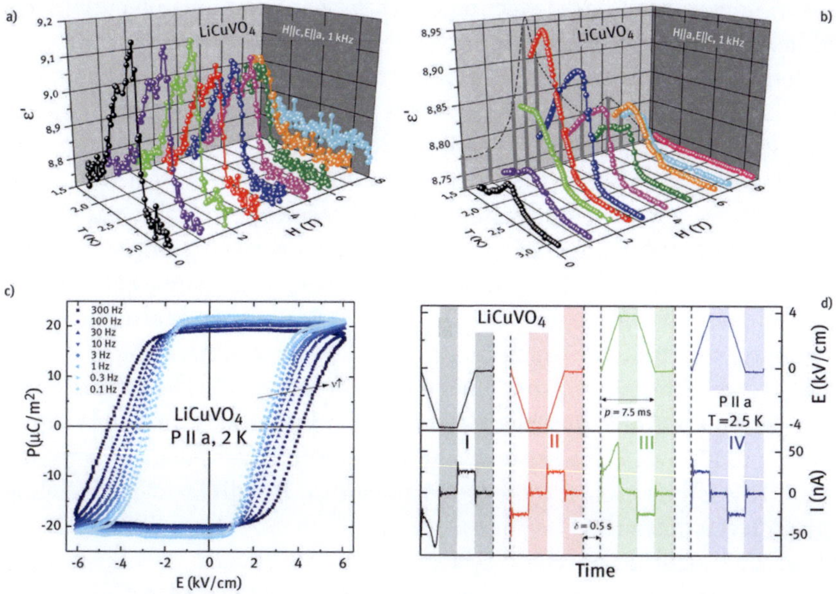

Figure 6.10: Dielectric constant at 1 kHz of a $LiCuVO_4$ single crystal as function of temperature and magnetic field, for (a) *E* along *a* and *H* along the ***c*** direction (non-switching case at *H* = 2.5 *T*) and (b) *E* along ***c*** and *H* along the *a* direction (switching case at *H* = 2.5 *T*) [20]. (c) Ferroelectric hysteresis loop at 2 K for various frequencies (the paraelectric background is subtracted for clarity reasons) [89]. (d) PUND measurement denoting the time-dependent excitation signal *E* and resulting current *I* for *P* parallel to the *a* direction at 2.5 K [20]. Polarization switching is evidenced by the spikes in the current response of the voltage pulses I and III as well as their absence in II and IV. *[(a), (b) and (d) Reprinted with permission from IOP Publishing, A. Ruff et al., J. Phys.: Condens. Matter **26**, 485901 (2014). (c) A. Ruff et al. Multiferroic Hysteresis Loop, Materials **10**, 1318 (2017) used in accordance with the Creative Commons Attribution (CC BY) license].*

three-dimensional ordering of its spiral magnetic structure at about 2.5 K. From these results, the switching and non-switching cases, which are expected on the basis of the theoretical symmetry predictions for multiferroic spiral magnets, are nicely confirmed. In addition, the direct observation of electric-field poling via PUND measurements demonstrates the ability to electrically control the chiral vector of the spin spiral in $LiCuVO_4$ and may indicate multiferroic domains of different helicities.

6.4 Conclusions and outlook

We have provided an overview of investigations of ferroelectric switching behaviour and of the dielectric properties and ferroelectric polarization in a broad temperature range as well as in applied magnetic fields for various typical multiferroic type I and type II systems. $BiFeO_3$ and hexagonal manganites are promising materials for applications in domain-wall-based electronics. They bear enormous potential for a new generation of materials exhibiting tuneable domain-wall effects. Their ferroelectric properties were investigated in a broad temperature and frequency range via dielectric spectroscopy. The analysis of the detected ferroelectric hysteresis loops paves the way to manipulate the polarization of these systems and thus the ferroelectric domains by an electric field. Charge-order-driven ferroelectricity in magnetite constitutes another route for multiferroic ordering, which is more interesting from a fundamental point of view. This system reveals charge ordering, which induces relaxor-ferroelectric behaviour, solving the discrepancy between the theoretically expected ferroelectricity and the frequent finding of a centrosymmetric structure. Finally, for the type I multiferroics, we briefly reviewed the dielectric properties of the multiferroic organic charge-transfer salt, κ-Cl. The formation of charge-order-driven electronic ferroelectricity seems to be the most plausible explanation pointing to a novel mechanism of charge-order-induced magnetic ordering and coupling. This is in contrast to the well-known type II multiferroics exhibiting the spin-driven mechanism [13].

The dielectric properties of type II systems with magnetically driven ferroelectricity, exhibiting a close coupling of ferroelectric and magnetic ordering, are briefly discussed for $TbMnO_3$ and $DyMnO_3$. They show a rather high polarization as well as the magnetic control of their improper ferroelectricity. For $LiCuVO_4$, constituting a prime example for a spin-driven mechanism [19], magnetic-field-dependent studies of its dielectric properties revealed the close coupling of magnetic order and ferroelectricity. So far, only few reports address the non-linear electrically driven polarization of spin-driven multiferroics. The results of hysteresis-loop measurements at low temperatures and applied magnetic fields demonstrate the electrical-field control of the spin helicity, which can be at least partly switched from clockwise to counterclockwise or vice versa. This further corroborates the presence of (multi)ferroic domains.

There are many further examples for multiferroic compounds in literature and it is impossible to provide a detailed discussion of the polarization and dielectric properties of all of them in the present article. Very current examples are the lacunar spinels [116–119], where ferroelectricity is driven by orbital order and local electric polarization was even found to be related to the occurrence of skymions [117], small whirl-like topological spin objects which may be used for advanced types of data storage devices.

However, for all these materials it turned out that the conductivity is of utmost importance for measuring and manipulating the polar properties on a macroscopic and microscopic level, respectively. The fascinating polar properties of a broad variety of magnetoelectric and multiferroic systems allow to envisage novel functionalities and may pave the way to the development of multifunctional magnetoelectrics for applications in novel electronic devices.

Funding: We acknowledge funding from the Deutsche Forschungsgemeinschaft (DFG) via the Transregional Collaborative Research Center TRR80 (Augsburg, Munich) and the BMBF via ENREKON 03EK3015.

References

[1] Fiebig M. Revival of the magnetoelectric effect. J Phys D Appl Phys. 2005;38:R123.
[2] Hill NA. Why are there so few magnetic ferroelectrics? J Phys Chem B. 2000;104:66–94.
[3] Eerenstein W, Mathur ND, Scott JF. Multiferroic and magnetoelectric materials. Nature. 2006;442:759.
[4] Fiebig M, Lottermoser T, Meier D, Trassin M. The evolution of multiferroics. Nat Rev Mater. 2016;1:16046.
[5] Catalan G, Seidel J, Ramesh R, Scott JF. Domain wall nanoelectronics. Rev Mod Phys. 2012;84:119.
[6] Meier D. Functional domain walls in multiferroics. J Phys Condens Matter. 2015;27:463003.
[7] Wang J, Neaton JB, Zheng H, Nagarajan V, Ogale SB, Liu B, et al. Epitaxial $BiFeO_3$ multiferroic thin film heterostructures. Science. 2003;299:1719.
[8] Schrettle F, Krohns S, Lunkenheimer P, Brabers VAM, Loidl A. Relaxor ferroelectricity and the freezing of short-range polar and charge order in magnetite. Phys Rev B. 2011;83:195109.
[9] Choi T, Horibe Y, Yi HT, Choi YJ, Wu W, Cheong S-W. Insulating interlocked ferroelectric and structural antiphase domain walls in multiferroic $YMnO_3$. Nat Mater. 2010;9:253.
[10] Ruff A, Li Z, Schafnitzel M, Krohns S, Dielectric properties and resource criticality aspects of hexagonal manganite. Ceramic Transactions. Volume **264**. Proceedings of the 12th pacific rim conference on ceramic and glass technology 2018.
[11] Lilienblum M, Lottermoser T, Manz S, Selbach SM, Cano A, Fiebig M. Ferroelectricity in the multiferroic hexagonal manganites. Nat Phys. 2015;11:1070.
[12] Ruff A, Li Z, Loidl A, Schaab J, Fiebig M, Cano A, et al. Frequency dependent polarisation switching in h-$ErMnO_3$. Appl Phys Lett. 2018;112:182908.
[13] Lunkenheimer P, Müller J, Krohns S, Schrettle F, Loidl A, Hartmann B, et al. Multiferroicity in an organic charge-transfer salt that is suggestive of electric-dipole-driven magnetism. Nat Mater. 2012;11:755.
[14] Khomskii D. Classifying multiferroics: mechanisms and effects. Physics. 2009;2:20.
[15] Kimura T, Goto T, Shintani H, Ishizaka K, Arima T, Tokura Y. Magnetic control of ferroelectric polarization. Nature. 2003;426:55.
[16] Kimura T, Lawes G, Goto T, Tokura Y, Ramirez AP. Magnetoelectric phase diagrams of orthorhombic $RMnO_3$ (R = Gd, Tb, and Dy). Phys Rev B. 2005;71:224425.
[17] Naito Y, Sato K, Yasui Y, Kobayashi Y, Kobayashi Y, Sato M. Ferroelectric transition induced by the Incommensurate magnetic ordering in $LiCuVO_4$. J Phys Soc Jpn. 2007;76:023708.
[18] Kimura T. Spiral magnets as magnetoelectrics. Annu Rev Mater Res. 2007;37:387.
[19] Tokura Y, Seki S. Multiferroics with spiral spin orders. Adv Mater. 2010;22:1554.
[20] Ruff A, Krohns S, Lunkenheimer P, Prokofiev A, Loidl A. Dielectric properties and electrical switching behavior of the spin-driven multiferroic $LiCuVO_4$. J Phys Condens Matter. 2014;26:485901.

[21] Lunkenheimer P, Krohns S, Riegg S, Ebbinghaus SG, Reller A, Loidl A. Colossal dielectric constant in transition-metal oxides. Eur Phys J Spec Top. 2009;180:61.
[22] Devonshire AF. Theory of ferroelectrics. Adv Phys. 1954;3:85.
[23] Cochran W. Crystal stability and the theory of ferroelectricity. Adv Phys. 1960;9:387.
[24] Jona F, Shirane G. Ferroelectric crystals. London: Pergamon Press, 1962.
[25] Blinc R, Zeks B. Soft modes in ferroelectrics and antiferroelectrics. Amsterdam: North Holland Publishing, 1974.
[26] Lines ME, Glass AM. Principles and applications of ferroelectrics and related materials. Clarendon: Oxford, 1977.
[27] Schrettle F, Krohns S, Lunkenheimer P, Hemberger J, Büttgen N, Krug von Nidda H-A, et al. Switching the ferroelectric polarization by external magnetic fields in the spin = 1/2 chain cuprate $LiCuVO_4$. Phys Rev B. 2008;77:144101.
[28] Cross LE. Relaxor ferroelectrics. Ferroelectrics. 1987;76:241.
[29] Samara GA. The relaxational properties of compositionally disordered ABO3 perovskites. J Phys Condens Matt. 2003;15:R367.
[30] Smolenski GA, Isupov VA, Agranovskaya AI, Popov SN. Ferroelectris with diffuse phase transitions. Sov Phys Solid State. 1961;2:2584.
[31] Lunkenheimer P, Schneider U, Brand R, Loidl A. Glassy dynamics. Contemp Phys. 2000;41:15.
[32] Viehland D, Jang SJ, Cross LE, Wuttig M. Freezing of the polarization fluctuations in lead magnesium niobate relaxors. J Appl Phys. 1990;68:2916.
[33] Levstik A, Kutnjak Z, Filipiĉ C, Pirc R. Glassy freezing in relaxor ferroelectric lead magnesium niobate. Phys Rev B. 1998;57:11204.
[34] Glazounov AE, Tagantsev AK. Direct evidence for Vögel–fulcher freezing in relaxor ferroelectrics. Appl Phys Lett. 1998;73:856.
[35] Lunkenheimer P, Kastner S, Köhler M, Loidl A. Temperature development of glassy -relaxation dynamics determined by broadband dielectric spectroscopy. Phys Rev E. 2010;81:051504.
[36] Sillescu H. Heterogeneity at the glass transition: a review. J Non-Cryst Solids. 1999;243:81.
[37] Ediger MD. Spatially heterogeneous dynamics in supercooled liquids. Annu Rev Phys Chem. 2000;51:99.
[38] Cole KS, Cole RH. Dispersion and absorption in dielectrics I. Alternating current characteristics. J Chem Phys. 1941;9:341.
[39] Davidson DW, Cole RH. Dielectric relaxation in glycerine. J Chem Phys. 1950;18:1417.
[40] Havriliak S, Negami S. A complex plane analysis of -dispersions in some polymers. J Polym Sci C. 1966;14:99.
[41] Viehland D, Jang SJ, Cross LE, Wuttig M. Deviation from curie-weiss behavior in relaxor ferroelectrics. Phys Rev B. 1992;46:8003.
[42] van Den Brink J, Khomskii D. Multiferroicity due to charge ordering. J Phys Condens Matter. 2008;20:434217.
[43] Lunkenheimer P, Bobnar V, Pronin AV, Ritus AI, Volkov AA, Loidl A. Origin of apparent colossal dielectric constants. Phys Rev B. 2002;66:052105.
[44] Emmert S, Wolf M, Gulich R, Krohns S, Kastner S, Lunkenheimer P, et al. Electrode polarization effects in broadband dielectric spectroscopy. Eur Phys J B. 2011;83:157.
[45] Scott JF. Ferroelectrics go bananas. J Phys Condens Matter. 2008;20:021001.
[46] Loidl A, Krohns S, Hemberger J, Lunkenheimer P. Bananas go paraelectric. J Phys Condens Matter. 2008;20:191001.

[47] Homes CC, Vogt T, Shapiro SM, Wakimoto S, Ramirez AP. Optical response of high-dielectric-constant perovskite-related oxide. Science. 2001;293:673.

[48] Lunkenheimer P, Fichtl R, Ebbinghaus SG, Loidl A. Non-intrinsic origin of the colossal dielectric constants in $CaCu_3Ti_4O_{12}$. Phys Rev B. 2004;70:172102.

[49] Dittl A, Krohns S, Sebald J, Schrettle F, Hemmida M, Krug von Nidda H-A, et al. On the magnetism of $Ln_{2/3}Cu_3Ti_4O_{12}$ (Ln = lanthanide). Eur Phys J B. 2011;79:391.

[50] Ikeda N, Ohsumi H, Ohwada K, Ishii K, Inami T, Kakurai K, et al. Ferroelectricity from iron valence ordering in the charge-frustrated system $LuFe_2O_4$. Nature. 2005;436:1136.

[51] Ren P, Yang Z, Zhu WG, Huan CHA, Wang L. Origin of the colossal dielectric permittivity and magnetocapacitance in $LuFe_2O_4$. J Appl Phys. 2011;109:074109.

[52] Niermann D, Waschkowski F, de Groot J, Angst M, Hemberger J. Dielectric properties of charge-ordered $LuFe_2O_4$ revisited: the apparent influence of contacts. Phys Rev Lett. 2012;109:016405.

[53] Ruff A, Krohns S, Schrettle F, Tsurkan V, Lunkenheimer P, Loidl A. Absence of polar order in $LuFe_2O_4$. Eur Phys J B. 2012;85:290.

[54] Scott JF. Ferroelectric memories, 1st ed. Springer Press: Berlin/Heidelberg, Germany, 2000.

[55] Choi YJ, Yi HT, Lee S, Huang Q, Kiryukhin V, Cheong S-W. Ferroelectricity in an ising chain magnet. Phys Rev Lett. 2008;100:047601.

[56] Van Aken BB, Palstra TTM, Filippetti A, Spaldin NA. The origin of ferroelectricity in magnetoelectric $YMnO_3$. Nat Mater. 2004;3:164.

[57] Iwara N, Kohn K. Magnetoelectric effect and rare earth magnetic ordering of $ErMnO_3$. Ferroelectrics. 1998;219:161.

[58] Schaab J, Skjaervo SH, Krohns S, Dai X, Holtz M, Cano A, et al. Electrical half-wave rectification at improper ferroelectric domain walls. Nat Nano. 2018;13:1028.

[59] Mostovoy M. Ferroelectricity in spiral magnets. Phys Rev Lett. 2006;96:067601.

[60] Sergienko IA, Dagotto E. Role of the dzyaloshinskii-moriya interaction in multiferroic perovskites. Phys Rev B. 2006;73:094434.

[61] Katsura H, Nagaosa N, Balatsky AV. Spin current and magnetoelectric effect in noncollinear magnets. Phys Rev Lett. 2005;95:057205.

[62] Catalan G, Scott JF. Physics and applications of bismuth ferrite. Adv Mater. 2009;21:2463.

[63] Moreau JM, Michel C, Gerson R, James WJ. Ferroelectric BiFeO3 X-ray and neutron diffraction study. J Phys Chem Sol. 1971;32:1315.

[64] Lu J, Schmidt M, Günther A, Schrettle F, Mayr F, Krohns S, et al. Magnetic susceptibility, heat capacity, electric polarization and dielectric constants of single crystalline $BiFeO_3$. Eur Phys J B. 2010;75:451.

[65] Lebeugle D, Colson D, Forget A, Viret M. Very large spontaneous electric polarization in $BiFeO_3$ single crystals at room temperature and its evolution under cycling fields. Appl Phys Lett. 2007;91:022907.

[66] Smith RT, Achenbach GD, Gerson R, James WJ. Dielectric properties of solid solutions of $BiFeO_3$ with $Pb(Ti,Zr)O_3$ at high temperature and high frequency. J Appl Phys. 1968;39:70.

[67] Fischer P, Polomska M, Sosnowska I, Szymanski M. Temperature dependence of the crystal and magnetic structures of $BiFeO_3$. J Phys C Solid State Phys. 1931;13:1980.

[68] Palewicz A, Przenioslo R, Sosnowska I, Hewat AW. Atomic displacements in $BiFeO_3$ as a function of temperature: neutron diffraction study. Acta Cryst B. 2007;63:537.

[69] Palewicz A, Sosnowska I, Przenioslo R, Hewat AW. $BiFeO_3$ crystal structure at low temperatures. Acta Phys Pol A. 2010;117:296.

[70] Krainik NN, Khuchua NP, Zhdanova VV, Evseev VA. Phase transitions in $BiFeO_3$. Sov Phys Solid State. 1966;8:654.

[71] Lunkenheimer P, Loidl A. Response of disordered matter to electromagnetic fields. Phys Rev Lett. 2003;91:207601.
[72] Meier D, Seidel J, Cano A, Delaney K, Kumagai Y, Mostovoy M, et al. Anisotropic conductance at improper ferroelectric domain walls. Nat Mater. 2012;11:284.
[73] Schaab J, Cano A, Lilienblum M, Yan Z, Bourret. E, Ramesh R, et al. Optimization of electronic domain-wall properties by aliovalent cation substitution. Adv Electron Mater. 2016;2:1500195.
[74] Lorenz B. Hexagonal Manganites—($rMnO_3$): class (I) Multiferroics with strong coupling of magnetism and ferroelectricity. ISRN Cond Mat Phys. 2013;497073:2013.
[75] Griffin SM, Lilienblum M, Delaney KT, Kumagai Y, Fiebig M, Spaldin NA. Scaling behavior and beyond equilibrium in the hexagonal manganites. Phys Rev X. 2012;2:041022.
[76] Holtz ME, Shapovalov K, Mundy JA, Chang CS, Yan Z, Bourret E, et al. Topological defects in hexagonal manganites: inner structure and emergent electrostatics. Nano Lett. 2017;17:5883.
[77] Ruff E, Krohns S, Lilienblum M, Meier D, Fiebig M, Lunkenheimer P, et al. Conductivity contrast and tunneling charge transport in the vortex-like ferroelectric domain patterns of multiferroic hexagonal $YMnO_3$. Phys Rev Lett. 2017;118:036803.
[78] Adem U, Mufti N, Nugroho AA, Catalan G, Nohed B, Palstra TTM. Dielectric relaxation in $YMnO_3$ single crystals. J Alloy Compd. 2015;638:228.
[79] Dey P, Natha TK, Nanda Goswami ML. Room temperature ferroelectric and ferromagnetic properties of multiferroics composites. Appl Phys Lett. 2007;90:162510.
[80] Ismailzade I, Kizhaev SA. DETERMINATION OF THE CURIE POINT OF THE FERROELECTRICS $YMnO_3$ AND $YbMnO_3$. Sov Phys Solid State. 1965;7:298.
[81] Krohns S, Lunkenheimer P, Ebbinghaus SG, Loidl A. Broadband dielectric spectroscopy on single-crystalline and ceramic $CaCu_3Ti_4O_{12}$. Appl Phys Lett. 2007;91:022910.
[82] Jonscher AK. The 'universal' dielectric response. Nature. 1977;267:673.
[83] Holstad TS, Evans DM, Ruff A, Smabraten DR, Schaab J, Tzschaschel C, et al. Electronic bulk and domain wall properties in B-Site doped hexagonal $ErMnO_3$. Phys Rev B. 2018;97:085143.
[84] Fennie CJ, Rabe KM. Ferroelectric transition in $YMnO_3$ from first principles. Phys Rev B. 2005;72:100103.
[85] Han MG, Zhu Y, Wu L, Aoki T, Volkov V, Wang X, et al. Ferroelectric switching dynamics of topological vortex domains in a hexagonal manganite. Adv Mater. 2013;25:2415.
[86] Yang KL, Zhang Y, Zheng SH, Lin L, Yan ZB, Liu J-M, et al. Electric field driven evolution of topological domain structure in hexagonal manganites. Phys Rev B. 2017;96:144103.
[87] Ishibashi Y, Orihara H. A theory of DE hysteresis loop. Integr Ferroelectr. 1995;9:57.
[88] Scott JF, Ross FM, Paz de Araujo CA, Scott MC, Huffman M. Structure and device characteristics of $SrBi_2Ta_2O_9$-based nonvolatile random-access memories. MRS Bull. 1996;21:33.
[89] Ruff A, Loidl A, Krohns S. Multiferroic Hysteresis loop. Mater. 2017;10:1318.
[90] Verwey EJW, Haayman PW. Electronic conductivity and transition point of magnetite Fe_3O_4. Physica. 1941;8:979.
[91] Verwey EJW, Haayman PW, Romeijn FC. Physical properties and cation arrangement of oxides with spinel structures II. electronic conductivity. J Chem Phys. 1947;15:181.
[92] Lopes AML, Araujo JP, Amaral VS, Correia JG, Tomioka Y, Tokura Y. New phase transition in the $Pr_{1-x}Ca_xMnO_3$ SYstem: evidence for electrical polarization in charge ordered manganites. Phys Rev Lett. 2008;100:155702.
[93] Wright JP, Attfield JP, Radaelli PG. Long range charge ordering in magnetite below the verwey transition. Phys Rev Lett. 2001;87:266401.

[94] Alexe M, Ziese M, Hesse D, Esquinazi P, Yamauchi K, Fukushima T, et al. Ferroelectric switching in multiferroic magnetite (Fe_3O_4) thin films. Adv Mater. 2009;21:4452.

[95] Tomić S, Dressel M. Ferroelectricity in molecular solids: a review of electrodynamic properties. Rep Prog Phys. 2015;78:096501.

[96] Lunkenheimer P, Loidl A. Dielectric spectroscopy on organic charge-transfer salts. J Phys Condens Matter. 2015;27:373001.

[97] Yoshimi K, Seo H, Ishibashi S, Brown SE. Tuning the magnetic dimensionality by charge ordering in the molecular TMTTF salts. Phys Rev Lett. 2012;108:096402.

[98] Giovannetti G, Nourafkan R, Kotliar G, Capone M. Correlation-driven electronic multiferroicity in $TMTTF_2 - X$ organic crystals. Phys Rev B. 2015;91:125130.

[99] Kanoda K. Mott Transition and Superconductivity in Q2D Organic Conductors, chapter 22. In: Lebed A, editor. The physics of organic superconductors and conductors. Springer: Berlin, 2008:623–642.

[100] Toyota N, Lang M, Müller J. Low-dimensional molecular metals. Berlin: Springer, 2007.

[101] Abdel-Jawad M, Terasaki I, Sasaki T, Yoneyama NN, Kobayashi N, Uesu Y, et al. Anomalous dielectric response in the dimer Mott insulator κ-(BEDT-TTF)$_2$Cu$_2$(CN)$_3$. Phys Rev B. 2010;82:125119.

[102] Hotta C. Quantum electric dipoles in spin-liquid dimer Mott insulator κ-(ET)$_2$Cu$_2$(CN)$_3$. Phys Rev B. 2010;82:241104(R).

[103] Li H, Clay RT, Mazumdar S. The paired-electron crystal in the two-dimensional frustrated quarter-filled band. J Phys Condens Matter. 2010;22:272201.

[104] Sedlmeier K, Elsässer S, Neubauer D, Beyer R, Wu D, Ivek T, et al. Absence of charge order in the dimerized κ-phase BEDT-TTF salts. Phys Rev B. 2012;86:245103.

[105] Dressel M, Grüner G, Carlson KD, Wang HH, Williams JM. Studies of the microwave resistivity of κ-(BEDT-TTF)$_2$Cu[N(CN)$_2$]Cl. Synth Met. 1995;70:927.

[106] Ito H, Ishiguro T, Kubota M, Saito G. Metal-nonmetal transition and superconductivity localization in the two-dimensional conductor κ-(BEDT-TTF)$_2$Cu[N(CN)$_2$]Cl under pressure. J Phys Soc Jpn. 1996;65:2987.

[107] Miyagawa K, Kawamoto A, Nakazawa Y, Kanoda K. Antiferromagnetic ordering and spin structure in the organic conductor, κ-(BEDT-TTF)$_2$Cu[N(CN)$_2$]Cl. Phys Rev Lett. 1995;75:1174.

[108] Pinterić M, Miljak M, Biškup N, Milat O, Aviani I, Tomić S, et al. Magnetic anisotropy and low-frequency dielectric response of weak ferromagnetic phase in κ-(BEDT-TTF)$_2$Cu[N(CN)$_2$]Cl, where BEDT-TTF is Bis(ethylenedithio)tetrathiafulvalene. Eur Phys J B. 1999;11:217.

[109] Tomić S, Pinterić M, Ivek T, Sedlmeier K, Beyer R, Wu D, et al. Magnetic ordering and charge dynamics in κ-(BEDT-TTF)$_2$Cu[N(CN)$_2$]Cl. J Phys Condens Matter. 2013;25:436004.

[110] Lang M, Lunkenheimer P, Müller J, Loidl A, Hartmann B, Hoang NH, et al. Multiferroicity in the Mott Insulating Charge-Transfer Salt κ – (BEDT – TTF)$_2$Cu[N(CN)$_2$]Cl. IEEE Trans Magn. 2014;50:2700107.

[111] Gati E, Fischer JKH, Lunkenheimer P, Zielke D, Köhler S, Kolb F, et al. Evidence for electronically driven ferroelectricity in a strongly correlated dimerized BEDT-TTF molecular conductor. Phys Rev Lett. 2018;120:247601.

[112] Lunkenheimer P, Hartmann B, Lang M, Müller J, Schweitzer D, Krohns S, et al. Electronic relaxor ferroelectricity in charge-ordered -(BEDT-TTF)$_2$I$_3$. Phys Rev B. 2015;91:245132.

[113] Kimura T, Ishihara S, Shintani H, Arima T, Takahashi KT, Ishizaka K, et al. Distorted perovskite with e^1_g configuration as a frustrated spin system. Phys Rev B. 2003;68:060403(R).

[114] Wang HW, Li CL, Yuan SL, Wang JF, Lu CL, Liu J-M. The crucial role of Mn spiral spin order in stabilizing the Dy–mn exchange striction in multiferroic $DyMnO_3$. Phys Chem Chem Phys. 2017;19:3706.

[115] Xiang HJ, Whangbo M-H. Density-functional characterization of the multiferroicity in spin spiral chain cuprates. Phys Rev Lett. 2007;99:257203.

[116] Singh K, Simon C, Cannuccia E, Lepetit M-B, Corraze B, Janod E, et al. Orbital-ordering-driven multiferroicity and magnetoelectric coupling in GeV_4S_8. Phys Rev Lett. 2014;113:137602.

[117] Ruff E, Widmann S, Lunkenheimer P, Tsurkan V, Bordács S, Kézsmárki I, et al. Multiferroicity and skyrmions carrying electric polarization in GaV_4S_8. Sci Adv. 2015;1:e1500916.

[118] Ruff E, Butykai A, Geirhos K, Widmann S, Tsurkan V, Stefanet E, et al. Polar and magnetic order in GaV_4Se_8. Phys Rev B. 2017;96:165–119.

[119] Geirhos K, Krohns S, Nakamura H, Waki T, Tabata Y, Kézsmárki I, et al. Orbital-order driven ferroelectricity and dipolar relaxation dynamics in multiferroic $GaMo_4S_8$. Phys Rev B. 2018;98:224306.

Marco Campanini, Rolf Erni and Marta D. Rossell

7 Probing local order in multiferroics by transmission electron microscopy

Abstract: The ongoing trend toward miniaturization has led to an increased interest in the magnetoelectric effect, which could yield entirely new device concepts, such as electric field-controlled magnetic data storage. As a result, much work is being devoted to developing new robust room temperature (RT) multiferroic materials that combine ferromagnetism and ferroelectricity. However, the development of new multiferroic devices has proved unexpectedly challenging. Thus, a better understanding of the properties of multiferroic thin films and the relation with their microstructure is required to help drive multiferroic devices toward technological application. This review covers in a concise manner *advanced* analytical imaging methods based on (scanning) transmission electron microscopy which can potentially be used to characterize complex multiferroic materials. It consists of a first broad introduction to the topic followed by a section describing the so-called phase-contrast methods, which can be used to map the polar and magnetic order in magnetoelectric multiferroics at different spatial length scales down to atomic resolution. Section 3 is devoted to electron nanodiffraction methods. These methods allow measuring local strains, identifying crystal defects and determining crystal structures, and thus offer important possibilities for the detailed structural characterization of multiferroics in the ultrathin regime or inserted in multilayers or superlattice architectures. Thereafter, in Section 4, methods are discussed which allow for analyzing local strain, whereas in Section 5 methods are addressed which allow for measuring local polarization effects on a length scale of individual unit cells. Here, it is shown that the ferroelectric polarization can be indirectly determined from the atomic displacements measured in atomic resolution images. Finally, a brief outlook is given on newly established methods to probe the behavior of ferroelectric and magnetic domains and nanostructures during in situ heating/electrical biasing experiments. These in situ methods are just about at the launch of becoming increasingly popular, particularly in the field of magnetoelectric multiferroics, and shall contribute significantly to understanding the relationship between the domain dynamics of multiferroics and the specific microstructure of the films providing important guidance to design new devices and to predict and mitigate failures.

This article has previously been published in the journal Physical Sciences Reviews. Please cite as: Campanini, M., Erni, R., Rossell, M. D. Probing local order in multiferroics by transmission electron microscopy *Physical Sciences Reviews* [Online] 2020, 5. DOI: 10.1515/psr-2019-0068

https://doi.org/10.1515/9783110582130-007

Keywords: (scanning) transmission electron microscopy, multiferroics, phase-contrast techniques, electron nanodiffraction, strain mapping, local ferroelectric polarization, in situ measurements

7.1 Introduction

Over the last decade, numerous experimental methods based on transmission electron microscopy have undergone quantum leaps. These abrupt advancements are mainly due to the significantly improved stability of microscope platforms and the availability of new electron optical devices. Aside from high performance spectrometers for electron energy-loss (EELS) and energy-dispersive X-ray spectroscopies (EDXS), fast and highly sensitive digital cameras based on CMOS technology have been introduced. These advanced cameras have replaced the inherently sluggish slow-scan CCD cameras which were not optimal, particularly for monitoring dynamics by in situ measurements. Moreover, novel detector geometries, electron monochromators and most notably aberration correctors, which can correct the intrinsic geometrical and ultimately also the chromatic aberrations of electromagnetic electron lenses, have significantly improved the information which can be gathered from atomic-resolution micrographs [1, 2]. Most noteworthy consequences of these technical improvements are that true atomic resolution micrographs of heavy and light atoms can be recorded in transmission and scanning transmission electron microscopy (TEM/STEM) approaching or exceeding half an Ångström resolution. In addition, atomic potentials can be mapped with high signal-to-noise ratios and with minimal blurring enabling the determination of atomic positions with precision in the picometer regime. New detection strategies make it possible to map electrostatic and magnetic fields with unprecedented resolutions. And from an analytical point of view, the high-performance spectrometers mentioned above enable the detection of highly sensitive chemical signals as well as electronic structure information by EELS or EDX of individual atomic columns and even single atoms [3]. Materials, their atomic structure and chemistry, can thus be characterized down to the atomic level. The main advantage of these electron microscopy-based methods over other techniques is that information can be extracted locally, at defect sites such as dislocation cores, at phase and domain boundaries and at other irregularities. For averaged information, diffraction methods and other techniques might provide a more precise, in terms of its statistical significance, but a less detailed picture.

When it comes to the characterization of the polar order in multiferroic materials, the full advantage of all the technical achievements can be exploited. Precise derivation of atomic positions based on high-resolution micrographs enables local and quantitative measurements of strain states, the distortion of unit cells or the (re-)arrangement of atoms within these unit cells. This is particularly important as the ferroelectric properties are (often) coupled to the symmetry of the unit cell and the relative positions of the atoms therein. However, for magnetic order in multiferroics the situation becomes a bit more intricate as the lenses used in medium and high-voltage transmission electron microscopes (60–300 kV) are made of magnetic fields, which in the objective lens (OL), i. e. the most important lens of the microscope, can reach values between about 1 and 2 T at normal operational conditions. Hence, the environmental magnetic field of the OL, wherein the specimen is immersed, prevents the study of the intrinsic domain structure of (ferro- and ferri-) magnetic materials. Operating the TEM with the OL turned off by using a lens either before or after the sample in order to focus the electron beam in STEM or TEM, respectively, allows for preserving the specimen's magnetic structure during the measurement, but comes at the expense of less resolution. These so-called Lorentz TEM or Lorentz STEM modi enable an image resolution in the range of typically 1 nm [4]. Though in general atomic resolution is not feasible in Lorentz modus, the resolution is still sufficient to derive critical information about the magnetic domain structure of the sample whose typical length scale typically exceeds the nanometer range.

So far, methods have been mentioned which improved the inspection of samples in the physical state it was prepared and transferred into the vacuum of the microscope column. Such retroactive measurements are important to learn about the basic states of matter but do not allow for finding out how the material had reached that state. Another branch of TEM applications, whose capabilities have started to flourish, covers so-called *in situ or in operando* measurements [2]. The goal of such measurements is to change the properties and/or the physical state of the sample in a controlled way in the microscope while the sample is under observation. One might argue that the small TEM samples, which are merely tens of nanometer thick, do not represent the state of matter as it adopts in a device. Nevertheless, such approaches allow for uncovering fundamental mechanisms, like, e. g. the change of the local atomic structure while temperature or electrical bias is varied or the moving mechanisms of domain walls and the identification of pinning centers. Though in situ transmission electron microscopy is not new at all, the recent success of these methods lies in the availability of special sample holders for MEMS (Micro-Electro-Mechanical Systems) chips upon which a small piece of the sample material, normally extracted and prepared by a focused ion beam (FIB) instrument, is mounted and contacted to the chip's electrodes. The advantages of the MEMS-based in situ technology over the conventional approach, i. e. where typically the state of disk-shaped samples of several millimeter diameter is being stimulated, lies in the fact that control and homogeneity over the physical state and the applied stimulus are enhanced. Moreover, because of the smaller volumes inspected in MEMS devices, the sample drift is minimal which is particularly important for heating experiments. In addition, cooling or heating rates can be achieved which are orders of magnitude higher than what is achievable by a conventional thermal system. Hence, aside from studying samples extracted from devices and test architectures, modern transmission electron microscopes go beyond simple imaging tools. The implementation of in situ techniques enabled by MEMS-based sample holders provides a physical test bench inside the microscope with the advantage that while the material is undergoing an induced transition, it can be inspected at a variety of length scales even down to the atomic scale where changes in the structure of individual unit cells can be observed.

While, of course, the study of ferromagnetic and ferroelectric materials, their defects and domains, is not new in transmission electron microscopy by using a variety of methods, such as CBED, conventional bright- and dark-field transmission electron microscopy or, e. g. Lorentz TEM, in this review we would like to focus on (i) new high-resolution electron microscopy techniques which can be used to characterize magnetoelectric multiferroics and which can be used to learn about the local order parameters in these materials, (ii) on new experimental strategies of measuring the macroscopic electrostatic and magnetic fields and (iii) on the new possibilities of in situ or in operando measurements. This review thus targets on introducing and discussing newly established techniques, whose potential might not have yet been fully explored, using specific examples for illustration.

7.2 S/TEM techniques for field mapping

In ferroic materials, the formation of an ordered state is usually associated with a symmetry breaking of the system [5]. Such symmetry breaking responsible for the ferroic order is conveniently characterized by an order parameter, which describes the degree of order of the microscopic entities relevant to define the ferroic state of the system. The most common examples of order parameters are the polarization P and the magnetization M, which respectively describe the degree of order of the microscopic electric dipoles and spins within the material.

In multiferroic materials the combination of different symmetry breakings leads to the coupling of order parameters, allowing for modifying one ferroic state by the application of a conjugated impulse. Among multiferroic materials, magnetoelectric multiferroics are materials in which ordered states of spins and electric dipoles coexist and are coupled, in such a way that the application of an external electric field can change the magnetic properties of the material and vice versa [6]. The complex phenomenology of magnetoelectric materials arises as a consequence of their peculiar morphology and structure, which display a wide range of features at different length scales including polar and magnetic domains, domain walls, vortices, etc. The investigation of ferroelectric and ferromagnetic materials is thus very challenging, involving different length scales that range from the meso- to the nanoscale.

Phase-contrast techniques in transmission electron microscopy offer the unique advantage of high spatial resolution and direct sensitivity to electrostatic and magnetic fields [7–10], allowing a multi-scale investigation of multiferroic materials. The qualitative and quantitative estimation of the order parameters describing the ferroic state and their evolution during the transformations – e. g. from the disordered to the ordered phase – is of fundamental importance for understanding the material properties and interpret the material behavior under applied external stimuli.

Recently, phase-contrast techniques have been successfully employed – usually in combination with in situ cryogenic capabilities – in the investigation of novel spin structures (see Chapter 4), including topological structures, such as magnetic skyrmions [11, 12], vortices and anti-vortices [13]. In this framework, phase-contrast techniques hold great promise for exploring dynamical magnetoelectric phenomena, as proven by state-of-the-art in situ experiments like e. g. [14, 15] and discussed in Chapter 9.

This subsection focuses on the application of phase-contrast techniques in transmission electron microscopy in the investigation of the fundamental aspects of multiferroic materials and their static properties, with a special emphasis on the study of domain configurations, mesoscopic polarization, and atomic electric fields.

7.2.1 The spatial resolution of S/TEM phase-contrast techniques

The properties of multiferroic materials can be investigated in transmission electron microscopy by means of so-called phase-contrast techniques, i. e. electron holography, Lorentz microscopy, and differential phase-contrast STEM. In the case of ferroelectric specimens, such studies can be carried out at a length scale ranging from the nano to the atomic scale, while for magnetic specimens the constraints due to the geometry of the OL limit the resolution to the nanometer range.

In common transmission electron microscopes, the OL works in the immersive mode, i. e. the sample is inserted in between the lens pole pieces where the magnetic field has its maximum value (about 2 T) [16]. The OL has a very important role in the image formation process in both TEM and STEM modes and its strong magnetic field allows proper electron focusing and the achievement of high-resolution imaging. Nevertheless, the strong magnetic field generated by the OL in the specimen plane is deleterious for the study of the nanoscale magnetism since it can induce a significant alteration of the magnetic state of the specimen during the experiments.

Therefore, phase-contrast techniques for the study of magnetic properties – either in TEM or STEM mode – have to be performed in field-free conditions and the corresponding operating mode of the microscope is called Lorentz mode. The field-free condition can be achieved by switching-off the OL and using a different focusing lens that is located behind or before the specimen plane in the case of TEM or STEM operation modes, respectively. In modern microscopes, the Lorentz lens or the condenser mini-lens are employed to perform this task. Due to the limited resolution typical of Lorentz mode, the phase-contrast analysis of magnetic specimens gives access to their mesoscale properties, i. e. the magnetic domain configuration, the domain walls shape and geometry, topological charges (e. g. skyrmions), the stray fields, etc. On the contrary, the investigation of ferroelectrics is not affected by this resolution limitation. The polar properties can be investigated at nanometer resolution – in order to study the local ferroelectric order, the domain wall nature, etc. – or at atomic resolution. Aberration-corrected STEM holds great potential for studying the electric fields at atomic resolution, making accessible a direct comparison between the atomistic structure of the material and its functional properties.

Here, it is worth mentioning that recently a new approach has been proposed in order to perform electron holography and Lorentz microscopy experiments on magnetic specimens without switching off the OL [17]. The method relies on compensating the magnetic fields generated by the upper and lower pole pieces at the specimen plane, by producing two equal magnetic fields but with an opposite sign. This special condition shall warrant a field-free environment for the sample without losing the resolving power provided by the OL, making this method very promising for the future development of magnetic field imaging with sub-nanometer resolution.

7.2.2 The electron beam sensitivity to electrostatic and magnetic fields

On transmitting an electron transparent sample, the illuminating electron wave undergoes both a change in amplitude and phase. While the change in amplitude is at the core of so-called amplitude contrast methods which are, e. g. exploited in bright-field or dark-field imaging where a particular Bragg diffracted beam is selected to form the image, the interaction of the electron beam with electrostatic and magnetic fields induces a change in the phase of the electron wavefunction – the Aharonov-Bohm effect [18] – given by:

$$\phi(x,y) = \phi_e + \phi_m = \sigma \int_{-\infty}^{+\infty} V(x,y,z)dz - \frac{e}{\hbar} \int_{-\infty}^{+\infty} A_z(x,y,z)dz \tag{7.1}$$

where σ is the interaction parameter equal to $\sigma = \lambda me/(2\pi\hbar^2)$, $\hbar$ the reduced Planck constant, V the electrostatic potential, and A_z the component of the magnetic vector potential parallel to the electron beam propagation direction. In the interaction constant, λ is the relativistic electron wavelength (that depends on the electron accelerating voltage), m is the electron relativistic mass, and e is the elementary charge. The integrals in eq. (7.1) are calculated along the electron trajectory, assuming the electron source and the detector plane at infinite distances from the specimen. In the following description, it is assumed that no other external electrostatic and magnetic fields exist, except for the one due to the specimen.

In the case the specimen does not display any ordered (ferroelectric/ferromagnetic) state, the only contribution to the phase shift of the electron beam is the one due to the mean inner electrostatic potential (V_{mip}) of the material and can be expressed as:

$$\phi_e(x,y) = \sigma V_{mip} t(x,y) \tag{7.2}$$

with t(x, y) the sample thickness at the lateral position (x, y). The mean inner potential V_{mip} is a material-dependent property that is defined as the volume average of the atomic electrostatic potentials of the specimen; accordingly, V_{mip} depends on the local composition, the density, and ionicity of the specimen [19].

In the case of non-ferroic specimens with a uniform composition ($V_{mip} = \text{const}$), the electrostatic phase shift is directly related to the thickness of the specimen. The mean inner potential term can thus be employed to estimate the sample thickness. On the other hand, in the case of ferroic specimens, the mean inner potential contribution needs to be subtracted to the total phase shift to investigate the properties of the ordered state. Strong phase changes due to thickness variations (as it occurs, for example, at the edges of the specimen) are undesirable and can lead to a misinterpretation of the experiments, especially in low-dimensional systems like nanoparticles (NPs), nanowires (NWs) and thin films.

In ferroelectric specimens, the polarization is responsible for an additional contribution to the electrostatic phase shift – that we will here call ϕ_P – which expresses the mesoscopic phase modulation of the electron wave due to the ferroelectric polarization. This contribution can be written as [20]:

$$\phi_P(x,y) = \frac{\sigma}{\varepsilon} \int_0^t \left[\int_0^{(x,y)} \mathbf{P}_\perp(x',y')dx'dy' \right] dz \tag{7.3}$$

With $\varepsilon = \varepsilon_0 \cdot \varepsilon_r$ the dielectric constant given by the product of the vacuum (ε_0) and the relative (ε_r) permittivity and $\mathbf{P}_\perp$ the component of the polarization orthogonal to the electron beam propagation direction. The use of the relative permittivity in eq. (7.3) is meant to include the depolarizing effect of free charges which reduce the polarization value of the system. Assuming a constant polarization across the thickness t, we obtain:

$$\phi_P(x,y) = \frac{\sigma t(x,y)}{\varepsilon} \left[\int_0^{(x,y)} P_\perp(x',y')dx'dy' \right] \tag{7.4}$$

Therefore, in the case of a specimen with a constant thickness (t(x,y) = const) and uniform composition ($V_{mip} = \text{const}$), the electrostatic phase shift becomes:

$$\phi_e(x,y) = \sigma V_{mip} t + \phi_P(x,y) = \phi_P(x,y) + \text{const} \tag{7.5}$$

The in-plane components $\left[P_x^\perp(x,y), P_y^\perp(x,y)\right]$ of the projected polarization can thus be retrieved from the phase image by means of the following expression:

$$\left[P_x^{\perp}(x,y), P_y^{\perp}(x,y)\right] = \frac{\varepsilon}{\sigma t}\nabla_{xy}\phi_P(x,y) \overset{V_{mip}\overset{*}{=}const}{\underset{t=const}{\longrightarrow}} \frac{\varepsilon}{\sigma t}\nabla_{xy}\phi_e(x,y) \tag{7.6}$$

where the symbol ∇_{xy} denotes the in-plane gradient operator given by $(\partial/\partial x, \partial/\partial y)$.

As mentioned above, in specimens of inhomogeneous thickness and/or variable composition the local changes in the mean inner potential can hamper the proper investigation of the specimen polarization. In such cases, the expression marked by the asterisk (*) in eq. (7.6) does not hold and the ferroelectric contribution to the phase has to be isolated.

In the case the specimen displays an ordered magnetic state, the magnetic contribution to the phase shift ϕ_m also needs to be considered. It can be shown that the magnetic phase shift is related to the magnetic flux (see scheme in Figure 7.1) by calculating the difference between ϕ_m values at two arbitrary points (x_1,x_1) and (x_2, x_2) in the phase image [10]:

$$\Delta\phi_m = \phi_m(x_1,y_1) - \phi_m(x_2,y_2) = -\frac{e}{\hbar}\int_{-\infty}^{+\infty} A_z(x_1,y_1,z)dz + \frac{e}{\hbar}\int_{-\infty}^{+\infty} A_z(x_2,y_2,z)dz \tag{7.7}$$

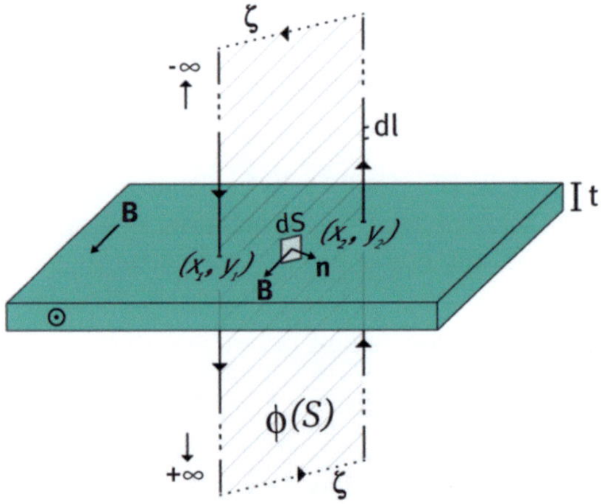

Figure 7.1: Schematic diagram of a uniformly magnetized thin film of constant thickness *t*. Φ is the magnetic flux through the surface S, bounded to the electron trajectories in (x_1,y_1) and (x_2,y_2).

Assuming that the magnetic potential generated by the specimen vanishes at infinite distance and no other transversal magnetic fields exist along the trajectories of the electrons, the amount given in eq. (7.7) can be alternatively expressed by a loop integral over a rectangular path, constituted by the two parallel electron trajectories – in the positions (x_1,y_1) and (x_2,y_2) – merged at infinite distances above and below the specimen plane by two segments (ζ) perpendicular to the trajectories themselves. The loop integral can be written as:

$$\Delta\phi_m = -\frac{e}{\hbar}\oint \mathbf{A}\cdot dl \tag{7.8}$$

Using the Stokes' theorem, the loop integral of the magnetic potential can be expressed by the magnetic flux (Φ) through the region of space (S) enclosed by the two electron trajectories (x_1,y_1) and (x_2,y_2):

$$\Delta\phi_m = \frac{e}{\hbar}\mathbf{B}\cdot\hat{\mathbf{n}}ds = \frac{\pi}{\phi_0}\Phi(S) \tag{7.9}$$

where $\phi_0 = h/2e = 2.07\cdot 10^{-15}Tm^2$ represents the "flux quantum" [10].

From the magnetic contribution to the phase shift, we can obtain the in-plane components of the magnetic induction $\left[B_x^P(x,y), B_y^P(x,y)\right]$ by calculating the gradient of ϕ_m [10, 21]:

$$\frac{\partial\phi_m(x,y)}{\partial x} = -\frac{e}{\hbar}\int\frac{\partial A_z(x,y,z)}{\partial x}dz = +\frac{e}{\hbar}B_y^P(x,y) \tag{7.10}$$

$$\frac{\partial\phi_m(x,y)}{\partial y} = -\frac{e}{\hbar}\int\frac{\partial A_z(x,y,z)}{\partial y}dz = -\frac{e}{\hbar}B_x^P(x,y)$$

where we replaced the partial derivatives of A_z using the expressions obtained by the magnetic potential definition:

$$B_x = \frac{\partial A_z(x,y)}{\partial y}, B_y = -\frac{\partial A_z(x,y)}{\partial x} \tag{7.11}$$

The formulas given in eq. (7.10) highlight the direct relationship between the magnetic phase shift and the in-plane components of the magnetic induction, which can be summarized as:

$$\left[B_y^P(x,y), -B_x^P(x,y)\right] = \frac{\hbar}{e}\nabla\phi_m(x,y) \tag{7.12}$$

7.2.3 Phase-contrast techniques in S/TEM

As previously described, the phase of the electron wave carries very useful information about the fields in the specimen. In conventional imaging techniques, however, we simply record an intensity distribution by collecting the electrons that impinge on a pixelated camera in the image plane (in TEM mode) or by integrating the scattered electrons over a certain angular range in the conjugated, i. e. diffraction, plane (in STEM mode). Unfortunately, in conventional imaging all the information directly related to the phase of the electrons is lost, being all the techniques solely sensitive to the intensity distribution of the scattered electrons. In order to study the electrostatic and magnetic fields in the specimen, we need to retrieve the phase shift imparted to the electron beam by the interaction with the sample. The techniques that allow retrieving the phase of the electron wave from intensity measurements are called phase-contrast techniques [22].

In particular, two different approaches allow mapping the specimen fields through phase-contrast techniques. The first one consists in retrieving the phase of the electron wavefunction by a reconstruction approach and calculating the fields by differentiating – using eqs. (7.6) and (7.12) – the phase of the wavefunction. The most common methods to measure the electron phase are off-axis electron holography and focal series reconstruction. The second approach is based on the direct measurement of the fields through DPC methods, from which the phase can be reconstructed upon integration. In this second category, DPC STEM with segmented-annular or pixelated detectors is gathering nowadays growing attention for its great potential, offering the advantages of high spatial-resolution, fast recording and parallel acquisition of different signals.

In the following, the basic principles of off-axis electron holography and DPC STEM are discussed, providing the reader with the fundamental tools to understand their application to ferroic materials investigations.

7.2.3.1 Off-axis electron holography

The full electron wavefront – in its amplitude and phase – can be retrieved by means of off-axis electron holography. This method consists in superimposing one electron beam that interacts with the specimen with another beam that propagates into the vacuum and acts as a reference wave [24]. After superimposing the two beams, we can record the interferometric pattern in the detector plane. In general, splitting of the beam is realized by partially illuminating the sample, meaning that one part of the beam, i. e. the object wave, passes through the specimen, while the other part, i. e. the reference wave, passes through vacuum. The superposition of the object and reference wave, respectively – is performed using the electrostatic biprism (also called Möllenstedt biprism [25]) placed behind the specimen plane. The superposition results in an interference pattern observable in the detector plane. A scheme of the electro-optical configuration commonly employed for electron holography experiments is sketched in Figure 7.2(a).

We can write the electron wavefunction in the image plane as:

$$\psi_i(\mathbf{r}) = A_i(\mathbf{r})e^{j\phi_i(\mathbf{r})} \tag{7.13}$$

where A_i and ϕ_i are respectively the electron wavefunction amplitude and phase in the image plane and j is the imaginary unit. The recorded intensity in the detector plane is thus given by:

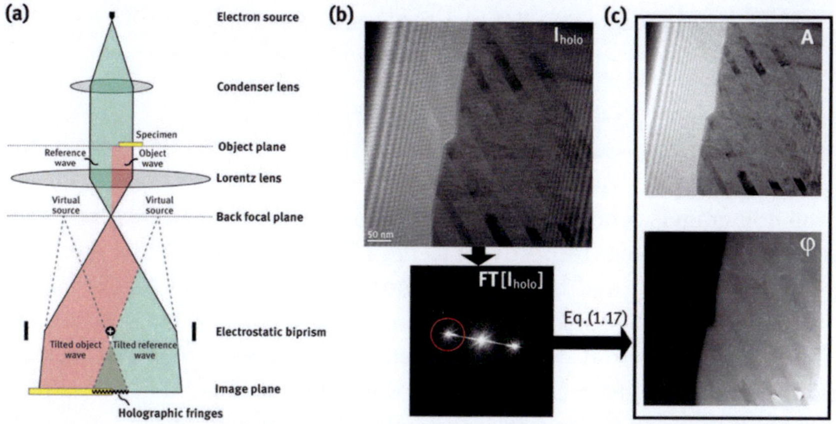

Figure 7.2: (a) Scheme of the conventional electron-optical configuration used for electron holography experiments (adapted from [23]). (b) Example of phase reconstruction from an electron holography experiment. One sideband is selected for the reconstruction of the amplitude and phase (c) through inverse Fourier transform operations.

$$I(\mathbf{r}) = |A_i(\mathbf{r})|^2 \tag{7.14}$$

As clearly visible in eq. (7.14), the information regarding the electron phase is lost in conventional imaging, as the intensity distribution is solely related to the wave-function amplitude.

In electron holography, we perform an interferometric recording that allows for solving the phase problem. In fact, if we calculate the intensity in the image plane after superimposing the reference and object beams, we obtain:

$$I_{holo}(\mathbf{r}) = \left|\psi_i(\mathbf{r}) + e^{2\pi j\, \mathbf{q_T} \cdot \mathbf{r}}\right|^2 = 1 + A_i^2(\mathbf{r}) + 2A_i(\mathbf{r})\cos[2\pi j\, \mathbf{q_T} \cdot \mathbf{r} + \phi_i] \tag{7.15}$$

where $\mathbf{q_T}$ is the wave-vector of the tilted reference beam. Eq. (7.15) shows that the intensity of an electron hologram is composed of three contributions, which are the unitary intensity of the reference image, the intensity of the image of the specimen, a modulated signal (given by the cosine term) that consists of interference fringes whose intensity maxima correspond to areas where the phase is the same for the object and reference beam. As an example, an electron hologram is shown in Figure 7.2(b).

The phase information is thus encoded in the intensity modulation appearing in the image and can be reconstructed by means of Fourier methods. In particular, taking the Fourier transform (FT) of the electron hologram, we obtain:

$$\mathrm{FT}[\mathrm{I}_{\mathrm{holo}}(\mathbf{r})] = \delta(\mathbf{q}) + \mathrm{FT}\left[\mathrm{A_i}^2(\mathbf{r})\right] + \underbrace{\delta(\mathbf{q}+\mathbf{q_T})\otimes\mathrm{FT}\left[\mathrm{A_i}(\mathbf{r})e^{j\phi_i(\mathbf{r})}\right]}_{(1)} + \underbrace{\delta(\mathbf{q}-\mathbf{q_T})\otimes\mathrm{FT}\left[\mathrm{A_i}(\mathbf{r})e^{-j\phi_{im}(\mathbf{r})}\right]}_{(2)} \tag{7.16}$$

The FT of the hologram is composed of two peaks centered at the origin of the reciprocal space – that correspond respectively to the FT of the reference and the FT of the image of the sample – and two peaks centered at $\mathbf{q} = \pm\,\mathbf{q_T}$ (see Figure 7.2(c)). These latter two peaks, called *sidebands* and marked by (1) and (2) in eq. (7.16), are equal to the FT of the electron wavefunction and its complex conjugated, respectively. The amplitude and phase can finally be restored from one of the two isolated sidebands by shifting it to the center of the image and computing the inverse FT [10, 26]. The amplitude and phase are thus given by:

$$\mathrm{A_i}(\mathbf{r}) = \{Re[\psi_i(\mathbf{r})] + Im\,[\psi_i(\mathbf{r})]\}^{\frac{1}{2}}, \phi_i(\mathbf{r}) = \tan^{-1}\left\{\frac{Im[\psi_i(\mathbf{r})]}{Re[\psi_i(\mathbf{r})]}\right\} \tag{7.17}$$

where we denoted the real and imaginary parts of the complex wavefunction with *Re* and *Im*, respectively.

In the holography experiment, the spatial and phase resolution of the hologram are respectively related to the spacing and contrast of the fringes, which can be tuned by varying several parameters including the biprism voltage and the magnification [24, 27]. Due to the opposite dependence of the contrast and spacing of the fringes on the biprism voltage, the experimental settings need to be properly chosen to achieve the desired spatial and phase resolutions. For the technical details about the optimal conditions (i. e. interference fringe spacing and signal-to-noise considerations) for the acquisition of electron holograms the reader is referred to more specific literature (see, e. g. [10, 28–32]).

In summary, electron holography experiments can be performed either at:

i. *medium-resolution*, for the investigation of the polar and magnetic properties of specimens. In this mode, the OL is switched off and the Lorentz lens is used instead to form the image. The achievable spatial resolution is limited by the aberration of the Lorentz lens and is usually in the nanometer range [23, 33].
ii. *high-resolution*, making use of the OL. This technique has been commonly used to improve the resolution of a high-resolution TEM image by the use of phase plates to correct for microscope lens aberrations [34–36].

However, it is important to recall that the investigation of magnetic materials has to be performed by medium-resolution electron-holography in Lorentz modus, which guarantees a *field-free* condition in the specimen plane, at the expense of the spatial resolution of the reconstructed phase.

Electron holography enables the retrieval of the total phase shift induced by the specimen to the electron beam. As previously discussed, the polar and magnetic properties of the specimen contribute to this phase with two different terms, which need to be separated. Depending on the terms that need to be addressed, this separation can be done by different methods. In the case of ferroelectric specimens, the ferroelectric phase shift can be separated from the mean inner potential contribution by performing the experiments respectively below and above the Curie temperature (T_C). In fact, above T_C the polar order is lost and the mean inner potential contribution is the only responsible for the phase shift. The phase difference of the two signals obtained respectively below and above T_C permits thus to isolate the phase shift due to the polar properties of the material.

In ferromagnetic specimens, their net magnetization induces a magnetic term in the phase shift of the electron wavefunction. The electrostatic contribution must be removed from the reconstructed phase image in order to retrieve this contribution. The most common way to achieve this separation is by the *time-reversal operation* of the electron beam [37] that is performed by recording two holograms of the same region of the specimen in normal (upside) and reversal (downside) positions. The sum and the difference of the reconstructed phase images can then be used to recover the electrostatic and the magnetic contributions to the phase, respectively. An alternative approach involves performing an in situ magnetization reversal to detect the phase of two states of the specimen with opposite magnetization [38]. The addition of the two phase-images gives access to the mean inner potential contribution to the phase. Magnetization reversal can be performed by using either a magnetizing TEM specimen holder or the magnetic field of the OL (eventually in combination with specimen tilt). This second approach is only applicable if the magnetization of the specimen can be perfectly reversed.

7.2.3.2 DPC STEM

In DPC STEM a convergent electron probe is used to make a raster scan of the specimen and the variations in the CBED pattern are detected by means of a segmented annular detector or a pixelated camera [39]. The technique was firstly ideated in the '70s, thanks to the pioneering works by Rose [40] and Dekkers [41]. Nowadays, DPC STEM is experiencing a renaissance thanks to the spread of commercially available segmented detectors and fast pixelated cameras. DPC STEM is attracting indeed a continuously growing interest by the scientific community thanks to its great potential, which allows contemporary different imaging modes by simultaneous acquisition of various signals – like e. g. high-angle annular dark-field (HAADF) and bright-field (BF) – the acquisition of secondary signals for spectroscopic analysis like EELS and EDXS. Additionally, the sub-Ångström resolving power achievable thanks to the new generation of aberration-corrected microscopes [42] enables to carry out these investigations at atomic resolution. Up to date, DPC in the Lorentz

STEM mode has been successfully employed to probe magnetic [15, 43–47] and electric fields [48–50] at the nanoscale, whereas DPC STEM at high-resolution has been employed to map electric fields at atomic resolution [51–54].

The DPC-STEM technique makes use of the deflection of the focused electron probe by a field in the sample to locally map this field. A scheme of the DPC principle is shown in Figure 7.3. If we describe classically, the propagation of the electron beam, the electrostatic and magnetic fields in the specimen plane induce a tilt of the electron probe in the specimen plane which leads to a shift of the bright field disk ("Ronchigram") in the diffraction plane. The DPC-STEM image is generated by performing a raster scan of the beam across the sample and recording for each scan position the off-centering of the transmitted disk. The variation of the disk position in the diffraction plane can be efficiently detected by a segmented detector. In particular, the off-centering of the disk induces an increase of the intensity impinging on certain detector segments at the expense of others, which are consequently less illuminated. The beam deflection angle can thus be estimated by measuring the differential contrast between opposite pairs of quadrants or – in a more generic segmented detector configuration, like e. g. the 16-quadrant configuration [57] – opposite sets of quadrants.

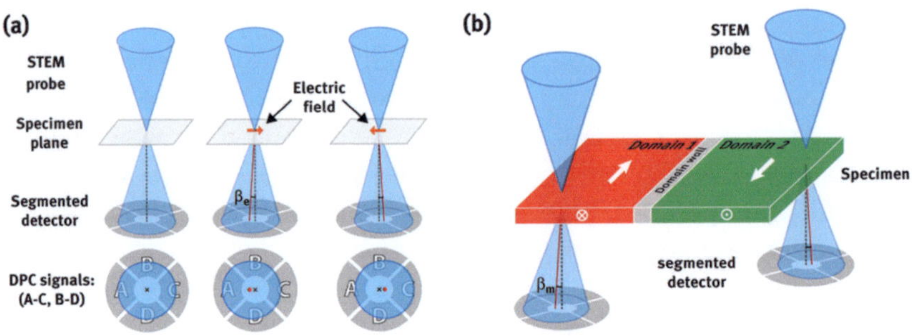

Figure 7.3: Scheme of the basic principles of DPC-STEM, performed using a segmented detector. (a) Electric and (b) magnetic field induced shift of the electron beam under the classical approximation for the electron propagation (adapted from [55, 56]).

From the classic treatment of the electron propagation, it is possible to calculate the deflection of the electron beam due to the Lorentz force in presence of an electric or magnetic field in the specimen plane (respectively β_e and β_m) [55, 58]:

$$\beta_e(\mathbf{r_P}) = \frac{e\lambda}{hv}\int_0^t \mathbf{E}_\perp(\mathbf{r_P}, z)dz,\ \beta_m(\mathbf{r_P}) = -\frac{e\lambda}{h}\int_0^t \mathbf{B}_\perp(\mathbf{r_P}, z) \times dz \tag{7.18}$$

where v is the electron velocity and the integrals are performed on the specimen thickness t($\mathbf{r_P}$) and $\mathbf{r_P}$ gives the position of the probe in the specimen plane. In the case of uniform electrostatic and magnetic fields, the relationships reduce to the handier forms:

$$\beta_e(\mathbf{r_P}) = \frac{e\lambda t(\mathbf{r_P})}{hv} E_\perp(\mathbf{r_P}),\ \beta_m(\mathbf{r_P}) = -\frac{e\lambda t(\mathbf{r_P})}{h} B_\perp(\mathbf{r_P}) \tag{7.19}$$

The deflections of the beam induced by electrostatic and magnetic fields can otherwise be calculated by an alternative approach. In particular, since DPC STEM does not measure the local phase shift but its gradient, eq. (7.19) can be derived by calculating the phase difference between electrons traveling along different trajectories and approximating the phase gradient with the finite difference among their phases [58].

It is interesting to note from eq. (7.19) that a local variation of the specimen thickness induces a change in the deflection angle. Therefore, the DPC-STEM technique can be used to directly measure the thickness variation of homogenous non-polar/non-magnetic specimens, where the specimen potential is solely given by the mean inner potential. In the case of ferroic materials, if the deflection angle caused by the effect that is to be measured is larger than the one caused by the thickness variations, the measurement will result with a good signal-to-noise ratio (the noise being the thickness variations at the area of interest) [58]. More generally, the reader should keep in mind that the thickness variations will generate some substructure in the signal of interest, which may lead to a misinterpretation of the experiments.

This simple model describing the electron beam deflection due to the Lorentz force allows for easily depicting the basic concepts behind the DPC-STEM technique. In reality, the quantification of fields in terms of a rigid shift of the Ronchigram can be inaccurate (Figure 7.4(a)) as a quantum mechanical treatment of the electron wave propagation is necessary (Figure 7.4(b)) [59, 60]. From a quantum mechanics

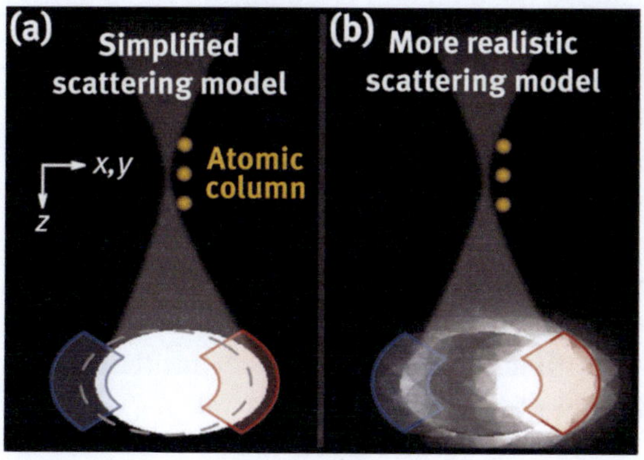

Figure 7.4: (a) Classical and (b) quantum-mechanical description of the effects of the electrostatic/magnetic fields on the DPC STEM signal. The quantum mechanical description includes dynamical effects that give rise to a complex redistribution of the intensity in the detector plane. Reproduced from [59].

point of view, the electrostatic and magnetic fields in the specimen plane induce a variation on the distribution of probability to find an electron at a certain position in the detector plane or equivalently – due to the Fourier reciprocity between the specimen and the Fraunhofer diffraction planes – with a certain momentum. The expectation value of the momentum operator **p** is given by [60]:

$$\langle \mathbf{p}_\perp \rangle(\mathbf{r_P}) = \int \langle \psi_\mathrm{D}(\mathbf{p}, \mathbf{r_P}) | \mathbf{p}_\perp | \psi_\mathrm{D}(\mathbf{p}, \mathbf{r_P}) \rangle \mathrm{d}p_x \mathrm{d}p_y = \mathbf{p}_\perp \mathrm{I}(\mathbf{p}_\perp, \mathbf{r_P}) \mathrm{d}p_x \mathrm{d}p_y \tag{7.20}$$

which corresponds to the center of mass (CoM) of the Ronchigram in the detector plane. In the formula given in eq. (7.20), $\mathbf{r_P}$ denotes the position of the convergent probe on the specimen, $\psi_\mathrm{D}(\mathbf{p}, \mathbf{r_P})$ is the electron wavefunction in the momentum space (correspondingly, in the detector plane), $\mathrm{I}(\mathbf{p}_\perp, \mathbf{r_P})$ the normalized intensity distribution in the diffraction plane. In the above expression, it is assumed that the beam intensity is constant over the specimen thickness.

If we aim at correlating the expectation value of the momentum in the detector plane with the fields in the specimen plane, we need to recall that in classical electrodynamics the Lorentz force is equal to the momentum transfer per time. According to Ehrenfest's theorem [61], we can write the quantum mechanical equivalent of the Lorentz force, which for the electrostatic field (the extension to the case of a magnetic field is straightforward) can be written as:

$$\mathbf{E}_\perp(\mathbf{r_P}) = -\langle \mathbf{p}_\perp \rangle(\mathbf{r_P}) . \frac{\mathrm{v}}{\mathrm{et}} \tag{7.21}$$

where $\mathbf{E}_\perp$ is the electric field projection average along the electron beam direction, v is the electron velocity, e the elementary charge, and t the specimen thickness. Although the rigid-shift model previously introduced does not accurately describe the electron propagation, the classical and the quantum mechanical treatment lead to similar results. In particular, from the quantum mechanical treatment, we obtain that the CoM of the Ronchigram – instead of the rigid shift of the beam – is linearly related to the projected local electric field in the sample (eq. (7.21)).

In order to make a proper estimation of the electric field, the expectation value of the momentum (as it is defined in eq. (7.20)) needs thus to be calculated. For this purpose, a pixel-array detector has to be used to record the full intensity distribution within the Ronchigram at each position of the STEM raster scan. This approach is commonly called 4D-STEM since it involves the acquisition of a two-dimensional (2D) diffraction image for each position of the raster scan – performed over a 2D grid of scan points across the region of interest CoM of the specimen. The acquisition of the entire Ronchigram also allows *a posteriori* integration of the scattered signal over different angular ranges of the CBED pattern to generate different signals, like e. g. bright-field (BF), annular bright-field (ABF), annular dark-field (ADF) [54, 62–64]. This very powerful feature makes 4D-STEM the most promising approach to combine differential phase-contrast and other imaging techniques in STEM mode, at the expense of a remarkable increase in the volume of generated data [65, 66]. Yet, one drawback of the currently existing 4D-STEM systems is the reduced scan speed which is given by the acquisition speed of the 2D camera. Current systems allow acquisition speeds of the order of 1000 frames per seconds [67, 68], which is two to three orders of magnitude slower than common scan speeds in STEM.

As already described, a position-sensitive detector – i. e. a segmented detector – can be alternatively employed for the qualitative detection of the momentum CoM displacement by measuring the increase or decrease of signal in pairs of opposite quadrants. This well-established technology makes use of an integrating detector, with the advantages of faster scan speeds, high signal-to-noise ratio, live imaging, and it currently represents the most promising method for routine applications. Nevertheless, it appears evident from the quantum mechanical perspective that despite this technology is very practical, its application for quantitative analysis requires additional care because the use of a segmented detector only gives a qualitative approximation of the momentum CoM by roughly sampling the scattering distribution variation.

It must be noted that eq. (7.21) was obtained under the assumption that the intensity of the STEM probe does not change significantly along the propagation direction. A proper description of the quantum mechanical problem would involve the integration over the three-dimensional (3D) fields in the specimen weighted with the local intensity of the electron beam [60]. However, such an assumption is verified in these special relevant cases [69]:

i. under medium resolution imaging conditions. In this case, the STEM probe size is increased beyond the unit cell size and the sample can be oriented as to minimize the dynamical scattering. The channeling effects are negligible. Under these conditions, specimens can often be described as pure phase objects (*phase object approximation*, POA) [70].
ii. under high-resolution imaging conditions – i. e. namely, zone-axis orientation and evaluation of large scattering angles – for specimens in the thickness range for which the *weak phase object approximation* (WPOA) holds. Such range extends to a few nanometers in the case of specimens composed by light atomic species, whereas it reduces to a few Ångstrom for medium and heavy elements [69, 71]. In this thickness range, the broadening of the electron beam during its propagation can be neglected.

Since the WPOA is a questionable approximation for real specimens (with an exception for 2D materials, like e. g. graphene), the correct interpretation of atomic resolution DPC/4D-STEM data is neither trivial nor immediate. Additionally, the atomic electric fields vary very strongly over distances equal to fractions of the Bohr radius, which are smaller than the typical lateral size of aberration-corrected STEM probes.

As a consequence, the interpretation of atomic resolution data requires comparison with simulations, as it is the case for other high-resolution S/TEM imaging techniques [59].

By fitting the CoM of the Ronchigram for each point in the raster scan, a projected electric field map can be obtained using eq. (7.21). Since the momentum transfer is linear with respect to the electric field, calculating its divergence we obtain:

$$\mathrm{div}\langle \mathbf{p}_\perp \rangle = -\frac{et}{v\varepsilon_0} e\rho_\perp \tag{7.22}$$

where $e\rho_\perp$ is the overall projected charge density, given by the sum of densities of protons and electrons. On the other hand, by calculating the inverse gradient we obtain:

$$grad^{-1}\langle \mathbf{p}_\perp \rangle = \frac{et}{v} V_\perp \tag{7.23}$$

being $V_\perp$ the projected electrostatic potential. The eq. (7.23) shows that when the WPOA is valid – i. e. the phase shift of the electron wavefunction is $\phi \approx \sigma V_\perp$ (see the conditions discussed previously) – the DPC and 4D-STEM are phase-contrast techniques in their integral perspective. Due to the finite size of the STEM probe, the experimentally measured electric field is the true electric field given in eq. (7.21) convolved with the electron probe, and similarly for the projected density of charge and projected potentials.

7.2.3.3 Phase-contrast data analysis and representation

As previously pointed out, the phase-contrast techniques allow for reconstructing the phase of the electron wavefunction and/or the projected in-plane fields. Therefore, the data obtained by phase-contrast techniques can be analyzed and plotted in different ways, depending on the features that need to be highlighted.

A very common way to represent the reconstructed phase, which gives visual information about the specimen field strength and direction, is by plotting the cosine of the phase [72]:

$$\mathrm{I}(\mathbf{r}) = \cos[\mathrm{n} \cdot \phi(\mathbf{r})] \tag{7.24}$$

eventually multiplying the phase by an integer n as an amplification factor. In the case $n = 1$, this representation gives the phase–contours for which the phase differs of 2π. The plot of phase-contours can be performed for both the electrostatic and magnetic phase contributions. Let us now analyze the meaning of the phase contours in both cases:

1. The contour lines obtained from the electrostatic contribution to the phase represent the electrostatic equipotential lines. The phase-contours are denser in

the regions where the potential is rapidly changing and the electric field direction is normal to the contours. Assuming a constant specimen thickness or that the contribution of the mean inner potential has been properly subtracted, the phase-contours represent the equipotential lines for the ferroelectric phase-shift and the polarization.
2. The contour lines obtained for the magnetic phase shift represent the variation of 2π of $\Delta\phi_m$. From eq. (7.9) it follows that the magnetic flux between two points laying on consecutive contours is given by $2\phi_0 = 414.10^{-15}\text{Tm}^2$ [10]. In this case, from eq. (7.12) it follows that the magnetic induction vector is tangent to the phase contours.

Starting from the electric or magnetic field components – i. e. a 2D map for each vector-component – a vector plot of the corresponding field is possible. Such plots are usually presented by using an HSV (H: hue, S: saturation, V: value) color map, that is an alternative representation of the RGB color scheme. In this color map, the color is related to the direction and orientation of the field, and the hue is related to its magnitude.

7.2.4 Investigations of electric and magnetic ordering in multiferroic materials

Here, we briefly discuss the application of electron holography and differential phase-contrast STEM to the investigation of the local order in ferroelectric and ferromagnetic specimens through a few selected examples.

7.2.4.1 Polar ordering in multiferroics

Electron holography has been successfully employed to investigate the polar state of ferroelectric and multiferroic thin films and nanocrystals. A ferroelectric material ($BaTiO_3$) is used in the following cases as an instructive example, though the same analyses are valid for multiferroic materials.

Matsumoto and coworkers [73] applied medium-resolution electron holography to a thin film of $BaTiO_3$ mapping the 90° domain structure of the polar phase. In order to extract the phase shift due to ferroelectricity, the authors performed the holography experiments at RT – a condition in which the specimen was in the polar phase – and above the Curie temperature (T_C) – i. e. without any polar order. Subtracting the phase shift obtained with the specimen heated above T_C from that obtained at RT allows for isolating the contribution intrinsic to ferroelectricity from those originated from the mean inner potential – e. g. thickness variation – or from electron-beam-induced specimen charging. Figure 7.5(a) shows the reconstructed phase in a region with the 90° ferroelectric domains. The reconstructed phase shows a weak contrast and faint stripes due to the ferroelectric domains are feebly visible, because of the thickness variation across the field of view. After subtracting

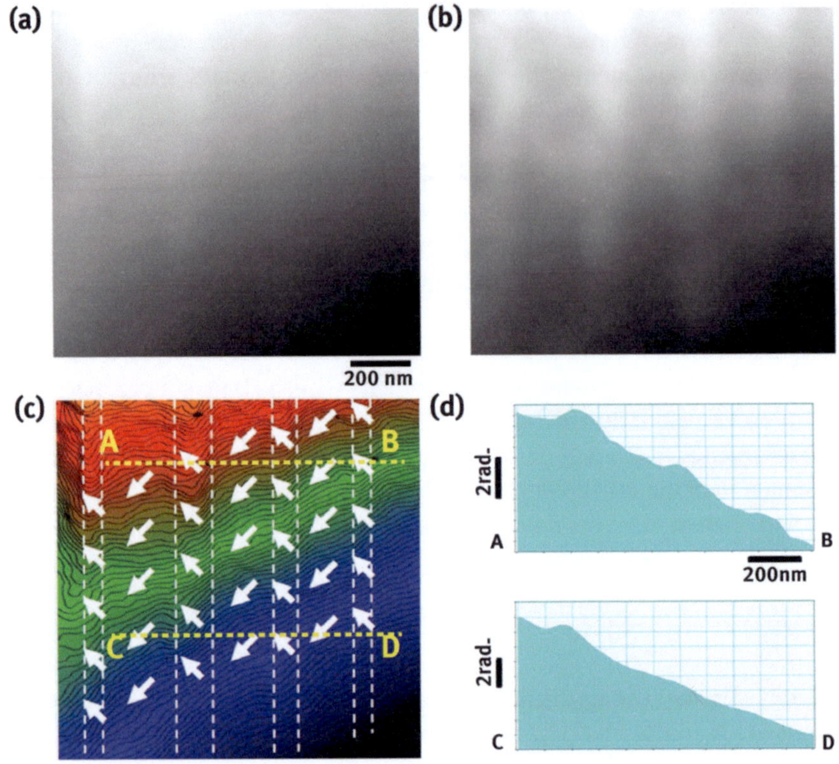

Figure 7.5: (a) Grayscale representation of the reconstructed phase of the 90° domain structure in ferroelectric $BaTiO_3$. (b) Difference between the reconstructed phases below and above T_C. (c) Color representation of the same phase difference overlaid with phase-contours and arrows representing the polarization. (d) Phase profiles along the segments AB and CD in (c). Reproduced from [73].

the contribution of the mean inner potential, the phase features due to the polar domains are immediately visible (Figure 7.5(b–d)).

Polking and coworkers [74] applied medium-resolution electron holography to ferroelectric $BaTiO_3$ nanocubes to investigate the effect of temperature and bias at the nanoscale. In particular, the RT polarized state of $BaTiO_3$ nanocubes appears visible as a phase variation at the [001] facet of the nanocubes (Figure 7.6(a)). Such a phase change disappears by heating the specimen above T_C (Figure 7.6(b)), where the contribution of the mean inner potential is the only one visible. Interestingly, a polar state (giving the same feature in the phase image) can be artificially induced by poling a nanocube with a W tip (Figure 7.4(c–f)).

The nanoscale ferroelectric domain configuration in $BaTiO_3$ thin films was successfully investigated also by medium-resolution DPC-STEM. Shibata and coworkers [51] applied the DPC technique performed using a four-quadrant segmented

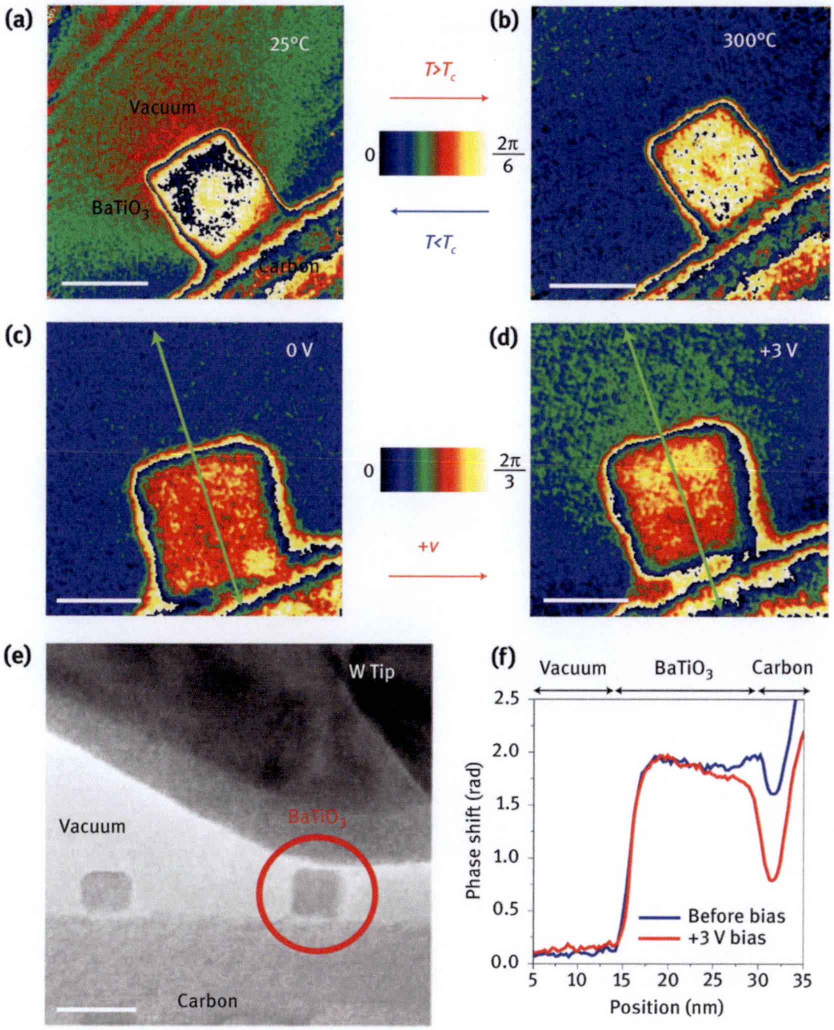

Figure 7.6: Phase images of a $BaTiO_3$ nanocube below (a) and above (b) the Curie temperature (403 K). The phase variation in (a) highlights a polar state with the polarization oriented along the [001] direction. Phase image of a different nanocube before (c) and after (d) electrical poling (+3 V). The poling induces a net polarization as visible from the variation of phase along the [001] direction (green lines). (e) TEM image of the poling experiments. (f) Phase profiles along the green lines in (c) and (d). Reproduced from [74].

detector and measuring the differential signal between opposite pairs of detector segments. Figure 7.7(a and b) show the $BaTiO_3$ unit cell distortion at the origin of the ferroelectric instability and the orientation of the specimen with respect to the segmented detector. The differential signal showed in Figure 7.7(c and d) highlights 90° ferroelectric domains (similar to the ones previously shown in Figure 7.5).

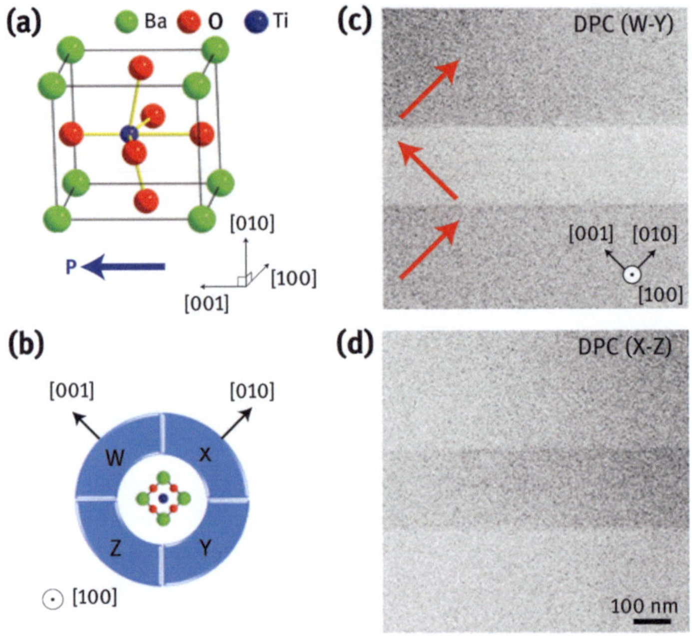

Figure 7.7: (a) Sketch of the $BaTiO_3$ unit cell illustrating the structural distortion at the origin of the ferroelectric polarization. (b) Schematic representation of the unit cell orientation with respect to the detector quadrants. (c,d) Medium-resolution differential phase-contrast signals showing a change in the contrast corresponding to 90° ferroelectric domains. The direction of the polarization is given by the red arrows in (c). Reproduced from [51].

As described in the previous section, DPC STEM has also been proven to be capable of atomic electric field mapping in high-resolution STEM mode. The correct interpretation of atomic electric field mapping is not straightforward, as it would require extensive use of image simulations to properly catch the dynamical effects which can strongly alter the momentum transfer in the reciprocal space. However, in ferroelectrics, the polarization typically varies very slowly (over distances comparable to the unit cell size) with respect to the atomic potentials, which are very strongly localized and peaked at the positions of the atomic nuclei. Since the polar field variation along the lateral size of the probe is very small, it will not affect significantly the dynamical scattering and its effect can be approximated as an additive low-frequency contribution superimposed to the strongly localized atomic fields.

As an example, we show the DPC-STEM analysis on a $SrRuO_3$ (SRO)/$BaTiO_3$ (BTO) bi-layer on a $SrTiO_3$ substrate. Figure 7.8(a) shows a HAADF-STEM image of the specimen, in which a dislocation is visible in the upper BTO layer. The dislocation acts as a pinning center for a domain wall separating two differently oriented polar domains, as visible from the high-resolution images shown in the insets. Figure 7.8(b) shows a rotation map obtained by Geometric Phase Analysis (GPA) (a

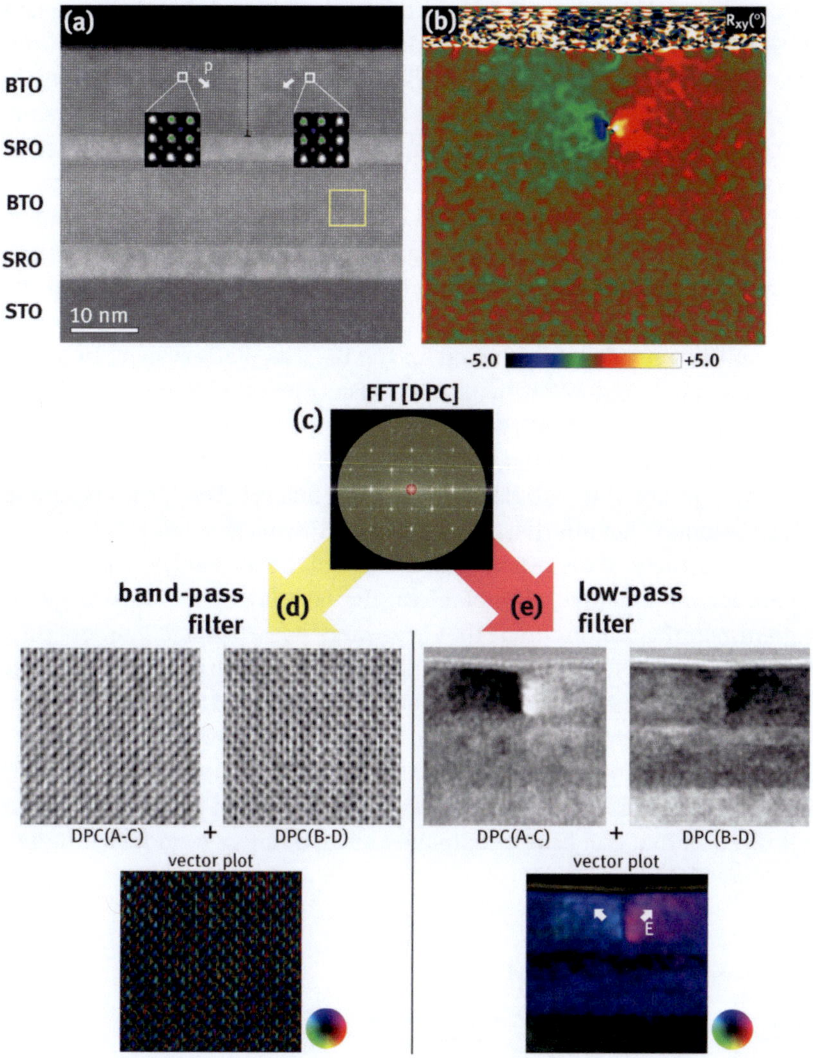

Figure 7.8: (a) HAADF-STEM image of a $SrRuO_3$ (SRO)/$BaTiO_3$ (BTO) bilayer on a $SrTiO_3$ substrate. A dislocation is visible in the upper BTO layer. In the inset, the atomically-resolved HAADF images at both the sides of the domain wall are shown. (b) Rotation map highlighting the dislocation in the upper BTO layer. (c) Fourier filtering of the DPC signal to separate low and medium frequencies contributions. Applying a band-pass filter with a frequency range of 0.69–8.8 nm^{-1} (d) we retrieve the atomic electric field (shown for the region marked by the yellow box in (a)), while applying a low pass filter that selects the frequency lower than 0.69 nm^{-1} the mesoscopic polarization is obtained.

detailed explanation of GPA is given in Section 7.4) highlighting the dislocation position at the interface between the two upper SRO/BTO layers. In order to separate the strongly localized atomic electric field from the mesoscopic polarization which varies at the domain wall between the two differently oriented domains, we can perform DPC STEM experiments at atomic resolution and afterward perform Fourier filtering of the data (Figure 7.8(c)). By applying a band-pass filter (as shown in Figure 7.8(d)), we can create an image with the sole contribution of the wavevectors in the range 0.69–8.8 nm^{-1}, which is equivalent to select the spatial frequencies in the range 1.45–11.2 Å. The obtained reconstructed field does not contain any information about the mesoscopic polarization and only the field of the atomic columns is visible in the map. On the contrary, if we apply a low-pass filter (selecting the wavevectors smaller than 0.69 nm^{-1}), we can retrieve the electric field due to the ferroelectric polarization.

A similar result was previously obtained for the $Bi_{0.8}Ca_{0.2}FeO_3$ system, in which spontaneously generated Ca-dopant fluctuations occurring in a layered fashion induce a complex polarization pattern with alternating polar (at the Ca-poor areas) and non-polar (Ca-rich areas) regions [56]. In Figure 7.9(a) a scheme of the polar pattern is shown, together with the HAADF and ABF images. The local change in the dopant composition is not effortlessly visible from the images, but it is linked to the alteration of the structural distortions responsible of the displacive ferroelectricity (see Ref [56]. for more details). Nonetheless, the layering of the dopant atoms generates a giant polarization gradient that is as large as ~ 70 μC cm^{-2} across only 3 nm. Interestingly, the DPC-STEM signal highlights asymmetries in the electric field components (Figure 7.9(b)). The asymmetries are clearly due to the superimposition of the much localized atomic electric field and the slowly varying polarization. Indeed, the polarization variation occurs on a length scale significantly larger with respect to the size of the aberration-corrected STEM probe, inducing a small beam deflection that can be approximated as an additive term just superimposed to the atomic field.

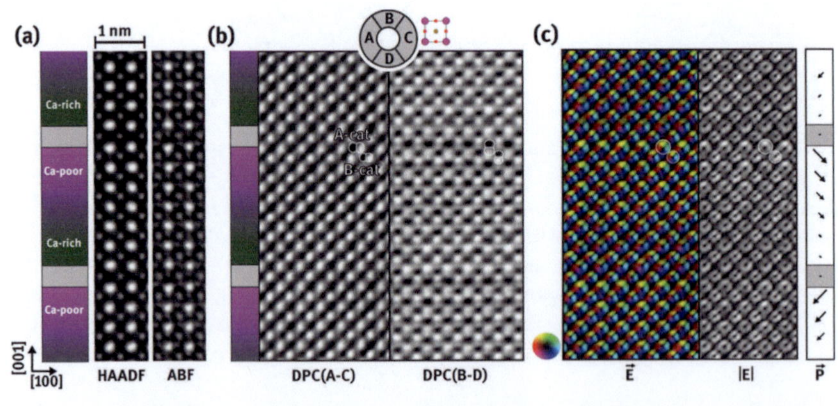

Figure 7.9: (a) Scheme of the Ca-dopant layering in a $Bi_{0.8}Ca_{0.2}FeO_3$ thin film, together with the HAADF and ABF signals. (b) DPC-STEM signals and (c) vector and amplitude plot of the projected electric field. Reproduced from [56].

7.2.4.2 Magnetic ordering in multiferroics

Electron holography has been widely used to study the magnetic domain configuration in ferromagnetic and multiferroic specimens, including thin films [75–79], nanowires [80–83] and NPs [84–86]. Here, we report as an example the application of electron holography to multiferroic Ni_2MnGa thin films and nanodisks.

Ni_2MnGa is a multiferroic ferromagnetic shape memory alloy in which a spontaneous deformation as high as 6–10 % can be induced by the application of a magnetic field. The ferroic properties of FMSAs arise as a consequence of coexistent ferromagnetic and ferroelastic (martensitic) phase transformations, and they give rise to a very peculiar nanostructure composed by twinned domains (usually called twin variants). The ferroelastic martensitic transformation results in martensite variants separated by mobile twin boundaries, which can be exploited to generate giant strains in FMSAs thin films and nanostructures. Figure 7.10(a) displays an electron hologram of a region showing twin variants oriented at 90° with respect to the MgO(001) substrate. The mean inner potential and the magnetic phase contributions can be separated acquiring the hologram before and after turning upside-down the specimen. In Figure 7.10(b) and 10(c) the electrostatic and the magnetic phase shift are respectively shown. By calculating the gradient of the magnetic phase shift we can retrieve the magnetic induction components using eq. (7.12). The

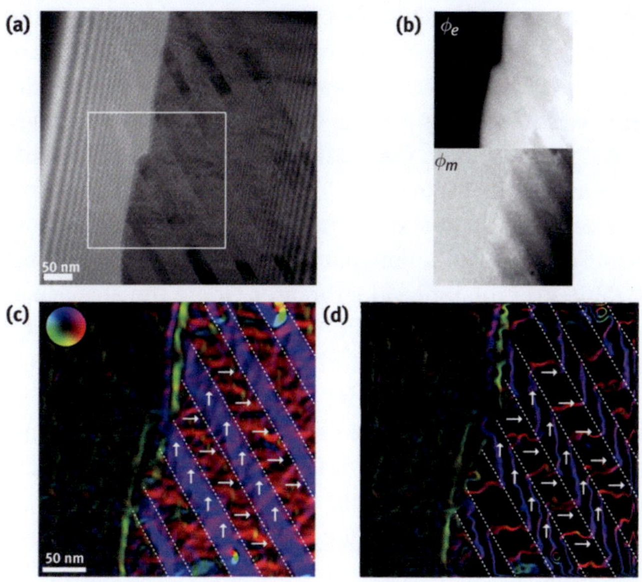

Figure 7.10: (a) 300 kV electron hologram acquired using a 200 V biprism bias (fringe spacing: 1.4 nm, fringe contrast: 20 %) on a Ni_2MnGa thin film in a region showing 90° twin variants. (b) Reconstructed electrostatic and magnetic phase shifts from the region marked by a white dashed box in (a). (c) Magnetic induction vector map and (d) magnetic induction flux lines of the 6x amplified phase. Adapted from [77].

magnetic induction color map is shown in Figure 7.10(d) and the phase contours of the 6x amplified phase are presented in Figure 7.10(d). Electron holography allows thus for reconstructing the magnetic domain configuration in the twinned martensite and to map the magnetic induction flux lines.

7.3 Electron nanodiffraction

Compared to the above discussed phase-contrast methods, a completely different approach to extract local information of ferroic materials is based on electron diffraction. The most common method to access local diffraction information in TEM is selected-area electron diffraction (SAED). However, should diffraction information be required for sample areas in the nanometer range, SAED is unsuitable. In contrast, so-called electron nanodiffraction uses a small electron probe to record electron diffraction patterns. Typically the electron probes for electron nanodiffraction have diameters ranging from a few tens of nanometers to less than 1 nm depending on the selected objective and condenser lens settings. Thus, several nanodiffraction techniques exist that exploit different convergence angles [87–89]. Nanoarea electron diffraction (NAED or NED) uses the objective prefield lens to form a small (nearly) *parallel* probe on the specimen [90, 91]. By inserting a small condenser aperture, the beam size may be as small as a few tens of nanometers. In this mode, the diffraction patterns consist of sharp diffraction spots (comparable to the ones obtained by a parallel beam in a TEM). Instead, in nanobeam electron diffraction (NBED or NBD) a *focused* probe is formed by placing the electron beam crossover after the second condenser lens far away from the front focal plane of the OL, while keeping the mini-condenser lens weakly excited [92, 93]. This gives a demagnified electron source image on the sample which can produce diffraction patterns from sub-nanometer regions. However, the diffraction patterns consist of small discs and, thus, NBED is only suitable for probing local structures where the absence of sharp diffraction spots is not critical, such as phase identification, orientation mapping and local strain measurements [88]. Schematic diagrams illustrating the electron paths in NAED and NBED modi are shown in Figure 7.11. By varying the excitation of the condenser lenses and apertures, the semi-convergence angle of the probe can be gradually changed between (nearly) parallel and convergent illumination. A gradual decrease of the convergence angle of the electron probe in NBED mode results in the NAED mode as a border case of vanishing convergence angle of the NBED mode. For each technique, the resulting electron probe and diffraction pattern from a $PbZr_{0.2}Ti_{0.8}O_3$ thin film acquired using a probe aberration-corrected (scanning) transmission electron microscope, with a more complex condenser lens system than the one illustrated in Figure 7.11, are given as insets.

Additionally, electron nanodiffraction patterns can be obtained in conjunction with STEM imaging using an annular dark-field detector with a low camera-length

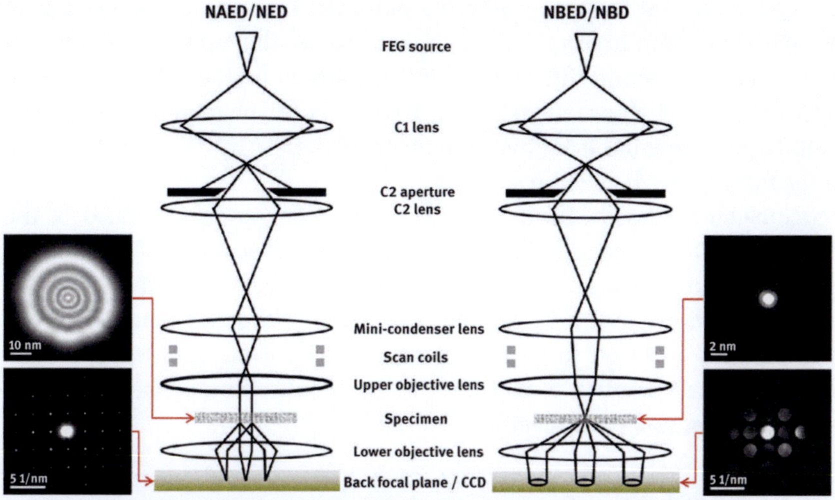

Figure 7.11: Ray diagrams showing the electron paths in NAED and NBED modes. For each technique, the resulting electron probe and diffraction pattern from a $PbZr_{0.2}Ti_{0.8}O_3$ thin film are shown as insets. Left: the experimental NAED probe (obtained with a semi-convergence angle <0.5 mrad) has a diameter of ~ 50 nm and was used to record the diffraction pattern with lattice reflections showing rather sharp diffraction spots. Right: the full-width-at-half-maximum of the NBED probe (obtained with a semi-convergence angle ~ 1.5 mrad) is 0.6 nm and was used to record the diffraction pattern consisting of small disks where the disk radius reflects the convergence angle of the electron probe. The ray diagrams are adapted from [87] and [91].

setting. This operation mode is known as scanning electron nanodiffraction (SEND) [88]. In its simplest form, SEND can be carried out by positioning the electron probe, based on the STEM image, at a selected area of interest and recording diffraction patterns of this area. Alternatively, SEND patterns can also be automatically acquired from a predefined list of probe positions by scanning along a line or across an area. Indeed, conceptually this approach is identical to the above-mentioned 4D-STEM, except for the diffraction-specific electron probe and the information extracted from the diffraction patterns. When SEND patterns are acquired across an area, they can be used to form an image. Thus, bright- and dark-field STEM images can be obtained from SEND by integrating the diffraction intensities of the direct beam and selected diffraction beams, respectively. In this manner, SEND works like a STEM. But beyond the simple integrated intensities, other information can be extracted off-line to form images. For instance, the diffraction patterns can be indexed and analyzed to achieve phase and crystal orientation mapping. Also, full 2D strain maps can be extracted by accurately measuring the position of the Bragg diffracted beams from each image pixel [94–99]. However, the non-uniform contrast present in the large diffraction discs (due to variations of thickness and bending across the specimen field of view) can result in an incorrect determination of their position and deliver results that are incorrect. One way to remove the systematic errors arising from the non-even intensity in the spots is to precess the electron beam around each sampling point. This approach known as nanobeam precession electron diffraction (NPED) leads to uniform illumination in the diffracted beams and allows the position of each spot to be more accurately determined [100]. Additionally, a higher convergence angle can be used to form a smaller probe for improved spatial resolution.

Both nanobeam modes, NBED and NPED, have been extensively used in the semiconductor industry to measure local strains in nanodevices, identify crystal defects and determine crystal structures. However, there are very few publications on electron nanodiffraction studies of ferroelectric and multiferroic materials [101]. Yet, the experiments reported to date suggest important possibilities for the detailed structural characterization of multiferroics in the ultrathin regime or inserted in multilayers or superlattice architectures.

7.4 Strain mapping techniques

Driven by the need of developing new semiconductor devices substantial progress has been made in the last decade in the advancement of transmission electron microscopy based techniques for strain mapping at high spatial resolution. Such developments are also very valuable to the research field of multiferroic magnetoelectric materials as their ground state can be tuned by epitaxial strain to adopt very different multiferroic phases exhibiting a number of fascinating phenomena. Thus, the exploration of epitaxial strain due to lattice mismatch is critical for the design and growth of epitaxial multiferroic thin films for next-generation power-efficient oxide electronics. Currently, the TEM techniques that provide the best spatial resolution (below 5 nm) and quantitative strain maps are: nanobeam electron diffraction (NBED) and its more advanced version NPED in scanning mode, dark-field electron holography (DFEH), high-resolution transmission electron microscopy (HRTEM or HREM) and high-angle annular dark-field scanning transmission electron microscopy (HAADF-STEM) [88, 99, 102].

7.4.1 Strain mapping from SEND techniques

The basic principles of NBED and NPED were introduced in the previous section. The main advantage of NPED over NBED is that a higher convergence angle can be used to form a smaller probe, thus, improving the lateral spatial resolution. Additionally, NPED is very insensitive to thickness variations in the specimen leading to more robust results. Hence, from these two reciprocal space techniques, NPED provides the best combination of spatial resolution (2–5 nm) and precision (0.02–0.05 %) [99]. However, both NBED and NPED require additional hardware control (an additional condenser lens and precession coils, respectively) and the data sets can easily contain thousands of diffraction patterns making off-line data processing lengthy and heavy.

7.4.2 Strain mapping from Dark Field Electron Holography

DFEH is a variant of off-axis electron holography introduced earlier in this review. In DFEH, an electron beam diffracted by a set of lattice planes is first selected using an objective aperture. Second, a biprism placed below the specimen is used to interfere the electron beams that have been diffracted by both the strained region of interest and a reference region, usually taken in the substrate. Thus, an interference pattern (or dark-field hologram) is obtained and used to calculate the relative phase of the diffracted beam between the two regions [103, 104]. The phase and exact location of a diffracted beam depend on the crystal spacing. The phase of a diffracted beam has an additional contribution given by $\varphi(\mathbf{r})=2\pi\,\mathbf{g}\cdot\mathbf{u}(\mathbf{r})$, where **g** is the scattering vector and **u(r)** the displacement field as a function of the position **r** [104]. Since the transmitted beam has $\mathbf{g}=\mathbf{0}$, it does not carry any information about the phase, and it is needed to select a diffracted beam for this type of holography. Hence, in order to do DFEH one needs to form a dark-field image [4] with a selected diffraction spot using an objective aperture, place the biprism between the strained and the reference area, and form an interference pattern within that dark-field image of the strained and the reference area. From this interference pattern, the strain map in the direction of the selected diffraction spot can be deduced. By repeating the same procedure for at least another linearly independent diffraction spot the full-field strains can be mapped.

For the measurement of strain, the specimens are required to be thin (<100 nm) and flat in order to ensure uniform diffraction conditions over the two regions. Any slight bending of the thin specimen is particularly prejudicial to obtain a homogeneous dark-field contrast [105]. Additionally, DFEH is often considered practically difficult because the intensity of dark-field holograms is low, and highly stable microscopes are needed for long acquisition times [106]. Nevertheless, DFEH is particularly attractive since strain maps with a large field of view (up to 1 µm) and high sensitivities (as low as 0.02–0.05 %) can be obtained [99, 107, 108]. Moreover, the DFEH technique excels in its speed of both data acquisition and processing of strain maps. As such, this technique has been extensively used for measuring strain in semiconductor devices and thin films [103, 105, 108]. But then again, there has been very limited research on quantitative measurements of displacement and strain fields by DFEH from multiferroic/ferroelectric films. Perhaps the most relevant example is the work by Denneulin et al. [109] on the deformation fields in epitaxial $Pb(Zr_{0.2}Ti_{0.8})O_3$ films containing 90° ferroelastic domains; see Figure 7.12. For the conditions used in their paper, they were able to measure the deformation

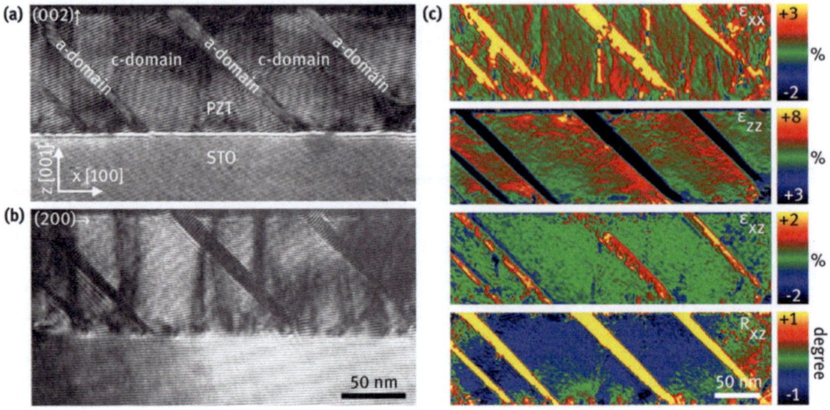

Figure 7.12: Local deformation gradients in an epitaxial $Pb(Zr_{0.2}Ti_{0.8})O_3$ film by DFEH. (a) (002) and (b) (200) dark-field electron holograms of a 100 nm thick film containing a- and c-domains. (c) From top to bottom: ε_{xx} in-plane, ε_{zz} out-of-plane, ε_{xz} shear deformation maps and R_{xz} rigid-body rotation map. Adapted from [109].

gradients in $Pb(Zr_{0.2}Ti_{0.8})O_3$ with a sensitivity of 0.1% and a spatial resolution of 6 nm over fields of view up to 500 nm. Thus, we anticipate that the ability of DFEH to measure strain over very large areas shall become very valuable when it comes to quantitatively measure strain in future nanoscale multiferroic devices and sensors.

7.4.3 Strain mapping from high-resolution images

HRTEM and HAADF-STEM allow imaging the structure of materials at the atomic scale. Additionally, with the development of aberration correctors, sub-Ångström resolution images can now be routinely obtained permitting the determination of atomic column positions with picometer precision. Then, by measuring the lattice distortions and/or local atomic displacements in the images quantitative strain maps can be extracted, provided a known reference area is present in the image. These techniques provide the best spatial resolution but at the expense of a limited field of view.

The key difference between HRTEM and HAADF-STEM lies in their modes of image formation. In TEM, an electron plane wave illuminates the specimen and the OL forms a magnified image, which is projected to the image plane. In the STEM mode, a finely focused electron probe is scanned across the specimen and the transmitted and/or scattered electrons from a localized volume of the specimen are collected on post-specimen (extended discs and/or annular) detectors as a function of raster position. TEM provides almost perfect coherent imaging whereas the STEM instrument can be configured for almost perfect incoherent imaging [110]. Hence, among the two techniques, there has been increased interest in measuring strain from HAADF-STEM images because peak intensities are registered at the atomic positions and are dependent on the atomic number (Z-contrast) over a large range of sample thicknesses, so that the images are directly interpretable. On the contrary, in HRTEM changes of specimen thickness and defocus can cause atomic columns to reverse contrast introducing artifacts in the deformation maps. Thus, atomically centered contrast in HRTEM is only accessible when the sample is at an optimal defocus and very thin. Moreover, on aberration-corrected microscopes, the aberrations (defocus, spherical aberrations) need to be balanced to optimize the phase-contrast that dominates the image contrast in HRTEM. Here, specimen preparation remains a practical challenge, especially when large fields of view are desired.

However, a major issue of using STEM imaging for strain measurements is the noise present in the images. As image intensities are acquired pixel by pixel, the recorded image is affected by relative movements of the electron probe and sample during acquisition, as well as systematic and random noises in the probe scan. Random scan noise is introduced due to sudden beam position changes (fly-back distortion) in conventional line-by-line raster scans, while systematic noise might arise from sample drift, environmental interferences, deviations from the saw-tooth deflecting voltage applied to the scan coils, or any systematic scan noise introduced during the scan fly-back time [89]. These are serious issues which can heavily degrade the quality of the STEM images and worsen the strain mapping precision. Hence, a stable environment (room conditions) and a stable microscope platform are essential for quantitative STEM measurements. For remaining instabilities, many approaches have been proposed to tackle the unwanted noise affecting STEM images. These can be grouped into those using reference data or those using frame-averaging [111].

Reference approaches assume that distortions remain constant between images recorded at different times [112, 113] or within whole scan-lines of an image [114]. However, as actual environmental distortions cannot be assumed to be reproducible, such methods often preserve scan-distortion artifacts in their corrected data [113–115]. This is illustrated in the example of Figure 7.13, where it is shown that residual scan-distortion artifacts are preserved in the corrected data.

Multi-frame acquisition approaches combined with post-processing have demonstrated significant improvements in precision and show an increased signal-to-noise ratio. These approaches include simple rigid-alignment [116], more advanced rigid-plus-affine registration of revolving image series [117], unconstrained non-rigid registration [118], and non-rigid registration using the "row-locking" technique [119]. Remarkably, sub-pm precision in the determination of atomic positions has been achieved by non-rigid registration and averaging of series of 20–30 short-exposure images [118, 119]. The resulting distortion-corrected images can be subsequently analyzed in order to obtain distortion and/or strain maps with high precision.

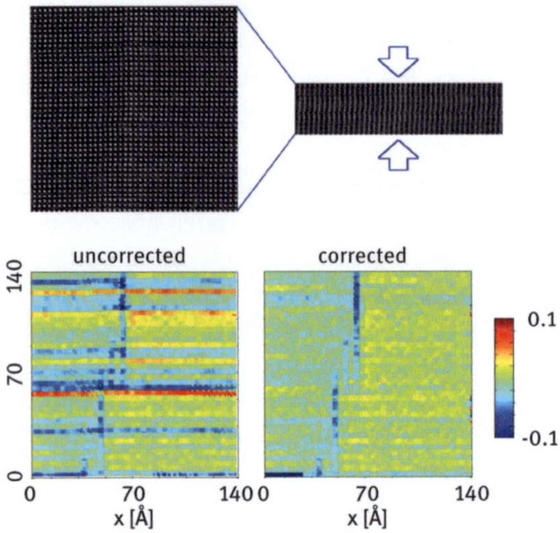

Figure 7.13: Effect of random scanning errors on the strain analysis of 109° domain walls in $BiFeO_3$. Top: HAADF-STEM image of a stepped domain wall running vertically across the whole field of view. If the image is vertically squashed, the atomic lattice no longer appears straight. Bottom: the unit-cell-wise shear strain ε_{xz} is depicted before and after removing the scanning errors, which are visible as characteristic horizontal streaks. Note that although most of the scan-distortion artifacts are removed, some are preserved in the corrected data. They are seen in blue and yellow horizontal strikes at both sides of the stepped domain wall. Adapted from [115].

An example of a noisy single HAADF-STEM image and the analogous distortion-corrected image obtained from averaging several images non-rigidly is given in Figure 7.14 and Figure 7.15. In Figure 7.14, a single HAADF-STEM image acquired with 20 μs dwell time of a $BiFeO_3$ thin film on a $La_{0.7}Sr_{0.3}MnO_3$ layer reveals the presence of two edge dislocations [79]. Here, the enlarged view (inset) of the area highlighted with a square clearly shows the presence of errors in the beam position along the fast scan axis (also known as "flags"). The errors in the beam position along the slow scan ("skips") lead to an extension or truncation of the image and are perceptible as horizontal stripes in the ε_{zz} and ε_{xz} strain and R_{xz} rotation maps. The analogous image obtained by non-rigid registration of 20 HAADF-STEM images acquired with 1 μs dwell time is shown in Figure 7.15. Both the enlarged view and the corresponding strain and rotation maps demonstrate that the flags and skips have been corrected resulting in a significant improvement in the strain precision.

Several methodologies have been described in the literature to obtain strain maps from HRTEM and HAADF-STEM images, either in Fourier space (GPA) or in real space (Peak Finding). They provide the best spatial resolution (1–5 nm) but a relatively low precision (0.2–0.4 %) as compared to the previously described deformation mapping techniques [99].

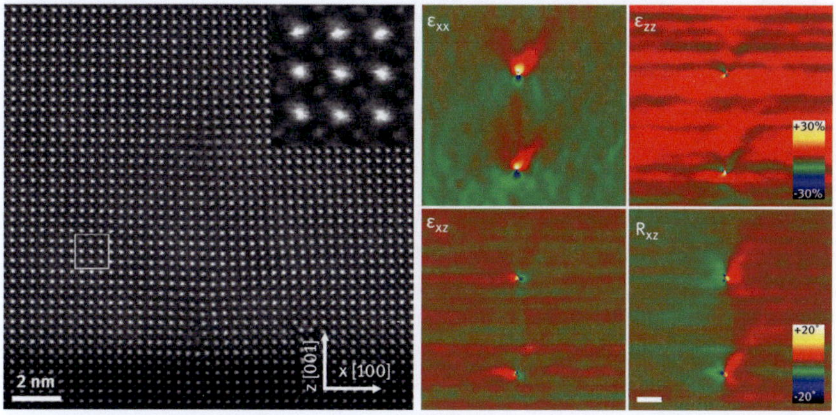

Figure 7.14: Single HAADF-STEM image acquired with 20 μs dwell time of a $BiFeO_3$ thin film on a $La_{0.7}Sr_{0.3}MnO_3$ layer revealing the presence of two edge dislocations at a distance of 8 nm. The inset is an enlarged view of the area highlighted with a square showing the errors in the beam position along the fast scan axis (flags). The errors in the beam position along the slow scan (skips) lead to an extension or truncation of the image and are perceptible as horizontal stripes in the ε_{zz} and ε_{xz} strain and R_{xz} rotation maps obtained by GPA. These fluctuations are absent in the ε_{xx} strain map because the scanning direction is parallel to the principal axis of the strain. The color bars give the change in strain and the rigid rotation angle in percent and degrees, respectively. The scale bar is 2 nm.

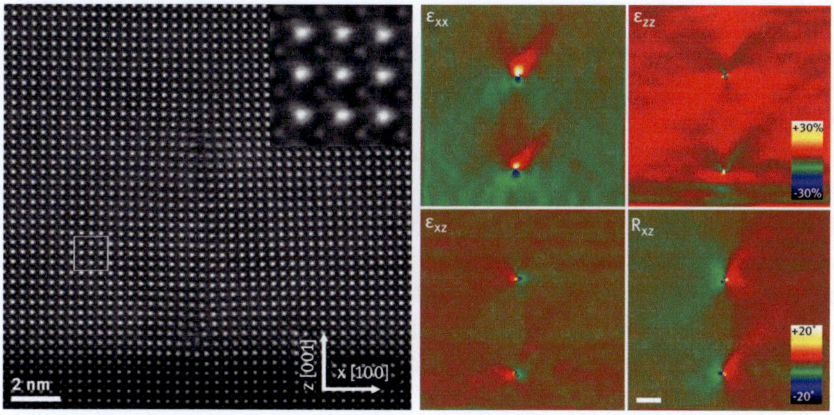

Figure 7.15: Non-rigid registration of 20 HAADF-STEM images acquired with 1 μs dwell time of the same area shown in Figure 7.14, and corresponding strain and rotation maps obtained by GPA. The inset is an enlarged view of the area highlighted with a square. Note that the flags and skips have been corrected. The color bars give the change in strain and the rigid rotation angle in percent and degrees, respectively. The scale bar is 2 nm.

Geometric Phase Analysis (GPA) works in the Fourier space and consists of measuring the local displacements of atomic planes by calculating the local Fourier components of the lattice fringes in a HRTEM or HAADF-STEM image [120]. This is done by centering a small aperture on a Bragg spot in the Fourier transform of the lattice image and performing an inverse Fourier transform. The phase component of the resulting complex image contains information about the local lattice displacements and the 2D displacement field can be derived by applying the method to two or more non-collinear Fourier components. Subsequently, the local strain components can be obtained by computing the derivative of the displacement field [121, 122]. The spatial resolution of the GPA technique is determined by the size of the mask used to extract the Bragg spot.

Peak-Finding methods work in real-space and take advantage of all information recorded in an atomic resolution HRTEM or HAADF-STEM image. They are based on locating all the intensity maxima registered at the atomic columns and superimposing a 2D reference lattice extrapolated from a distortion-free region of the material in order to calculate the local discrete displacement field at each node. Then, by derivation of the calculated displacement field, the strain field is obtained [114, 123]. This methodology works very well for the analysis of pseudomorphic heterostructures but the automatic image girding usually fails at defective regions, such as dislocations cores. At these positions human intervention is typically needed to correctly construct the lattice. An alternative real space algorithm based on the detection of pairs of intensity maxima was proposed by Galindo et al. [124]. The algorithm, known as Peak Pairs Analysis (PPA) [125], works on a filtered image by locating pairs of peaks along two predefined basis vectors. From the deformation of each pair across the image, the displacement field can be reconstructed and used to calculate strain maps. PPA exhibits a good behavior at heavily distorted areas, for instance at interfaces of plastically relaxed systems, without user's intervention. However, PPA fails when lattice peaks are not clearly detected due to static tilts of atomic columns near specific dislocations or grain boundaries. In these cases, GPA is recommended [124].

The TEM techniques introduced in this section are very valuable to the research field of multiferroics as the nanoscale strain gradients are known to influence different functional properties of the films [79, 126–130]. Thus, if properly exploited, the epitaxial strain can be used as an important factor to tailor the macroscopic response toward desired properties, such as ferroic transition temperatures, magnetic order and ferroelectric polarization.

7.5 Ferroelectric polarization mapping from high-resolution images

Aside from local strain measurements, atomic-resolution micrographs also allow for assessing the local symmetry of individual unit cells and thus for indirectly measuring the local polarization of the sample (see Chapter 6). The ferroelectric polarization can be determined from the atomic displacements measured in atomic resolution images. Earlier reports on polarization mapping mostly used HRTEM techniques such as negative spherical aberration (Cs) condition [131] or exit wave reconstruction of focal series [74]. However, in the last few years HAADF-STEM has emerged as the preferred technique for mapping atomic polar displacements because the samples can be thicker, which helps to preserve the native domain structure, and the interpretation of the image contrast is straightforward and reliable [56, 79, 115, 132–139]. Mapping of atomic polar displacements from HAADF-STEM images was successfully applied to reveal various domain wall configurations [115, 132–135, 140], flux-closure domain patterns [136], ferroelectric vortex domains [137, 138], polarization gradients [56, 141], as well as interface and surface reconstructions in ferroelectric thin films [142]; see Figure 7.16. In these examples, the ferroelectric films present the *pseudocubic perovskite-type structure* with general formula ABO_3, where A and B are cations of different size and O is the anion bonding to both. In the perovskite structure, the larger A cations lie in the center of 12 oxygen ions in dodecahedral coordination, and the smaller B cations are octahedrally coordinated. The polar displacement resulting from the off-centering of the cations with respect to the oxygen octahedra induces spontaneous polarization. However, in the HAADF-STEM images, the light elements such as oxygen are usually undetectable when the compounds contain relatively heavier atoms; then polarization maps are effectively calculated from the relative displacements of the two cation sublattices. As the B cations are displaced along with the oxygen octahedra, the displacement vector can be determined by measuring the polar displacement of the B position in the image plane from the center of its four A neighbors. Thus, the displacement vector as the manifestation of the ferroelectric polarization in ABO_3 points toward the center of the negative oxygen ions and can be used to determine the polarization vector. In the polarization maps derived from HAADF-STEM images, the polarization vectors are typically plotted opposite to the polar displacement of the B cations. An example of a polarization map, superimposed on a HAADF-STEM image of a $Bi_{0.8}Ca_{0.2}FeO_3$ thin film, calculated from the displacement of the Fe cation with respect to the CoM of the Bi cell, is shown in Figure 7.17(a).

A more straightforward way of obtaining polarization maps can be attained, in principle, by using annular bright-field (ABF) STEM imaging as it allows for simultaneously visualizing both heavy and light atomic columns in a single image [143–148]. This imaging mode uses an annular detector spanning an angular range within the direct scattering cone, which is defined by the probe-forming aperture of the focused electron beam [143, 147]. In ABF-STEM the image contrast is mostly dominated by elastically scattered electrons and the images show, for suitable defocus values, a negative absorptive-type signal at all (light and heavy) column locations over a large range of thicknesses [147, 149]. Therefore, this imaging mode has turned out to become an important imaging mode for the quantitative structure determination of systems composed of a mixture of light and heavy atoms, such as

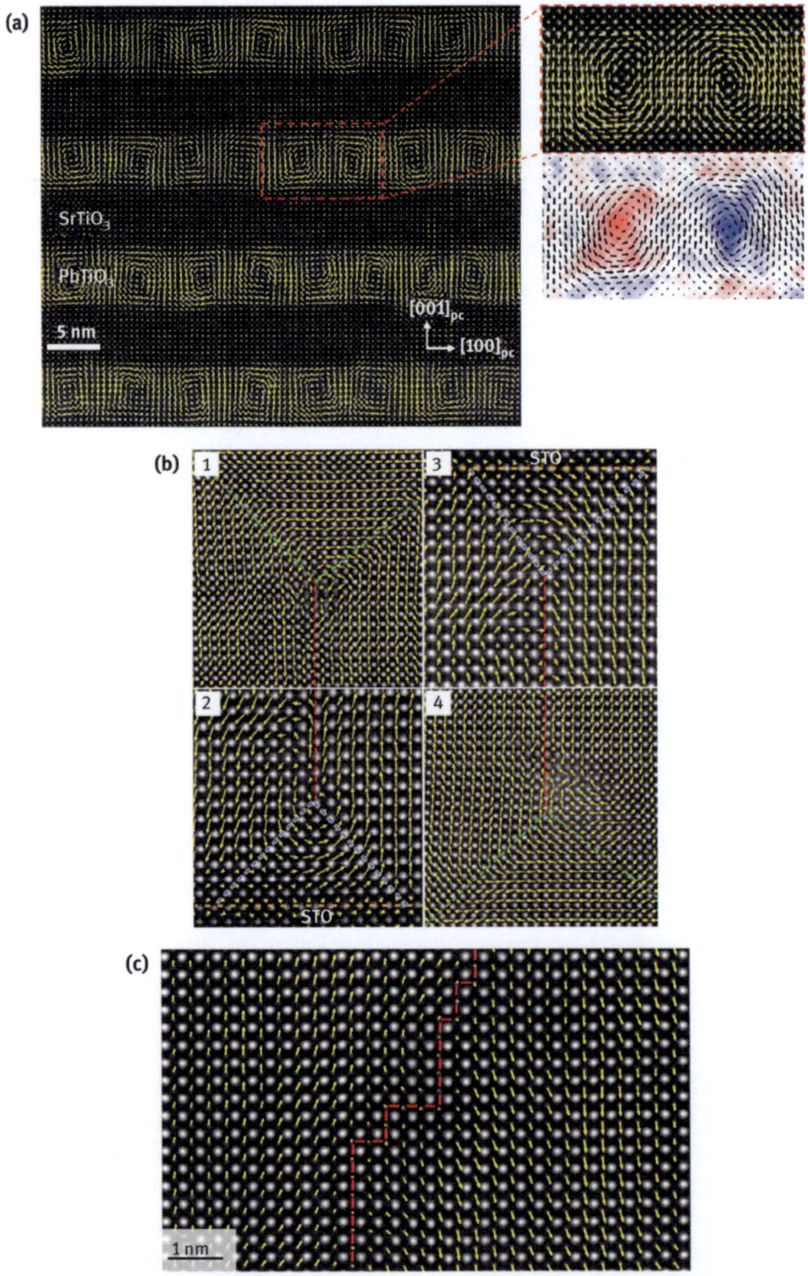

Figure 7.16: Mapping of atomic polar displacements from HAADF-STEM images. (a) Ferroelectric vortex domains in a $(SrTiO_3)_{10}/(PbTiO_3)_{10}$ superlattice [138]. The curl of the polar displacement $(\nabla \times P)_{[010]}$ is plotted for a vortex-antivortex pair with a red/blue color scale; clockwise (negative) is blue and anticlockwise (positive) is red. (b) Flux-closure domains in strained $PbTiO_3$ films [136]. (c) 180° domain wall in a $PbZr_{0.2}Ti_{0.8}O_3$ film displaying a mixed Ising–Néel type character [135].

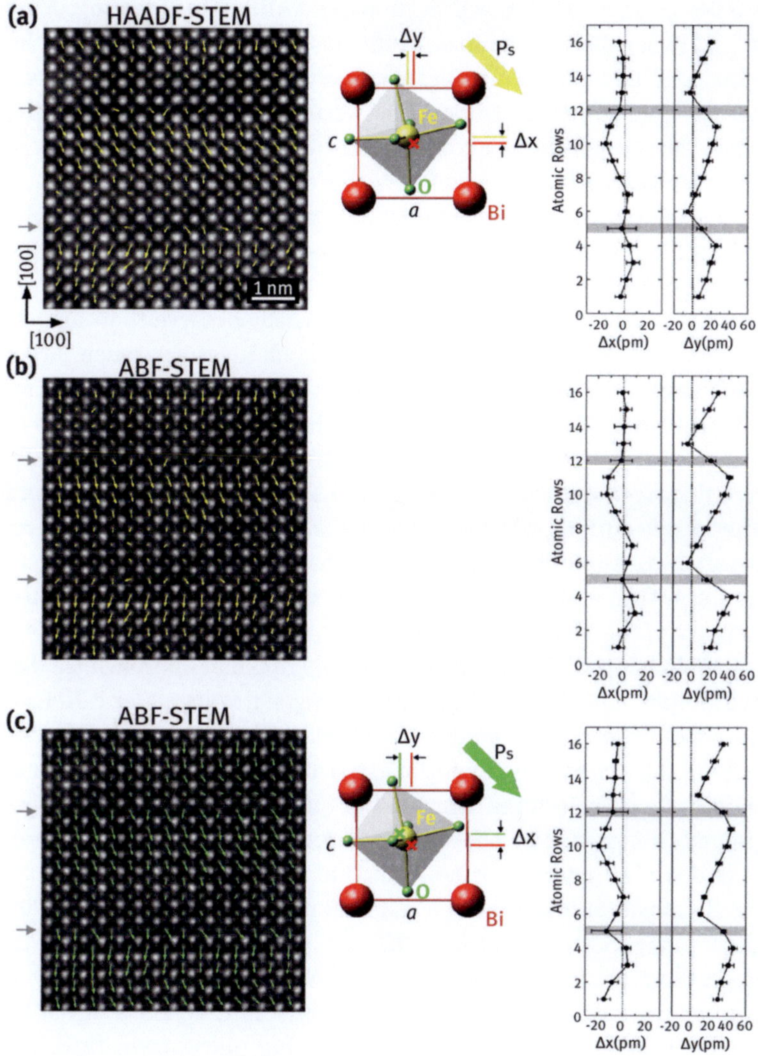

Figure 7.17: Periodic giant polarization gradients in a $Bi_{0.8}Ca_{0.2}FeO_3$ thin film. Polarization maps superimposed to (a) an averaged HAADF-STEM image and (b) the simultaneously recorded inverted ABF-STEM image calculated from the displacement of the Fe cation with respect to the CoM of the Bi cell (marked with a red cross in the model). (c) Polarization map superimposed to the inverted ABF-STEM image calculated from the relative displacements of the CoM of the oxygen octahedron and the CoM of the Bi cell (marked with green and red crosses in the model, respectively). As indicated in the models, the polarization vectors are plotted opposite to the polar displacements of the B cations and O anions. The corresponding profiles along the [001] direction of the in-plane (Δx) and out-of-plane (Δy) components of the atomic displacements are shown at the right side. The gray rectangles indicate the position of two oxygen deficient layers. Adapted from [56].

oxides and perovskites [56, 139, 150]. However, unlike HAADF-STEM imaging, ABF-STEM images are highly sensitive to sample mistilt [147, 151–153]. Specifically, the misalignment of the sample results in significant column-type dependent shifts, especially for thicker specimens. Recent work on the accuracy of ABF-STEM imaging revealed that a small sample mistilt of about 6 mrad relative to the electron beam causes an artificial displacement of 11.9 pm between the oxygen and cation columns for cubic $SrTiO_3$ [152]. Artifacts of this magnitude can thus result in the depiction of a deformed unit cell preventing the reliable quantification of lateral atomic displacements based on ABF-STEM images. This is exemplified in Figure 7.17. By comparing the profiles in Figure 7.17(a and b), one can easily observe that the out-of-plane (Δy) relative displacements of the two cation sublattices are much larger for the ABF-STEM image (~ 47 pm) than for the HAADF-STEM image (~ 30 pm). Further, the profile in Figure 7.17(c), calculated from the relative displacements of the CoM of the oxygen octahedron and the CoM of the Bi cell, is shifted laterally away from the zero-displacement axis as a result of a small sample mistilt around the [100] axis. When using ABF-STEM imaging for polarization mapping it is therefore extremely important to align the sample's zone axis to the optical axis as accurately as possible, i. e. within a precision of tilt of <1–2 mrad, which is typically the limit of the tilt accuracy of the TEM goniometer [151].

For a practical application of the above outlined experimental approach we discuss a non-trivial example, namely the multiferroic *hexagonal manganites* h-$RMnO_3$ (R = Sc, Y, Dy-Lu), which belongs to a family of materials attracting much interest, particularly because of the observation of antiphase structural domains clamped to ferroelectric domain walls [154–160] (as discussed in Chapter 3). The structural domains, when arranged in a particular phase relation forming a clover-leaf pattern, result in the appearance of structural vortices, which in turn induce magnetic vortices [161]. Additionally, it was shown that in hexagonal manganites ferroelectric domain walls can carry a magnetic moment [161–163].

The high-temperature paraelectric $RMnO_3$ structure displays $P6_3/mmc$ symmetry and consists of corner-linked MnO_5 trigonal bipyramid layers alternating along the c-axis with R layers. At the symmetry-lowering phase transition, polarization emerges as a result of periodic tilts of the MnO_5 bipyramids and displacements of the R ions along the c axis; see Figure 7.18. This periodic $\sqrt{3}\times\sqrt{3}\times 1$ lattice distortion, which is driven by the condensation of a zone boundary K_3 phonon, triples the crystallographic unit cell and is often referred to as the trimerization transition. Thus, the R ions arrange in the form of sinusoidal waves and display either a ↑↑↓ pattern with positive (+) polarization or a ↓↓↑ pattern with negative (–) polarization. Each polarization pattern exhibits three possible permutations resulting in three distinct φ structural domains (α, β, and γ). Therefore, the low-temperature ferroelectric phase, $P6_3cm$, presents a total of six structural domains, labeled α^+, β^-, γ^+, α^-, β^+, γ^-. By driving the system through this structural transition, topological defects are

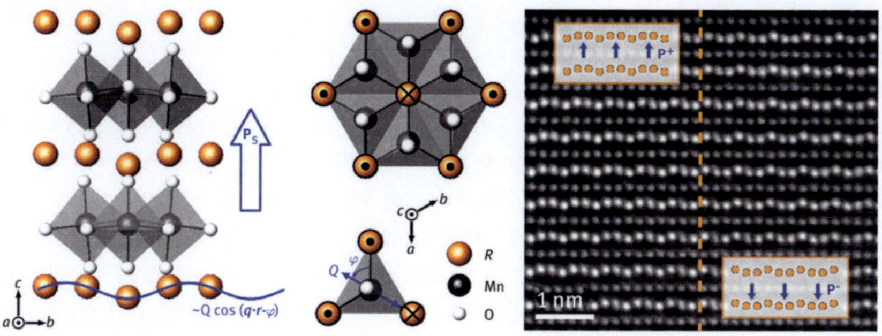

Figure 7.18: Ferroelectric order in hexagonal manganites. The side and top views of the $P6_3cm$ unit cell show the lattice trimerization **Q** with tilt amplitude Q (in this case away from the common center) and azimuthal tilt angle φ of the MnO_5 bipyramids, and the resulting sinusoidal corrugation pattern of the R atoms. For clarity, the oxygen atomic displacements for one of the three bipyramids constituting the tilt are indicated. The HAADF-STEM image corresponds to a neutral domain wall in $YMnO_3$ (marked by the broken line) separating two neighboring domains with opposite polarization. The insets show the Y displacements observed at each side of the domain wall.

created in the form of discrete vortices around which φ changes monotonically in a clockwise or counterclockwise fashion. The core region is typically about 3 nm in diameter and exhibits nearly zero polarization [164, 165].

The crystal-lattice trimerization can be fitted by using the model proposed by Holtz et al. [165]:

$$u = Q\cos(\boldsymbol{q}\cdot\boldsymbol{r} - \varphi)$$

where $\boldsymbol{q}$ is the wavevector of the K_3 mode, $\boldsymbol{r}$ the position vector of the R atom, Q the amplitude of the sinusoidal wave and φ its phase, with discrete values given by $\varphi = n\pi/3$ (n = 0, 1, . . ., 5). The primary order parameter of the structural trimerization is hence given by $\mathbf{Q} = (Q\cos\varphi, Q\sin\varphi)$. Noteworthy, the two parameters Q and φ are connected to remarkable physical quantities: Q is related to the maximum displacement (1.5 x Q) between R atoms, and φ is associated with the six possible structural domains generated by the possible permutations of the ↑↑↓ and ↓↓↑ patterns.

From Landau theory, the polarization P can be directly calculated from the values of Q and φ from the relation [165]:

$$\mathbf{P} \sim Q^3\cos 3\varphi$$

Therefore, from the fitted values of Q and φ, the polarization can be calculated. The paired charge density due to the material polarization can then be calculated using the equation:

$$\rho_P = -\nabla \cdot \mathbf{P}$$

and allows a direct visualization of the charged domain walls.

As an example, Figure 7.19 shows a HAADF-STEM image of $YMnO_3$, where the displacement pattern of the Y atoms changes gradually across three 180° domain walls. Thus, four distinct ferroelectric domains can be identified. Additionally, the charge density map reveals that, due to the non-straight nature of the domain walls, in some areas the polarization is not parallel to the wall, so that $\nabla \cdot \mathbf{P} \neq 0$. This example demonstrates that by carefully applying the above methodology critical structural information about complex multiferroics can be obtained, on length scales which are difficult to access by other methods.

7.6 New possibilities of in situ measurements

The success of in situ TEM measurements has been facilitated in recent years by the development of a number of technologies. Progress in Micro-Electro-Mechanical Systems (MEMS) has led to the development of dedicated TEM holders providing different stimuli (such as heating, electrical biasing, mechanical force, etc.) or a combination of multiple stimuli at once. The investigations can moreover be carried out in vacuum, gas or liquid environment. Additionally, the introduction of scanning/transmission electron microscopes with sub-Ångström resolution as well as fast and sensitive detection cameras capable of high speed (with more than 1000 frames per second) support the direct observation of dynamic phenomena in situ down to the atomic scale.

This section covers the latest developments in in situ TEM instrumentation which are of relevance to characterize magnetoelectric multiferroics. In particular, we describe here recent advances of in situ MEMS-based holders for heating and electrical biasing experiments and provide relevant literature on in situ observations of domain structure evolution and phase transitions.

7.6.1 Heating stages

For many years the only available in situ heating stages for TEM were furnace type holders made from tantalum or Inconel, and designed to accommodate standard 3 mm disk specimens. These holders can reach 1200 °C (when using tantalum), but require the use of circulating water to cool down the specimen rod and to maintain the targeted temperature at the specimen. Hence, these traditional systems suffer from slow response times, (comparably) inaccurate temperatures and severe sample instability complicating atomic resolution imaging.

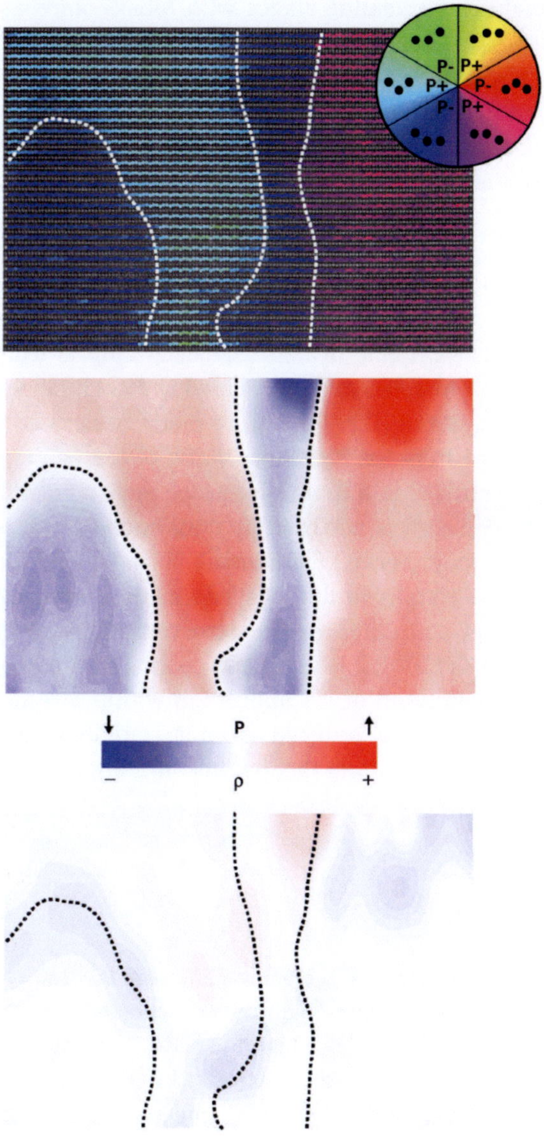

Figure 7.19: Charged domain walls in $YMnO_3$. Top: HAADF-STEM image with a color overlay of the trimerization phase. The displacement pattern of the Y atoms changes gradually across the 180° domain walls. Middle: plot of the polarization. Red indicates polarization up and blue polarization down, according to the color bar. Bottom: charge density map. Positively and negatively charged walls are indicated in red and blue, respectively.

To address these issues, microfabricated heating stages with MEMS chips as disposable cartridges were developed [166, 167] and have become commercially available. The MEMS chips are typically made of silicon as the main body and present a window which is covered with a free-standing non-electron transparent membrane consisting of an electrical insulator material (usually silicon carbide or silicon nitride) [167–170]. The membrane is offered with either electron transparent windows made of amorphous silicon nitride or carbon (for nanostructures) or through-holes (for FIB lamella applications or 2D materials). The cartridges incorporate integrated circuitry which is used to pass an electrical current and, thus, to produce localized resistive heating directly on a small region of the chip of about 0.1 mm^2. The Protochips systems [171] utilize ceramic heaters and each chip is calibrated by the supplier (using an optical pyrometer) and is provided with a chip-specific calibration file. Instead, the other systems [172, 173] utilize metal heaters and the temperature is accurately determined from the resistance value of the heating element by using a separate pair of electrodes in a four-point probe measurement. Precise temperature stabilization (<0.1 °C/min) is achieved by continuously monitoring the resistance in a closed control loop, hence guaranteeing accuracy and reproducibility of the in situ experiments. Thanks to the localized heating and the small thermal mass involved, low power consumption (few tens of milliwatts) is achieved enabling temperatures as high as 1300 °C, heating and cooling in the millisecond range, minor sample displacement, and drift stabilization within few minutes instead of hours. Thus, with some care heat-induced specimen modifications can be recorded at atomic resolution in real time from the first instance after a heat change [174–176].

7.6.2 Electrical stages

In early studies, instruments and attachments for in situ biasing TEM were laboratory-made systems [177–183]. Practically, these experiments required complicated sample preparation strategies and it was difficult to estimate the magnitude of the actual electric field applied to the imaging area of the specimen. In the early 2000s, the commercialization by Nanofactory Instruments of TEM holders with an integrated scanning probe made in situ biasing experiments much easier [184–188]. These in situ holders use a conductive nanoprobe that is brought into contact to the specimen by using a precise piezo-driven manipulator. In particular, for polarization switching experiments of ferroelectric thin films, cross-sectional specimens are required. Thus, an electric field can be induced along the film normal by applying a voltage between the nanoprobe (contacted at the film surface) and a bottom electrode (contacted with silver paste at the conductive substrate or buffer layer); see Figure 7.20(a). A typical ferroelectric switching process observed in situ using a conductive nanoprobe is shown in Figure 7.20(b and c). In this example, the nucleation and growth of a P[00–1] domain

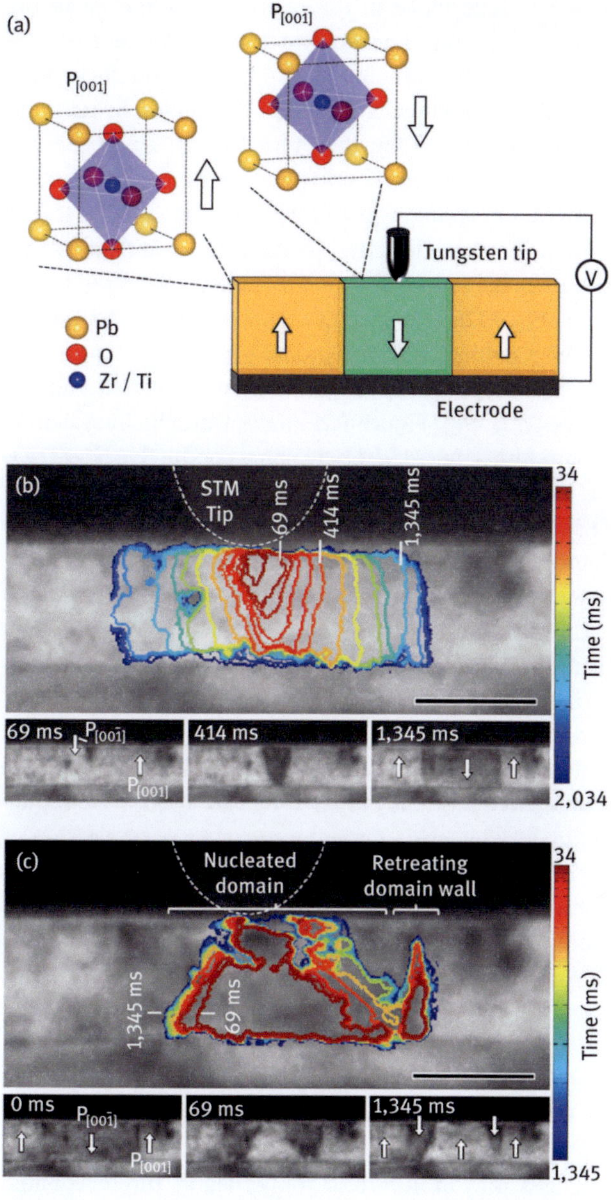

Figure 7.20: In situ TEM observations of polarization switching using a conductive nanoprobe. (a) Schematics of the switching experiment in the TEM. Polarization switching is induced by an applied electric field oriented along the film normal between a surface probe in direct contact with the $PbZr_{0.2}Ti_{0.8}O_3$ film surface and a planar bottom electrode. (b) Growth of a P[00−1] domain during the initial switching cycle. (c) Reverse switching under reverse bias results in the nucleation of P[001] domains at the bottom. The lines are contour plots of the domain outline versus time. Selected corresponding TEM images are shown below each panel. Scale bars are 100 nm. Reproduced from [189].

in a $PbZr_{0.2}Ti_{0.8}O_3$ film is observed to take place in the region below the tip of the probe during the initial switching cycle, while reverse switching under reverse bias results in the nucleation of multiple P[001] domains at the bottom interface [189]. While much research on the switching mechanisms has been conducted by using in situ conductive nanoprobe systems [184, 187–190], they face two serious disadvantages. First, it is not possible to examine ferroelectric switching over large areas under a uniform electric field. Instead, an inhomogeneous high field is confined to the region below the tip of the probe [187–189]. Second, additional piezoelectric and flexoelectric coupling effects on the spontaneous polarization might arise from the force contact of the probe against the film surface [191, 192]. This suggests that the employed geometry, where the film surface is in direct contact with the probe, is probably not entirely reliable. As an alternative, a sandwich-type capacitor geometry with bottom and top electrodes would be a better choice [193, 194]. Figure 7.21 shows the switching behavior of an epitaxial $PbZr_{0.2}Ti_{0.8}O_3$ thin film sandwiched between a conductive Nb-$SrTiO_3$ substrate and a top electrode consisting of Au and Pt in a capacitor-type geometry [193]. The series of dark-field TEM images acquired at distinct external biases demonstrates that by using this configuration a uniform electric field is attained over a very large area. Moreover, this geometry ensures that the electrical boundary conditions closely resemble those of an actual ferroelectric capacitor [195]. Here, also, advances in MEMS technology for miniaturized TEM specimen holders shall bring new opportunities for more consistent in situ biasing studies. Similarly to the MEMS heating chips, this approach uses prefabricated circuitry developed by lithography on silicon chips containing thin silicon nitride membranes. A cross-sectional slice of the ferroelectric film sandwiched between two conductive layers is prepared by FIB. Then, the conductive layers are connected to the circuitry by writing metal interconnects by electron or ion-beam-induced deposition within a SEM or FIB [196]. Notably, MEMS-based holders and chips fitted with four or more contacts offer the unique additional capability to enable simultaneous in situ heating and biasing experiments, thus, opening up new fields of research for in situ electron microscopy.

For all the reasons explained above, MEMS-based TEM heating holders are becoming increasingly popular. This is reflected in the exponential growth in the number of publications, and material conference sessions and workshops on in situ TEM organized over the last decade. In particular, it is expected that the MEMS-based in situ technology shall become standard for characterizing ferroelectric materials enabling a whole new level of insight into multiferroic properties at unprecedented length scales. For example, in situ thermal TEM experiments shall be used to study the ferroelectric phase transition in thin films and to investigate the mechanisms controlling the Curie temperature in the low thickness range, such as clamping of the film distortions exerted by the substrate. These experiments performed at the atomic scale should give direct insight into the ferroelectric ordering and help clarify the displacement and order-disorder polarization mechanisms. Dynamic measurements of the polarization reversal by applying electric fields through the connected top and

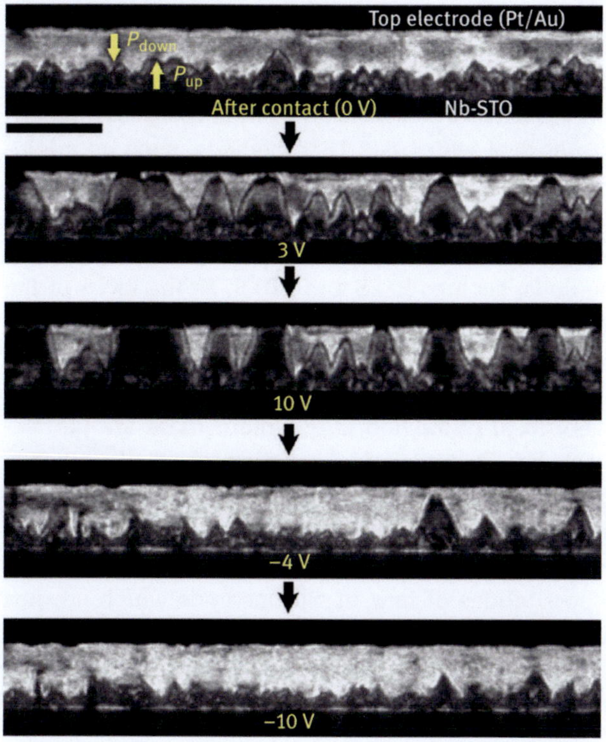

Figure 7.21: In situ TEM observations of polarization switching using a capacitor-type geometry. The ferroelectric $PbZr_{0.2}Ti_{0.8}O_3$ thin film is sandwiched between a conductive Nb-$SrTiO_3$ substrate and a top electrode consisting of Au and Pt. The series of dark-field TEM images show the domain-switching behavior induced by a series of external biases. The scale bar is 200 nm. Reproduced from [193].

bottom electrodes will be crucial to study in situ the polarization switching process and domain wall propagation at the atomic scale. In situ electron holography, at forward and reverse bias conditions, shall reveal valued information about the charge redistribution. Also, in situ biasing experiments shall allow for observing the response of magnetic domains to the applied electric field, which is essential to understanding the multiferroic behavior. In conclusion, MEMS-based TEM holders shall contribute significantly to understanding the relationship between the domain dynamics of multiferroics and the specific microstructure of the films providing important guidance to design new devices and to predict and mitigate failures.

7.7 Summary

This review focused on introducing advanced methods based on (scanning) transmission electron microscopy which can be used to characterize complex multiferroic materials. The review did not deal with basic TEM methods, such as bright-field/dark-field imaging, selected area diffraction, which, of course, are important for the study of materials, in general, and undoubtedly need to be considered as the fundament upon which new and advanced methods build up. Neither did the review deal with analytical methods, such as EELS and EDXS, as the focus of the present review was on analytical imaging methods. After introducing the topic, so-called phase-contrast methods were first discussed which can be used to map electrostatic and magnetic fields in TEM and STEM modus. Diffraction methods were then covered which provide very local diffraction information; these are particularly useful to gather structural data in materials consisting of small domains. Thereafter, methods were discussed which allow for analyzing local strain, whereas in the subsequent section methods were introduced which allow for measuring local polarization effects on a length scale of individual unit cells. Finally, a brief outlook was given about the possibilities of new in situ methods which are just about at the launch of becoming more popular, particularly in the field of magnetoelectric multiferroics.

References

[1] Erni R. Aberration-corrected imaging in transmission electron microscopy. London: Imperial College Press, 2014.

[2] Carter CB, Williams DB. Transmission electron microscopy: diffraction, imaging, and spectrometry. Switzerland: Springer International Publishing, 2016.

[3] Ramasse QM, Seabourne CR, Kepaptsoglou D-M, Zan R, Bangert U, Scott AJ. Probing the bonding and electronic structure of single atom dopants in graphene with electron energy loss spectroscopy. Nano Lett. 2013;13:4989.

[4] Williams DB, Carter CB. Transmission electron microscopy. United States: Springer, 2009.

[5] Newnham RE. Properties of materials: anisotropy, symmetry, structure. New York, United States: Oxford University Press, 2005

[6] Fiebig M, Lottermoser T, Meier D, Trassin M. The evolution of multiferroics. Nat Rev Mater. 2016;1:16046.

[7] Matteucci G, Missiroli G, Nichelatti E, Migliori A, Vanzi M, Pozzi G. Electron holography of long-range electric and magnetic fields. J Appl Phys. 1991;69:1835.

[8] Rau WD, Schwander P, Baumann FH, Höppner W, Ourmazd A. Two-dimensional mapping of the electrostatic potential in transistors by electron holography. Phys Rev Lett. 1999;82:2614.

[9] Cumings J, Zettl A, McCartney MR, Spence JC. Electron holography of field-emitting carbon nanotubes. Phys Rev Lett. 2002;88:056804.

[10] Kasama T, Dunin-Borkowski RE, Beleggia M. Electron holography of magnetic materials. In: Monroy F, editor(s). Holography-Different fields of application, 2011 ed. London, United Kingdom: IntechOpen.

[11] Nagaosa N, Tokura Y. Topological properties and dynamics of magnetic skyrmions. Nat Nanotechnol. 2013;8:899.

[12] Pollard SD, Garlow JA, Yu J, Wang Z, Zhu Y, Yang H. Observation of stable Néel skyrmions in cobalt/palladium multilayers with Lorentz transmission electron microscopy. Nat Commun. 2017;8:1.

[13] Phatak C, Petford-Long AK, Heinonen O. Direct observation of unconventional topological spin structure in coupled magnetic discs. Phys Rev Lett. 2012;108:067205.

[14] Rubiano Da Silva N, Möller M, Feist A, Ulrichs H, Ropers C, Schäfer S. Nanoscale mapping of ultrafast magnetization dynamics with femtosecond Lorentz microscopy. Phys Rev X. 2018;8:31052.

[15] Fu X, Pollard SD, Chen B, Yoo BK, Yang H, Zhu Y. Optical manipulation of magnetic vortices visualized in situ by Lorentz electron microscopy. Sci Adv. 2018;4:eaat3077.

[16] Rodríguez LA, Magén C, Snoeck E, Gatel C, Marín L, Serrano-Ramón L, et al. Quantitative in situ magnetization reversal studies in Lorentz microscopy and electron holography. Ultramicroscopy. 2013;134:144.

[17] Shibata N, Kohno Y, Nakamura A, Morishita S, Seki T, Kumamoto A, et al. Atomic resolution electron microscopy in a magnetic field free environment. Nat Commun. 2019;10:2308.

[18] Aharonov Y, Bohm D. Significance of electromagnetic potentials in the quantum theory. Phys Rev. 1959;115:485.

[19] Winkler F, Tavabi AH, Barthel J, Duchamp M, Yucelen E, Borghardt S, et al. Quantitative measurement of mean inner potential and specimen thickness from high-resolution off-axis electron holograms of ultra-thin layered WSe2. Ultramicroscopy. 2017;178:38.

[20] Lichte H, Reibold M, Brand K, Lehmann M. Ferroelectric electron holography. Ultramicroscopy. 2002;93:199.

[21] Beleggia M, Kasama T, Dunin-Borkowski RE. The quantitative measurement of magnetic moments from phase images of nanoparticles and nanostructures—I. Fundamentals. Ultramicroscopy. 2010;110:425.

[22] Reimer L, Kohl H. Scattering and phase-contrast. Springer Ser Opt Sci. 2008;36:193.

[23] Bonevich JE, Pozzi G, Tonomura A. Electron holography of electromagnetic fields. Introduction to electron holography. Boston, MA, United States: Springer, 1999.

[24] Lichte H, Lehmann M. Electron holography—basics and applications. Reports Prog Phys. 2008;71:016102.

[25] Möllenstedt G, Düker H. Beobachtungen und Messungen an Biprisma-Interferenzen mit Elektronenwellen. Zeitschrift Für Phys. 1956;145:377.

[26] Völkl E, Lehmann M. The reconstruction of off-axis electron holograms. Introduction to electron holography. Boston, MA, United States: Springer, 1999.

[27] Lehmann M, Lichte H. Tutorial on off-axis electron holography. Microsc Microanal. 2002;8:447.

[28] Röder F, Lubk A, Wolf D, Niermann T. Noise estimation for off-axis electron holography. Ultramicroscopy. 2014;144:32.

[29] Frost BG, Matteucci G. On recording, processing and interpretation of low magnification electron holograms. Introduction to electron holography. Boston, MA, United States: Springer, 1999.

[30] de Ruijter WJ, Weiss JK. Detection limits in quantitative off-axis electron holography. Ultramicroscopy. 1993;50:269.

[31] Cooper D, Truche R, Rivallin P, Hartmann J-M, Laugier F, Bertin F, et al. Medium resolution off-axis electron holography with millivolt sensitivity. Appl Phys Lett. 2007;91:143501.

[32] Harscher A, Lichte H. Experimental study of amplitude and phase detection limits in electron holography. Ultramicroscopy. 1996;64:57.

[33] Phatak C, Bain JA, Zhu JG, De Graef M. Aberration corrected Lorentz microscopy for perpendicular magnetic recording media. Microsc Microanal. 2008;14:832.
[34] Lehmann M, Lichte H, Geiger D, Lang G, Schweda E. Electron holography at atomic dimensions—present state. Mater Charact. 1999;42:249.
[35] Lichte H. Electron image plane off-axis holography of atomic structures. In: Mulvey T, Sheppard C, editor(s). Advances in optical and electron microscopy Vol. 12. San Diego, CA, United States: Academic Press Inc, 1991:25.
[36] Rau WD, Lichte H. High resolution off-axis electron holography. Introduction to electron holography. Boston, MA, United States: Springer, 1999.
[37] Tonomura A, Matsuda T, Endo J, Arii T, Mihama K. Holographic interference electron microscopy for determining specimen magnetic structure and thickness distribution. Phys Rev B. 1986;34:3397.
[38] Kasama T, Dunin-Borkowski RE, Scheinfein MR, Tripp SL, Liu J, Wei A. Off-axis electron holography of self-assembled Co nanoparticle rings. MRS Proc. 2007;1026:1003.
[39] Lazić I, Bosch EG. Analytical review of direct STEM imaging techniques for thin samples. Adv Imaging Electron Phys. 2017;199:75.
[40] Rose H. Phase-contrast in scanning transmission electron microscopy. Optik (Stuttg). 1974;39:416.
[41] Dekkers NH, Lang HD. Differential phase-contrast in a STEM. Optik (Stuttg). 1974;41:452.
[42] Erni R, Rossell MD, Kisielowski C, Dahmen U. Atomic-resolution imaging with a sub-50-pm electron probe. Phys Rev Lett. 2009;102:96101.
[43] Chapman JN, Batson PE, Waddell EM, Ferrier RP. The direct determination of magnetic domain wall profiles by differential phase-contrast electron microscopy. Ultramicroscopy. 1978;3:203.
[44] Chapman JN, Morrison GR, Jakubovics JP, Taylor RA. Determination of domain wall structures in thin foils of a soft magnetic alloy. J Magn Magn Mater. 1985;49:277.
[45] Sannomiya T, Haga Y, Nakamura Y, Nittono O, Takahashi Y. Observation of magnetic structures in Fe granular films by differential phase-contrast scanning transmission electron microscopy. J Appl Phys. 2004;95:214.
[46] O'Shea KJ, MacLaren DA, McGrouther D, Schwarzbach D, Jungbauer M, Hühn S, et al. Nanoscale mapping of the magnetic properties of (111)-oriented La0.67Sr0.33MnO3. Nano Lett. 2015;15:5868.
[47] Chen C, Li H, Seki T, Yin D, Sanchez-Santolino G, Inoue K, et al. Direct determination of atomic structure and magnetic coupling of magnetite twin boundaries. ACS Nano. 2018;12:2662.
[48] Brown HG, Shibata N, Sasaki H, Petersen TC, Paganin DM, Morgan MJ, et al. Measuring nanometre-scale electric fields in scanning transmission electron microscopy using segmented detectors. Ultramicroscopy. 2017;182:169.
[49] Haas B, Rouvière J-L, Boureau V, Berthier R, Cooper D. Direct comparison of off-axis holography and differential phase-contrast for the mapping of electric fields in semiconductors by transmission electron microscopy. Ultramicroscopy. 2019;198:58.
[50] Shibata N, Findlay SD, Sasaki H, Matsumoto T, Sawada H, Kohno Y, et al. Imaging of built-in electric field at a p-n junction by scanning transmission electron microscopy. Sci Rep. 2015;5:10040.
[51] Shibata N, Findlay SD, Kohno Y, Sawada H, Kondo Y, Ikuhara Y. Differential phase-contrast microscopy at atomic resolution. Nat Phys. 2012;8:611.
[52] Shibata N, Seki T, Sánchez-Santolino G, Findlay SD, Kohno Y, Matsumoto T, et al. Electric field imaging of single atoms. Nat Commun. 2017;8:15631.

[53] Campanini M, Ciprian R, Bedogni E, Mega A, Chiesi V, Casoli F, et al. Lorentz microscopy sheds light on the role of dipolar interactions in magnetic hyperthermia. Nanoscale. 2015;7:7717.
[54] Hachtel JA, Idrobo JC, Chi M. Sub-Ångstrom electric field measurements on a universal detector in a scanning transmission electron microscope. Adv Struct Chem Imaging. 2018;4:10.
[55] Krajnak M, McGrouther D, Maneuski D, O'Shea V, McVitie S. Pixelated detectors and improved efficiency for magnetic imaging in STEM differential phase-contrast. Ultramicroscopy. 2016;165:42.
[56] Campanini M, Erni R, Yang CH, Ramesh R, Rossell MD. Periodic giant polarization gradients in doped BiFeO3 thin films. Nano Lett. 2018;18:717.
[57] Shibata N, Findlay SD, Matsumoto T, Kohno Y, Seki T, Sánchez-Santolino G, et al. Direct visualization of local electromagnetic field structures by scanning transmission electron microscopy. Acc Chem Res. 2017;50:1502.
[58] Zweck J, Schwarzhuber F, Wild J, Galioit V. On detector linearity and precision of beam shift detection for quantitative differential phase-contrast applications. Ultramicroscopy. 2016;168:53.
[59] Close R, Chen Z, Shibata N, Findlay SD. Towards quantitative, atomic-resolution reconstruction of the electrostatic potential via differential phase-contrast using electrons. Ultramicroscopy. 2015;159:124.
[60] Müller K, Krause FF, Béché A, Schowalter M, Galioit V, Löffler S, et al. Atomic electric fields revealed by a quantum mechanical approach to electron picodiffraction. Nat Commun. 2014;5:5653.
[61] Ehrenfest P. Bemerkung über die angenäherte Gültigkeit der klassischen Mechanik innerhalb der Quantenmechanik. Zeitschrift Für Phys. 1927;45:455.
[62] Yang H, Jones L, Ryll H, Simson M, Soltau H, Kondo Y, et al. 4D STEM: high efficiency phase-contrast imaging using a fast pixelated detector. J Phys Conf Ser. 2015;644:012032.
[63] Chen Z, Weyland M, Ercius P, Ciston J, Zheng C, Fuhrer MS, et al. Practical aspects of diffractive imaging using an atomic-scale coherent electron probe. Ultramicroscopy. 2016;169:107.
[64] Fang S, Wen Y, Allen CS, Ophus C, Han GG, Kirkland AI, et al. Atomic electrostatic maps of 1D channels in 2D semiconductors using 4D scanning transmission electron microscopy. Nat Commun. 2019;10:1127.
[65] Jesse S, Chi M, Belianinov A, Beekman C, Kalinin SV, Borisevich AY, et al. Big data analytics for scanning transmission electron microscopy ptychography. Sci Rep. 2016;6:26348.
[66] Belianinov A, Vasudevan R, Strelcov E, Steed C, Yang SM, Tselev A, et al. Big data and deep data in scanning and electron microscopies: deriving functionality from multidimensional data sets. Adv Struct Chem Imaging. 2015;1:6.
[67] Tate MW, Purohit P, Chamberlain D, Nguyen KX, Hovden R, Chang CS, et al. High dynamic range pixel array detector for scanning transmission electron microscopy. Microsc Microanal. 2016;22:237.
[68] Faruqi AR, McMullan G. Direct imaging detectors for electron microscopy. Nucl Instrum Methods Phys Res A. 2018;878:180.
[69] Lubk A, Zweck J. Differential phase-contrast: an integral perspective. Phys Rev A. 2015;91:023805.
[70] Lubk A, Wolf D, Lichte H. The effect of dynamical scattering in off-axis holographic mean inner potential and inelastic mean free path measurements. Ultramicroscopy. 2010;110:438.
[71] Rose H. Nonstandard imaging methods in electron microscopy. Ultramicroscopy. 1976;2:251.
[72] Shindo D, Murakami Y. Electron holography of magnetic materials. J Phys D: Appl Phys. 2008;41:183002.

[73] Matsumoto T, Koguchi M, Suzuki K, Nishimura H, Motoyoshi Y, Wada N. Ferroelectric 90° domain structure in a thin film of BaTiO3 fine ceramics observed by 300kV electron holography. Appl Phys Lett. 2008;92:72902.
[74] Polking MJ, Han M-G, Yourdkhani A, Petkov V, Kisielowski CF, Volkov VV. Ferroelectric order in individual nanometre-scale crystals. Nat Mater. 2012;11:700.
[75] Chapman JN. The investigation of magnetic domain structures in thin foils by electron microscopy. J Phys D Appl Phys. 1984;17:623.
[76] Körner M, Röder F, Lenz K, Fritzsche M, Lindner J, Lichte H, et al. Quantitative imaging of the magnetic configuration of modulated nanostructures by electron holography. Small. 2014;10:5161.
[77] Ranzieri P, Campanini M, Fabbrici S, Nasi L, Casoli F, Cabassi R, et al. Achieving giant magnetically induced reorientation of martensitic variants in magnetic shape-memory Ni-Mn-Ga films by microstructure engineering. Adv Mater. 2015;27:4760.
[78] Gatel C, Warot-Fonrose B, Biziere N, Rodríguez LA, Reyes D, Cours R, et al. Inhomogeneous spatial distribution of the magnetic transition in an iron-rhodium thin film. Nat Commun. 2017;8:15703.
[79] Agrawal P, Campanini M, Rappe A, Liu S, Grillo V, Hébert C, et al. Structure and properties of edge dislocations in BiFeO3. Phys Rev Mater. 2019;3:034410.
[80] Akhtari-Zavareh A, Carignan LP, Yelon A, Ménard D, Kasama T, Herring R, et al. Off-axis electron holography of ferromagnetic multilayer nanowires. J Appl Phys. 2014;116:23902.
[81] Cantu-Valle J, Díaz Barriga-Castro E, Vega V, García J, Mendoza-Reséndez R, Luna C, et al. Quantitative magnetometry analysis and structural characterization of multisegmented cobalt-nickel nanowires. J Magn Magn Mater. 2015;379:294.
[82] Reyes D, Biziere N, Warot-Fonrose B, Wade T, Gatel C. Magnetic configurations in Co/Cu multilayered nanowires: evidence of structural and magnetic interplay. Nano Lett. 2016;16:1230.
[83] Rodríguez LA, Bran C, Reyes D, Berganza E, Vázquez M, Gatel C, et al. Quantitative nanoscale magnetic study of isolated diameter-modulated FeCoCu nanowires. ACS Nano. 2016;10:9669.
[84] Snoeck E, Gatel C, Lacroix LM, Blon T, Lachaize S, Carrey J, et al. Magnetic configurations of 30 nm iron nanocubes studied by electron holography. Nano Lett. 2008;8:4293.
[85] Gatel C, Bonilla FJ, Meffre A, Snoeck E, Warot-Fonrose B, Chaudret B, et al. Size-specific spin configurations in single iron nanomagnet: from flower to exotic vortices. Nano Lett. 2015;15:6952.
[86] Varón M, Beleggia M, Kasama T, Harrison RJ, Dunin-Borkowski RE, Puntes VF, et al. Dipolar magnetism in ordered and disordered low-dimensional nanoparticle assemblies. Sci Rep. 2013;3:3.
[87] Zuo J-M, Tao J. Scanning electron nanodiffraction and diffraction imaging. Scanning transmission electron microscopy. New York: Springer, 2011.
[88] Zuo JM, Spence JC. Advanced transmission electron microscopy. New York: Springer, 2017.
[89] Kuhn M, Cea S, Zhang J, Wormington M, Nuytten T, De Wolf I, et al. Transistor strain measurement techniques and their applications. In: Ma Z, Seiler DG, editor(s). Metrology and diagnostic techniques for nanoelectronics. New York, United States: Pan Stanford Ltd., 2016.
[90] Gao M, Zuo JM, Twesten RD, Petrov I, Nagahara LA, Zhang R. Structure determination of individual single-wall carbon nanotubes by nanoarea electron diffraction. Appl Phys Lett. 2003;82:2703.
[91] Zuo JM, Gao M, Tao J, Li BQ, Twesten R, Petrov I. Coherent nano-area electron diffraction. Microsc Res Tech. 2004;64:347.

[92] Armigliato A, Frabboni S, Gazzadi GC. Electron diffraction with ten nanometer beam size for strain analysis of nanodevices. Appl Phys Lett. 2008;93:161906.
[93] Béché A, Rouvière JL, Clément L, Hartmann JM. Improved precision in strain measurement using nanobeam electron diffraction. Appl Phys Lett. 2009;95:123114.
[94] Chen LC, Chen FR, Kai JJ, Chang L, Ho JK, Jong CS, et al. Microstructural investigation of oxidized Ni/Au ohmic contact to p-type GaN. J Appl Phys. 1999;86:3826.
[95] Okamoto H, Hokazono A, Adachi K, Yasutake N, Itokawa H, Okamoto S, et al. In situ doped embedded-SiGe source/drain technique for 32 nm node p-channel metal-oxide-semiconductor field-effect transistor. Jpn J Appl Phys. 2008;47:2564.
[96] Toda A, Nakamura H, Fukai T, Ikarashi N. Channel strain in advanced complementary metal-oxide-semiconductor field effect transistors measured using nano-beam electron diffraction. Jpn J Appl Phys. 2008;47:2496.
[97] Favia P, Popovici M, Eneman G, Wang G, Bargallo-Gonzalez M, Simoen E, et al. Nano-beam diffraction: crystal structure and strain analysis at the nanoscale. ECS Transactions. 2010;33:205.
[98] Kim SW, Byeon DS, Jang H, Koo SM, Lee HJ, Ko DH. Strain characterization of fin-shaped field effect transistors with SiGe stressors using nanobeam electron diffraction. Appl Phys Lett. 2014;105:83104.
[99] Cooper D, Denneulin T, Bernier N, Béché A, Rouvière JL. Strain mapping of semiconductor specimens with nm-scale resolution in a transmission electron microscope. Micron. 2016;80:145.
[100] Vincent R, Midgley PA. Double conical beam-rocking system for measurement of integrated electron diffraction intensities. Ultramicroscopy. 1994;53:271.
[101] Miao S, Zhang X, Zhu J. Direct compositional analysis of ordered domains in pure and La-Doped Pb(Mg1/3Nb2/3)O3 by analytical electron microscopy techniques. J Am Ceram Soc. 2004;84:2091.
[102] Béché A, Rouvière JL, Barnes JP, Cooper D. Strain measurement at the nanoscale: comparison between convergent beam electron diffraction, nano-beam electron diffraction, high resolution imaging and dark field electron holography. Ultramicroscopy. 2013;131:10.
[103] Hÿtch M, Houdellier F, Hüe F, Snoeck E. Nanoscale holographic interferometry for strain measurements in electronic devices. Nature. 2008;453:1086.
[104] Hÿtch MJ, Houdellier F, Hüe F, Snoeck E. Dark-field electron holography for the measurement of geometric phase. Ultramicroscopy. 2011;111:1328.
[105] Denneulin T, Cooper D, Rouvière JL. Practical aspects of strain measurement in thin SiGe layers by (004) dark-field electron holography in Lorentz mode. Micron. 2014;62:52.
[106] Béché A, Rouvière JL, Barnes JP, Cooper D. Dark field electron holography for strain measurement. Ultramicroscopy. 2011;111:227.
[107] Hüe F, Hÿtch M, Houdellier F, Snoeck E, Claverie A. Strain mapping in MOSFETS by high-resolution electron microscopy and electron holography. Mater Sci Eng B Solid-State Mater Adv Technol. 2008;154–155:221.
[108] Cooper D, Barnes JP, Hartmann JM, Béché A, Rouvière JL. Dark field electron holography for quantitative strain measurements with nanometer-scale spatial resolution. Appl Phys Lett. 2009;95:53501.
[109] Denneulin T, Wollschläger N, Everhardt AS, Farokhipoor S, Noheda B, Snoeck E, et al. Local deformation gradients in epitaxial $Pb(Zr_{0.2}Ti_{0.8})O_3$ layers investigated by transmission electron microscopy. J Phys Condens Matter. 2018;30:215701.
[110] Pennycook SJ. Scanning transmission electron microscopy: Z-contrast imaging. Characterization of materials. New Jersey, United States: Wiley, 2012.

[111] Jones L, Wenner S, Nord M, Ninive PH, Løvvik OM, Holmestad R, et al. Optimising multi-frame ADF-STEM for high-precision atomic-resolution strain mapping. Ultramicroscopy. 2017;179:57.
[112] Sanchez AM, Galindo PL, Kret S, Falke M, Beanland R, Goodhew PJ. Quantitative strain mapping applied to aberration-corrected HAADF images. Microsc Microanal. 2006;12:285.
[113] Sanchez AM, Galindo PL, Kret S, Falke M, Beanland R, Goodhew PJ. An approach to the systematic distortion correction in aberration-corrected HAADF images. J Microsc. 2006;221:1.
[114] Zuo JM, Shah AB, Kim H, Meng Y, Gao W, Rouvière JL. Lattice and strain analysis of atomic resolution Z-contrast images based on template matching. Ultramicroscopy. 2014;136:50.
[115] Lubk A, Rossell MD, Seidel J, He Q, Yang SY, Chu YH, et al. Evidence of sharp and diffuse domain walls in BiFeO3 by means of unit-cell-wise strain and polarization maps obtained with high resolution scanning transmission electron microscopy. Phys Rev Lett. 2012;109:047601.
[116] Couillard M, Radtke G, Botton GA. Strain fields around dislocation arrays in a Σ9 silicon bicrystal measured by scanning transmission electron microscopy. Philos Mag. 2013;93:1250.
[117] Sang X, LeBeau JM. Revolving scanning transmission electron microscopy: correcting sample drift distortion without prior knowledge. Ultramicroscopy. 2014;138:28.
[118] Yankovich AB, Berkels B, Dahmen W, Binev P, Sanchez SI, Bradley SA, et al. Picometre-precision analysis of scanning transmission electron microscopy images of platinum nanocatalysts. Nat Commun. 2014;5:4155.
[119] Jones L, Yang H, Pennycook TJ, Marshall MS, Van Aert S, Browning ND, et al. Smart Align—a new tool for robust non-rigid registration of scanning microscope data. Adv Struct Chem Imaging. 2015;1:8.
[120] Hÿtch MJ, Snoeck E, Kilaas R. Quantitative measurement of displacement and strain fields from HREM micrographs. Ultramicroscopy. 1998;74:131.
[121] Hÿtch MJ, Putaux J-L, Pénisson J-M. Measurement of the displacement field of dislocations to 0.03 Å by electron microscopy. Nature. 2003;423:270.
[122] Johnson CL, Hÿtch MJ, Buseck PR. Nanoscale waviness of low-angle grain boundaries. Proc Natl Acad Sci. 2004;101:17936.
[123] Du K, Rau Y, Jin-Phillipp NY, Phillipp F. Lattice distortion analysis directly from high resolution transmission electron microscopy images—the {LADIA} program package. J Mater Sci. 2002;18:135.
[124] Galindo PL, Kret S, Sanchez AM, Laval JY, Yáñez A, Pizarro J, et al. The Peak Pairs algorithm for strain mapping from HRTEM images. Ultramicroscopy. 2007;107:1186.
[125] Galindo P, Pizarro J, Molina S, Ishizuka K. High resolution peak measurement and strain mapping using peak pairs analysis. Microsc Anal. 2009;23:23.
[126] Rossell MD, Erni R, Prange MP, Idrobo J-C, Luo W, Zeches RJ, et al. Atomic structure of highly strained BiFeO3 thin films. Phys Rev Lett. 2012;108:047601.
[127] Zeches RJ, Rossell MD, Zhang JX, Hatt AJ, He Q, Yang C-H, et al. A strain-driven morphotropic phase boundary in BiFeO3. Science. 2009;326:977.
[128] Agrawal P, Guo J, Yu P, Hébert C, Passerone D, Erni R, et al. Strain-driven oxygen deficiency in multiferroic SrMnO3 thin films. Phys Rev B. 2016;94:104101.
[129] Zhang JX, He Q, Trassin M, Luo W, Yi D, Rossell MD, et al. Microscopic origin of the giant ferroelectric polarization in tetragonal-like BiFeO3. Phys Rev Lett. 2011;107:147602.
[130] Guo JW, Wang PS, Yuan Y, He Q, Lu JL, Chen TZ, et al. Strain-induced ferroelectricity and spin-lattice coupling in SrMnO3 thin films. Phys Rev B. 2018;97:235135.

[131] Jia CL, Urban KW, Alexe M, Hesse D, Vrejoiu I. Direct observation of continuous electric dipole rotation in flux-closure domains in ferroelectric Pb(Zr,Ti)O3. Science. 2011;331:1420.

[132] Wang WY, Tang YL, Zhu YL, Xu YB, Liu Y, Wang YJ, et al. Atomic level 1D structural modulations at the negatively charged domain walls in BiFeO3 films. Adv Mater Interfaces. 2015;2:1500024.

[133] Lubk A, Rossell MD, Seidel J, Chu YH, Ramesh R, Hÿtch MJ, et al. Electromechanical coupling among edge dislocations, domain walls, and nanodomains in BiFeO3 revealed by unit-cell-wise strain and polarization maps. Nano Lett. 2013;13:1410.

[134] Li L, Gao P, Nelson CT, Jokisaari JR, Zhang Y, Kim SJ, et al. Atomic scale structure changes induced by charged domain walls in ferroelectric materials. Nano Lett. 2013;13:5218.

[135] De Luca G, Strkalj N, Manz S, Bouillet C, Fiebig M, Trassin M. Nanoscale design of polarization in ultrathin ferroelectric heterostructures. Nat Commun. 2017;8:1419.

[136] Tang YL, Zhu YL, Ma XL, Borisevich AY, Morozovska AN, Eliseev EA, et al. Observation of a periodic array of flux-closure quadrants in strained ferroelectric PbTiO3 films. Science. 2015;348:547.

[137] Nelson CT, Winchester B, Zhang Y, Kim SJ, Melville A, Adamo C, et al. Spontaneous vortex nanodomain arrays at ferroelectric heterointerfaces. Nano Lett. 2011;11:828.

[138] Yadav AK, Nelson CT, Hsu SL, Hong Z, Clarkson JD, Schlepuëtz CM, et al. Observation of polar vortices in oxide superlattices. Nature. 2016;530:198.

[139] Gao P, Liu HJ, Huang YL, Chu YH, Ishikawa R, Feng B, et al. Atomic mechanism of polarization-controlled surface reconstruction in ferroelectric thin films. Nat Commun. 2016;7:11318.

[140] Campanini M, Trassin M, Ederer C, Erni R, Rossell MD. Buried in-plane ferroelectric domains in Fe-doped single-crystalline Aurivillius thin films. ACS Appl Electron Mater. 2019;1:1019.

[141] Damodaran AR, Pandya S, Qi Y, Hsu SL, Liu S, Nelson C, et al. Large polarization gradients and temperature-stable responses in compositionally-graded ferroelectrics. Nat Commun. 2017;8:14961.

[142] Jin L, Xu PX, Zeng Y, Lu L, Barthel J, Schulthess T, et al. Surface reconstructions and related local properties of a BiFeO3 thin film. Sci Rep. 2017;7:39698.

[143] Okunishi E, Ishikawa I, Sawada H, Hosokawa F, Hori M, Kondo Y. Visualization of light elements at ultrahigh resolution by STEM annular bright field microscopy. Microsc Microanal. 2009;15:164.

[144] Ishikawa R, Okunishi E, Sawada H, Kondo Y, Hosokawa F, Abe E. Direct imaging of hydrogen-atom columns in a crystal by annular bright-field electron microscopy. Nat Mater. 2011;10:278.

[145] Oshima Y, Sawada H, Hosokawa F, Okunishi E, Kaneyama T, Kondo Y, et al. Direct imaging of lithium atoms in LiV2O4 by spherical aberration-corrected electron microscopy. J Electron Microsc (Tokyo). 2010;59:457.

[146] Findlay SD, Shibata N, Sawada H, Okunishi E, Kondo Y, Yamamoto T, et al. Robust atomic resolution imaging of light elements using scanning transmission electron microscopy. Appl Phys Lett. 2009;95:191913.

[147] Findlay SD, Shibata N, Sawada H, Okunishi E, Kondo Y, Ikuhara Y. Dynamics of annular bright field imaging in scanning transmission electron microscopy. Ultramicroscopy. 2010;110:903.

[148] Findlay SD, Saito T, Shibata N, Sato Y, Matsuda J, Asano K, et al. Direct imaging of hydrogen within a crystalline environment. Appl Phys Express. 2010;3:116603.

[149] Okunishi E, Sawada H, Kondo Y. Experimental study of annular bright field (ABF) imaging using aberration-corrected scanning transmission electron microscopy (STEM). Micron. 2012;43:538.

[150] Guzmán R, Maurel L, Langenberg E, Lupini AR, Algarabel PA, Pardo JA, et al. Polar-graded multiferroic SrMnO3 thin films. Nano Lett. 2016;16:2221.
[151] Kim YM, Pennycook SJ, Borisevich AY. Quantitative comparison of bright field and annular bright field imaging modes for characterization of oxygen octahedral tilts. Ultramicroscopy. 2017;181:1.
[152] Gao P, Kumamoto A, Ishikawa R, Lugg N, Shibata N, Ikuhara Y. Picometer-scale atom position analysis in annular bright-field STEM imaging. Ultramicroscopy. 2018;184:177.
[153] Zhou D, Müller-Caspary K, Sigle W, Krause FF, Rosenauer A, van Aken PA. Sample tilt effects on atom column position determination in ABF-STEM imaging. Ultramicroscopy. 2016;160:110.
[154] Choi T, Horibe Y, Yi HT, Choi YJ, Wu W, Cheong SW. Insulating interlocked ferroelectric and structural antiphase domain walls in multiferroic YMnO3. Nat Mater. 2010;9:253.
[155] Kumagai Y, Spaldin NA. Structural domain walls in polar hexagonal manganites. Nat Commun. 2013;4:1540.
[156] Han MG, Zhu Y, Wu L, Aoki T, Volkov V, Wang X, et al. Ferroelectric switching dynamics of topological vortex domains in a hexagonal manganite. Adv Mater. 2013;25:2415.
[157] Zhang S, Gilbert I, Nisoli C, Chern G-W, Erickson MJ, O'Brien L, et al. Crystallites of magnetic charges in artificial spin ice. Nature. 2013;500:553.
[158] Zhang QH, Tan GT, Gu L, Yao Y, Jin CQ, Wang YG, et al. Topology breaking of the vortex in multiferroic Y0.67Lu0.33MnO3. Appl Phys Lett. 2014;105:12902.
[159] Tian L, Wang Y, Ge B, Zhang X, Zhang Z. Direct observation of interlocked domain walls and topological four-state vortex-like domain patterns in multiferroic YMnO3 single crystal. Appl Phys Lett. 2015;106:112903.
[160] Li J, Li J, Chiang FK, Chen Z, Ma C, Chu MW, et al. Homotopy-theoretic study & atomic-scale observation of vortex domains in hexagonal manganites. Sci Rep. 2016;6:28047.
[161] Artyukhin S, Delaney KT, Spaldin NA, Mostovoy M. Landau theory of topological defects in multiferroic hexagonal manganites. Nat Mater. 2014;13:42.
[162] Geng Y, Lee N, Choi YJ, Cheong SW, Wu W. Collective magnetism at multiferroic vortex domain walls. Nano Lett. 2012;12:6055.
[163] Das H, Wysocki AL, Geng Y, Wu W, Fennie CJ. Bulk magnetoelectricity in the hexagonal manganites and ferrites. Nat Commun. 2014;5:2998.
[164] Zhang Q, Tan G, Gu L, Yao Y, Jin C, Wang Y, et al. Direct observation of multiferroic vortex domains in YMnO3. Sci Rep. 2013;3:2741.
[165] Holtz ME, Shapovalov K, Mundy JA, Chang CS, Yan Z, Bourret E, et al. Topological defects in hexagonal manganites: inner structure and emergent electrostatics. Nano Lett. 2017;17:5883.
[166] Zhang M, Olson EA, Twesten RD, Wen JG, Allen LH, Robertson IM, et al. In situ transmission electron microscopy studies enabled by microelectromechanical system technology. J Mater Res. 2005;20:1802.
[167] Allard LF, Bigelow WC, Jose-Yacaman M, Nackashi DP, Damiano J, Mick SE. A new MEMS-based system for ultra-high-resolution imaging at elevated temperatures. Microsc Res Tech. 2009;72:208.
[168] Creemer JF, Santagata F, Morana B, Mele L, Alan T, Iervolino E, et al. An all-in-one nanoreactor for high-resolution microscopy on nanomaterials at high pressures. Proc IEEE Int Conf Micro Electro Mech Syst. 2011;1103.
[169] Mele L, Santagata F, Pandraud G, Morana B, Tichelaar FD, Creemer JF, et al. Wafer-level assembly and sealing of a MEMS nanoreactor for in situ microscopy. J Micromech Microeng. 2010;20:85040.

[170] Morana B, Santagata F, Mele L, Mihailović M, Pandraud G, Creemer JF, et al. A silicon carbide MEMS microhotplate for nanomaterial characterization in TEM. Proc IEEE Int Conf Micro Electro Mech Syst. 2011;380.

[171] Protochips. Available at: https://www.protochips.com/products/fusion/. Accessed 2019.

[172] DENSsolutions. Available at: https://denssolutions.com/products/wildfire/. Accessed 2019.

[173] Hummingbird. Available at: http://hummingbirdscientific.com/products/heating-biasing/. Accessed 2019.

[174] He K, Robertson AW, Fan Y, Allen CS, Lin YC, Suenaga K, et al. Temperature dependence of the reconstruction of zigzag edges in graphene. ACS Nano. 2015;9:4786.

[175] Harmand JC, Patriarche G, Glas F, Panciera F, Florea I, Maurice JL, et al. Atomic step flow on a nanofacet. Phys Rev Lett. 2018;121:166101.

[176] Ryu GH, France-Lanord A, Wen Y, Zhou S, Grossman JC, Warner JH. Atomic structure and dynamics of self-limiting sub-nanometer pores in monolayer WS2. ACS Nano. 2018;12:11638.

[177] Yamamoto N, Yagi K, Honjo G. Electron microscopic studies of ferroelectric and ferroelastic Gd2(MoO4)3. IV. Polarization reversal and field induced phase transformation. Phys Status Solidi. 1980;62:657.

[178] Snoeck E, Normand L, Thorel A, Roucau C. Electron microscopy study of ferroelastic and ferroelectric domain wall motions induced by the in situ application of an electric field in BaTiO3. Phase Transitions. 1994;46:77.

[179] Randall CA, Barber DJ, Whatmore RW, Groves P. A TEM study of ordering in the perovskite, Pb (Sc1/2Ta1/2)O3. J Mater Sci. 1986;21:4456.

[180] Tan X, He H, Shang J-K. In situ transmission electron microscopy studies of electric-field-induced phenomena in ferroelectrics. J Mater Res. 2005;20:1641.

[181] He H, Tan X. Electric-field-induced transformation of incommensurate modulations in antiferroelectric Pb0.99Nb0.02[(Zr1-xSnx)1-yTiy]0.98O3. Phys Rev B. 2005;72:024102.

[182] Qu W, Zhao X, Tan X. Evolution of nanodomains during the electric-field-induced relaxor to normal ferroelectric phase transition in a Sc-doped Pb(Mg1/3Nb2/3)O3 ceramic. J Appl Phys. 2007;102:84101.

[183] Sato Y, Hirayama T, Ikuhara Y. Real-time direct observations of polarization reversal in a piezoelectric crystal: Pb(Mg1/3Nb2/3)O3-PbTiO3 studied via in situ electrical biasing transmission electron microscopy. Phys Rev Lett. 2011;107:187601.

[184] Nelson CT, Gao P, Jokisaari JR, Heikes C, Adamo C, Melville A, et al. Domain dynamics during ferroelectric switching. Science. 2011;334:968.

[185] Chang H, Kalinin SV, Yang S, Yu P, Bhattacharya S, Wu PP, et al. Watching domains grow: in situ studies of polarization switching by combined scanning probe and scanning transmission electron microscopy. J Appl Phys. 2011;110:52014.

[186] Zhang JX, Xiang B, He Q, Seidel J, Zeches RJ, Yu P, et al. Large field-induced strains in a lead-free piezoelectric material. Nat Nanotechnol. 2011;6:98.

[187] Gao P, Britson J, Jokisaari JR, Nelson CT, Baek SH, Wang Y, et al. Atomic-scale mechanisms of ferroelastic domain-wall-mediated ferroelectric switching. Nat Commun. 2013;4:2791.

[188] Gao P, Britson J, Nelson CT, Jokisaari JR, Duan C, Trassin M, et al. Ferroelastic domain switching dynamics under electrical and mechanical excitations. Nat Commun. 2014;5:3801.

[189] Gao P, Nelson CT, Jokisaari JR, Baek SH, Bark CW, Zhang Y, et al. Revealing the role of defects in ferroelectric switching with atomic resolution. Nat Commun. 2011;2:591.

[190] Gao P, Nelson CT, Jokisaari JR, Zhang Y, Baek SH, Bark CW, et al. Direct observations of retention failure in ferroelectric memories. Adv Mater. 2012;24:1106.

[191] Chen L, Ouyang J, Ganpule CS, Nagarajan V, Ramesh R, Roytburd AL. Formation of 90° elastic domains during local 180° switching in epitaxial ferroelectric thin films. Appl Phys Lett. 2004;84:254.

[192] Lu H, Bark CW, Esque De Los Ojos D, Alcala J, Eom CB, Catalan G, et al. Mechanical writing of ferroelectric polarization. Science. 2012;336:59.

[193] Han MG, Marshall MS, Wu L, Schofield MA, Aoki T, Twesten R, et al. Interface-induced nonswitchable domains in ferroelectric thin films. Nat Commun. 2014;5:4693.

[194] Lee JK, Shin GY, Song K, Choi WS, Shin YA, Park SY, et al. Direct observation of asymmetric domain wall motion in a ferroelectric capacitor. Acta Mater. 2013;61:6765.

[195] Scott JF. Applications of modern ferroelectrics. Science. 2007;315:954.

[196] Zhang DL, Huang WC, Chen ZW, Zhao WB, Feng L, Li M, et al. Structure evolution and multiferroic properties in cobalt doped Bi4NdTi3Fe1-xCoxO15-Bi3NdTi2Fe1-xCoxO12-δ intergrowth Aurivillius compounds. Sci Rep. 2017;7:43540.

David Szaller, Alexey Shuvaev, Alexander A. Mukhin,
Artem M. Kuzmenko and Andrei Pimenov

8 Controlling of light with electromagnons

Abstract: Magnetoelectric coupling in multiferroic materials opens new routes to control the propagation of light. The new effects arise due to dynamic magnetoelectric susceptibility that cross-couples the electric and magnetic fields of light and modifies the solutions of Maxwell equations in media. In this paper, two major effects will be considered in detail: optical activity and asymmetric propagation. In case of optical activity the polarization plane of the input radiation rotates by an angle proportional to the magnetoelectric susceptibility. The asymmetric propagation is a counter-intuitive phenomenon and it represents different transmission coefficients for forward and backward directions. Both effects are especially strong close to resonance frequencies of electromagnons, i. e. excitations in multiferroic materials that reveal simultaneous electric and magnetic character.

Keywords: Magnetoelectricity, Multiferroics, Spectroscopy, Electromagnons, Optical Activity

8.1 Introduction

Magnetoelectric coupling in multiferroics leads to appearance of new excitations that carry both, magnetic and electric dipole moments. These excitations are expected due to mixing of magnetic and electric order parameters in these materials. Thus, original excitations – magnons and phonons – are getting “coloured”, i. e. magnons can be excited by an ac electric field and the phonons start to carry a magnetic dipole moment.

The existence of magnetoelectric excitations was pointed out already long time ago by Smolenskii and Chupis [1], which was followed by further investigations especially in the Russian literature [2–4]. The general idea of new excitations is rather simple: as long as the magnetoelectric coupling is absent, the magnons and phonons are independent and show no cross-coupling. The inclusion of the coupling into the equations of motion mixes the eigenmodes of the magnetic and electric subsystems thus leading to magnetoelectric excitations. Unfortunately, in most cases and for the materials investigated at that time the coupling was mainly due to relativistic corrections and therefore rather small. Typical estimates of the order of the effects gave the

This article has previously been published in the journal Physical Sciences Reviews. Please cite as: Szaller, D., Shuvaev, A., Mukhinn, A. A., Kuzmenko, A. M., Pimenov, A. Controlling of light with electromagnons *Physical Sciences Reviews* [Online] 2020, 5. DOI: 10.1515/psr-2019-0055

https://doi.org/10.1515/9783110582130-008

values, e. g. for the magnetoelectrically induced frequency shifts not more than $\Delta\omega/\omega{\sim}0.01$, thus making the experimental observation nearly impossible.

The revival of the magnetoelectric effect [5] has led to discovery of new material systems with strong coupling of electricity and magnetism. For example, orthorhombic rare-earth (R) manganites with general formula $RMnO_3$ were materials where unusually strong magnetoelectric excitations have been detected and they were termed electromagnons. After first observation [6] in $GdMnO_3$ and $TbMnO_3$ the list of compounds with electromagnons has grown rapidly [7–11], see Refs. [12, 13] for review. Recently, electromagnons with electric dipole contribution as large as $\Delta\varepsilon \sim 40$ have been reported in multiferroic borates [14].

In several cases, including those described in this paper, the electromagnons originate from the eigenmodes of the ordered magnetic structure. If the magnetic symmetry of the structure allows the magnetoelectric coupling, static electric polarization arises simultaneously with magnetic ordering. One possible configuration is illustrated in Figure 8.1(a) showing a cycloidal magnetic ordering (see also the paper by T. Kurumaji, Chapter 4). The cycloidal ordering breaks the inversion symmetry leading to an electric polarization in the form [15, 16]:

$$\mathbf{P}_0 \sim \mathbf{e}_{ij} \times (\mathbf{S}_i \times \mathbf{S}_j). \tag{8.1}$$

Here, $\mathbf{S}_i$ and $\mathbf{S}_j$ are the neighbour Mn^{3+} spins within the spin cycloid and $\mathbf{e}_{ij}$ is the vector connecting them (see Figure 8.1(a)). In eq. (8.1), the electric polarization is directly coupled to the magnetic order opening the magnetoelectric excitation channel. In this case, the magnons can be excited by the electric field of light. As mentioned above, due to the weakness of the Dzyaloshinskii–Moriya (DM) mechanism, these electromagnons in multiferroic manganites are difficult to observe. In Figure 8.1(b), this excitation is hardly seen in direct transmission (blue points). In order to detect the DM electromagnon in $DyMnO_3$, either magnetoelectric channel (black squares and green line) or purely magnetic channel (Figure 8.1(c)) can be used. However, in several examples, including $SmFe_3(BO_3)_4$ discussed below, even the "direct" electromagnon may be of substantial intensity.

Especially in several new multiferroic systems the microscopic mechanism of Heisenberg exchange (HE) striction may lead to strong electromagnons. In this case, the coupling is proportional to the scalar product of the neighbour spins $S_i \cdot S_j$ and is orders of magnitude stronger than the relativistic DM mechanism. In $DyMnO_3$ (Figure 8.1(b)) as in several other multiferroic manganites the HE electromagnon is a zone-boundary magnon that gets an electric dipole moment due to simultaneous existence of the DM coupling. Not going into the details of the specific models [11, 18, 19] we note here that in addition to the exchange striction mechanism a cycloidal magnetic structure is necessary to make the electromagnons observable by optical spectroscopy. Surprisingly, to realize the cycloidal magnetic structure the DM mechanism is essential. Therefore, in some sense the interaction

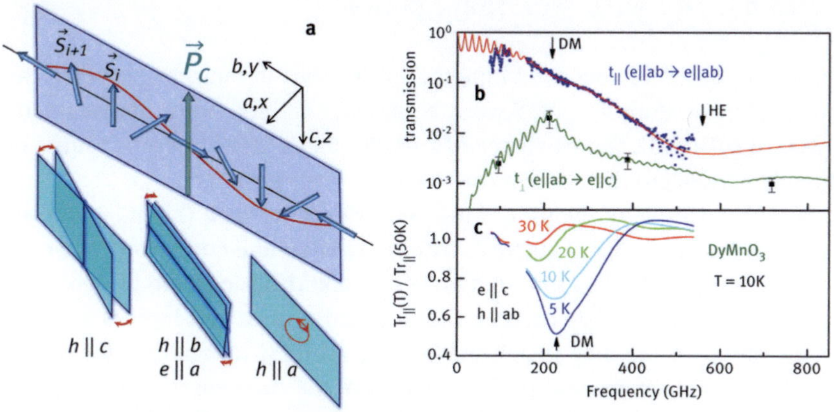

Figure 8.1: Illustration of electromagnons in $DyMnO_3$ as example. (a) Electromagnon of the ordered magnetic structure that breaks the space inversion symmetry. Shown is an example of cycloidal structure realized in multiferroic manganites. The static polarization and the magnetoelectric character of the electromagnon mode with $h \| b$ and $e \| a$ is due to inverse Dzyaloshinskii–Moriya (DM) coupling [15, 16]. (b, c) Simultaneous observation [17] of the DM electromagnon and strong Heisenberg exchange (HE) electromagnon in $DMnO_3$. Experiment: blue points – transmission in parallel polarizers, black squares – transmission in crossed polarizers that is sensitive to magnetoelectric susceptibility only, solid lines in (c) – transmission in the geometry sensitive to magnetic channel only ($h \| b$, see panel (a). Solid lines in (b) – model calculations assuming simultaneous presence of both electromagnons. Adapted from Ref. [17].

of both mechanisms leads to the dynamic magnetoelectric coupling. In such cases, two types of electromagnons may be expected: strong "Heisenberg" electromagnon and much weaker DM electromagnon. Indeed, in $TbMnO_3$ and $DyMnO_3$ (Figure 8.1) both excitations could be reliably observed and separated in the spectra [11, 17, 20].

Here, we review possible routes to control light using the magnetoelectric excitations. We demonstrate that magnetoelectricity can do much more than only giving electric dipole moment to the magnons. The appearance of magnetoelectric terms in Maxwell equations leads to completely new phenomena like optical activity or asymmetric light propagation that will be considered in the following sections. These effects arise because the magnetoelectic susceptibility opens new ways in controlling light. The remaining of this paper is organized as follows. In Section 8.2, we connect optical activity with magnetoelectric excitations (Section 8.2.1) and present two examples of experimental observation of the polarization rotation due to magnetoelectric effect. We review the data on two intensively studied multiferroics $SmFe_3(BO_3)_4$ (Section 8.2.2) and $Ba_2CoGe_2O_7$ (Section 8.2.3), both presenting an easy-plane type antiferromagnetic ground state strongly coupled to the electric polarization. In Section 8.3, the effect of directional anisotropy due to magnetoelectric effect is discussed, again using two examples, $SmFe_3(BO_3)_4$ (Section 8.3.1) and $Ba_2CoGe_2O_7$ (Section 8.3.2). Sections 8.4 and 8.5 give the Summary and Outlook, respectively.

8.2 Optical activity with electromagnons

In this section, we describe how the magnetoelectric terms in multiferroics can lead to the polarization rotation of light. As can be already derived from the considerations of the sum rules [21], the existence of a static magnetoelectric effect must be accompanied by a nonzero dynamic magnetoelectric susceptibility.

We start with a simple derivation of the connection between the magnetoelectric susceptibility and the optical activity. Within experimental conditions, the optical activity may be measured as the rotation of the polarization plane of light after passing through the sample. We assume that a linearly polarized light is incident on the sample and only the magnetoelectric susceptibility is responsible for the rotation. Even in the simplified case the calculation of the transmission through the magnetoelectric sample is a difficult task. In real experiments, the calculations are done numerically, utilizing, e. g. the Berreman approach [14, 22, 23] or the Müller matrix formalism [24]. However, in the approximation of thin sample useful relations between polarization rotation and dynamic magnetoelectric susceptibilities can be obtained as we demonstrate below.

8.2.1 Polarization rotation via magnetoelectric effect

The rotation of the polarization plane as a result of the magnetoelectric susceptibility may be calculated in a simple way in the approximation of thin sample. Within this approximation, the plane waves propagating within the sample and the exponential coordinate dependence $\sim \exp(\pm ikz)$ is reduced for $kz \ll 1$ to linear terms only: $\exp(\pm ikz) \approx (1 \pm ikz)$. Here, k is the wave-vector that results from the solution of the Maxwell equations within the sample and z is the propagation direction. In addition, we simplify the geometry assuming the plane-parallel sample with the wave propagating perpendicular to the surface and neglecting possible effects of asymmetric propagation (see Section 8.3). The usefulness of the present approximation is a linear coordinate dependence of all fields within the sample.

We consider the linear polarization state of the incident light with oscillating electric field $E_0 \| x$-axis. After interaction with the sample, the reflected and transmitted waves become elliptical in general case, that can be presented as a sum of two linearly polarized waves with polarizations along the x- and y-axes, respectively. The geometry of the problem and the notations are given in Figure 8.2. In a first step, the third Maxwell equation:

$$\oint \mathbf{E}d\mathbf{s} = -\frac{\partial}{\partial t}\int\int \mathbf{B}d\mathbf{A}, \tag{8.2}$$

is applied to the path **1** shown in Figure 8.2, taking into account the path integration along the sample surface only. Here, **B** stands for the magnetic flux density. In this case, the left-hand side of eq. (8.2) per unit length is reduced to $-E_0 - E_r^p + E_t^p$. The surface integral on the right hand side assuming the $\sim \exp(-i\omega t)$ time dependence can be written as:

$$-\frac{\partial}{\partial t}\int\int BdA = i\omega dB_y.$$

Here, d is the sample thickness and B_y is the averaged value of the magnetic field across the sample. In the present approximation, it may be rewritten as a half sum of the values at the left and right sample surfaces. This useful simplification will be applied to all components of the fields.

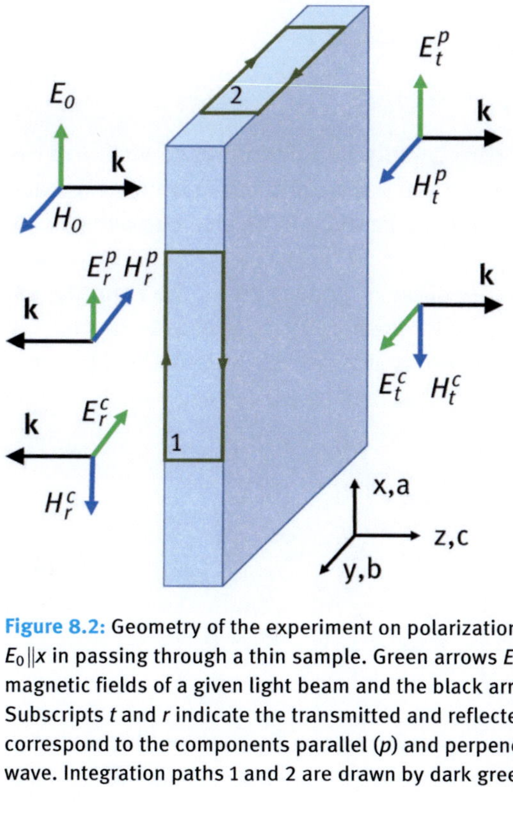

Figure 8.2: Geometry of the experiment on polarization rotation of a linearly polarized wave with $E_0 \| x$ in passing through a thin sample. Green arrows E and blue arrows H denote the electric and magnetic fields of a given light beam and the black arrow **k** shows the propagation direction. Subscripts t and r indicate the transmitted and reflected waves, respectively. Superscripts correspond to the components parallel (p) and perpendicular (c) to the polarization of the incident wave. Integration paths 1 and 2 are drawn by dark green.

In a similar way, eq. (8.2) is applied to the path **2** in Figure 8.2 and then the fourth Maxwell equation:

$$\oint \mathbf{H}d\mathbf{s} = \frac{\partial}{\partial t}\int\int \mathbf{D}d\mathbf{A},$$

is applied to both paths in Figure 8.2, where **H** stands for the magnetic field and **D** is the electric displacement field. As a result, the following set of four equations is obtained:

$$\begin{aligned} -E_0 - E_r^p + E_t^p &= i\omega d B_y \\ -E_r^c - E_t^c &= -i\omega d B_x \\ H_r^c - H_t^c &= -i\omega d D_y \\ H_0 - H_r^p - H_t^p &= -i\omega d D_x. \end{aligned} \tag{8.3}$$

The averaged fields **D** and **B** are connected to **E** and **H** inside the sample via the material equations:

$$\begin{aligned} \mathbf{B} &= \mu_0 \hat{\mu} \mathbf{H} + \sqrt{\varepsilon_0 \mu_0} \hat{\chi}^{me} \mathbf{E} \\ \mathbf{D} &= \sqrt{\varepsilon_0 \mu_0} \hat{\chi}^{em} \mathbf{H} + \varepsilon_0 \hat{\varepsilon} \mathbf{E}. \end{aligned} \tag{8.4}$$

Here, μ_O and ε_O stand for the vacuum permeability and permittivity, while $\hat{\mu}$, $\hat{\varepsilon}$ denote the relative magnetic permeability and electric permittivity tensors of the material. The $\hat{\chi}^{me}$ and $\hat{\chi}^{em}$ tensors are the magnetoelectric and inverse magnetoelectric susceptibilities of the sample.

Unfortunately, if we take the susceptibilities in general form, the resulting expressions will be still too complicated. In order to simplify the problem, we take the example of $SmFe_3(BO_3)_4$. As has been shown by microscopic calculations [14], the simplified susceptibilities in case of $SmFe_3(BO_3)_4$ may be written as:

$$\begin{aligned} \hat{\chi}^m(\omega) &= \begin{pmatrix} 0 & 0 & 0 \\ 0 & \chi_{yy}^m & \chi_{yz}^m \\ 0 & \chi_{zy}^m & \chi_{xx}^m \end{pmatrix} \quad \hat{\chi}^{me}(\omega) = \begin{pmatrix} 0 & 0 & 0 \\ 0 & \chi_{yy}^{me} & 0 \\ 0 & \chi_{zy}^{me} & 0 \end{pmatrix} \\ \hat{\chi}^{em}(\omega) &= \begin{pmatrix} 0 & 0 & 0 \\ 0 & \chi_{yy}^{em} & \chi_{yz}^{em} \\ 0 & 0 & 0 \end{pmatrix} \quad \hat{\chi}^{e}(\omega) = \begin{pmatrix} \chi_{xx}^e & 0 & 0 \\ 0 & \chi_{yy}^e & 0 \\ 0 & 0 & \chi_{zz}^e \end{pmatrix} \end{aligned} \tag{8.5}$$

Here, $\hat{\chi}^m$ and $\hat{\chi}^e$ are the magnetic and electric susceptibilities of the material. Compared to Ref. [25], we neglect here the terms which are not resonant at the electromagnon frequency.

Because the electromagnetic waves outside the sample are transversal, the z components B_z and D_z are zero. Applying the boundary conditions of the normal components of the fields, $D_z = const$, $B_z = const$ and taking into account the linearity within the sample, we see that B_z and D_z are zero everywhere. This allows us to exclude the z-components of the **H** and **E** fields from material equations, eqs. (8.4) and (8.5). As a result, four "effective" boundary conditions for the transverse components of the fields inside the sample are obtained.

The remaining four equations are necessary to connect the fields across the sample. They are obtained taking into account that the transversal components of **H** and **E** are continuous at the sample surfaces. Again, the fields inside the sample are calculated within the present approximation as an average of the fields at two surfaces. The resulting equations can be written as:

$$\begin{aligned} E_0 + E_r^p + E_t^p &= 2E_x \\ E_t^c - E_r^c &= 2E_y \\ -H_r^c - H_t^c &= 2H_x \\ H_0 - H_r^p + H_t^p &= 2H_y. \end{aligned} \tag{8.6}$$

Finally, the electric and magnetic fields for the plane waves in vacuum are connected via $H_{r,t}^{c,p} = E_{r,t}^{c,p}/Z_0$ where Z_o is the impedance of the free space.

Excluding the **E**, **H**, **D**, **B** fields inside the sample from the obtained array of 12 equations the following solution for the polarization rotation (θ) and ellipticity (η) of a sample with d thickness is easily calculated for the linear incident polarization $E_0 \| x$:

$$tan(\theta + i\eta) = E_t^c/E_t^p \approx -\frac{i\omega d}{2}\tilde{\chi}_{yy}^{em}. \tag{8.7}$$

The exact solution within the present approximation is given by:

$$tan(\theta + i\eta) = E_t^c/E_0 \approx t_c = -\frac{i\omega d}{2}\frac{\tilde{\chi}_{yy}^{em}}{\left(1 - \frac{i\omega d\hat{\mu}_{yy}}{2Z_0}\right)\left(1 + \frac{i\omega d Z_0 \tilde{\varepsilon}_y}{2}\right)} \tag{8.8}$$

$$t_p = E_t^p/E_0 = \frac{1}{2}\left[\frac{1 + i\omega d Z_0 \varepsilon_x/2}{1 - i\omega d Z_0 \varepsilon_x/2} + \frac{1 + \frac{1 + i\omega d\hat{\mu}_{yy}}{2Z_0}}{1 - \frac{1 + i\omega d\hat{\mu}_{yy}}{2Z_0}}\right] \approx 1 \tag{8.9}$$

The notations in eqs. (8.8) and (8.9) are:

$$\tilde{\chi}_{yy}^{em} = \chi_{yy}^{em} - \frac{\chi_{yz}^{em}\chi_{zy}^{m}}{1 + \chi_{zz}^{m}},$$

$$\hat{\mu}_{yy} = \tilde{\mu}_{yy} + \frac{\tilde{\chi}_{yy}^{me}\tilde{\chi}_{yy}^{em}}{\left(\frac{2}{i\omega d Z_0} - \tilde{\varepsilon}_y\right)} \approx \tilde{\mu}_{yy} \approx 1,$$

$$\tilde{\mu}_{yy} = 1 + \chi_{yy}^{m} - \frac{\chi_{yz}^{m}\chi_{zy}^{m}}{1 + \chi_{zz}^{m}} \approx 1,$$

$$\tilde{\varepsilon}_y = 1 + \chi_{yy}^{e} - \frac{\chi_{yz}^{em}\chi_{zy}^{me}}{1 + \chi_{zz}^{m}}, \text{ and}$$

$$\varepsilon_x = \varepsilon_{x,\infty} + \chi_{xx}^{e}.$$

Here, $\varepsilon_{x,\infty}$ is the high-frequency permittivity, and t_p, t_c stand for the transmission in the parallel and perpendicular polarization, respectively. Eq. (8.7) demonstrates that the diagonal term of the magnetoelectric susceptibility in a reasonable approximation is responsible for the polarization rotation.

On a similar manner, the polarization rotation in the perpendicular geometry $E_0 \| y$ can be obtained and up to the leading term it is given by:

$$tan(\theta + i\eta) = E_t^c / E_t^p \approx -\frac{i\omega d}{2} \tilde{\chi}_{yy}^{me}, \tag{8.10}$$

with

$$\tilde{\chi}_{yy}^{me} = \chi_{yy}^{me} - \frac{\chi_{zy}^{me} \chi_{yz}^{m}}{1 + \chi_{zz}^{m}},$$

being the renormalized magnetoelectric susceptibility.

We note that the physical mechanism of the thin sample approximation corresponds to the interaction of the electromagnetic waves with two sample surfaces. Therefore, strictly speaking, in eqs. (8.7) and (8.10), the effect of the sample volume is underestimated. In the opposite approximation [26], one neglects the interaction with surfaces and considers the propagation in the bulk only. In this case, the polarization rotation is again proportional to the diagonal term of the magnetoelectric susceptibility $\chi_{yy}^{me,\,em}$ with an additional $\pi/2$ phase shift compared to eqs. (8.7) and (8.10).

8.2.2 Optical activity in samarium ferroborate

The mechanisms described in the previous subsection can be demonstrated experimentally in $SmFe_3(BO_3)_4$. As shown in Figure 8.3, samarium ferroborate contains two interacting localized magnetic subsystems given by Sm^{3+} and Fe^{3+} ions. The iron subsystem orders antiferromagnetically below $T_N = 34$ K with an easy-plane magnetic structure oriented perpendicularly to the trigonal *c*-axis. Although the Sm^{3+} moments play an important role in the magnetoelectric properties of $SmFe_3(BO_3)_4$, they do not order up to the lowest temperatures. $SmFe_3(BO_3)_4$ has a non-centrosymmetric trigonal structure with R32 space group [27].

Static electric polarization in multiferroic ferroborates can be explained by symmetry arguments and by taking into account that Fe^{3+} moments are oriented antiferromagnetically within the crystallographic *ab*-plane [28, 29]. The terms governing the ferroelectric polarization (**P**) along the *a*- and *b*-axis (or *x*- and *y*-axis) are given by:

$$P_x \sim L_x^2 - L_y^2, \quad P_y \sim -2L_xL_y. \tag{8.11}$$

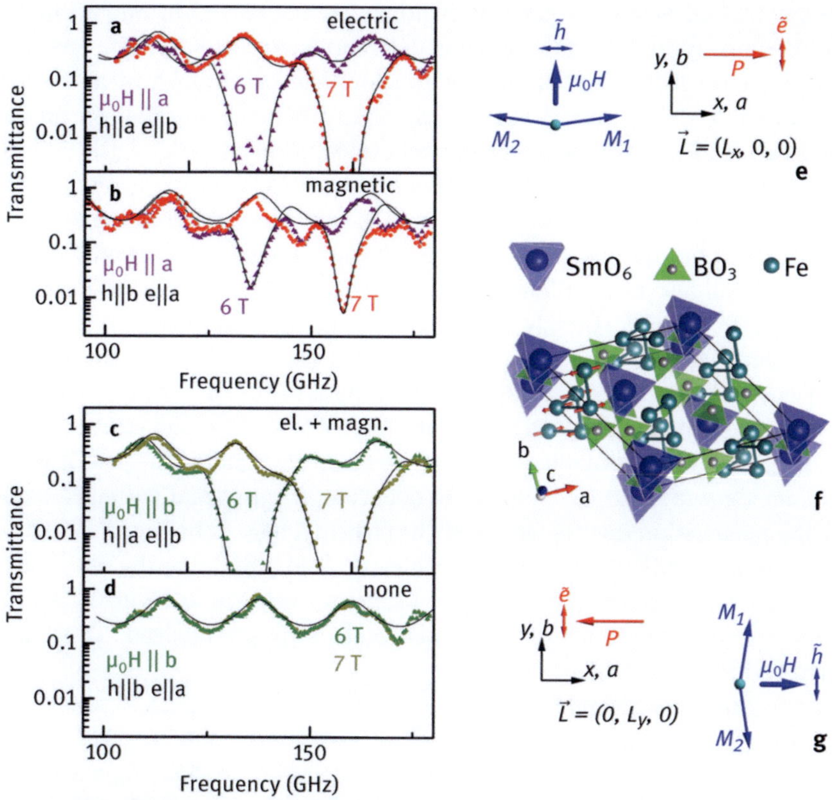

Figure 8.3: Magnetic structure and the excitation of electromagnon in $SmFe_3(BO_3)_4$ in terahertz spectra [14]. (a, b) Transmittance spectra for external magnetic field $\mu_0\mathbf{H}\|a$, where the electromagnon can be selectively excited either by electric (a) or by magnetic (b) field of light. (c, d) Geometry with magnetic field $\mu_0\mathbf{H}\|b$ where the electromagnon is either excited simultaneously in both channels (c) or is silent (d). The orientation of two iron sublattices in external magnetic fields ($\mathbf{M}_1$ and $\mathbf{M}_2$) is shown schematically for two relevant geometries: $\mu_0\mathbf{H}\|a$ (e) and $\mu_0\mathbf{H}\|b$ (g). The excitation conditions for the electromagnon are indicated by $\tilde{e}$ (electric field) and $\tilde{h}$ (magnetic field). Panel (f) shows the crystal structure of the magnetoelectric borates.

Here, $\mathbf{L} = \mathbf{M}_1 - \mathbf{M}_2$ is the antiferromagnetic vector with $\mathbf{M}_1$ and $\mathbf{M}_2$ being the magnetic moments of two magnetic Fe^{3+} sublattices. More details of the symmetry analysis of the static magnetoelectric effects in $SmFe_3(BO_3)_4$ can be found in Refs. [28–30].

Eq. (8.11) allows us to understand the behaviour of the static and dynamic properties in external magnetic fields. As shown in Figure 8.3(e), static magnetic field along the y-axis stabilizes the magnetic configuration with $L_y = 0$ and $L_x \neq 0$. In agreement with eq. (8.11) in this case the static polarization is oriented parallel to the a-axis (Figure 8.3(g)). For magnetic fields along the x-axis and above the spin flop value $L_x = 0$ and $L_y \neq 0$ which leads to antiparallel orientation of electric polarization with respect to the a-axis.

The magnetic and electric excitation channels are connected in $SmFe_3(BO_3)_4$ because of direct coupling of antiferromagnetism and ferroelectricity. The low-frequency magnetoelectric mode of interest corresponds to oscillations of antiferromagnetic moment **L** in the easy *ab*-plane [31]. It can be excited either by an ac electric field with $\mathbf{e}\perp\mathbf{P}$ or by an ac magnetic field $\mathbf{h}\perp\mathbf{M}$ (see Figure 8.3(e,g)). Here **P** is the static electric polarization and $\mathbf{M}\|\mu_0\mathbf{H}$ is a weak field-induced ferromagnetic moment with $\mathbf{M}\perp\mathbf{L}$. Because electric polarization is directly coupled to the antiferromagnetic order, it becomes possible to excite the spin oscillations not only by an ac magnetic field but also by an alternating electric field as well. The excitation conditions of the electromagnon strongly differ for two magnetic configurations given in Figure 8.3. In particular, the excitation conditions of the electromagnon for the configuration in Figure 8.3(a) are given by $\mathbf{e}\|b$ and for Figure 8.3(b) by $\mathbf{h}\|b$, respectively. Therefore, for an ab-cut of $SmFe_3(BO_3)_4$ one can selectively excite either electric ($\mathbf{e}\|b$) or magnetic ($\mathbf{h}\|b$) component of the electromagnon simply by rotating the polarization plane of the incident radiation. However, we note that in agreement with eqs. (8.7) and (8.10), the polarization rotation angle will be closely similar in both geometries.

The excitation conditions for the static magnetic field parallel to the *b*-axis are shown in Figure 8.3(g) and the corresponding spectra are given in Figure 8.3(c,d). In this geometry, the conditions of directional anisotropy are realised. This case will be discussed in Section 8.3.1.

Full spectral and polarization analysis of the data in Figure 8.3(a,b) allows us to obtain the angle of the polarization rotation in these experiments [14]. The results are shown in Figure 8.4. In these experiments, a polarization rotation angle exceeding 120 degrees is obtained for a sample with thickness of $\sim$2 mm. We stress out that in agreement with eqs. (8.7) and (8.10) this rotation arises purely from dynamic magnetoelectric susceptibilities $\chi_{yy}^{me,em}$, eq. (8.5) which are intrinsic for the electromagnon in $SmFe_3(BO_3)_4$.

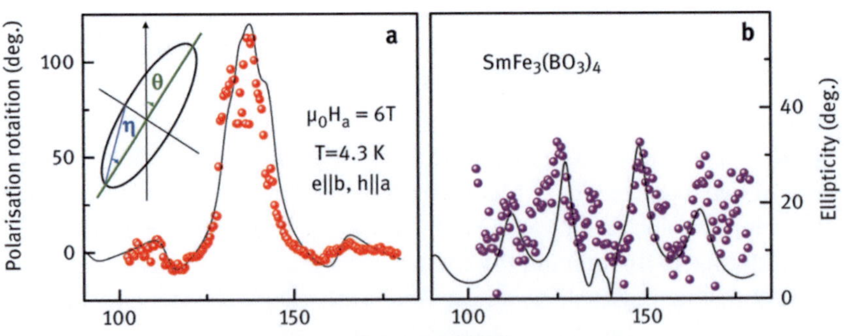

Figure 8.4: Optical activity with electromagnon in $SmFe_3(BO_3)_4$ (Ref [14]). (a) Polarization rotation (θ). The inset shows the definition of rotation angle and ellipticity. (b) Ellipticity (η). Symbols – experiment, solid line – model calculations.

8.2.3 Optical activity in $Ba_2CoGe_2O_7$

In this section we present another example of polarization rotation by electromagnons in $Ba_2CoGe_2O_7$. $Ba_2CoGe_2O_7$, as shown in Figure 8.5(a,b), crystallizes in the non-centrosymmetric, tetragonal $P\bar{4}2_1m$ symmetry [32, 33] with square-lattice layers of the magnetic Co^{2+} cations perpendicular to the [001] (tetragonal) axis of the crystal. The nearest neighbours are connected along the [110] and $[1\bar{1}0]$ directions. The Co^{2+} ions are surrounded by tetrahedral oxygen cages, which are bridged by GeO_4 tetrahedra, while Ba^{2+} spacer ions separate the layers. The $P\bar{4}2_1m$ space group has 2_1 two-fold screw axes along the [100] and [010] directions, while the (110) and $(1\bar{1}0)$ planes correspond to the m mirror planes.

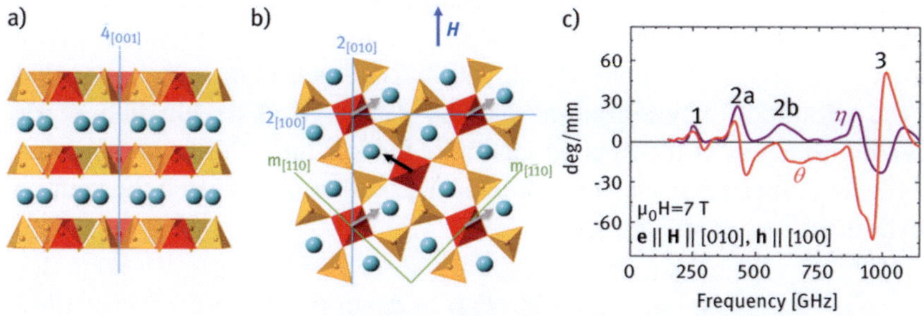

Figure 8.5: $P\bar{4}2_1m$ crystal structure and natural optical activity of $Ba_2CoGe_2O_7$ as reproduced from Ref. [41]. Co^{2+} ions are surrounded by red tetrahedral oxygen cages, GeO_4 tetrahedra are orange and Ba^{2+} spacer ions are shown by blue spheres. At the corners of the tetrahedral cages oxygen atoms are omitted for clarity. (a) 3D image of the layered crystal structure viewed from the [100] direction, (b) the tetragonal (001) magnetic easy plane of the crystal. Symmetries of the crystal structure are presented in (b). Blue lines indicate rotation symmetries around the line, while green lines are projections of mirror planes. Subscripts are used to label the axes of rotations and the mirror plane normals. Gray and black arrows represent the chessboard-like canted antiferromagnetic magnetic ordering in magnetic field (blue arrow) along [010] at $T < T_N$. (c) The optical rotation θ (red), and ellipticity η (purple) spectra in $\mu_0 H_{[010]} = 7\,T$ magnetic field [41], with incident polarization $\mathbf{e} \parallel [010]$, $\mathbf{h} \parallel [100]$. The signal of the electromagnon excitations is marked with numbers: 1 indicates the quasi-ferromagnetic, 2a and 2b the split quasi-antiferromagnetic, and 3 the spin-stretching [42] mode.

The tetrahedral local symmetry of the oxygen cage results in the quenching of the orbital moment and thus in the $S = 3/2$ spin-only moment of the Co^{2+} ion. At low temperatures ($T < T_N \approx 7\,K$), the magnetic moments order antiferromagnetically. The two-sublattice antiferromagnetic state shows an easy-plane character, where spins lie within the (001) plane, due to the single-ion anisotropy of the Co sites [32].

The magnetoelectric behaviour of $Ba_2CoGe_2O_7$ in the static limit has been extensively investigated both experimentally, [34–36] and theoretically, [37–39]. The observed magnetoelectric properties can be understood in the frame of the spin-dependent metal-ligand p–d hybridization model [40]. Namely, the hybridization of the $2p$ orbitals of the surrounding oxygen cage with the Co^{2+} $3d$ orbitals depends on the direction of the Co magnetic moment [34, 35] and the emergent onsite polarization can be described [34] as:

$$\mathbf{P} \sim \sum_{i=1}^{4} (\mathbf{S}\delta_i)^2 \delta_i, \tag{8.12}$$

where the contributions of the four Co–O bonds are summed up. Here, δ_i vectors stand for the unit vectors pointing from the Co^{2+} ion towards each neighbouring oxygen. In a basic picture, the more parallel a given bond to the Co moment is, the stronger the bond gets. Thus, the Co^{2+} ion shifts towards that oxygen atom, generating a local electric polarization. As evident from eq. (8.12), **P** is independent of the sign of spins **S**, as should be expected based on time-reversal invariance of the electric polarization.

In moderate external magnetic fields, when the magnetization can be considered to be proportional to the applied field, within a classical spin model, the orientation of the sublattice magnetizations can be easily given as function of the magnitude and direction of the applied magnetic field. Within this approximation, the magnetically induced electric polarization can be determined using the spin-dependent hybridization model [34, 40]:

$$P_{[100]} = A_\perp \left(h_\perp - \sqrt{1-h_\perp^2} \right) h_z \sqrt{1-h_z^2} \sin\phi, \tag{8.13}$$

$$P_{[010]} = A_\perp \left(h_\perp - \sqrt{1-h_\perp^2} \right) h_z \sqrt{1-h_z^2} \cos\phi, \tag{8.14}$$

$$P_{[001]} = A_z \left(h_\perp^2 - h_\perp \sqrt{1-h_\perp^2} - 12 \right) (1-h_z^2) \sin(2\phi). \tag{8.15}$$

Here, $h_\perp = H\sin\theta / H_\perp^{Sat}$ and $h_z = H\cos\theta / H_z^{Sat}$ represent the magnetic field (H) relative to the corresponding saturation fields in the tetragonal plane ($H_\perp^{Sat}$) and along the tetragonal axis (H_z^{Sat}), respectively. Azimuthal (θ) and polar (ϕ) angles of the field are measured from the [001] and [100] axes, respectively. The phenomenological coefficients A_z and $A_\perp$ scale the magnetoelectric coupling for polarization perpendicular and along the tetragonal axis, respectively.

Magnetic field applied along the [100] or [010] axis breaks the mirror-plane symmetries of the crystal, and the magnetic point group symmetry decreases to $2_{[100]}2'_{[010]}2'_{[001]}$ and $2'_{[100]}2_{[010]}2'_{[001]}$, respectively. Thus, in the non-chiral $Ba_2CoGe_2O_7$ crystal, the chirality can be induced by magnetic field pointing in appropriate directions [41]. Opposed to the chemical chirality of molecules originating from the chiral distribution of static charges, in our case the specific pattern of the magnetic moments on the non-centrosymmetric lattice cause the breaking of mirror symmetries. The molecular chirality is encoded into the arrangement of atoms, thus the handedness of the molecule cannot be reversed. In contrast, by turning the magnetic field from [100] to [010] the handedness of $Ba_2CoGe_2O_7$ can be switched, since the ground states of the two cases are related by the $m_{[110]}$ mirror-plane symmetry [41].

The chirality of $Ba_2CoGe_2O_7$ can be proven by measuring the natural optical activity. Indeed, considerable rotation of the incoming linear polarization state of light was observed, as reproduced from Ref. [41] in Figure 8.5(c). The sign of the optical rotation and ellipticity does not change when reversing the magnetic field, while they change sign for a rotation of the field from [100] to [010], which operation changes the handedness of chirality. In the intermediate orientation of the field, when the mirror plane is restored, the effect disappears [41].

Considering the experimental geometry in Figure 8.5(c), the observed rotation means an induced oscillating electric polarization along [100] by an oscillating magnetic field along the same direction. However, based on eq. (8.13), no electric polarization is expected when the magnetic moments lie in the (001) plane, giving $\chi^{me}_{[100][100]}(\omega = 0) = 0$ in the static limit. The effect can only persist in the dynamic range, corresponding to a time-reversal symmetric $\chi^{me}_{[100][100]}(\omega)$ [41].

In contrast to to the previously discussed case of $SmFe_3(BO_3)_4$, where the optical activity could be connected to a change in the polarization in the static limit, the optical activity of $Ba_2CoGe_2O_7$ has no overall static contribution in the studied geometry. This is also evident from the derivative-like asymmetric line shape of the optical rotation angle θ in Figure 8.5(c), while for $SmFe_3(BO_3)_4$ the line shape is symmetric. From the symmetry point of view, the effect is symmetric under time reversal for $Ba_2CoGe_2O_7$, while for $SmFe_3(BO_3)_4$ it is antisymmetric, forbidding/allowing an antisymmetric static contribution, respectively.

8.3 Directional anisotropy via the magnetoelectric coupling

Different propagation/absorption of particles or waves along opposite directions, termed as directional anisotropy, is a strikingly counterintuitive phenomenon. This effect is non-reciprocal, i. e. it is not symmetric with respect to the interchange of the source and detector. Specially in electrodynamics, Helmholtz's reciprocity theorem [43–45] is a fundamental principle of geometrical optics, stating that counter-propagating light beams follow identical paths due to the equal refractive indices of the opposite directions. However, in that case the medium is implicitly assumed to be invariant under time and space reversal [46]. When violating these fundamental symmetries, propagation along opposite directions can be different. Although the spatial symmetry can be reduced by an external electric field and the time-reversal symmetry can be broken by an external magnetic field, the experiments show that internal fields, e. g. non-centrosymmetric charge order (electric polarization or chiral structure) combined with magnetization are more effective in producing directional anisotropy. Namely, the relative difference in the studied quantity (absorption, phase shift, emitted or diffracted intensity) between forward and backward propagation is only in the order of $10^{-8}-10^{-2}$ for centrosymmetric, non-magnetic media placed in external electric and magnetic fields. However, the spontaneous symmetry breaking magnetism and electric polarization or chirality of multiferroics can result in relative non-reciprocal effects reaching the order of unity [25, 41, 47]. This optical non-reciprocity, when appearing in the absorption, is termed as directional dichroism, while in the phase shift it is known as directional birefringence.

Besides the ground state, strong directional anisotropy in the optical properties also imposes certain conditions on the excitation spectrum. Namely, in the lowest-order process directional anisotropy is produced by the magnetoelectric coupling, requiring simultaneously magnetic- and electric-dipole resonances. Optical directional anisotropy can be viewed as the constructive/destructive addition of oscillating electric polarization components generated by the electric and magnetic fields of light, where the sign of the interference is opposite for the forward/backward directions. This means that the dimensionless magnetic and electric oscillator strengths of the magnetoelectric excitations must be equal to the full suppression of the transmission in one direction and leaving the sample transparent for the opposite light propagation [47]. Considering that electric excitations are usually by orders of magnitude stronger than magnetic ones, the ideal candidate to observe one-way transparency is an originally magnetic excitation dressed with an electric character, for instance the both magnetic- and electric dipole active electromagnons in multiferroic crystals.

Static magnetoelectric coupling and directional dichroism are closely related, and their connection can be mathematically formulated by using the Kramers–Kronig [48, 49] relations in the static limit [21]:

$$\chi_{\gamma\delta}^{me}(0)=\frac{c}{2\pi}\mathcal{P}\int_0^{\infty}\frac{\Delta\alpha(\omega)}{\omega^2}d\omega. \tag{8.16}$$

Based on the resulting sum rule, the off-diagonal zero-frequency magnetoelectric properties are predominantly generated by the directional dichroism of low-frequency excitations, such as electromagnons, which is due to enhancement of the contribution of absorption difference, $\Delta\alpha$, at low frequencies via the ω^2 denominator. In the sum rule c denotes the speed of light in vacuum, while $\mathcal{P}$ is the Cauchy principal value integral.

Considering the symmetry of the experimental geometry, all realizations of the directional dichroism up to now belong to two classes, which are termed as optical magnetoelectric effect (OME) [50] or toroidal dichroism [51] and magneto-chiral dichroism (MChD) [52]. However, rigorous symmetry analysis [53] allows other possibilities as well, which have not been observed so far.

In the case of OME the light beam propagates parallel/antiparallel to the vector product of the polarization and the magnetization, which defines the $\mathbf{T}=\mathbf{P}\times\mathbf{M}$ toroidal moment [5]. Since $\mathbf{T}$ is a well-defined static quantity which has identical symmetry properties to the $\mathbf{k}$ wave vector of the light beam, the system can distinguish between propagation along/opposite to $\mathbf{T}$. In other words, none of the symmetries of the system can connect $+\mathbf{k}$ with $-\mathbf{k}$, because that symmetry would also annihilate **T**. **T** can also be defined locally, i. e. it can be finite without macroscopic magnetization, thus the OME can be utilized to distinguish between antiferromagnetic domains of time-reversed magnetic and identical polar structure [54]. We will present the one-way transparency by OME on the example of the electromagnon in $SmFe_3(BO_3)_4$ in Section 8.3.1. Interestingly, due to its strong magnetoelectric coupling, $SmFe_3(BO_3)_4$ also offers another way to control absorption of light via the magnetoelectric effect. In the static limit, by applying electric field, the population of magnetic domains can be tuned and through the selection rules of purely magnetic excitations the optical response can be changed drastically [55].

MChD can arise in chiral media when light travels parallel/antiparallel to the magnetization. With the use of the same argument as above, one can see that none of the symmetries can connect $+\mathbf{k}$ with $-\mathbf{k}$, since in chiral systems, i. e. systems having only rotational symmetries, the **M** and **k** vectors have the same transformation properties. The one-way transparency by MChD will be presented in Section 8.3.2 on the example of the multiferroic melilite, $Ba_2CoGe_2O_7$.

Finally, directional anisotropy in the microwave range due to skyrmions in multiferroics is discussed by M. Mochizuki, Chapter 9.

8.3.1 One-way transparency by optical magnetoelectric effect in samarium ferroborate

As was demonstrated in Figure 8.3(c,g), in the geometry with external magnetic fields $\mathbf{H}\|b$-axis the electromagnon in $SmFe_3(BO_3)_4$ is excited simultaneously via electric and magnetic channels: $h\|a$, $e\|b$. In this case the magnetoelectric coupling starts to distinguish between two possible propagation directions. This effect arises because electric polarization, $\mathbf{P}\|a$, is oriented perpendicular to the induced magnetization $\mathbf{M}\|\mathbf{H}\|b$. Consequently, $SmFe_3(BO_3)_4$ gets a toroidal [5] moment, $\mathbf{T}=\mathbf{P}\times\mathbf{M}$, which may be parallel or antiparallel to the propagation direction [56]. The toroidal moment allows the existence of strong directional birefringence in $SmFe_3(BO_3)_4$.

The susceptibility matrices for this geometry can be written in a simplified form [25]:

$$\begin{aligned}
\hat{\chi}^{m}(\omega) &= \begin{pmatrix} \chi^{m}_{xx} & 0 & \chi^{m}_{xz} \\ 0 & \chi^{m}_{yy} & 0 \\ -\chi^{m}_{xz} & 0 & \chi^{m}_{xx} \end{pmatrix} & \hat{\chi}^{me}(\omega) &= \begin{pmatrix} 0 & \chi^{me}_{xy} & 0 \\ 0 & 0 & 0 \\ 0 & \chi^{me}_{zy} & 0 \end{pmatrix} \\
\hat{\chi}^{em}(\omega) &= \begin{pmatrix} 0 & 0 & 0 \\ \chi^{me}_{xy} & 0 & -\chi^{me}_{zy} \\ 0 & 0 & 0 \end{pmatrix} & \hat{\chi}^{e}(\omega) &= \begin{pmatrix} \chi^{e}_{xx} & 0 & 0 \\ 0 & \chi^{e}_{yy} & 0 \\ 0 & 0 & \chi^{e}_{zz} \end{pmatrix}.
\end{aligned} \tag{8.17}$$

The set of basic equations to obtain the electromagnetic waves for the case of directional anisotropy (nonreciprocal transmission) is discussed by M. Mochizuki, Chapter 9. Here, we reproduce only the solutions of the Maxwell equations. For a linearly polarized electromagnetic wave with $e \| y$, $\mathbf{h} \| x$ propagating along the z-direction the complex refractive indexes of the forward (n^+) and backward (n^-) solutions are different and they are given by:

$$n^{\pm} = \sqrt{\tilde{\varepsilon}_y \tilde{\mu}_x} \pm \tilde{\alpha}_{xy}. \tag{8.18}$$

Here, $\tilde{\alpha}_{xy} = \chi_{xy}^{me} - \chi_{zy}^{me}\chi_{xz}^{m}/\mu_z$, $\tilde{\varepsilon}_y = \varepsilon_y + (\chi_{zy}^{me})^2/\mu_z$, and $\tilde{\mu}_x = \mu_x + (\chi_{xz}^{m})^2/\mu_z$ are the renormalized material parameters; $\varepsilon_y = 1 + \chi_{yy}^{e}$ is the dielectric permittivity along the y-axis, $\mu_x = 1 + \chi_{xx}^{m}$ and $\mu_z = 1 + \chi_{zz}^{m}$ are the magnetic permeabilities along the x-axis and z-axis, respectively. The solution for the perpendicular polarization with $e \| x$, $\mathbf{h} \| y$ is trivial and can be written as $n^{\pm} = \sqrt{\varepsilon_x \mu_y}$. Importantly, two possible solutions for the propagating wave are linearly polarized. Therefore, no polarization rotation is expected for the geometry with $\mathbf{H} \| b$ and the two linear polarizations do not mix. This is in contrast to a related geometry with $\mathbf{H} \| a$ revealing optical activity which has been discussed in Section 8.2.2. Figure 8.6 presents the experimental results on directional birefringence and directional dichroism in $SmFe_3(BO_3)_4$.

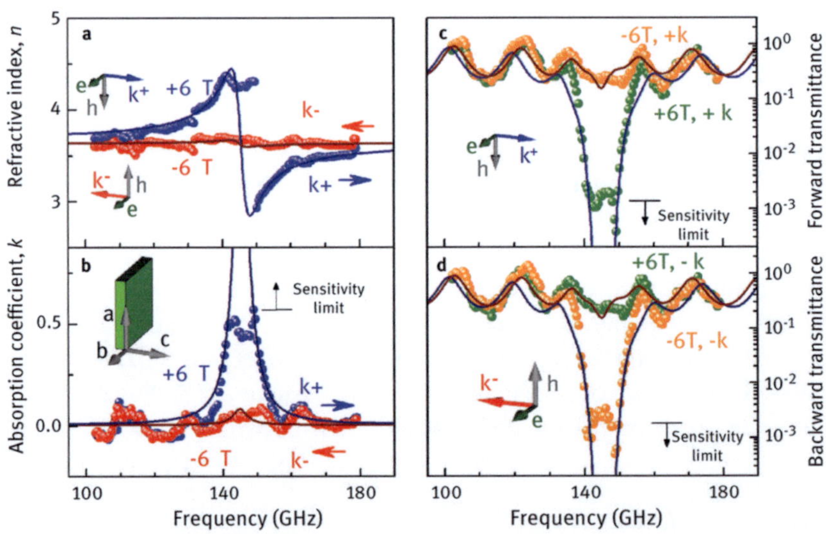

Figure 8.6: Nonreciprocal transmission in $SmFe_3(BO_3)_4$. Left panels: (a) Refractive index in $SmFe_3(BO_3)_4$ for forward (blue) and backward (red) propagation of the millimetre-wave radiation. (b) Absorption coefficient in $SmFe_3(BO_3)_4$ for forward and backward propagation. In this experiments, the inversion of the propagation is realized by the inversion of external magnetic field. Right panels: demonstration of the non-reciprocity by explicit exchange of source and detector. (c) Forward transmission for two opposite directions of external magnetic field. (d) Backward transmission. The geometry of the experiment is given in the inset. Symbols – experiment, solid lines – fits assuming a Lorentzian form of the electromagnon [25].

8.3.2 One-way transparency by magneto-chiral dichroism in multiferroic $Ba_2CoGe_2O_7$

According to the magnetoelectric sum rule in eq. (8.16), the static magnetoelectric coupling in eqs. (8.13)–(8.15) is a precursor of the direction-dependent absorption – directional dichroism – appearing in the low-frequency spin excitations of $Ba_2CoGe_2O_7$. Indeed, $Ba_2CoGe_2O_7$ was the first compound exhibiting considerable directional dichroism in the THz frequency range [57]. Here we focus on the one-way transparency observed in the MChD geometry [21, 41, 47], when the chirality of the system is induced by the magnetic field along the [100] or [010] axes.

In Figure 8.7(b) and (e), following Refs. [21, 41], we review the MChD spectra of $Ba_2CoGe_2O_7$. The clear difference in the absorption of waves propagating parallel and

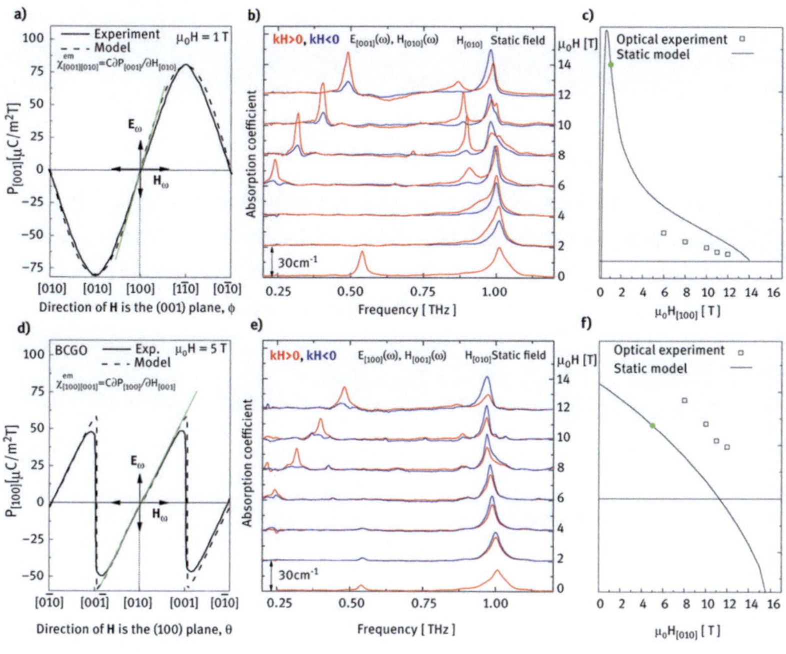

Figure 8.7: Static and dynamic magnetoelectric properties of $Ba_2CoGe_2O_7$ (BCGO). Change of the electric polarization (P) when rotating the magnetic field (H) in the (001) (a) and (100) (d) planes, respectively. Experimental results [34] and model calculations using eqs. (8.13) and (8.15) are presented by solid and dashed line, respectively. Here, the magnetoelectric susceptibility is indicated by the slope of green lines. Arrows labelled by $\mathbf{H}^\omega$ and $\mathbf{E}^\omega$ show the magnetic and electric components of light in the analogous optical experiment. The absorption measurements [21, 41] presented in (b)/(e) correspond to the static experiments in (a)/(d). Propagation along/opposite to the magnetic field is indicated by red/blue lines. The spectra are vertically shifted proportionally to the magnetic field. (c) and (f) show the field dependences of static magnetoelectric susceptibilities resulted from static (lines) and optical experiments (symbols). Green dots in (c) and (f) correspond to the slope of the green lines in (a) and (d), respectively. Reproduced from Ref. [21].

antiparallel to the magnetic field is observable both for the quasi-ferromagnetic mode ($\omega \sim H$) and for the exotic spin-stretching excitation [42] ($\omega \approx 1\,\mathrm{THz}$). A reference experiment [41] with magnetic field and light propagation along the [110] direction showed no directional dichroism, corresponding to the restored $m'_{[110]}$ and $m_{[1\bar{1}0]}$ mirror symmetries. Moreover, repeating the measurement [47] for $\mathbf{k} \parallel \mathbf{H} \parallel [010]$ resulted in a directional dichroism of opposite sign and equal magnitude as compared to the $\mathbf{k} \parallel \mathbf{H} \parallel [100]$ case, proving the magneto-chiral origin of the effect.

Using the sum rule in eq. (8.16), the directional dichroism spectra can be connected to the static magnetoelectric experiments [34]. Figure 8.7(d) shows $P_{[100]}(\theta)$, as reproduced from Ref. [34], where θ is the angle between the magnetic field and the [001] axis. In the first-order approximation, the tilting of the magnetic field from the [010] direction by $\delta\theta$ can be considered as an additional transversal field $\delta\mathbf{H} = (0, 0, H\sin\delta\theta)$, thus:

$$\begin{aligned}\chi^{em}_{[100][001]}\left(H_{[010]} = H\right) &= c\frac{\partial P_{[100]}}{\partial H_{[001]}}\bigg|_{H_{[010]}=H} \approx \frac{c}{H}\frac{\partial P_{[100]}}{\partial \theta}\bigg|_{H_{[010]}=H} \\ &\approx \lim_{\delta\to 0+}\frac{2c}{\pi H}P_{[100]}\left(H_{[0\delta 1]} = H\right).\end{aligned} \tag{8.19}$$

Here, $\lim_{\delta\to 0+} P_{[100]}\left(H_{[0\delta 1]} = H\right)$ is equivalent to the amplitude of the $P_{[100]}(\theta)$ function. In the analogous optical experiment $\mathbf{H} \parallel [010]$, while the oscillating magnetic field is parallel to $\delta\mathbf{H} \parallel [001]$ and the electric field of light oscillates along [100]. The corresponding THz-frequency absorption spectra are shown for the two opposite propagation directions (red and blue lines) in Figure 8.7(e), where the directional dichroism can be described by $\Delta\alpha(\omega) = \frac{4\omega}{c}\Im\chi^{em}_{[100][001]}(\omega)$.

Similarly, the $P_{[001]}(\phi)$ curve in Figure 8.7(a), as taken from Ref. [34], can be used to calculate $\chi^{em}_{[001][010]}$ in the static limit:

$$\begin{aligned}\chi^{em}_{[001][010]}\left(H_{[100]} = H\right) &= c\frac{\partial P_{[001]}}{\partial H_{[010]}}\bigg|_{H_{[100]}=H} \approx \frac{c}{H}\frac{\partial P_{[001]}}{\partial \phi}\bigg|_{H_{[100]}=H} \\ &\approx \frac{2c}{H}P_{[001]}\left(H_{[110]} = H\right),\end{aligned} \tag{8.20}$$

for $\mathbf{H} \parallel [100]$, where $P_{[001]}\left(H_{[110]} = H\right)$ is the amplitude of $P_{[001]}(\phi)$. In the optical range, $\mathbf{H} \parallel [100]$, oscillating magnetic and electric fields along [010] and [001] result in directional dichroism corresponding to $\chi^{em}_{[001][010]}$.

Figure 8.7(c) and (f) shows the comparison between the field dependencies of magnetoelectric tensor elements obtained from the static and optical experiments. While the order of magnitude and the tendency in the field dependence is similar, there is a significant difference in the numerical values, which can be attributed to the different experimental conditions (temperature, sample growth, uncertainty in the geometrical parameters of the sample, etc.) [21].

8.4 Summary

The appearance of the magnetoelectric term in the Maxwell equations leads to novel possibilities to control the propagation of light. The new effects are present in several multiferroics because of internal coupling between electric and magnetic order. The electromagnons, new electromagnetic excitations in multiferroics, extend the magnetoelectric coupling to the dynamic regime. Due to resonance-like character of the electromagnons, new optical effects may be especially strong close to the resonance frequency.

In this contribution, we considered two major optical effects due to dynamic magnetoelectric coupling: optical activity and directional anisotropy. As has been demonstrated in Section 8.2, in case of multiferroics the optical activity may be reduced to magnetoelectric susceptibility and in a simple approximation is even proportional to it. As two characteristic examples, the polarization rotation of $SmFe_3(BO_3)_4$ and $Ba_2CoGe_2O_7$ have been considered, both demonstrating large effects close to the electromagnon frequencies.

Depending on the magnetic symmetry, the existence of the magnetoelectric term may lead to another important effect, the directional anisotropy. Besides symmetry considerations, direct solution of the Maxwell equations allows us to look for a difference in forward and backward propagation. Directional anisotropy arises when two related eigenmodes, corresponding to opposite propagation directions, reveal different propagation constants. Here, we demonstrated that both, $SmFe_3(BO_3)_4$ and $Ba_2CoGe_2O_7$, reveal strong directional anisotropy under certain experimental conditions.

8.5 Outlook

In spite of the richness of optical effects reported in multiferroic materials, several questions remain open. One example of an unexplored optical effect is given by the gyrotropic birefringence, i. e. kind of polarization rotation that reverses sign both, under time inversion and under space inversion. Presently, a classification of possible optical effects and their connection to specific terms in the magnetoelectric susceptibility is still missing. This classification should also include the symmetry analysis of the magnetoelectric coefficients and suggest magnetic structures for possible realizations.

Both, optical activity and asymmetric transmission described above are sensitive to the polarization state of the radiation. That is, these effects can be observed basically for one solution of the Maxwell equations only. The other solution remains in several cases unaffected. Therefore, especially for the case of asymmetric transmission it is highly desired to look for magnetoelectric materials with polarization-independent directional anisotropy.

Ordered magnetic structures that are necessary for the optical effect exist mainly at rather low temperatures, or at high magnetic fields [58]. This is a disadvantage with respect to possible application of multiferroics in optical communication. In the static regime, present success to push the phase transitions in multiferroics above the room temperature is remarkable. In case of dynamical properties, except for the weak directional dichroism of $BiFeO_3$ [59], the room temperature effects still have to be found.

Acknowledgements: This work was supported by the Russian Science Foundation (16-12-10531) and by the Austrian Science Funds (W 1243, I 2816-N27, I 1648-N27).

References

[1] Smolenskii GA, Chupis IE. Ferroelectromagnets. Sov Phys Usp. 1982;25:475.
[2] Krivoruchko VN, Yablonskii DA. Antiferroelectric resonance in noncentrosymmetric multi-sublattice magnets. Sov Phys JETP. 1988;67:1886.
[3] Eremenko VV, Kharchenko NF, Litvinenko YG, Naumenko VM. Magneto-optics and spectroscopy of antiferromagnets. New York: Springer, 1992.
[4] Turov EA. Purely antiferromagnetic oscillatory mode in a two-sublattice ferromagnetic phase. J Exp Theor Phys Lett. 2001;73:87–9.
[5] Fiebig M. Revival of the magnetoelectric effect. J Phys D: Appl Phys. 2005;38:R123.
[6] Pimenov A, Mukhin AA, Ivanov VY, Travkin VD, Balbashov AM, Loidl A. Possible evidence for electromagnons in multiferroic manganites. Nat Phys. 2006;2:97.
[7] Sushkov AB, Valdes Aguilar R, Park S, Cheong S-W, Drew HD. Electromagnons in multiferroic YMn2O5 and TbMn2O5. Phys Rev Lett. 2007;98:027202.
[8] Takahashi Y, Kida N, Yamasaki Y, Fujioka J, Arima T, Shimano R, et al. Evidence for an electric-dipole active continuum band of spin excitations in multiferroic TbMnO3. Phys Rev Lett. 2008;101:187201.
[9] Kida N, Ikebe Y, Takahashi Y, He JP, Kaneko Y, Yamasaki Y, et al. Electrically driven spin excitation in the ferroelectric magnet DyMnO3. Phys Rev B. 2008;78:104414.
[10] Pimenov A, Loidl A, Mukhin AA, Travkin VD, Ivanov VY, Balbashov AM. Terahertz spectroscopy of electromagnons in $Eu1 - xYxMnO3$. Phys Rev B. 2008;77:014438.
[11] Valdes Aguilar R, Mostovoy M, Sushkov AB, Zhang CL, Choi YJ, Cheong S-W, et al. Origin of electromagnon excitations in multiferroic RMnO3. Phys Rev Lett. 2009;102:047203.
[12] Tokura Y, Seki Sh, Nagaosa N. Multiferroics of spin origin. Rep Prog Phys. 2014;77:076501.
[13] Dong Sh, Liu J-M, Cheong S-W, Ren Zh. Multiferroic materials and magnetoelectric physics: symmetry, entanglement, excitation, and topology. Adv Phys. 2015;64:519–626.
[14] Kuzmenko AM, Shuvaev A, Dziom V, Pimenov A, Schiebl M, Mukhin AA, et al. Giant gigahertz optical activity in multiferroic ferroborate. Phys Rev B. 2014;89:174407.
[15] Katsura H, Nagaosa N, Balatsky AV. Spin current and magnetoelectric effect in noncollinear magnets. Phys Rev Lett. 2005;95:057205.
[16] Mostovoy M. Ferroelectricity in spiral magnets. Phys Rev Lett. 2006;96:067601.
[17] Shuvaev A, Dziom V, Pimenov A, Schiebl M, Mukhin AA, Komarek AC, et al. Electric field control of terahertz polarization in a multiferroic manganite with electromagnons. Phys Rev Lett. 2013;111:227201.

[18] Miyahara S, Furukawa N. Theory of electric field induced one-magnon resonance in cycloidal spin magnets. ArXiv 0811.4082, 2008.
[19] Mochizuki M, Furukawa N, Nagaosa N. Theory of electromagnons in the multiferroic Mn perovskites: the vital role of higher harmonic components of the spiral spin order. Phys Rev Lett. 2010;104:177206.
[20] Shuvaev AM, Travkin VD, Ivanov VY, Mukhin AA, Pimenov A. Evidence for electroactive excitation of the spin cycloid in TbMnO3. Phys Rev Lett. 2010;104:097202.
[21] Szaller D, Bordács S, Kocsis V, Rõõm T, Nagel U, Kézsmárki I. Effect of spin excitations with simultaneous magnetic- and electric-dipole character on the static magnetoelectric properties of multiferroic materials. Phys Rev B. 2014;89:184419.
[22] Berreman DW. Optics in stratified and anisotropic media – 4x4-matrix formulation. J Opt Soc Am. 1972;62:502.
[23] Shuvaev AM, Astakhov GV, Brüne C, Buhmann H, Molenkamp LW, Pimenov A. Terahertz magneto-optical spectroscopy in HgTe thin films. Semicond Sci Technol. 2012;27:124004.
[24] Stanislavchuk TN, Kang TD, Rogers PD, Standard EC, Basistyy R, Kotelyanskii AM, et al. Synchrotron radiation-based far-infrared spectroscopic ellipsometer with full Mueller-matrix capability. Rev Sci Instr. 2013;84:023901.
[25] Kuzmenko AM, Dziom V, Shuvaev A, Pimenov A, Schiebl M, Mukhin AA, et al. Large directional optical anisotropy in multiferroic ferroborate. Phys Rev B. 2015;92:184409.
[26] Mukhin AA. Unpublished data.
[27] Vasiliev AN, Popova EA. Rare-earth ferroborates RFe3(BO3)4. Low Temp Phys. 2006;32:735–47.
[28] Zvezdin AK, Krotov SS, Kadomtseva AM, Vorob'ev GP, Popov YF, Pyatakov AP, et al. Magnetoelectric effects in gadolinium iron borate GdFe3(BO3)4. JETP Lett. 2005;81:272–6.
[29] Zvezdin AK, Vorob'ev GP, Kadomtseva AM, Popov YF, Pyatakov AP, Bezmaternykh LN, et al. Magnetoelectric and magnetoelastic interactions in NdFe3(BO3)4 multiferroics. JETP Lett. 2006;83:509–14.
[30] Popov AI, Plokhov DI, Zvezdin AK. Quantum theory of magnetoelectricity in rare-earth multiferroics: Nd, Sm, and Eu ferroborates. Phys Rev B. 2013;87:024413.
[31] Kuz'menko AM, Mukhin AA, Ivanov VY, Kadomtseva AM, Bezmaternykh LN. Effects of the interaction between R and Fe modes of the magnetic resonance in RFe3(BO3)4 rare-earth iron borates. JETP Lett. 2011;94:294–300.
[32] Zheludev A, Sato T, Masuda T, Uchinokura K, Shirane G, Roessli B. Spin waves and the origin of commensurate magnetism in Ba2CoGe2O7. Phys Rev B. 2003;68:024428.
[33] Hutanu V, Sazonov A, Murakawa H, Tokura Y, Náfrádi B, Chernyshov D. Symmetry and structure of multiferroic Ba2CoGe2O7. Phys. Rev. B 2011;84:212101.
[34] Murakawa H, Onose Y, Miyahara S, Furukawa N, Tokura Y. Ferroelectricity induced by spin-dependent metal-ligand hybridization in Ba2CoGe2O7. Phys Rev Lett. 2010;105:137202.
[35] Murakawa H, Onose Y, Miyahara S, Furukawa N, Tokura Y. Comprehensive study of the ferroelectricity induced by the spin-dependent *d-p* hybridization mechanism in Ba2*X*Ge2O7 (*X* = Mn Co and Cu). Phys Rev B. 2012;85:174106.
[36] Yi HT, Choi YJ, Lee S, Cheong S-W. Multiferroicity in the square-lattice antiferromagnet of Ba2CoGe2O7. Appl Phys Lett. 2008;92:212904.
[37] Yamauchi K, Barone P, Picozzi S. Theoretical investigation of magnetoelectric effects in Ba2CoGe2O7. Phys Rev B. 2011;84:165137.
[38] Perez-Mato JM, Ribeiro JL. On the symmetry and the signature of atomic mechanisms in multiferroics: the example of $Ba_2CoGe_2O_7$. Acta Crystallogr Sect A. 2011;67:264–8.
[39] Toledano P, Khalyavin DD, Chapon LC. Spontaneous toroidal moment and field-induced magnetotoroidic effects in Ba2CoGe2O7. Phys Rev B. 2011;84:094421.

[40] Arima T. Ferroelectricity induced by proper-screw type magnetic order. J Phys Soc Jpn. 2007;76:073702.
[41] Bordacs S, Kezsmarki I, Szaller D, Demko L, Kida N, Murakawa H, et al. Chirality of matter shows up via spin excitations. Nat Phys. 2012;8:734–8.
[42] Penc K, Romhányi J, Rõõm T, Nagel U, Antal Á, Fehér T, et al. Spin-stretching modes in anisotropic magnets: spin-wave excitations in the multiferroic Ba2CoGe2O7. Phys Rev Lett. 2012;108:257203.
[43] Stokes GG. On the perfect Blackness of the Central Spot in Newton's Rings, and on the Verification of Fresnel's Formula for the intensities of Reflected and Reflacted Rays. In: Cambridge Library Collection – Mathematics vol. 2. Cambridge: Cambridge University Press, 2009:89–103.
[44] von Helmholtz H. Handbuch der physiologischen Optik, vol. 1, 1st ed. Leipzig: Leopold Voss, 1856.
[45] Born M, Wolf E. Principles of optics: electromagnetic theory of propagation, interference and diffraction of light. Cambridge: Cambridge University Press, 2000.
[46] Kuzmenko AM, Dziom V, Shuvaev A, Pimenov A, Szaller D, Mukhin AA, et al. Sign change of polarization rotation under time or space inversion in magnetoelectric $YbAl_3(BO_3)_4$. Phys Rev B. 2019;99:224417.
[47] Kézsmárki I, Szaller D, Bordács S, Kocsis V, Tokunaga Y, Taguchi Y, et al. One-way transparency of four-coloured spin-wave excitations in multiferroic materials. Nat Commun. 2014;5:3203.
[48] Kronig RL. On the theory of dispersion of X-rays. J Opt Soc Am. 1926;12:547–57.
[49] Kramers HA. Diffusion of light by atoms. Atti Congr Int Fisici. 1927;2:545–57.
[50] Arima T. Magneto-electric optics in non-centrosymmetric ferromagnets. J Phys: Condens Matter. 2008;20:434211.
[51] Bordács S, Kocsis V, Tokunaga Y, Nagel U, Rõõm T, Takahashi Y, et al. Unidirectional terahertz light absorption in the pyroelectric ferrimagnet CaBaCo4O7. Phys Rev B. 2015;92:214441.
[52] Rikken GL, Raupach E. Observation of magneto-chiral dichroism. Nature. 1997;390:493.
[53] Szaller D, Bordács S, Kézsmárki I. Symmetry conditions for nonreciprocal light propagation in magnetic crystals. Phys Rev B. 2013;87:014421.
[54] Kocsis V, Penc K, Rõõm T, Nagel U, Vít J, Romhányi J, et al. Identification of antiferromagnetic domains via the optical magnetoelectric effect. Phys Rev Lett. 2018;121:057601.
[55] Kuzmenko AM, Szaller D, Kain T, Dziom V, Weymann L, Shuvaev A, et al. Switching of magnons by electric and magnetic fields in multiferroic borates. Phys Rev Lett. 2018;120:027203.
[56] Tokura Y. Multiferroics – toward strong coupling between magnetization and polarization in a solid. J Magn Magn Mater. 2007;310:1145–50.
[57] Kézsmárki I, Kida N, Murakawa H, Bordács S, Onose Y, Tokura Y. Enhanced directional dichroism of terahertz light in resonance with magnetic excitations of the multiferroic Ba2CoGe2O7 oxide compound. Phys Rev Lett. 2011;106:057403.
[58] Viirok J, Nagel U, Rõõm T, Farkas DG, Balla P, Szaller D, et al. Directional dichroism in the paramagnetic state of multiferroics: A case study of infrared light absorption in Sr2CoSi2O7 at high temperatures. Phys Rev B. 2019;99:014410.
[59] Kézsmárki I, Nagel U, Bordács S, Fishman RS, Lee JH, Yi HT, et al. Optical diode effect at spin-wave excitations of the room-temperature multiferroic BiFeO3. Phys Rev Lett. 2015;115:127203.

Masahito Mochizuki

9 Dynamical magnetoelectric phenomena of skyrmions in multiferroics

Abstract: Magnetic skyrmions, nanoscopic spin vortices carrying a quantized topological number in chiral-lattice magnets, are recently attracting great research interest. Although magnetic skyrmions had been observed only in metallic chiral-lattice magnets such as B20 alloys in the early stage of the research, their realization was discovered in 2012 also in an insulating chiral-lattice magnet Cu_2OSeO_3. A characteristic of the insulating skyrmions is that they can host multiferroicity, that is, the noncollinear magnetization alignment of skyrmion induces electric polarizations in insulators with a help of the relativistic spin-orbit interaction. It was experimentally confirmed that the skyrmion phase in Cu_2OSeO_3 is indeed accompanied by the spin-induced ferroelectricity. The resulting strong magnetoelectric coupling between magnetizations and electric polarizations can provide us with a means to manipulate and activate magnetic skyrmions by application of electric fields. This is in sharp contrast to skyrmions in metallic systems, which are driven through injection of electric currents. The magnetoelectric phenomena specific to the skyrmion-based multiferroics are attracting intensive research interest, and, in particular, those in dynamical regime are widely recognized as an issue of vital importance because their understanding is crucial both for fundamental science and for technical applications. In this article, we review recent studies on multiferroic properties and dynamical magnetoelectric phenomena of magnetic skyrmions in insulating chiral-lattice magnet Cu_2OSeO_3. It is argued that the multiferroic skyrmions show unique resonant excitation modes of coupled magnetizations and polarizations, so-called electromagnon excitations, which can be activated both magnetically with a microwave magnetic field and electrically with a microwave electric field. The interference between these two activation processes gives rise to peculiar phenomena in the gigahertz regime. As its representative example, we discuss a recent theoretical prediction of unprecedentedly large nonreciprocal directional dichroism of microwaves in the skyrmion phase of Cu_2OSeO_3. This phenomenon can be regarded as a one-way window effect on microwaves, that is, the extent of microwave absorption changes significantly when its incident direction is reversed. This dramatic effect was indeed observed by subsequent experiments. These studies demonstrated that the multiferroic skyrmions can be a promising building block for microwave devices.

This article has previously been published in the journal Physical Sciences Reviews. Please cite as: Mochizuki, M. Dynamical magnetoelectric phenomena of skyrmions in multiferroics *Physical Sciences Reviews* [Online] 2020, 5. DOI: 10.1515/psr-2019-0017

https://doi.org/10.1515/9783110582130-009

Keywords: skyrmion, electromagnon, nonreciprocal directional dichroism, magnetoelectric coupling, chiral magnetic insulator

9.1 Introduction

Noncollinear spin textures in magnets have continuously attracted intense research interest because they often exhibit intriguing physical phenomena and useful device functions through coupling to charge degree of freedom [1, 2]. Domain walls and magnetic vortices in ferromagnets can be driven through injection of electric currents via the spin transfer torque mechanism, which leads to their possible application to spintronics devices such as race-track memory. Magnetic spirals in insulating magnets often induce ferroelectric polarization **P** via relativistic spin–orbit interactions [3–5], and resulting multiferroicity hosts rich magnetoelectric phenomena due to the cross-correlation of magnetizations and polarizations [6].

In addition to these magnetic structures, magnetic skyrmions, vortex-like swirling spin textures characterized by a quantized topological invariant, are recently attracting a great deal of interest [7, 9–11]. Originally, the concept of skyrmion was proposed by Tony Skyrme in 1960s as a solution of the nonlinear sigma model in three dimensions to account for stability of baryons in nuclear physics [12, 13]. The original skyrmion proposed by Skyrme comprises vector fields pointing in all directions wrapping a sphere like a hedgehog as shown in Figure 9.1(a). In 1980s, the emergence of skyrmions as vortex-like magnetic textures in ferromagnets without spatial inversion symmetry was theoretically predicted by Bogdanov and collaborators [14, 15]. This magnetic skyrmion is regarded as a stereo-projection of the original hedgehog skyrmion onto a two-dimensional plane [Figure 9.1(b)]. It was also predicted that skyrmions are often crystallized to form a triangular lattice in magnets, which is called skyrmion crystal [Figure 9.1(c)].

The emergence of skyrmion crystals was experimentally confirmed in a metallic B20 compound MnSi under an external magnetic field in 2009 by a small-angle neutron scattering experiment [17]. Subsequently, real-space images of the swirling skyrmion spin structures and the triangular skyrmion crystal were observed by a Lorentz transmission electron microscopy for thin-plate specimens of $Fe_{1-x}Co_xSi$ [18]. The skyrmion crystal phase was also observed in other B20 compounds such as FeGe, $Mn_{1-x}Fe_xGe$, and MnSi by small-angle neutron scatterings [19–23] and Lorentz transmission electron microscopies [24–27]. These compounds have a chiral cubic crystal structure, which belongs to the $P2_13$ space group.

In the B20 compounds, a strong Dzyaloshinskii–Moriya interaction is active because the spatial inversion symmetry is broken [28, 29]. This interaction favors a rotating alignment of magnetizations with 90° rotation angle with a fixed handedness (spin chirality) and thus competes with the ferromagnetic exchange interaction which favors a parallel alignment of magnetizations. A competition between these

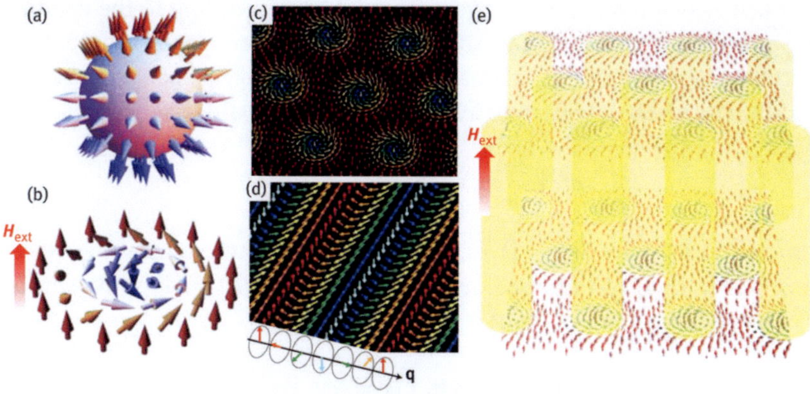

Figure 9.1: (a) Schematic illustrations of (a) the hedgehog-type skyrmion with vector fields or spins pointing in all directions wrapping a sphere, (b) the vortex-type magnetic skyrmion which can be regarded as a stereo-projection of the hedgehog skyrmion onto a plane, (c) the helimagnetic state realized as a consequence of the competition between the ferromagnetic exchange interaction and the Dzyaloshinskii–Moriya interaction in chiral-lattice magnets, and (d) the skyrmion crystal composed of hexagonally packed vortex-type skyrmions in chiral-lattice magnets under an external magnetic field. (e) Three-dimensional structure of a magnetic skyrmion crystal, in which the vortex-type skyrmions are stacked along an external magnetic field $\boldsymbol{H}_{ext}$ to form hexagonally aligned vortex tubes.

two interactions results in the formation of helical spin structure with moderate spin-rotation angles [see Figure 9.1(d)] when the external magnetic field is absent.

When a magnetic field is applied at certain temperatures, this helimagnetic state changes into a skyrmion crystal. A triangular skyrmion crystal appears on a plane normal to the applied magnetic field. Magnetizations constituting each skyrmion point antiparallel to the magnetic field at its center, whereas at the periphery, they point parallel to the magnetic field so as to increase a gain of the energy associated with the Zeeman interaction [see Figure 9.1(b)]. The magnetizations gradually rotate upon propagating along radial directions from the center to the periphery in each skyrmion. Such vortex-like textures are stacked along the magnetic field to form hexagonally packed three-dimensional tubes [Figure 9.1(e)].

A typical size of skyrmion in chiral-lattice magnets is of nanometric order ranging from 5 to 100 nm, which is determined by a ratio D/J with D and J being coupling constants of the Dzyaloshinskii–Moriya interaction and the ferromagnetic exchange interaction, respectively. This ratio varies depending on materials. The skyrmion diameter becomes smaller in a material with larger D/J because a stronger Dzyaloshinskii–Moriya interaction results in more rapid rotation of magnetizations along the radial direction.

Magnetic skyrmions in metallic B20 compounds have turned out to show specific transport phenomena and spintronics functions upon injection of electric currents. For example, magnetizations constituting a skyrmion work as a fictitious magnetic field acting on conduction electrons via exchange coupling to the conduction-electron spins, which causes a Hall effect of the conduction electrons (topological Hall effect) [30–34]. Note that a summation of the solid angles spanned by three neighboring magnetizations over a skyrmion take a quantized value as

$$\int_{\mathrm{uc}} d^2r \left(\frac{\partial \boldsymbol{m}}{\partial x} \times \frac{\partial \boldsymbol{m}}{\partial y} \right) \cdot \boldsymbol{m} = 4\pi Q \tag{9.1}$$

in the continuum form. Here $Q(=\pm 1)$ is a quantized invariant called topological charge, and $\boldsymbol{m}$ is a normalized classical magnetization vector. This formula can be rewritten in a discretized form on a square lattice as

$$\sum_{i}^{\mathrm{uc}} \boldsymbol{m}_i \cdot (\boldsymbol{m}_{i+\hat{x}} \times \boldsymbol{m}_{i+\hat{y}}) = 4\pi Q \tag{9.2}$$

The integration and the summation in the above equations are performed over a unit cell (uc) of the skyrmion lattice, which corresponds to one single-layer skyrmion. As its consequence, the skyrmion behaves as one magnetic flux quantum acting on conduction electrons. Another interesting phenomenon is the current-driven motion of skyrmions [35–40]. Like other noncollinear magnetic textures such as domain walls and magnetic vortices, skyrmions can be driven by injection of electric current. Here the electric current exerts torque to magnetizations of skyrmions via the angular-momentum transfer from the conduction-electron spins, which is referred to as the spin-transfer torque. It has turned out that the skyrmion can be driven with ultra-low electric current densities. Surprisingly, its threshold value turned out to be 10^5–10^6 A/m^2 [35, 36, 39], which is five or six orders of magnitude smaller than the threshold current density required to drive ferromagnetic domain walls. Since the discovery of these properties, skyrmions have attracted great interest from the viewpoint of spintronics applications to magnetic memory devices.

In the early stage of research, observations of the skyrmions and the skyrmion crystals had been limited to metallic B20 compounds such as MnSi, FeGe, and $Fe_{1-x}Co_xSi$. However, in 2009, the first example of skyrmion-hosting insulating magnet Cu_2OSeO_3 was discovered [41–43]. Formations of skyrmions and skyrmion crystals in this compound were observed by a Lorentz transmission electron microscopy for thin-film specimens [41] and small-angle neutron-scattering experiments for bulk specimens [42, 44].

The crystal structure of Cu_2OSeO_3 [Figure 9.2(a)] belongs to the chiral cubic $P2_13$ space group, which is identical to that of the B20 compounds. However, the coordination of atoms in Cu_2OSeO_3 is quite different from that in the B20 compounds. The Cu_2OSeO_3 crystal contains two inequivalent Cu^{2+} sites with different oxygen coordination. One is surrounded by a square pyramid of oxygen ligands, whereas the other is surrounded by a trigonal bipyramid. The nominal ratio between the former and the latter Cu^{2+} ions is 3:1.

The magnetic structure of Cu_2OSeO_3 [Figure 9.2(b)] consists of tetrahedra composed of four Cu^{2+} (S=1/2) ions at their apexes. On each tetrahedron, a three-up and one-down type ferrimagnetic spin configuration is realized below $T_C \sim 58$K [45–47]. For bulk samples, the skyrmion crystal phase appears only as a small pocket (so-called A phase) on the verge of the phase boundary between the paramagnetic phase and the helimagnetic phase in the phase diagram in plane of temperature T and magnetic field H [see Figure 9.2(c)]. On the other hand, the area of the skyrmion phase spreads over a wide T–H range for thin-film samples [Figure 9.2(d)]. Magnetic modulation periods in the helimagnetic and the skyrmion crystal states are approximately 630 Å, which is much longer than the crystallographic lattice constant of ~8.9 Å.

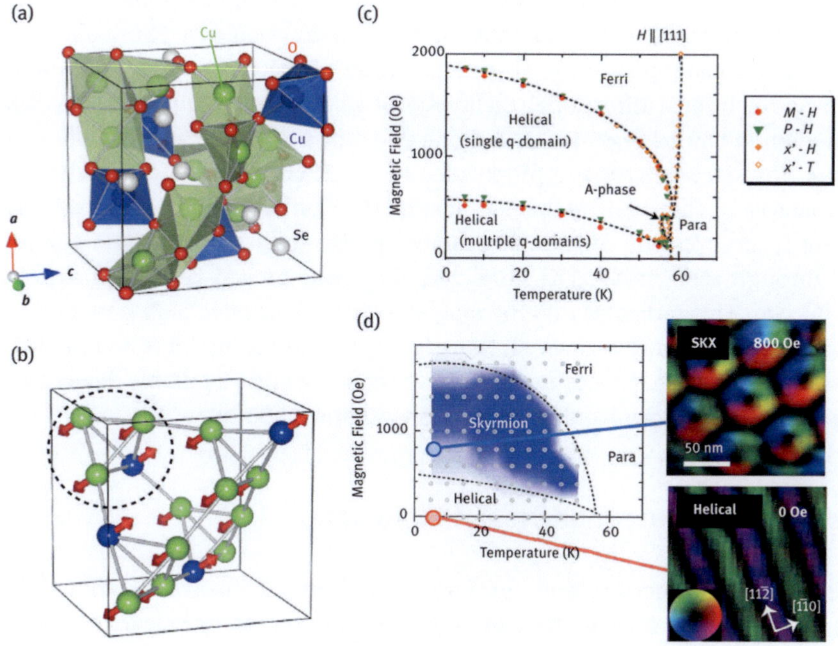

Figure 9.2: (a) Crystal structure of chiral magnetic insulator Cu_2OSeO_3 which belongs to the chiral cubic $P2_13$ symmetry group with broken spatial inversion symmetry. (b) Magnetic structure of Cu_2OSeO_3 as a network of octahedra composed of four Cu^{2+} ions with spin $S = 1/2$ at their apexes. Three-up and one-down spin configuration is realized in each octahedra (indicated by a dashed circle) resulting in the ferrimagnetic order. (c) Experimental phase diagram in plane of temperature T and magnetic field H for a bulk specimen [41]. (d) Experimental T-H phase diagram for a 100-nm-thick thin-plate specimen [41], in which the skyrmion crystal phase spreads over a wide area and appears even at the lowest temperatures.

It is widely known that noncollinear magnetic orders like cycloidal and helical spin states in insulating magnets often induce electric polarizations via the relativistic spin–orbit interactions [3–6]. We can therefore expect multiferroic nature for the insulating skyrmion phase in Cu_2OSeO_3. In contrast, the skyrmion phase in metallic systems like the B20 alloys cannot induce eletric polarizations, and thus the systems remain to be non-multiferroic. Indeed, it was experimentally confirmed that the skyrmion phase in Cu_2OSeO_3 hosts ferroelectric polarization [41, 43]. Thus the magnetoelectric responses specific to skyrmions are anticipated in this material. The electromagnon excitation is one of the most important phenomena [48–50]. When a magnetic order couples to electric polarizations, spin-wave modes or magnon modes can be excited through activating the electric polarizations using an AC electric-field component of electromagnetic wave, which can be regarded as coupled oscillations of magnetizations and polarizations (see also Chapter 8).

These electrically activated collective excitations called electromagnons give rise to rich physical phenomena and optical/microwave device functions. In particular, it was theoretically predicted that an interference between two excitation channels of electromagnons in the skyrmion crystal phase causes a unprecedentedly large nonreciprocal directional dichroism at microwave frequencies [51, 52]. This phenomenon can be regarded as a diode effect or a one-way window effect of microwaves, that is, a microwave propagates in a certain direction can penetrate through a sample of chiral-lattice magnet hosting the skyrmion phase, whereas that propagating in an opposite direction becomes significantly absorbed and cannot penetrate through the sample. The predicted effect was indeed confirmed experimentally [53, 54]. This indicates that the magnetic skyrmions have high potential as a building block of microwave devices. Namely this phenomenon offers an opportunity to realize microwave devices much smaller than currently available devices by replacing constituent circuits and wave guides with one material.

9.2 Magnetoelectric properties of Cu_2OSeO_3

Before discussing the electromagnon excitations and the microwave directional dichroism in the skyrmion crystal phase of Cu_2OSeO_3, it is necessary to learn the magnetoelectric properties of this material. In fact, a presence or an absence of the magnetism-induced ferroelectricity and, if any, a direction of the ferroelectric polarization in the skyrmion crystal phase of Cu_2OSeO_3 depend on an orientation of the external (static) magnetic field $\boldsymbol{H}$, which can be understood from symmetry argument [41, 43]. The argument is applicable generally to skyrmion crystals in magnets with cubic crystal symmetries. The crystal structure of Cu_2OSeO_3 belongs to a nonpolar space group $P2_13$, and thus does not host ferroelectricity originally. In addition, the spin structure of the skyrmion crystal as well as those of the conical and ferromagnetic (ferrimagnetic) spin states are not polar, either. However, a combination of the crystal and magnetic symmetries renders the system polar and allows the emergence of ferroelectric polarization.

As shown in Figure 9.3(a), the crystal structure of Cu_2OSeO_3 has four three-fold rotation axes (**3** axes) along ⟨111⟩, and three two-fold screw axes (**2_1** axes) along ⟨100⟩. On the other hand, the magnetic structure of skyrmion crystal has a six-fold rotation axis (**6** axis) along the external magnetic field ***H***, and two-fold rotation axes followed by time reversal (**$2'$** axes) in the plane normal to ***H*** as shown in Figure 9.3(b). When a skyrmion spin texture is formed on a cubic crystal lattice of

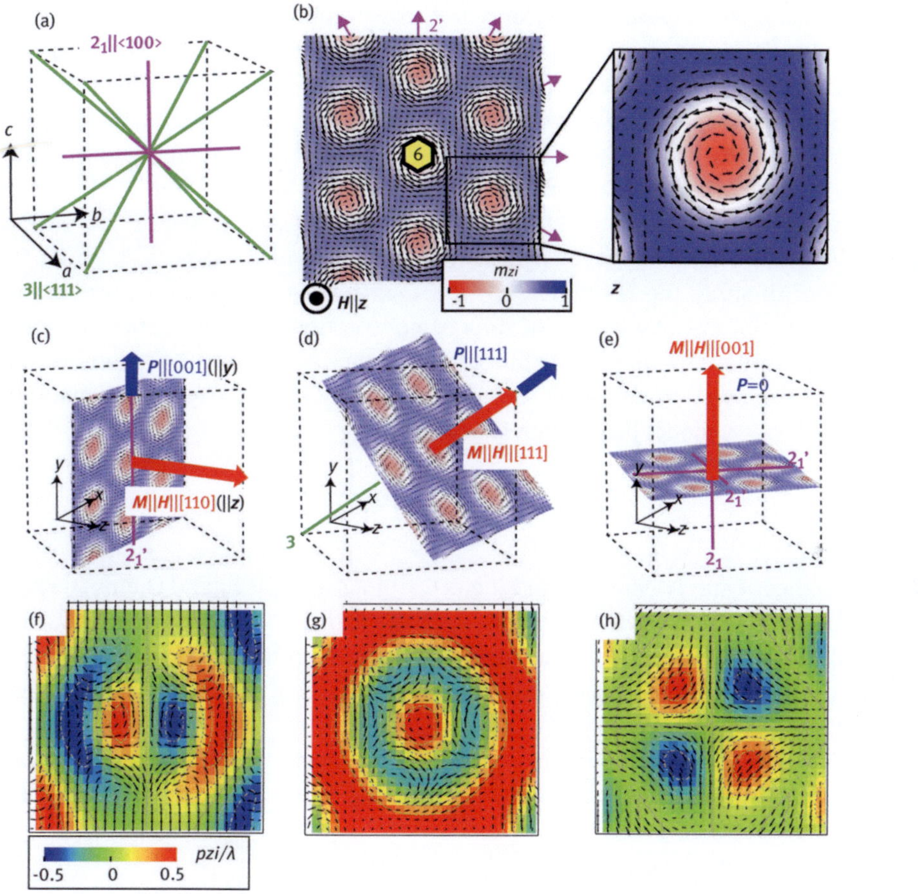

Figure 9.3: (a) Symmetry axes in the crystal structure of Cu_2OSeO_3, i. e. four three-fold rotation axes (**3** ∥ ⟨111⟩) and three two-fold screw axes (**2_1** ∥ ⟨100⟩). (b) Symmetry axes of the skyrmion crystal, i. e. a six-fold rotation axis (**6** ∥ ***H***) and six two-fold rotation axes followed by the time reversal (**$2'$** ⊥ ***H***). (c) Under the application of ***H*** ∥[110], only the **$2_1'$** axis along [001] survives as a symmetry axis, which allows the emergence of ferroelectric polarization ***P*** ∥[001]. (d) Under the application of ***H*** ∥[111], only the **3** axis along [111] survives as a symmetry axis, which allows the emergence of ***P*** ∥[111]. (e) Under the application of ***H*** ∥[001], one **2_1** axis and two **$2_1'$** axes survive as symmetry axes, which forbidden the emergence of ***P***. (f)–(h) Spatial distributions of local electric polarizations $\boldsymbol{p}_i$ for various orientations of ***H***, i. e. (f) ***H*** ∥[110], (g) ***H*** ∥[111], and (h) ***H*** ∥[001]. The in-plane (out-of-plane) components of $\boldsymbol{p}_i$ are displayed by arrows (colors).

Cu_2OSeO_3, most of the symmetry elements disappear. Eventually the system can be polar depending on the orientation of $\boldsymbol{H}$.

Considering the fact that skyrmion crystals appear always on a plane normal to $\boldsymbol{H}$, we discuss three cases with different $\boldsymbol{H}$ orientations shown in Figure 9.3(c)–(e). When a skyrmion crystal sets in under $\boldsymbol{H}\,\|[110]$ [see Figure 9.3(c)], only a $\mathbf{2}_1{}'$ axis ($\|[001]$) normal to $\boldsymbol{H}$ survives. Consequently, the system becomes polar along [001], and the emergence of ferroelectric polarization $\boldsymbol{P}\,\|[001]$ is allowed. Likewise, in the case of $\boldsymbol{H}\,\|[111]$ [see Figure 9.3(d)], only a **3** axis parallel to $\boldsymbol{H}$ remains unbroken, which allows the emergence of $\boldsymbol{P}\,\|\,\boldsymbol{H}\,\|[111]$. In contrast, in the case of $\boldsymbol{H}\,\|[001]$ [see Figure 9.3(e)], an orthogonal arrangement of the three $\mathbf{2}_1{}'$ axes along $\langle 001\rangle$ remains, and thus no ferroelectric polarization can be expected. Importantly, these arguments hold not only for the skyrmion crystal state but also for helimagnetic and ferrimagnetic states. Namely, the ferroelectric polarization $\boldsymbol{P}\,\|[001]$ and $\boldsymbol{P}\,\|[111]$ can emerge under $\boldsymbol{H}\,\|[110]$ and $\boldsymbol{H}\,\|[111]$, respectively, whereas $\boldsymbol{P}$ should be absent under $\boldsymbol{H}\,\|[001]$ for these magnetic states.

In order to study the magnetoelectric responses of multiferroic materials, we need to know not only behaviors of the net polarization $\boldsymbol{P}$ but also those of local electric polarizations $\boldsymbol{p}_i$ as its constituents. The distribution and behaviors of the local polarizations $\boldsymbol{p}_i$ in Cu_2OSeO_3 are described by the spin-dependent metal-ligand hybridization model [55], which is expressed by

$$\boldsymbol{p}_i \propto \sum_j (\boldsymbol{e}_{ij}\cdot\boldsymbol{m}_i)^2 \boldsymbol{e}_{ij} \tag{9.3}$$

Here the indices i and j denote the magnetic Cu^{2+} ions and the ligand O ions, respectively. The normalized directional vector $\boldsymbol{e}_{ij}$ connects the ith Cu^{2+} ion and jth ligand O ion. This formula expresses that a single local magnetization $\boldsymbol{m}_i$ on the ith Cu^{2+} ion induces the local electric polarization $\boldsymbol{p}_i$ through hybridizing with surrounding ligand O ions.

On the other hand, according to a symmetry argument in the Ginzburg–Landau expansions, the local electric polarization $\boldsymbol{p}_i = (p_{ia}, p_{ib}, p_{ic})$ is expressed by using local magnetization $\boldsymbol{m}_i$ as,

$$\boldsymbol{p}_i = (p_{ia}, p_{ib}, p_{ic}) = \lambda(m_{ib}m_{ic}, m_{ic}m_{ia}, m_{ia}m_{ib}). \tag{9.4}$$

Here a, b and c are the orthogonal axes of the cubic ($P2_13$) setting. The magnetoelectric coefficient λ for Cu_2OSeO_3 is evaluated as $\lambda = 5.64\times 10^{-27}\mu$Cm from experimental data of the ferroelectric polarization [51]. Note that eqs. (9.3) and (9.4) have totally different forms. However, we can confirm that the spatial distribution of local polarizations $\boldsymbol{p}_i$ calculated from eq. (9.3) using experimental coordinate data of Cu^{2+} and O ions in Cu_2OSeO_3 perfectly coincides with that calculated from eq. (9.4). We show calculated spatial maps of the local polarizations $\boldsymbol{p}_i$ for different $\boldsymbol{H}$ orientations with respect to the crystallographic axes in Figure 9.3(f)–(h).

9.3 Spin model and phase diagram

Magnetism in chiral magnets is described by the following continuum spin model as long as magnetic orders with sufficiently slow spatial and temporal variations are concerned [56]:

$$\begin{aligned}\mathcal{H} = \int d\boldsymbol{r}\Bigg[&\frac{J}{2a}(\nabla \boldsymbol{m})^2 + \frac{D}{a^2}\boldsymbol{m}\cdot(\nabla\times\boldsymbol{m}) - \frac{g\mu_B\mu_0}{a^3}\boldsymbol{H}\cdot\boldsymbol{m} \\ &+ \frac{A_1}{a^3}(m_x^4 + m_y^4 + m_z^4) \\ &+ \frac{A_2}{2a}[(\nabla_x m_x)^2 + (\nabla_y m_y)^2 + (\nabla_z m_z)^2]\Bigg].\end{aligned} \tag{9.5}$$

Here the first and the second terms describe the ferromagnetic exchange interaction and the Dzyaloshinskii–Moriya interaction, respectively, while the third term depicts the Zeeman coupling to an external magnetic field $\boldsymbol{H}$. Two types of magnetic anisotropies allowed in crystals with cubic symmetry are also taken into account by the fourth and fifth terms, but they turned out to play only a minor role as long as realistic small values of A_1 and A_2 are considered. In order to treat this continuum spin model numerically, it is convenient to divide the space into cubic (square) meshes, which gives a classical Heisenberg model on the cubic (square) lattice [57]. The Hamiltonian of this lattice spin model is given by,

$$\begin{aligned}\mathcal{H}_0 = &-J\sum_{<i,j>}\boldsymbol{m}_i\cdot\boldsymbol{m}_j - D\sum_{i,\hat{\gamma}=\hat{x},\hat{y},\hat{z}}\boldsymbol{m}_i\times\boldsymbol{m}_{i+\hat{\gamma}}\cdot\hat{\gamma} \\ &-g\mu_B\mu_0\boldsymbol{H}\cdot\sum_i\boldsymbol{m}_i,\end{aligned} \tag{9.6}$$

We treat a tetrahedron composed of four Cu^{2+} (S=1/2) ions with three-up and one-down collinear spins as a magnetic unit in Cu_2OSeO_3, and describe this four-spin assembly by a classical magnetization vector $\boldsymbol{m}_i$ whose norm m is unity.

We choose the ratio $D/J = 0.09$, which gives a skyrmion diameter of ~99 sites for the skyrmion crystal phase. If we assume that the distance between adjacent tetrahedra is ~5 Å, this number corresponds to the skyrmion diameter of ~50 nm in agreement with the observation in the Lorentz transmission electron microscopy for Cu_2OSeO_3 [41]. The skyrmions and the helimagnetic states discussed here have long spatial periodicities, and thus these magnetic states are nearly decoupled from the background lattice structure. This fact justifies the theoretical treatment based on the spin model on the simple cubic lattice without considering the complicated real crystal structure.

Theoretical phase diagrams of this lattice spin model have turned out to reproduce the experimental phase diagrams of Cu_2OSeO_3 as well as those of the B20 compounds [57, 58]. Figure 9.4 shows a ground-state phase diagram of this lattice spin

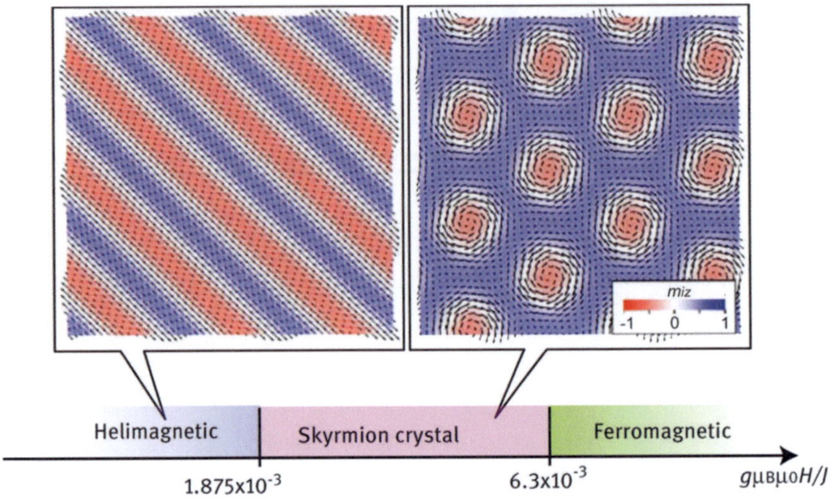

Figure 9.4: Ground-state phase diagram of the classical Heisenberg model given by eq. (9.6) for a thin-plate specimen of chiral-lattice magnet as a function of magnetic field applied normal to the plane.

model in two dimensions as a function of the external magnetic field H_z normal to the plane [59].We find that the skyrmion crystal phase appears in a moderate field-strength region ranging $1.875\times10^{-3} < |g\mu_B\mu_0 H_z/J| < 6.3\times10^{-3}$, sandwiched by the helimagnetic phase (low-field phase) and the ferrimagnetic phase (high-field phase). This theoretical phase diagram well reproduces a feature of the experimental phase diagram for thin-film samples of Cu_2OSeO_3 shown in Figure 9.2(d), where the helimagnetic, skyrmion crystal, and ferrimagnetic phases successively appear as a magnetic field increases at the lowest temperatures. For the ferromagnetic exchange parameter J = 3 meV relevant for Cu_2OSeO_3, these critical fields, respectively, correspond to 486 Oe and 1632 Oe, which coincide well with the experimental values of 500 Oe and 1800 Oe indicated by the experimental phase diagram in Figure 9.2(d).

Magnetic modulation period or spiral pitch of the helimagnetic state and the skyrmion crystal state are determined by a ratio of the two interactions D/J. From a saddle-point equation of the energy written as a function of the spin-rotation angle θ between $\boldsymbol{m}_i$ and $\boldsymbol{m}_{i+\hat{x}}$ (or between $\boldsymbol{m}_i$ and $\boldsymbol{m}_{i+\hat{y}}$), we obtain a relation $\tan\theta = D/(\sqrt{2}J)$. Accordingly, the periodicity λ_{HL} for the helimagnetic state propagating along <110> direction is given by $\lambda_{\mathrm{HL}} = \sqrt{2}\pi a/\theta$. The periodicity of the skyrmion crystal λ_{SkX} becomes approximately $\lambda_{\mathrm{SkX}} = (2/\sqrt{3})\lambda_{\mathrm{HL}}$.

Magnetization dynamics of a skyrmion crystal and other magnetic structures in chiral-lattice magnets are studied using the Landau–Lifshitz–Gilbert (LLG) equation. The equation is given by,

$$\frac{d\boldsymbol{m}_i}{dt} = -\gamma \boldsymbol{m}_i \times \boldsymbol{H}_i^{\mathrm{eff}} + \alpha_{\mathrm{G}} \left(\boldsymbol{m}_i \times \frac{d\boldsymbol{m}_i}{dt} \right). \tag{9.7}$$

The first term describes the gyrotropic motion of local magnetization $\boldsymbol{m}_i$ around the effective local magnetic field $\boldsymbol{H}_i^{\mathrm{eff}}$ where γ is the gyromagnetic ratio. The second term is the Gilbert-damping term where α_{G} is the Gilbert-damping coefficient, which is phenomenologically introduced to describe dissipative nature of the magnetization dynamics. Because of this term, each local magnetization gyrating around the effective local magnetic field $\boldsymbol{H}_i^{\mathrm{eff}}$ is eventually forced to point along the $\boldsymbol{H}_i^{\mathrm{eff}}$ through losing the energy and reducing the gyration radius. The effective field $\boldsymbol{H}_i^{\mathrm{eff}}$ can be calculated from the Hamiltonian $\mathcal{H}=\mathcal{H}_0+\mathcal{H}'(t)$ as

$$\boldsymbol{H}_i^{\mathrm{eff}} = -\frac{1}{\gamma\hbar}\frac{\partial \mathcal{H}}{\partial \boldsymbol{m}_i}. \tag{9.8}$$

Here $\mathcal{H}_0$ is the lattice-spin Hamiltonian for chiral-lattice magnets given by eq. (9.6), while $\mathcal{H}'(t)$ describes time-dependent external fields.

9.4 Microwave nonreciprocal directional dichroism

As argued in the above section, the magnetic skyrmions in insulating magnets attain multiferroic nature and thus exhibit collective excitations which can be activated not only magnetically by AC magnetic field $\boldsymbol{H}^{\omega}$ but also electrically by AC electric field $\boldsymbol{E}^{\omega}$ because of the magnetoelectric coupling [48–50]. The collective modes are characterized as coupled oscillations of magnetizations and polarizations, which are referred to as electromagnons. In particular, when the ferroelectric polarization $\boldsymbol{P}$ and the net magnetization $\boldsymbol{M}$ are orthogonal to each other as in the case of $\boldsymbol{H} \parallel [110]$ [see Figure 9.3(c)], the $\boldsymbol{H}^{\omega}$ and $\boldsymbol{E}^{\omega}$ components of an electromagnetic wave propagating parallel or antiparallel to $\boldsymbol{P} \times \boldsymbol{M}$ can activate the same electromagnon modes.

Such simultaneous magnetic and electric activities of electromagnons in this specific configuration can be seen in the dynamical magnetic susceptibility $\chi_{\alpha\beta}^{\mathrm{mm}}(\omega)$ and the dynamical dielectric susceptibility $\chi_{\alpha\beta}^{\mathrm{ee}}(\omega)$ for the skyrmion crystal state. They are, respectively, given by

$$\chi_{\alpha\beta}^{\mathrm{mm}}(\omega) = \frac{\Delta M_{\alpha}^{\omega}}{\mu_0 H_{\beta}^{\omega}} \tag{9.9}$$

and

$$\chi_{\alpha\beta}^{\mathrm{ee}}(\omega) = \frac{\Delta P_{\alpha}^{\omega}}{\epsilon_0 E_{\beta}^{\omega}}. \tag{9.10}$$

Here α and β represent the orthogonal axes x, y, z where the z-axis is defined to be parallel to the static external magnetic field $\boldsymbol{H} \parallel [110]$. These susceptibilities can be numerically calculated by tracing the time evolutions of the net magnetization $\boldsymbol{M}(t) = \frac{g\mu_B}{N}\sum_i \boldsymbol{m}_i(t)$ and the ferroelectric polarization $\boldsymbol{P}(t) = \frac{1}{N}\sum_i \boldsymbol{p}_i(t)$ after applying a short pulse of magnetic field or electric field as a perturbation $\mathscr{H}'(t)$ uniformly to the whole system. We can obtain ΔM_α^ω and ΔP_α^ω by performing the Fourier transformation of the time profiles $\Delta\boldsymbol{M}(t) = \boldsymbol{M}(t) - \boldsymbol{M}(0)$ and $\Delta\boldsymbol{P}(t) = \boldsymbol{P}(t) - \boldsymbol{P}(0)$.

Figure 9.5(a) shows imaginary parts of the calculated dynamical susceptibilities $\mathrm{Im}\chi_{\alpha\beta}^{\mathrm{mm}}(\omega)$ and $\mathrm{Im}\chi_{\alpha\beta}^{\mathrm{ee}}(\omega)$. We can find that the collective modes active to AC magnetic field seen as peaks in the spectrum of $\mathrm{Im}\chi_{\alpha\beta}^{\mathrm{mm}}$ can be found also in the spectrum of $\mathrm{Im}\chi_{\alpha\beta}^{\mathrm{ee}}$. More concretely, the spectrum of $\mathrm{Im}\chi_{yy}^{\mathrm{mm}}$ with $\boldsymbol{y} \parallel [001]$ in Figure 9.5(a) has a couple of resonance peaks, whereas the spectrum of $\mathrm{Im}\chi_{zz}^{\mathrm{ee}}$ with $\boldsymbol{z} \parallel [110]$ in Figure 9.5(b) also has peaks at the same frequencies. This indicates that the in-plane AC magnetic field $\boldsymbol{H}^\omega(\perp \boldsymbol{H})$ and the out-of-plane AC electric field $\boldsymbol{E}^\omega(\parallel \boldsymbol{H})$ activate the same collective modes. On the other hand, both $\mathrm{Im}\chi_{zz}^{\mathrm{mm}}$ in Figure 9.5(a) and $\mathrm{Im}\chi_{yy}^{\mathrm{ee}}$ in Figure 9.5(b) have a single peak at the same frequency, indicating that the out-of-plane AC magnetic field $\boldsymbol{H}^\omega(\parallel \boldsymbol{H})$ and the in-plane AC electric field $\boldsymbol{E}^\omega(\perp \boldsymbol{H})$ activate the same collective mode.

A couple of rotation modes are activated by the in-plane AC magnetic field $\boldsymbol{H}^\omega \parallel \boldsymbol{y}(\perp \boldsymbol{H})$ or the out-of-plane AC electric field $\boldsymbol{E}^\omega \parallel \boldsymbol{z}(\parallel \boldsymbol{H})$ [59]. In these modes, all the skyrmions in the skyrmion crystal uniformly circulate, which occur in a counterclockwise (clockwise) fashion for the lower-lying (higher-lying) mode. Figure 9.5(c) shows spatiotemporal maps of the local magnetizations $\boldsymbol{m}_i$ (left panels) and the induced local electric polarizations $\boldsymbol{p}_i$ (right panels) for the counterclockwise rotation mode. It is interesting that these kinds of rotawtional dynamics of skyrmions can be driven by lineally polarized microwave fields. On the other hand, a breathing mode is activated by the out-of-plane AC magnetic field $\boldsymbol{H}^\omega \parallel \boldsymbol{z}(\parallel \boldsymbol{H})$ or the in-plane AC electric field $\boldsymbol{E}^\omega \parallel \boldsymbol{y}(\perp \boldsymbol{H})$, in which all the skyrmions in the skyrmion crystal uniformly expand and shrink in an oscillatory manner as shown in Figure 9.5(d) [59, 60].

Behaviors of the net magnetization $\boldsymbol{M}$ and the ferroelectric polarization $\boldsymbol{P}$ for each mode are shown in the bottom panels of Figure 9.5(c) and (d). For the counterclockwise rotational mode, the angle between $\boldsymbol{P}$ and $\boldsymbol{M}$, which is originally 90°, becomes larger and smaller in an oscillatory manner. On the other hand, for the breathing mode, simultaneous elongation and shrinkage of $\boldsymbol{P}$ and $\boldsymbol{M}$ occur oscillatorily, while keeping the angle between them at 90°.

These simultaneous electric and magnetic activities of the electromagnon excitations are nothing but a source of the nonreciprocal directional dichroism of microwaves in the multiferroic skyrmion crystal phase [51]. Starting from Maxwell's equations, we can derive a relation for electromagnetic waves:

$$\boldsymbol{H}^\omega \parallel \boldsymbol{K}^\omega \times \boldsymbol{E}^\omega. \tag{9.11}$$

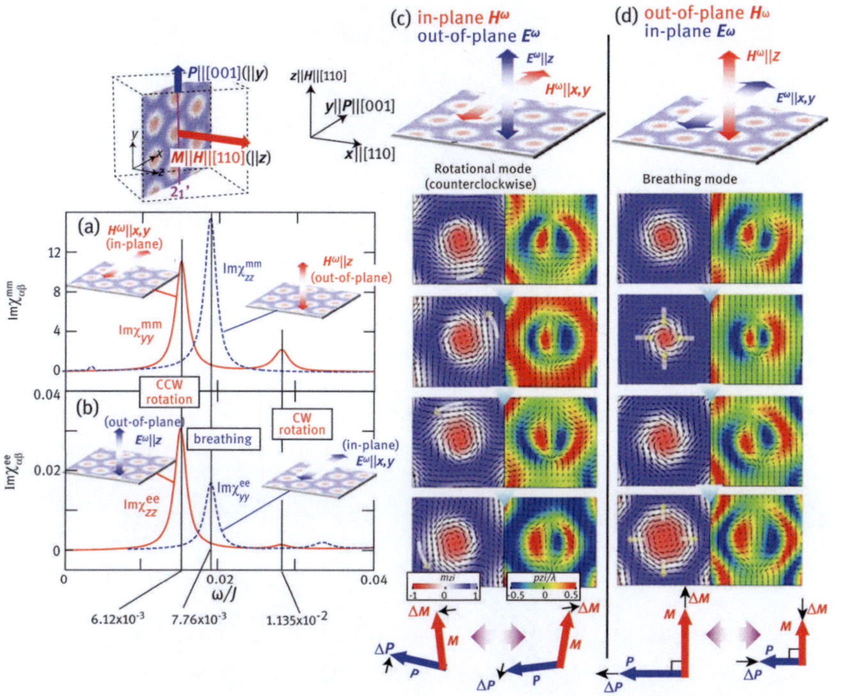

Figure 9.5: (a) [(b)] Imaginary parts of the dynamical magnetic [dielectric] susceptibilities $\mathrm{Im}\chi^{\mathrm{mm}}_{\alpha\beta}(\omega)$ [$\mathrm{Im}\chi^{\mathrm{ee}}_{\alpha\beta}(\omega)$] calculated for the skyrmion crystal phase under $\boldsymbol{H} \parallel [110](\parallel \boldsymbol{z})$ [51]. $\mathrm{Im}\chi^{\mathrm{mm}}_{yy}$ in (a) and $\mathrm{Im}\chi^{\mathrm{ee}}_{zz}$ in (b) have resonance peaks at the same frequencies, indicating that both the in-plane AC magnetic field $\boldsymbol{H}^\omega \parallel \boldsymbol{y}$ and the out-of-plane AC electric field $\boldsymbol{E}^\omega \parallel \boldsymbol{z}$ excite common collective modes. On the other hand, $\mathrm{Im}\chi^{\mathrm{mm}}_{zz}$ in (a) and $\mathrm{Im}\chi^{\mathrm{ee}}_{yy}$ in (b) also have a resonance peak at the same frequency, indicating that both the out-of-plane AC magnetic field $\boldsymbol{H}^\omega \parallel \boldsymbol{z}$ and the in-plane AC electric field $\boldsymbol{E}^\omega \parallel \boldsymbol{y}$ excite a common collective mode. (c), (d) Spatiotemporal dynamics of magnetizations $\boldsymbol{m}_i$ (left panels) and polarizations $\boldsymbol{p}_i$ (right panels) for electromagnon excitations in the skyrmion crystal under $\boldsymbol{H} \parallel [110]$ ($\parallel \boldsymbol{z}$) [51]. (a) Counterclockwise rotational mode excited by the in-plane AC magnetic field $\boldsymbol{H}^\omega \perp \boldsymbol{H}$ or by the out-of-plane AC electric field $\boldsymbol{E}^\omega \parallel \boldsymbol{H}$. (b) Breathing mode excited by the out-of-plane AC magnetic field $\boldsymbol{H}^\omega \parallel \boldsymbol{H}$ or by the in-plane AC electric field $\boldsymbol{E}^\omega \perp \boldsymbol{H}$. Dynamical behaviors of the net magnetization $\boldsymbol{M}$ and the ferroelectric polarization $\boldsymbol{P}$ are shown for each mode in the bottom panels.

This equation indicates that a propagation direction of the electromagnetic wave or the sign of its wave vector $\boldsymbol{K}^\omega$ governs a relative relationship between $\boldsymbol{H}^\omega$ and $\boldsymbol{E}^\omega$ orientations. This relationship will be opposite when the sign of $\boldsymbol{K}^\omega$ is reversed.

Accordingly, in the case of a linearly polarized microwave with $\boldsymbol{H}^\omega \parallel [001](\perp \boldsymbol{M})$ and $\boldsymbol{E}^\omega \parallel [110](\perp \boldsymbol{P})$ in Figure 9.6(b), the $\boldsymbol{H}^\omega$ and $\boldsymbol{E}^\omega$ components of a microwave propagating in $[1\bar{1}0]$ $(-\boldsymbol{x})$ direction cooperatively excite an oscillation of the angle between $\boldsymbol{P}$ and $\boldsymbol{M}$ so as to intensely activate the rotation mode [see bottom panels of Figure 9.5(c)]. In turn, when the microwave is propagating in the opposite direction, that is, $[\bar{1}10]$ $(+\boldsymbol{x})$ direction, the $\boldsymbol{H}^\omega$ and $\boldsymbol{E}^\omega$ components contribute destructively to

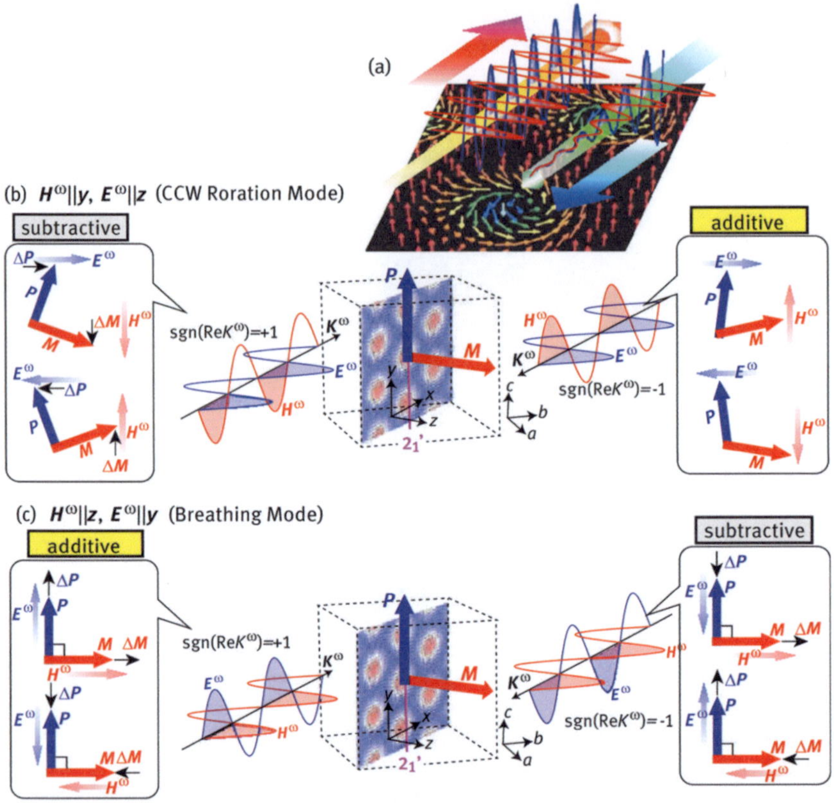

Figure 9.6: (a) Schematic illustration of the nonreciprocal directional dichroism of electromagnetic waves induced by the skyrmion crystal. An electromagnetic wave irradiating in a certain direction to a material is significantly absorbed, while that in the opposite direction is not absorbed so much and thus can penetrate through the material. (b), (c) Microwave-polarization configurations with which the nonreciprocal directional dichroism occurs in the skyrmion crystal phase with net magnetization $\boldsymbol{M}$ $||[110](||\ \boldsymbol{z})$ and ferroelectric polarization $\boldsymbol{P}$ $||[001](||\ \boldsymbol{y})$. (a) $\boldsymbol{H}^{\omega}$ $||[001](||\ \boldsymbol{y})$ and $\boldsymbol{E}^{\omega}$ $||[001](||\ \boldsymbol{z})$. The $\boldsymbol{H}^{\omega}$ and $\boldsymbol{E}^{\omega}$ components contribute in an additive way to the excitation of rotational modes for a microwave propagating in the negative ($-\boldsymbol{x}$ or $[1\bar{1}0]$) direction, whereas in a subtractive way for a microwave propagating in the positive ($+\boldsymbol{x}$ or $[\bar{1}10]$) direction. Consequently the absorption intensity of the former microwave with sign(ReK^{ω}) = − 1 becomes larger than that of the latter one with sign(ReK^{ω}) = + 1. (b) $\boldsymbol{H}^{\omega}$ $||[110](||\ \boldsymbol{z})$ and $\boldsymbol{E}^{\omega}$ $||[001](||\ \boldsymbol{z})$. In this case, a microwave propagating in the $+\boldsymbol{x}$ direction is absorbed significantly through intensely exciting the breathing mode as compared to that propagating in the $-\boldsymbol{x}$ direction.

activation of the rotation mode. Consequently, the microwave with $\boldsymbol{K}^{\omega}$ $||$ $-\boldsymbol{x}$ strongly excites the spin wave resonance and, thereby, is strongly absorbed, whereas the microwave with $\boldsymbol{K}^{\omega}$ $||$ $+\boldsymbol{x}$ excites the spin wave resonance only weakly and, thus, is absorbed only little. In this way, the intensity of the microwave absorption differs depending on its incident direction.

In addition to the rotation mode considered for the microwave polarization in Figure 9.6(b), the interference between the magnetic and electric activation channels of the breathing mode can also give rise to the nonreciprocal directional dichroism of microwave when the microwave polarization is $\boldsymbol{H}^{\omega}$ $\|[110](\| \boldsymbol{M})$ and $\boldsymbol{E}^{\omega}$ $\|[001](\| \boldsymbol{P})$ as shown in Figure 9.6(c). In this case, the $\boldsymbol{H}^{\omega}$ and $\boldsymbol{E}^{\omega}$ components of a microwave propagating in the $+\boldsymbol{x}$ direction cooperatively induce length oscillations of $\boldsymbol{P}$ and $\boldsymbol{M}$ so as to intensely activate the breathing mode, whereas those of a microwave propagating in the $-\boldsymbol{x}$ direction do not [see bottom panels of Figure 9.5(d)]. Consequently, the microwave with $\boldsymbol{K}^{\omega} \parallel +\boldsymbol{x}$ is strongly absorbed, whereas the microwave with $\boldsymbol{K}^{\omega} \parallel -\boldsymbol{x}$ is weakly absorbed. Again, the absorption intensity becomes different depending on the incident direction of the microwave.

The nonreciprocal propagation of electromagnetic waves can be formulated starting with Maxwell's equations for materials with finite net magnetization $\boldsymbol{M}$ and ferroelectric polarization $\boldsymbol{P}$. The equations are given by,

$$\mathrm{rot}\boldsymbol{E} = -\frac{\partial \boldsymbol{B}}{\partial t} \tag{9.12}$$

$$\mathrm{rot}\boldsymbol{B} = \frac{\partial \boldsymbol{D}}{\partial t} \tag{9.13}$$

where

$$\boldsymbol{B} = \mu_0(\hat{\mu}^{\infty}\boldsymbol{H} + \boldsymbol{M}), \tag{9.14}$$

$$\boldsymbol{D} = \epsilon_0\hat{\epsilon}^{\infty}\boldsymbol{E} + \boldsymbol{P} \tag{9.15}$$

$$\boldsymbol{P} = \epsilon_0\hat{\chi}^{\mathrm{ee}}\boldsymbol{E} + \hat{\chi}^{\mathrm{em}}\sqrt{\epsilon_0\mu_0}\boldsymbol{H} \tag{9.16}$$

$$\boldsymbol{M} = \mu_0\hat{\chi}^{\mathrm{mm}}\boldsymbol{H} + \hat{\chi}^{\mathrm{me}}\sqrt{\frac{\epsilon_0}{\mu_0}}\boldsymbol{E} \tag{9.17}$$

A key ingredient of the nonreciprocal directional dichroism is a fact that the magnetoelectric susceptibilities $\hat{\chi}^{\mathrm{em}}$ and $\hat{\chi}^{\mathrm{me}}$ become finite owing to the magnetoelectric coupling in multiferroics, which gives the $\boldsymbol{H}$-linear contribution to the ferroelectric polarization $\boldsymbol{P}$ and the $\boldsymbol{E}$-linear contribution to the net magnetization $\boldsymbol{M}$. We employ the following expressions for the AC electric field and the AC magnetic field:

$$\boldsymbol{E}_{\mathrm{ac}} = \boldsymbol{E}^{\omega}\exp[\mathrm{i}(\boldsymbol{K}^{\omega}\cdot\boldsymbol{r} - \omega t)] \tag{9.18}$$

$$\boldsymbol{H}_{\mathrm{ac}} = \boldsymbol{H}^{\omega}\exp[\mathrm{i}(\boldsymbol{K}^{\omega}\cdot\boldsymbol{r} - \omega t)] \tag{9.19}$$

Inserting these expressions into eqs. (9.12) and (9.13), we obtain the following equations for the Fourier components,

$$\omega\boldsymbol{B}^{\omega} = \boldsymbol{K}^{\omega}\times\boldsymbol{E}^{\omega}, \tag{9.20}$$

$$-\omega \boldsymbol{D}^{\omega} = \boldsymbol{K}^{\omega} \times \boldsymbol{H}^{\omega} \tag{9.21}$$

where

$$\boldsymbol{B}^{\omega} = [\hat{\mu}^{\infty} + \hat{\chi}^{\mathrm{mm}}(\omega)]\mu_0 \boldsymbol{H}^{\omega} + \hat{\chi}^{\mathrm{me}}(\omega)\sqrt{\epsilon_0\mu_0}\boldsymbol{E}^{\omega} \tag{9.22}$$

$$\boldsymbol{D}^{\omega} = [\hat{\epsilon}^{\infty} + \hat{\chi}^{\mathrm{ee}}(\omega)]\epsilon_0 \boldsymbol{E}^{\omega} + \hat{\chi}^{\mathrm{em}}(\omega)\sqrt{\epsilon_0\mu_0}\boldsymbol{H}^{\omega} \tag{9.23}$$

Solving these simultaneous equations in terms of $\boldsymbol{K}^{\omega}$ and considering the relations,

$$K^{\omega} = \frac{\omega N(\omega)}{c}, \tag{9.24}$$

we can derive an expression of the complex refractive index

$$N(\omega) = n(\omega) + i\kappa(\omega). \tag{9.25}$$

for respective electromagnetic-wave polarizations as follows:
Case 1: $\boldsymbol{H}^{\omega} \parallel \boldsymbol{y}$ and $\boldsymbol{E}^{\omega} \parallel \boldsymbol{z}$

$$N(\omega) \sim \sqrt{[\epsilon_{zz}^{\infty} + \chi_{zz}^{\mathrm{ee}}(\omega)][\mu_{yy}^{\infty} + \chi_{yy}^{\mathrm{mm}}(\omega)]} - \frac{\mathrm{sign}(K^{\omega})}{2}[\chi_{yz}^{\mathrm{me}}(\omega) + \chi_{zy}^{\mathrm{em}}(\omega)], \tag{9.26}$$

Case 2: $\boldsymbol{H}^{\omega} \parallel \boldsymbol{z}$ and $\boldsymbol{E}^{\omega} \parallel \boldsymbol{y}$

$$N(\omega) \sim \sqrt{[\epsilon_{yy}^{\infty} + \chi_{yy}^{\mathrm{ee}}(\omega)][\mu_{zz}^{\infty} + \chi_{zz}^{\mathrm{mm}}(\omega)]} + \frac{\mathrm{sign}(K^{\omega})}{2}[\chi_{zy}^{\mathrm{me}}(\omega) + \chi_{yz}^{\mathrm{em}}(\omega)]. \tag{9.27}$$

The complex refractive index $N(\omega)$ describes how the material responds to an irradiated electromagnetic wave. In particular, its imaginary part $\kappa(\omega)$ is called extinction coefficient, which represents to what extent the material absorbs the electromagnetic wave. More concretely, the absorption coefficient $\alpha(\omega)$ is related with $\kappa(\omega)$ as,

$$\alpha(\omega) = \frac{2\omega\kappa(\omega)}{c} = \frac{2\omega}{c}\mathrm{Im}N(\omega). \tag{9.28}$$

Note that the expressions of $N(\omega)$ given by eqs. (9.26) and (9.27) contain a term including a sign of K^{ω}. Through this sign dependence of $N(\omega)$, the absorption coefficient $\alpha(\omega)$ eventually depends on the incident direction of electromagnetic wave.

Magnitudes of the nonreciprocal directional dichroism of electromagnetic waves for respective polarization configurations are given by,
Case 1

$$\Delta\alpha(\omega) = \alpha_{-}(\omega) - \alpha_{+}(\omega) = \mathrm{Im}[\chi_{yz}^{\mathrm{me}}(\omega) + \chi_{zy}^{\mathrm{em}}(\omega)] \tag{9.29}$$

Case 2

$$\Delta\alpha(\omega) = \alpha_{+}(\omega) - \alpha_{-}(\omega) = \mathrm{Im}[\chi_{zy}^{\mathrm{me}}(\omega) + \chi_{yz}^{\mathrm{em}}(\omega)] \tag{9.30}$$

Here $\alpha_+(\omega)$ and $\alpha_-(\omega)$ denote the absorption coefficients for electromagnetic waves propagating in the positive and negative x directions, respectively. Note that the absorption intensity $I(\omega)$ is related with the absorption coefficient $\alpha(\omega)$ as

$$I(\omega) = I_0 \exp[-\alpha(\omega)d], \tag{9.31}$$

with d being the sample thickness.

The above argument indicates that for the quantitative calculation of the complex refractive index $N(\omega)$ and the absorption coefficient $\alpha_\pm(\omega)$, it is necessary to calculate not only the dynamical magnetic and dielectric susceptibilities $\hat{\chi}^{\mathrm{mm}}$ and $\hat{\chi}^{\mathrm{ee}}$ but also the dynamical magnetoelectric susceptibilities $\hat{\chi}^{\mathrm{em}}$ and $\hat{\chi}^{\mathrm{me}}$, which are given by,

$$\chi^{\mathrm{em}}_{\alpha\beta}(\omega) = \frac{\Delta P^\omega_\alpha}{\sqrt{\epsilon_0\mu_0}H^\omega_\beta}, \tag{9.32}$$

and

$$\chi^{\mathrm{me}}_{\alpha\beta}(\omega) = \sqrt{\frac{\mu_0}{\epsilon_0}}\frac{\Delta M^\omega_\alpha}{E^\omega_\beta}. \tag{9.33}$$

These susceptibilities describe the response of ferroelectric polarization $\boldsymbol{P}$ to AC magnetic field $\boldsymbol{H}^\omega$ and the response of net magnetization $\boldsymbol{M}$ to AC electric field $\boldsymbol{E}^\omega$, respectively.

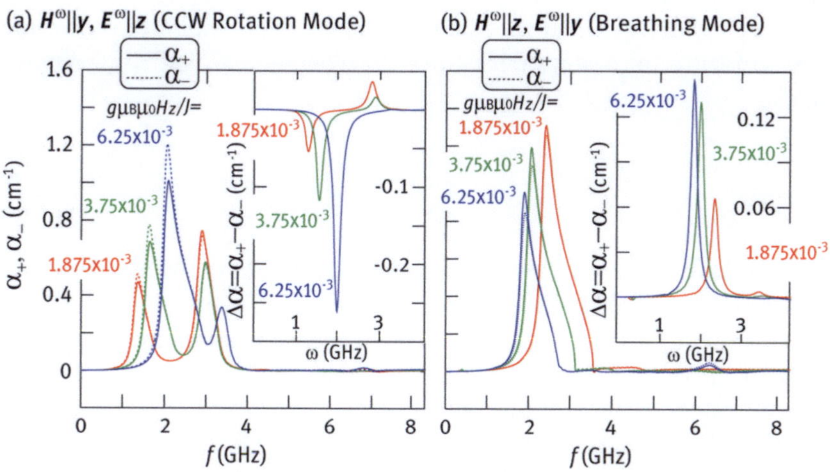

Figure 9.7: Calculated microwave absorption spectra for the skyrmion crystal phase realized in a (two-dimensional) thin-plate specimen of Cu_2OSeO_3 under the external magnetic field $\boldsymbol{H}$ ||[110] [51]. Here α_+ and α_- are the microwave absorption coefficients for positive and negative incident directions, that is, sign(ReK^ω) = +1 and sign(ReK^ω) = −1, respectively. The difference $\Delta\alpha$ represents a magnitude of the directional dichroism. The spectra in (a) are for the microwave polarization of $\boldsymbol{H}^\omega$ || [001] (|| $\boldsymbol{y}$) and $\boldsymbol{E}^\omega$ || [110] (|| $\boldsymbol{z}$) where the rotational modes are excited. On the other hand, the spectra in (b) are for the microwave polarization of $\boldsymbol{H}^\omega$ || [110] (|| $\boldsymbol{z}$) and $\boldsymbol{E}^\omega$ || [001] (|| $\boldsymbol{y}$) where the breathing mode is excited.

Figure 9.7(a) and (b) show the calculated microwave absorption coefficients $\alpha_+(\omega)$ and $\alpha_-(\omega)$ as well as their difference $\Delta\alpha(\omega)$ as functions of the microwave angular frequency ω for two different microwave polarizations. The spectra are calculated for various value of the external magnetic field $\boldsymbol{H}_z$.

In the case of the microwave polarization $\boldsymbol{H}^\omega \parallel y$ and $\boldsymbol{E}^\omega \parallel z$ with which the rotation modes are excited [see Figure 9.7(a)], the spectra have double peaks corresponding to the lower-lying counterclockwise rotation mode and the higher-lying clockwise rotation mode. The difference of absorption coefficient $\Delta\alpha(\omega)$ increases as the static external magnetic field H_z increases, which reaches 0.25 cm^{-1} at maximum. This value corresponds to relative difference $\Delta\alpha/\alpha_{\text{av}} = 2(\alpha_+ - \alpha_-)/(\alpha_+ + \alpha_-)$ of ~20%.

On the other hand, when the microwave polarization is $\boldsymbol{H}^\omega \parallel y$ and $\boldsymbol{E}^\omega \parallel z$ with which the breathing mode is excited [see Figure 9.7(b)], the spectra have a single peak. The difference of the absorption coefficient $\Delta\alpha(\omega)$ again increases as H_z increases and reaches 0.14 cm^{-1} at maximum, which corresponds to the relative difference of ~10%.

Here it should be mentioned that emergence of such gigantic nonreciprocal directional dichroism is quite rare in any frequency range of electromagnetic wave [61–66]. In particular, there has been no report for the GHz regime. This is because usual multiferroic materials based on rather simple noncollinear magnetism with a short modulation period tend to have a large spin-wave gap and resulting higher resonance frequencies typically at THz regime. On the contrary, the skyrmions and other topological spin textures with a considerably long modulation period tend to have a smaller spin-wave gap and specific resonance modes, which provides rare opportunities to realize interesting microwave phenomena and useful microwave functionalities.

9.5 Summary

In this article, we have discussed the electromagnon excitations of the magnetic skyrmions in an insulating chiral-lattice magnet Cu_2OSeO_3 and unprecedentedly large nonreciprocal directional dichroism at microwave frequencies caused by the interferences between magnetic and electric activation channels of the electromagnon excitations. For practical application of this phenomenon to microwave devices such as microwave isolators, it is necessary to further enhance the effect. For this purpose, search for novel skyrmion-based multiferroic materials with a larger ferroelectric polarization or a stronger magnetoelectric coupling is needed. In fact, Cu_2OSeO_3 had been an only example of the skyrmion-hosting magnetic insulators. However, efforts to seek new materials have been continuously made, and lacunar spinel vanadates GaV_4S_8 and GaV_4Se_8 have been recently discovered as a new class of multiferroic skyrmion materials [67, 68]. In addition, it was recently reported that Neel-type skyrmions in magnetic insulator $VOSe_2O_5$ with a tetragonal polar crystal structure host multiferroicity [69] (see also Chapter 4). Nonetheless, we have only a few examples of the skyrmion-hosting magnetic insulators currently, and thus the research should be continued. One important direction of the research is to seek chiral-lattice multiferroics with ferroelectric polarization induced by the inverse Dzyaloshinskii–Moriya mechanism [4, 5] because this mechanism tends to induce large ferroelectric polarization as compared to the spin-dependent metal-ligand hybridization mechanism [55] in Cu_2OSeO_3. Another promising direction is the artificial designing of large magnetoelectric susceptibilities and intense electromagnon resonances by using metamaterials, thin films, and synthetic nanomaterials to realize the strong microwave magnetoelectric responses.

Acknowledgements: This work was supported by JSPS KAKENHI (Grants No. 17H02924, No. 16H06345, and No. 19H00864), Waseda University Grant for Special Research Projects (Project Nos. 2018K-257 and 2019C-253), and JST PRESTO (Grant No. JPMJPR132A).

References

[1] Nagaosa N, Tokura Y. Emergent electromagnetism in solids. Phys Scr. 2012;T146:014020–014020.
[2] Mochizuki M. Current-driven dynamics of skyrmions. In: Seidel Jan, editor. Topological structures in ferroic materials domain walls, skyrmions and vortices, 1st ed. Springer Series in Materials Science Vol. 228. Switzerland: Springer International Publishing, 2016:55–81.
[3] Kimura T, Goto T, Shintani H, Ishizaka K, Arima T, Tokura Y. Magnetic control of ferroelectric polarization. Nature. 2003;426:55–8.
[4] Katsura H, Nagaosa N, Balatsky AV. Spin current and magnetoelectric effect in noncollinear magnets. Phys Rev Lett. 2005;95.
[5] Mostovoy M. Ferroelectricity in spiral magnets. Phys Rev Lett. 2006;96.
[6] Kimura T. Spiral magnets as magnetoelectrics. Annu Rev Mater Res. 2007;37:387–413.
[7] Pfleiderer C. Surfaces get hairy. Nat Phys. 2011;7:673–74.
[8] Nagaosa N, Tokura Y. Topological properties and dynamics of magnetic skyrmions. Nat Nanotech. 2013;8:899–911.
[9] Fert A, Cros V, Sampaio J. Skyrmions on the track. Nat Nanotech. 2013;8:152–6.
[10] Seki S, Mochizuki M. Skyrmions in magnetic materials, 1st ed. SpringerBriefs in Physics. Switzerland: Springer International Publishing, 2015.
[11] Mochizuki M, Seki S. Dynamical magnetoelectric phenomena of multiferroic skyrmions. J Phys: Condens Matter. 2015;27:503001–503001.
[12] Skyrme TH. A non-linear field theory. Proc R Soc London Ser A Math Phys Sci. 1961;260:127–38.
[13] Skyrme TH. A unified field theory of mesons and baryons. Nucl Phys. 1962;31:556–69.
[14] Bogdanov AN, Yablonskii DA. Thermodynamically stable "vortices" in magnetically ordered crystals. Sov Phys JETP. 1989;68:101–3.
[15] Bogdanov A, Hubert A. Thermodynamically stable magnetic vortex states in magnetic crystals. J Magn Magn Mater. 1994;138:255–69.
[16] Rößler UK, Bogdanov AN, Pfleiderer C. Spontaneous skyrmion ground states in magnetic metals. Nature. 2006;442:797–801.

[17] Muhlbauer S, Binz B, Jonietz F, Pfleiderer C, Rosch A, Neubauer A, Georgii R, Boni P. Skyrmion lattice in a chiral magnet. Science. 2009;323:915–9.
[18] Yu XZ, Onose Y, Kanazawa N, Park JH, Han JH, Matsui Y, Nagaosa N, Tokura Y, et al. Real-space observation of a two-dimensional skyrmion crystal. Nature. 2010;465:901–4.
[19] Pappas C, Lelièvre-Berna E, Falus P, Bentley PM, Moskvin E, Grigoriev S, Fouquet P, Farago B. Chiral paramagnetic Skyrmion-like phase in MnSi. Phys Rev Lett. 2009;102.
[20] Pfleiderer C, et al. Skyrmion lattices in metallic and semiconducting B20 transition metal compounds. J Phys: Condensed Matter. 2010;22:164207–164207.
[21] Münzer W, et al. Skyrmion lattice in the doped semiconductorFe1–xCoxSi. Phys Rev B. 2010;81.
[22] Adams T, et al. Long-range crystalline nature of the Skyrmion lattice in MnSi. Phys Rev Lett. 2011;107.
[23] Grigoriev SV, et al. Chiral properties of structure and magnetism inMn1–xFexGeCompounds: when the left and the right are fighting, who wins? Phys Rev Lett. 2013;110.
[24] Yu XZ, Kanazawa N, Onose Y, Kimoto K, Zhang WZ, Ishiwata S, Matsui Y, Tokura Y, et al. Near room-temperature formation of a skyrmion crystal in thin-films of the helimagnet FeGe. Nat Mater. 2010:106–109.
[25] Tonomura A, Yu X, Yanagisawa K, Matsuda T, Onose Y, Kanazawa N, Park HS, Tokura Y, et al. Real-space observation of Skyrmion lattice in helimagnet MnSi thin samples. Nano Lett. 2012;12:1673–77.
[26] Shibata K, Yu XZ, Hara T, Morikawa D, Kanazawa N, Kimoto K, Ishiwata S, Matsui Y, Tokura Y, et al. Towards control of the size and helicity of skyrmions in helimagnetic alloys by spin–orbit coupling. Nat Nanotech. 2013;8:723–8.
[27] Morikawa D, Shibata K, Kanazawa N, Yu XZ, Tokura Y. Crystal chirality and skyrmion helicity in MnSi and (Fe, Co)Si as determined by transmission electron microscopy. Phys Rev B. 2013;88.
[28] Dzyaloshinsky I. A thermodynamic theory of “weak” ferromagnetism of antiferromagnetics. J Phys Chemi Solids. 1958;4:241–55.
[29] Moriya T. Anisotropic Superexchange Interaction and Weak Ferromagnetism. Phys Rev. 1960;120:91–8.
[30] Binz B, Vishwanath A. Chirality induced anomalous-Hall effect in helical spin crystals. Phys B: Conden Matter. 2008;403:1336–40.
[31] Neubauer A, Pfleiderer C, Binz B, Rosch A, Ritz R, Niklowitz PG, Böni P, et al. Topological hall effect in the APhase of MnSi. Phys Rev Lett. 2009;102.
[32] Zang J, Mostovoy M, Han JH, Nagaosa N. Dynamics of Skyrmion crystals in metallic thin films. Phys Rev Lett. 2011;107.
[33] Kanazawa N, Onose Y, Arima T, Okuyama D, Ohoyama K, Wakimoto S, Kakurai K, Ishiwata S, Tokura Y, et al. Large topological hall effect in a short-period helimagnet MnGe. Phys Rev Lett. 2011;106.
[34] Schulz T, Ritz R, Bauer A, Halder M, Wagner M, Franz C, Pfleiderer C, Everschor K, Garst M, Rosch A, et al. Emergent electrodynamics of skyrmions in a chiral magnet. Nat Phys. 2012;8:301–4.
[35] Jonietz F, et al. Spin transfer torques in MnSi at ultralow current densities. Science. 2010;330:1648–51.
[36] Yu XZ, Kanazawa N, Zhang WZ, Nagai T, Hara T, Kimoto K, Matsui Y, Onose Y, Tokura Y, et al. Skyrmion flow near room temperature in an ultralow current density. Nat Commun. 2012:988.
[37] Everschor K, Garst M, Duine RA, Rosch A. Current-induced rotational torques in the skyrmion lattice phase of chiral magnets. Phys Rev B. 2011;84.
[38] Everschor K, Garst M, Binz B, Jonietz F, Mühlbauer S, Pfleiderer C, Rosch A, et al. Rotating skyrmion lattices by spin torques and field or temperature gradients. Phys Rev B. 2012;86.

[39] Iwasaki J, Mochizuki M, Nagaosa N. Universal current-velocity relation of skyrmion motion in chiral magnets. Nat Commun. 2013;4.
[40] Iwasaki J, Mochizuki M, Nagaosa N. Current-induced skyrmion dynamics in constricted geometries. Nat Nanotech. 2013;8:742–7.
[41] Seki S, Yu XZ, Ishiwata S, Tokura Y. Observation of Skyrmions in a multiferroic material. Science. 2012;336:198–201.
[42] Seki S, Kim JH, Inosov DS, Georgii R, Keimer B, Ishiwata S, Tokura Y, et al. Formation and rotation of skyrmion crystal in the chiral-lattice insulator Cu2OSeO3. Phys Rev B. 2012;85.
[43] Seki S, Ishiwata S, Tokura Y. Magnetoelectric nature of skyrmions in a chiral magnetic insulator Cu2OSeO3. Phys Rev B. 2012;86.
[44] Adams T, Chacon A, Wagner M, Bauer A, Brandl G, Pedersen B, Berger H, Lemmens P, Pfleiderer C, et al. Long-wavelength helimagnetic order and Skyrmion lattice phase in Cu2OSeO3. Phys Rev Lett. 2012;108.
[45] Bos Jan-Willem G, Colin Claire V, Palstra Thomas TM. Magnetoelectric coupling in the cubic ferrimagnetCu2OSeO3. Phys Rev B. 2008;78. DOI:10.1103/PhysRevB.78.094416.094416.
[46] Belesi M, Rousochatzakis I, Wu HC, Berger H, Shvets IV, Mila F, Ansermet JP. Ferrimagnetism of the magnetoelectric compound Cu2OSeO3 probed by 77Se NMR. Phys Rev B. 2010;82.
[47] Belesi M, Philippe T, Rousochatzakis I, Wu HC, Berger H, Granville S, Shvets IV, Ansermet JPh. Magnetic properties of the magnetoelectric compound Cu2OSeO3: Magnetization and 77Se NMR study. J Phys: conf ser. 2011;303:012069–012069.
[48] Smolenskiĭ GA, Chupis IE. Ferroelectromagnets. Sov Phys Usp. 1982;25:475–93.
[49] Pimenov A, Mukhin AA, Ivanov VY, Travkin VD, Balbashov AM, Loidl A. Possible evidence for electromagnons in multiferroic manganites. Nat Phys. 2006;2:97–100.
[50] Katsura H, Balatsky AV, Nagaosa N. Dynamical magnetoelectric coupling in helical magnets. Phys Rev Lett. 2007;98.
[51] Mochizuki M, Seki S. Magnetoelectric resonances and predicted microwave diode effect of the skyrmion crystal in a multiferroic chiral-lattice magnet. Phys Rev B. 2013;87.
[52] Mochizuki M. Microwave magnetochiral effect in Cu2OSeO3. Phys Rev Lett. 2015;114.
[53] Okamura Y, Kagawa F, Mochizuki M, Kubota M, Seki S, Ishiwata S, Kawasaki M, Onose Y, Tokura Y, et al. Microwave magnetoelectric effect via skyrmion resonance modes in a helimagnetic multiferroic. Nat Commun. 2013;4.
[54] Okamura Y, Kagawa F, Seki S, Kubota M, Kawasaki M, Tokura Y. Microwave magnetochiral dichroism in the chiral-lattice magnet Cu2OSeO3. Phys Rev Lett. 2015;114.
[55] Arima T. Ferroelectricity induced by proper-screw type magnetic order. J Phys Soc Jpn. 2007;76:073702–073702.
[56] Bak P, Jensen MH. Theory of helical magnetic structures and phase transitions in MnSi and FeGe. J Phys C: Solid State Phys. 1980;13:L881–L885.
[57] Yi SD, Onoda S, Nagaosa N, Han JH. Skyrmions and anomalous Hall effect in a Dzyaloshinskii-Moriya spiral magnet. Phys Rev B. 2009;80.
[58] Buhrandt S, Fritz L. Skyrmion lattice phase in three-dimensional chiral magnets from Monte Carlo simulations. Phys Rev B. 2013;88.
[59] Mochizuki M. Spin-wave modes and their intense excitation effects in Skyrmion crystals. Phys Rev Lett. 2012;108.
[60] Petrova O, Tchernyshyov O. Spin waves in a skyrmion crystal. Phys Rev B. 2011;84.
[61] Jung JH, Matsubara M, Arima T, He JP, Kaneko Y, Tokura Y. Optical magnetoelectric effect in the polar GaFeO3 ferrimagnet. Phys Rev Lett. 2004;93.
[62] Saito M, Taniguchi K, Arima T. Gigantic optical magnetoelectric effect in CuB2O4. J Phys Soc Jpn–J. 2008;77:013705–013705.

[63] Saito M, Ishikawa K, Taniguchi K, Arima T. Magnetic control of crystal chirality and the existence of a large magneto-optical dichroism effect in CuB2O4. Phys Rev Lett. 2008;101.
[64] Takahashi Y, Shimano R, Kaneko Y, Murakawa H, Tokura Y. Magnetoelectric resonance with electromagnons in a perovskite helimagnet. Nat Phys. 2011;8:121–5.
[65] Bordács S, et al. Chirality of matter shows up via spin excitations. Nat Phys. 2012;8:734–8.
[66] Miyahara S, Furukawa N. Theory of magnetoelectric resonance in two-dimensional S= 3/2 antiferromagnet Ba2CoGe2O7 via spin-dependent metal-ligand hybridization mechanism. J Phys Soc Jpn. 2011;80:073708–073708.
[67] Kézsmárki I, et al. Néel-type skyrmion lattice with confined orientation in the polar magnetic semiconductor GaV4S8. Nat Mater. 2015;14:1116–22.
[68] Ruff E, Widmann S, Lunkenheimer P, Tsurkan V, Bordács S, Kézsmárki I, Loidl A, et al. Multiferroicity and skyrmions carrying electric polarization in GaV4S8. Sci Adv. 2015;1: e1500916–e1500916.
[69] Kurumaji T, Nakajima T, Ukleev V, Feoktystov A, Arima T, Kakurai K, Tokura Y, et al. Néel-type Skyrmion lattice e in the tetragonal polar magnet VOSe2O5. Phys Rev Lett. 2017;119.

Julien Varignon, Nicholas C. Bristowe, Eric Bousquet
and Philippe Ghosez

10 Magneto-electric multiferroics: designing new materials from first-principles calculations

Abstract: In parallel with the revival of interest for magneto-electric multiferroic materials in the beginning of the century, *first-principles* simulations have grown incredibly in efficiency during the last two decades. Density functional theory calculations, in particular, have so become a must-have tool for physicists and chemists in the multiferroic community. While these calculations were originally used to support and explain experimental behaviour, their interest has progressively moved to the design of novel magneto-electric multiferroic materials. In this article, we mainly focus on oxide perovskites, an important class of multifunctional material, and review some significant advances to which contributed *first-principles* calculations. We also briefly introduce the various theoretical developments that were at the core of all these advances.

Keywords: Density Functional Theory, multiferroism, magnetism, ferroelectricity, perovskites, magneto-electricity

Although the magneto-electric effect, that allows the control of a magnetization (polarization) with an electric (magnetic) field and conversely, was proposed at the end of the nineteenth century by Pierre Curie [1], it only really started to be studied during the 1960s and 1970s, after Cr_2O_3 was proposed as good candidate by Landau and Lifshitz [2]. With our lives being more and more dependent on technology, the identification of multifunctional materials handling simultaneously large arrays of information has become a real challenge for producing low energy consumption devices. Magneto-electric compounds thus received a considerable renewal of interest in the early 2000s [3] due to their potential for improved and new technological applications [4–7].

Before *first-principles* calculations became possible, identification of magneto-electric materials was typically guided by symmetry and thermodynamic considerations. According to thermodynamic requirements, the linear magneto-electric effect α is bound by the electric (χ^e) and magnetic (χ^m) susceptibilities through the formula α^2 $\chi^e.\chi^m$ [8]. Since ferroelectric (ferromagnetic) materials usually exhibit the largest electric (magnetic) susceptibilities, it is often expected that the linear magneto-electric effect α should be large in multiferroics, *i. e.* materials combining ferromagnetism and ferroelectricity.

This article has previously been published in the journal Physical Sciences Reviews. Please cite as: Varignon, J., Bristowe, N. C., Bousquet, E., Ghosez, P. Magneto-electric multiferroics: designing new materials from first-principles calculations Physical Sciences Reviews [Online] 2020, 5. DOI: 10.1515/psr-2019-0069

https://doi.org/10.1515/9783110582130-010

Nevertheless, the relationship between α, χ^e and χ^m is just an upper bound and there is in fact no guarantee that multiferroic materials will provide the largest magneto-electric effect. Strong coupling between the orders is an additional key ingredient.

The interest in multiferroic materials also owes its revival to the strong developments of *first-principles* simulations, most notably to Density Functional Theory (DFT). In the 1990s, DFT became the standard platform for studying ferroelectricity in ABO_3 materials and it provided a deep understanding of the origin of ferroelectricity in $BaTiO_3$ and $PbTiO_3$ [9]. Although originally restricted to explain experimental observations, the microscopic understanding enabled by DFT became a key tool for guiding experiments. For instance, DFT studies clarified ferroelectric and piezoelectric properties [9, 10] or finite-size effects appearing in thin films [11–13]. As we will see in the examples tackled in this chapter, interest in magneto-electric compounds focused on ferroelectric magnets and thus, it was natural to direct DFT studies of multiferroics. At the same time, strong advances in computer efficiency were also beneficial for enabling fast, reliable and large-scale studies.

As before for ferroelectricity, DFT initially served as a platform to explain the origin of multiferroism in systems such as in $BiFeO_3$ [14] (lone pair mechanism [15]), hexagonal $YMnO_3$ [16, 17] (lattice improper ferroelectricity, Chapter 3) or $TbMnO_3$ [18] (magnetic improper ferroelectricity, Chapter 4). The numerous successes encountered by *first-principles* simulations established DFT as a powerful tool to design strategies for engineering magneto-electric multiferroic materials. Among all magneto-electric multiferroic materials, many studies considered ABO_3 perovskites, a well-known class of multifunctional materials of technological relevance. These compounds adopt an aristotype cubic structure where the A cations are located at the corner of the cell, the B cations are at its centre and are surrounded each by O_6 octahedra which form a corner sharing network. This structure is very attractive since it can accommodate various chemical elements on the A and B sites of the structure, including magnetic elements. According to the A-to-B cation size mismatch, measured through the Goldschmidt tolerance factor [19] $\frac{t = r_A + r_O}{\sqrt{2}(r_B + r_O)}$ where r is the ionic radius of the different ions, the structure can develop several lattice distortions such as O_6 rotations that in turn can alter the magnetic properties. Since the most popular ferroelectrics are perovskite oxides, it was straightforwardly used by theorists and experimentalists as a natural playground for engineering new properties and functionalities. Finally, developments of experimental techniques such as Pulse Laser Deposition or Molecular Beam Epitaxy allowing atomic-scale deposition of oxides as thin films, superlattices or nanostructures also favoured interest in oxide perovskites. In this article, we will therefore focus on this broad and promising family of compounds although, in practice, much more compounds can be multiferroic.

It is worth noticing that the choice of ABO_3 materials for engineering multiferroism was not *a priori* so judicious. Ferroelectricity and magnetism are expected to be mutually exclusive in ABO_3 materials, a fact formulated in terms of the "d^0" rule by Spaldin [20]. The rule relies on the fact that ferroelectricity in prototypical ferroelectrics such as

$BaTiO_3$ is related to *O 2p* – *B 3d* hybridization and requires empty *d* levels while magnetism requires partially filled *d*-states. The popular multiferroic material $BiFeO_3$ navigates around this rule with ferroelectricity held by the Bi cations at the A-site through a lone pair mechanism and magnetism carried by Fe cations at the B-site [14]. This has motivated the search of materials in which ferroelectricity and magnetism are driven by a different cation such as the double perovskite Bi_2FeCrO_6 [21–23]. Later in this chapter, we will see that in some cases magnetism and polar displacements can nevertheless be held by the same cation [24].

In this article, we first briefly discuss the key theoretical developments that made possible the prediction of novel multiferroic compounds using DFT simulations. We then highlight some strategies seeded by *first-principles* simulations that have been identified for engineering magneto-electric compounds. In fact, all these strategies typically rely on the basic concept of turning a non-polar magnet to a ferroelectric magnet. Namely, we will focus on the use of strain engineering, octahedra connectivity, lattice mode couplings, electronic or magnetic driven ferroelectrics (*i. e.* type II multiferroics) and interfacial systems based on an interface between a ferroelectric and a (anti-)ferromagnet.

10.1 Theoretical aspects

DFT is a powerful computational method aimed at solving the electronic problem. It is based on the Hohenberg and Kohn theorems [25] and the Kohn-Sham ansatz [26], that allows to find the ground state of the N many-body problem, associated with the wave function $|\psi(\vec{r})>$ depending on 3N degrees of freedom, from the sole knowledge of the electronic density $\rho(\vec{r})$ that only depends on three spatial degrees of freedom. In its formulation, DFT is an exact theory. However, the exact form of the functional linking the energy to the density is unknown and must therefore be approximated. During the years, various approximate functionals have been developed which were successful for a wide variety of systems. Conceptualized in the sixties, DFT became a powerful computational tool during the eighties thanks to the advent of efficient supercomputers. Acknowledging this theoretical breakthrough, Walter Kohn was awarded the Nobel Prize in chemistry in 1998. We guide the reader to the textbook of R. Martin for a detailed and comprehensive description of DFT [27] as we will just provide the basic developments that have allowed studies of magneto-electric compounds with *first-principles* simulations. We describe below the standard methodologies used to model and compute ferroelectric, magnetic and magneto-electric properties.

10.1.1 Computing ferroelectric properties

In ferroelectrics, a fundamental quantity is the spontaneous electric polarization, defined in most textbooks as the dipole moment per unit cell. In truly periodic infinite solids as those considered by imposing Born-Von Karman boundary conditions, this quantity is however ill-defined (*i. e.* dependent on the arbitrary choice of the unit cell) which prevented practical computation of the polarization for a long time. The breakthrough arose in 1992 with the modern theory of the polarization established by King-Smith and Vanderbilt as well as Resta who formulated the polarization as a Berry phase [28–31]. Independently, derivatives of the polarization such as the Born effective charges or piezoelectric constants are also accessible by linear response using density functional perturbation theory [30–32]. More recently, methods have also been developed to deal with periodic solids under finite electric or displacement fields [33–35].

10.1.2 Modelling magnetic systems

To study magneto-electric coupling, one obviously has to deal with magnetism. In this specific case, the DFT formulated by Kohn and Sham (Kohn-Sham DFT) can be easily extended to magnetic systems by treating separately spin up and spin down channels. One substitutes the scalar density $\rho(\vec{r})$ by a 2-by-2 $\bar{\bar{n}}(\vec{r})$ density matrix according to the following formula:

$$\bar{\bar{n}}(\vec{r}) \Rightarrow \frac{1}{2}\begin{pmatrix} \rho(\vec{r})+m_z(\vec{r}) & m_x(\vec{r})-im_y(\vec{r}) \\ m_x(\vec{r})+im_y(\vec{r}) & \rho(\vec{r})-m_z(\vec{r}) \end{pmatrix} \tag{10.1}$$

where $m(\vec{r})$ is the magnetic density. In addition to the scalar density $\rho(\vec{r})$ minimizing the energy, the spin-density $\bar{\bar{n}}(\vec{r})$ can also relax the magnetization both in magnitude and direction. At this stage, one must distinguish two cases: (i) the collinear approach reducing the spin polarized density to diagonal components, which is the most widely adopted methodology due to its more tractable computational effort (only $m_z(\vec{r})$ is considered as a scalar and thus spins can only relax in magnitude) and (ii) the non-collinear approach that considers the full 2×2 $\bar{\bar{n}}(\vec{r})$ density matrix that is computationally heavier, due to the presence of several local minima in the energy landscape as a function of relative spin directions (see Section 10.1.3 below).

By including the spin-orbit interaction, one fully considers the coupling between spin and orbital degrees of freedom and has access to spin canting and magnetocrystalline anisotropy. The spin-orbit coupling also allows to access to the orbital magnetization, which has been reached only very recently though a modern theory of magnetization [36–40], analogous to the Berry phase of the theory of polarization.

10.1.3 Approximation of exchange and correlation effects

Although being in principle an exact theory of the ground state, DFT has some limitations in its practical implementation: it relies on an approximation of exchange and correlations effects, which often possess a spurious self-interaction problem (an electron is interacting with its own potential, potentially yielding undesired electron delocalization phenomena). The most common approximations to exchange and correlations are the Local Density Approximation (LDA) and Generalized Gradient Approximation (GGA, adding an additional term to LDA that depends on the density gradients) that have been successful in describing metals and band insulators. However, magneto-electric multiferroic materials involve magnetic cations, thus possessing partially filled *d* (or *f*) shells (*e.g.* for multiferroics based on *3d* elements for instance) and *de facto* a much more localized electronic structure. Standard approximations are usually not appropriate for studying such materials.

At the time of the revival of magneto-electric compounds, one had to use other methods to better treat electron-electron repulsion and cancel the spurious self-interaction term. The most widely adopted method is DFT + U, involving two empirical parameters U and J, corresponding to on-site Coulomb interactions and intra-site spin-exchange, respectively [41]. Both parameters are captured through a Hubbard-like model. We emphasize here that although the terminology "+U" is reminiscent of the Hubbard model, DFT is a single determinant theory with a static mean-field treatment of electron interactions, *i. e.* there are no dynamical correlation effects, and U is totally different to the interelectronic U parameter appearing in the Hubbard model. Typically, one usually fits these U and J parameters by adjusting their value in order to reach the best agreement with known experimental data, although methodologies exist to self-consistently evaluate these values [42, 43]. However, the method is not always fully predictive and one might sometimes rescale the computed parameters [44]. Two different implementations of DFT + U exist, with U and J entering either as different parameters [45] or as a single effective $U_{eff} = U\text{-}J$ parameter [46]. While the latter methodology is usually suited for collinear simulations, within few exceptions [47], both U and J parameters are important when dealing with non-collinear magnetic structures as J acts on the spin canting as demonstrated by Bousquet and Spaldin [48].

In the search of methods alleviating the use of empirical parameters, hybrid functionals also appeared as valuable methodologies. They combine LDA/GGA functionals with a part of exact Hartree–Fock exchange (thus forming "hybrid functionals") in order to better reproduce exchange and correlation effects and cancel the spurious self-interaction term. These functionals are usually providing a better description of structural aspects and band gaps over standard LDA and GGA functionals. The most famous hybrid functionals used for studying multiferroic materials are B3LYP [49] and HSE [50] and its updates. One must also highlight the B1WC functional that was developed to properly reproduce the properties of ferroelectric

ABO_3 materials [51]. Although more computationally demanding than DFT + U methods, these hybrid functionals encountered many successes in modelling multiferroic materials [52–56]. Alternatively, they can be used as a starting point for extracting DFT + U parameters [54, 57]. The very latest developments of meta-GGA functionals (adding an additional term depending on the second derivative of the density to GGA functionals), such as the SCAN functional [58], are producing insulating phases in most *3d* transition metal oxide perovskites and related compounds [60, 61], and thus would be an invaluable method for studying multiferroism in the future.

We end this section by emphasizing that many alternative methods exist for studying multiferroic materials. For instance, Filippetti and Spaldin have developed a simplified Self-Interaction Correction method [62]. Alternatively, a post Hartree–Fock method [63] (Interaction Configuration) has been used to accurately compute the evolution of the magnetic exchange integrals as a function of an applied electric field in oxides with complex magnetic structures [64, 65]. Finally, one can also invoke the most complex GW method including many body effects which was used to benchmark results on $BiFeO_3$ [52].

10.1.4 Computing the contribution to magneto-electric coupling

Now that we have exposed the basic methodologies and developments involved in the study of multiferroic compounds, we can focus more precisely on magneto-electric coupling. For the magneto-electric coupling, one usually refers to the magneto-electric tensor which is a 2 by 2 matrix whose components (i,j) are given by the second derivative of the free energy F with respect to applied electric $\vec{E}$ and magnetic $\vec{H}$ fields:

$$\alpha_{i,j} - \frac{1}{\Omega_0}\frac{\partial^2 F}{\partial E_i \partial H_j} = \frac{\partial P_i}{\partial H_j} = \frac{\partial M_j}{\partial E_i} \tag{10.2}$$

where Ω_0 is the unit cell volume and P_i (M_i) is the polarization (magnetization) along the i axis. Within a Born Oppenheimer approximation, one can decompose the magneto-electric coupling into three contributions corresponding to electronic (α_{el}, pure electronic response at fixed atomic positions and lattice parameters), ionic (α_{ionic}, additional effect of atomic relaxation) and strain (α_{strain}, additional effect from lattice relaxation) parts. Let us remind here that for a magnetic system, the magnetization has two contributions originating from spin (labelled S) and orbital (labelled O), and thus the magneto-electric coupling is expected to have in total six contributions. To the best of our knowledge, the latest developments of DFT have calculated five out of six components of α [66–70]: only α^O_{strain} remains elusive.

Although revival of multiferroics began in the early 2000s, computation of the different components of the magneto-electric tensor only started in 2008. Again, boosted by experience with ferroelectric phenomena, Iñiguez proposed magnetic effective charges Z^m, analogous of Born effective charges Z^e, both given by the change of magnetization (polarization) with respect to an atomic displacement $\vec{u}$ [67]:

$$Z^m_{i,j} = \Omega_0 \frac{\partial M_i}{\partial u_j} \tag{10.3}$$

$$Z^e_{i,j} = \Omega_0 \frac{\partial P_i}{\partial u_j} \tag{10.4}$$

Using these quantities and by noting an analogy with the dielectric tensor, Iñiguez easily expressed the ionic part of the magnetoelectric coupling by the following formula:

$$\alpha^S_{ion,\, i,j} = \frac{1}{\Omega_0} \sum_{n=1}^{N_{IR}} \frac{Z^e_{n,i} Z^m_{n,j}}{K^2_n} \tag{10.5}$$

where K_n are the eigenvalues of the force constant matrix and the sum runs over all infrared modes. K_n, Z^e and Z^m are directly accessible by DFT simulations and hence calculation of α^S_{ion} is rather straightforward. It was then extended to the orbital contribution in 2012 [70]. In the same spirit, Wojdel and Iñiguez extended the formalism to the strain part of the magneto-electric coupling in 2009 [68]:

$$\alpha^S_{strain,\, i,j} = \frac{1}{\Omega_0} \sum_{m,n=1}^{3} e_{i,m} \left(C^{-1}\right)_{mn} h_{j,n} \tag{10.6}$$

where e and h are the piezoelectric and piezomagnetic tensors, respectively, and C is the elastic constant matrix.

In 2011, Bousquet *et al.* identified another pathway for computing the magnetoelectric coefficients [66]: they proposed to evaluate the change of polarization under an applied magnetic field. The authors modelled the magnetic field $\vec{H}$ using a Zeeman term $\bar{\bar{\Delta}} V_{Zeeman}(\vec{r})$ that is added to the DFT scheme. It is represented by a very simple 2 by 2 matrix, reminiscent of the density matrix $\bar{\bar{n}}(\vec{r})$, and it takes the elegant expression:

$$\bar{\bar{\Delta}} V_{Zeeman}(\vec{r}) \Rightarrow \frac{-g}{2} \mu_B \mu_0 \begin{pmatrix} H_z & H_x - iH_y \\ H_x + iH_y & -H_z \end{pmatrix} \tag{10.7}$$

By computing the change of polarization for different finite values of magnetic field, one can directly access the magneto-electric coupling components. Most notably, this technique has the advantage of giving access to the separate ionic and electronic parts of α^S (*i. e.* the electronic part is given at fixed lattice parameters and atomic positions while the ionic part is accessible once the structure is relaxed). While it was usually assumed that it is negligible in most cases, Bousquet *et al.* showed that α^S_{el} is about 25 % of α^S_{ion} in the reference magnetoelectric material Cr_2O_3

[66], an antiferromagnetic (AFM) insulator below $T_N = 307$ K adopting a centrosymmetric *R-3c* space group but whose magnetic structure allows a magnetoelectric effect. Going further, Ricci and Bousquet reported that α^S_{el} can be as large as α^S_{ion} in the troilite phase of FeS [71], which is attributed to the presence of a metal-insulator phase transition in this compound. In their paper, Ricci and Bousquet also reported that diagonal magnetoelectrics (compounds in which the magnetoelectric tensor reduces to diagonal components only) can be seen as being (electrically) polarized in opposite directions for each spin channel (up and down) such that the difference of the polarization contribution of each channel is non-zero (while the sum is exactly zero, FeS being non-polar). This can also be seen as the integration of the magnetic monopolar term (*e. g.* $\int_r m(\vec{r}).\vec{r}.d\vec{r}$), also coined magnetoelectric monopolarization [72]. A full *first-principles* Berry phase formulation and Wannier representation of the magnetoelectric monopole has been developed by Thöle, Fechner and Spaldin [73], which paves the road to systematically compute this original property of magnetoelectric crystals. Finally, the strategy developed by Bousquet and Spaldin was employed in 2016 by Tillack, Yates and Radaelli to disentangle the different contributions – cycloidal, longitudinal and chiral terms – to the magnetoelectric tensor in Cr_2O_3 in the low-field and spin-flop phases [74].

Later on, the computation of magnetic properties as a function of an applied electric field, thanks to the emergence of the modern theory of magnetization, enabled access to the orbital part to the magneto-electric coupling. Again, using Cr_2O_3 as the prototypical example, Malashevich and co-workers showed that the orbital contribution to α (both electronic and ionic parts) are smaller than the spin contribution. However, Scaramucci *et al.* found distinct results in $LiFePO_4$ with rather similar spin and orbital contributions to α_{ion}[70]. So far, and to the best of our knowledge, the orbital contribution to the strain part of the magneto-electric coupling has not been computed.

Similarly to the monopolarization in "diagonal magnetoelectrics", another exotic property can be associated to non-diagonal magnetoelectrics: the bulk magnetic toroidal moment [72]. Identically to the electric toroidal moment, the magnetic (or spin) toroidal moment can be described by $\sum_i \vec{r}_i.\vec{m}_i$, where $\vec{m}_i$ is the magnetic moment at position $\vec{r}_i$. This magnetic toroidal moment is directly associated to the antisymmetric part of the magnetoelectric tensor and it breaks both space and time reversal symmetries together. A new primary ferroic order (on the top of the ferroelectric one, which breaks the space inversion only, of the ferromagnetic one, which breaks the time reversal only and of the ferroelastic one, which is invariant under both space and time reversal), *i. e.* ferrotoroidic, could be defined where the toroidal moment would be an independent order parameter [75, 76]. Recently, Gao, Vanderbilt and Xiao have developed a microscopic theory of spin toroidization for periodic crystal [77] but its implementation in a DFT code is still lacking.

10.1.5 Accessing finite temperature properties

Finally, DFT simulations are performed at 0 K. Complementary methods have been involved in order to alleviate this restriction imposed by DFT simulations. Using Monte-Carlo simulations combined with DFT calculations, Mostovoy *et al.* have considered the spin fluctuations coming from thermal effects [78]. They showed that these exchange-striction effects can break the inversion centre in Cr_2O_3 (that is otherwise centrosymmetric) and induce a polarization, resulting in a magneto-electric effect in the material. Additionally, they demonstrated that the contribution of the spin-orbit interaction to α is one order of magnitude smaller than the exchange-striction effects. These results are supported by quantum chemistry calculations of Varignon *et al.* on $YMnO_3$ that showed that the modification of magnetic exchange integrals under an applied electric field (*i. e.* somehow indicative of the strength of the magneto-electric coupling) are mainly governed by striction effects rather than by the spin-orbit interaction [65].

Finally, *second-principles* methods are constantly on the rise to grant access to finite temperature properties. For instance, effective Hamiltonians of Bellaiche and co-workers gave access to the magneto-electric coefficients and to the origin of the spin-spiral of $BiFeO_3$ [79, 80]. Such a work relies on the effective Hamiltonian approach, initially developed for ferroelectrics during the 90's [81, 82]. Further generalization of effective Hamiltonians has been proposed in order to treat explicitly all structural degrees of freedom from the construction of first-principles based effective atomic potentials [83], as implemented in SCALE-UP [www.secondprinciples.unican.es/spdft] and MULTIBINIT [www.abinit.org] software. More recently, the method has even been extended in order to reintroduce electronic degrees of freedom [84] and perform second-principles simulations at finite temperature while considering explicitly lattice and electron degrees of freedom. This opens new perspectives for the modelling of magneto-electric multiferroics at the mesoscale and finite temperatures.

10.2 Strain engineering

10.2.1 Inducing ferroelectricity in magnets

The advent of experimental growth techniques allowing atomically-controlled deposition of epitaxial oxides thin films has enabled promising pathways for engineering new physics and phenomena with ABO_3 oxides. In 2004, the identification of room-temperature ferroelectricity under moderate tensile epitaxial strain in $SrTiO_3$, an otherwise quantum paraelectric material in bulk, offered promising perspectives for multiferroism [85]. It became straightforward to use such a strategy to induce ferroelectricity in magnetic materials.

This strategy is universal and relies on the coupling between the ferroelectric polarization P and the homogeneous strain η. In the high-symmetry cubic cell of ABO_3 compounds, the lowest-order $P-\eta$ coupling term to the free energy takes the form $g\eta P^2$, so that we can write:

$$F(P,\eta) \propto (a+g\eta)P^2 + bP^4 + \varsigma\eta^2 \propto a_{eff}P^2 + bP^4 + \varsigma\eta^2 \tag{10.8}$$

where a, b, c, ς and g are coefficients. For non-polar compounds in bulk (*i. e.* $\eta = 0$), the coefficient a is positive, and the energy as a function of a polar displacement P is characterized by a single well energy potential whose minimum is located at $P = 0$ (red curve of Figure 10.1). For a “proper ferroelectric” in bulk, the coefficient a is negative and the system is then characterized by a double well potential whose minimum is now located at $P \neq 0$ (*i. e.* $+P_S$ and $-P_S$) (blue curve in Figure 10.1) [87]. For a non-polar material in thin film form, the effect of the epitaxial strain η is to renormalize the coefficient of eq. (10.8) into a_{eff}. For η sufficiently large and with appropriate sign, a_{eff} can become negative and produce a ferroelectric behaviour. Thus, turning a non-polar compound to a ferroelectric is *a priori* feasible for any material *via* strain engineering whatever the sign of the coefficient g is: if $g > 0$, one would use compressive strain (*i. e.* $\eta < 0$) and if $g < 0$, one involves tensile strain (*i. e.* $\eta > 0$). However, the strength of the $P-\eta$ coupling is different for each material and not necessarily large so that being already at the verge of a ferroelectric transition (*i. e.* having a polar phonon mode at relatively low frequency) is often a prerequisite to destabilize the system at experimentally achievable epitaxial strains.

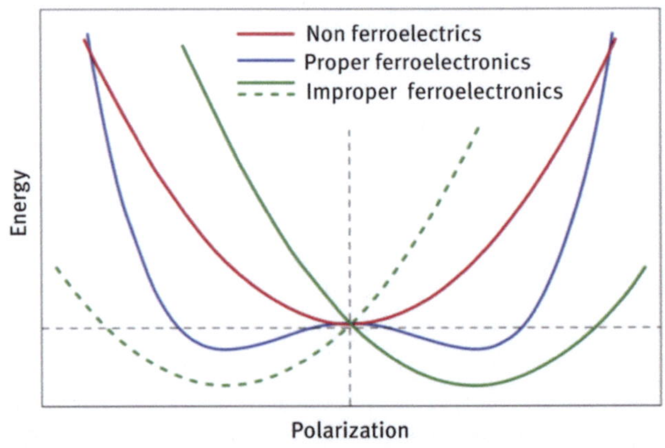

Figure 10.1: Energy potential as a function of a polar displacement. Figure extracted from Ref. [86].

This powerful strategy encountered many successes and most notably, it has been employed to demonstrate that magnetism and ferroelectricity are not mutually exclusive but can coexist in $CaMnO_3$ [24]. In bulk, $CaMnO_3$ is a G-type AFM insulator adopting a *Pnma* symmetry characterized by the usual $a^-b^+a^-$ oxygen cage rotations in Glazer's notation [88] (see Figure 10.2(d) for sketched of magnetic orders). However, on the basis of *first-principles* simulations, Bhattacharjee, Bousquet and Ghosez have identified in the cubic phase that $CaMnO_3$ does exhibit both weak ferroelectric and large oxygen cage rotation imaginary phonon frequencies ω – one recalls here that the curvature of the energy E is proportional to ω^2. While the ferroelectric mode instability is suppressed by the rotations in bulk, they proposed that appropriate strain engineering can make $CaMnO_3$ ferroelectric. This scenario was verified experimentally in 2012 by applying a 2.3 % tensile strain to the material [90]. This result attracted some attention since it was showing for the first time that both ferroelectricity and magnetism can be held by the B site cations [24], thus breaking somewhere the "d^0" rule.

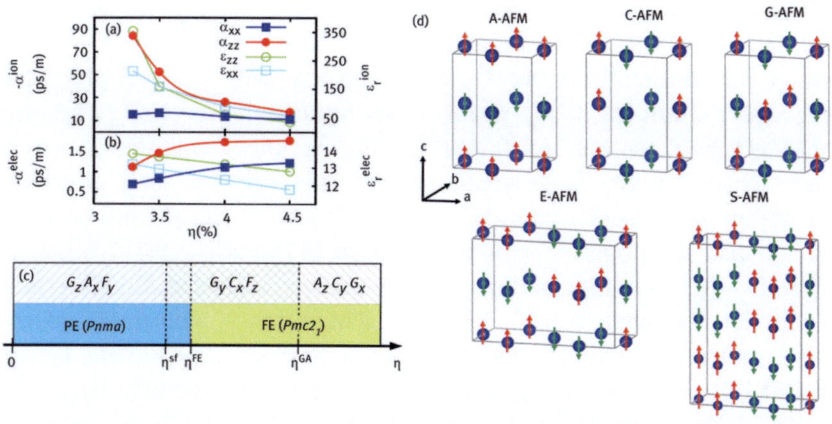

Figure 10.2: Multiferroic properties of $CaMnO_3$ under strain engineering. xx (squares) and zz (circles) components of the magneto-electric (filled symbols) and dielectric (open symbols) constant versus the epitaxial strain for ionic (a) and electronic (b) contributions. (c) Structural, ferroelectric and magnetic properties of $CaMnO_3$ versus epitaxial strain. $\eta^{sf} = 2.6\,\%$ is the strain value needed for inducing a spin-flop transition, $\eta^{FE} = 3.2\,\%$ corresponds to the strain required for the non-polar – ferroelectric transition and at $\eta^{GA} = 4.6\,\%$, the magnetic order switches from a G-type to a A-type. (d) Sketched of magnetic orders appearing in orthorhombic perovskites. The orthorhombic structure corresponds to a $(\sqrt{2}a_c, \sqrt{2}a_c, 2a_c)$ cubic cell where a_c is the cubic cell lattice parameter. Panels (a), (b) and (c) are adapted from Ref. [89].

Being ferroelectric and magnetic does not necessarily imply that there is a large magneto-electric effect in the material. The existence of a linear magneto-electric effect in strained $CaMnO_3$ was verified by Bousquet and Spaldin in 2011 [89]. In any material adopting a *Pnma* symmetry with a G-type AFM order, a small canting of spins is allowed thus permitting weak ferromagnetism along the z axis (see Ref. [91] for a detailed description of non-collinear magnetism). Nevertheless, the inversion centre existing in the *Pnma* symmetry forbids any magneto-electric effect [92]. By using strain engineering, Bousquet and Spaldin have shown that $CaMnO_3$ adopts a *Pmc2₁* ferroelectric phase at 3.2 % of tensile epitaxial strain (see Figure 10.2(c)). Although the easy axis rotates with respect to the *Pnma* symmetry at 2.6 % of strain, a linear magneto-electric effect is allowed with ionic and electronic components larger than that exhibited by Cr_2O_3 (see Figure 10.2(a) and 10.2(b)) [89]. For tensile strains larger than 4.6 %, another magnetic phase transition from a G-AFM to a A-AFM order at constant structural symmetry occurs but a linear magneto-electric effect is not anymore allowed (Figure 10.2(c) and 10.2(d)). This symmetry analysis is totally general and may apply to any material adopting a *Pnma* symmetry, which is the most common symmetry adopted by ABO_3 perovskites. A recent group theory analysis led by Senn and Bristowe has enumerated the possible couplings between polarization and magnetism in ABO_3 perovskite oxides [93].

In addition to unlocking ferroelectricity in some otherwise non-polar materials in bulk, strain engineering can also unveil unexpected ferroelectric phases in perovskites. For instance, it has led to a rich ferroelectric phase diagram as a function of the applied epitaxial strain in $BiFeO_3$, including a super-tetragonal phase for highly compressive strain [94, 95] (T phase briefly discussed in Section 10.2.3, see chapter on $BiFeO_3$ by Bibes *et al.* for further details). Another "super-ferroelectric" phase is also achieved under large tensile in ABO_3 ferroelectrics, such as $PbTiO_3$ [96, 97], or otherwise non-polar compounds in bulk such as $SrTiO_3$ and $BaMnO_3$ [98]. Strain can also be applied to superlattices such as $(SrCoO_3)_1/(SrTiO_3)_1$ [99] or $(PbTiO_3)_1/(SrTiO_3)_1$ [100] for engineering various ferroelectric phases. Finally, strain engineering is not restricted to ABO_3 materials. For instance, Bousquet and co-workers have proposed that it can yield ferroelectricity in simple magnetic binary oxides like EuO [101] or multiferroicity in fluorite ABF_3 compounds [102, 103].

10.2.2 Inducing magnetism in ferroelectrics

Additionally to induce ferroelectricity, Spaldin and co-workers have unveiled that strain also favours the formation of oxygen vacancies in oxides films such as in $CaMnO_3$ [104, 105] (the inverse effect was also shown theoretically and experimentally in $PrVO_3$ thin films [106]). This additional lever offers nice perspectives for unlocking new functionalities in oxides and for engineering multiferroism [107]. For instance, on the basis of DFT calculations, Xu *et al.* have shown that strain can favour the appearance of defects in the ferroelectric non-magnetic $PbTiO_3$, giving rise to magnetism and thus to multiferroism [108].

10.2.3 Structural softness

Equations (10.5) and (10.6) proposed by Iñiguez and co-workers clearly suggest a route to design compounds with large magneto-electric effects: vanishingly small values of K_n or C will produce diverging behaviours for α [68] and thus large magnetoelectric response. This is coined as a “structural softness”. Since the dielectric constant ε_{ion} also diverges, positioning the material close to a ferroelectric phase transition may result in the enhancement of α. Iñiguez and Wojdel proposed to use the rhombohedral (R) and tetragonal (T) polymorphs of $BiFeO_3$ as a proof of concept. As a function of the applied compressive epitaxial strain, they found that the transition from R to T phase occurs at 4.4 %. Nevertheless, the authors observed a strong enhancement of the magneto-electric coefficient for the R phase at 6 % strain (see Figure 10.3), although not the ground state at this strain level, which is ascribed to a phonon with a vanishingly small force-constant.

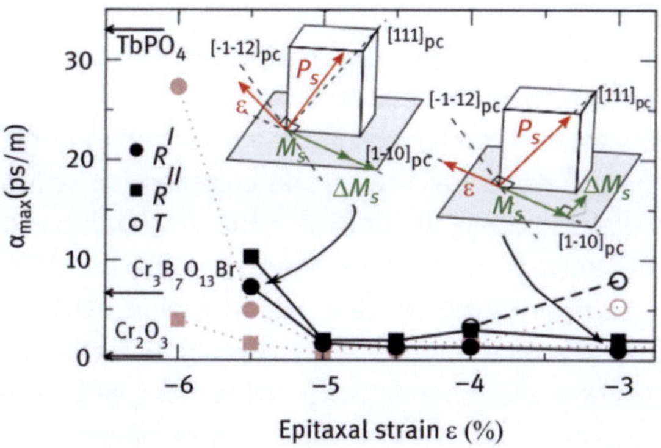

Figure 10.3: Magneto-electric response of the rhombohedral R (filled symbols) and tetragonal T (open symbols) phases as a function of the epitaxial strain. Figure extracted from Ref. [109].

10.2.4 Phase competitions through spin-lattice coupling

In magnetic materials, exchange interactions are directly affected by changes of structural aspects such as modifications of B-O bond lengths and/or B-O-B bond angles. Consequently, they are sensitive to the optical phonon modes appearing at the zone centre. The frequency ω associated with these phonon modes are in turn sensitive to spin orders. This phenomenon is coined as “spin-lattice coupling” and the frequency ω can be written as [110–112]:

$$\omega \propto \omega_{PM} + \gamma \vec{S_i}.\vec{S_j} \tag{10.9}$$

where ω_{PM} is the frequency in the spin-disordered paramagnetic (PM) phase, $\langle \vec{S_i}.\vec{S_j} \rangle$ is the nearest neighbour spin-spin correlation function in the cell and γ is the spin-lattice constant. We can simply express the free energy as a function of the polar P and the ferromagnetic L (antiferromagnetic M) order parameters:

$$F(P,L) \propto +\gamma' P^2 L^2 \tag{10.10}$$

$$F(P,M) \propto -\gamma P^2 M^2 \tag{10.11}$$

It appears that, similarly to strain in eq. (10.8), the magnetic order can produce a significant renormalization of a ferroelectric phonon frequency. In this case however, the coupling is bi-quadratic (see also Section 10.4.3 for a similar example of lattice mode coupling) so that induce polarization will mandatorily require a positive coefficient. In such a case, starting from a non-polar material, a sufficiently large γ can produce a sizable renormalization of the zone-centre mode frequencies and induce ferroelectricity in the material.

This mechanism can be coupled to strain engineering to bring the material close to a phase competition between two magnetic states. This is the strategy employed by Fennie and Rabe in $EuTiO_3$, an AFM material adopting a non-polar cubic structure at all temperatures [112]. As a function of the biaxial compressive strain applied to the material, the authors observed that the critical value to destabilize a ferroelectric phonon mode is smaller for a ferromagnetic state than for the AFM configuration (see Figure 10.4(a)). It follows that there is a narrow region between 0.9 % and 1.2 % of biaxial strain where the material possesses competing ferroelectric-ferromagnetic (FE-FM) and antiferromagnetic-paraelectric (AFM-PE) (see Figure 10.4(b)). Locating the material in this narrow compressive strain region offers a control of magnetic properties with an electric field: applying a magnetic field aligns the spins ferromagnetically (FM), and thus the material becomes a ferroelectric. Conversely, applying an electric field will force a ferromagnetic alignment of the spins. We emphasize that the ground state of $EuTiO_3$ remains AFM-PE until 1.2 % of biaxial strain (Figure 10.4(b)). The existence of the FE-FM phase at larger strain values was confirmed experimentally [113] although a tensile strain was involved – we highlight that the effect of the epitaxial strain is similar for compressive and tensile strain, the polar axis is just rotated by 90° between the two strain states.

Although a nice proof of concept, the low Néel temperature of $EuTiO_3$ ($T_N \propto$ 5.5 K) hinders practical use of the material. Going from materials involving *4f* elements to materials implying *3d* elements is a good alternative. Lee and Rabe have proposed that $SrMnO_3$, an AFM material with a T_N around 250 K in bulk, exhibits a large spin-phonon coupling and adopts a ferromagnetic-ferroelectric ground state at large epitaxial strain [114]. Large spin-phonon coupling was also observed in materials with magnetism carried by both *4f* and *3d* elements [115]. Finally, strong coupling between magnetism and lattice degrees of freedom is not restricted to ABO_3 materials with a perovskite structure. It has been also observed in hexagonal materials such as $YMnO_3$ [116, 117] or in MnF_2 [118].

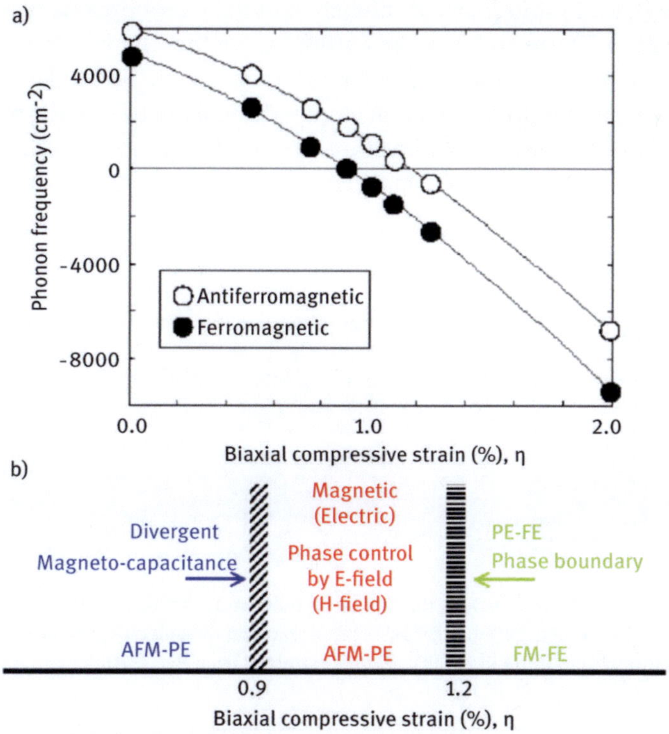

Figure 10.4: (a) Phonon frequency of a ferroelectric mode as a function of the epitaxial strain for a ferromagnetic (filled circles) and an antiferromagnetic (open symbols). (b) Phase diagram of $EuTiO_3$ as a function of the epitaxial strain. PE and FM stand for paraelectric and ferromagnetic, respectively. Figures extracted from Ref. [112].

10.3 Oxide interfaces

When computing the phonon dispersion curves of most ABO_3 compounds in their ideal high symmetry cubic phase, one often identifies several unstable phonon frequencies associated with polar zone-centre modes. This is the case, for instance, in $CaTiO_3$ (see Figure 10.5(a)) [119]. However, one observes other unstable phonon modes associated with oxygen cage rotations, also called antiferrodistortive motions, located at the M and R points of the Brillouin zone – $a^0b^+a^0$ and $a^-b^0c^-$ rotations in Glazer's notation [88], respectively. They usually produce larger energy gains than the unstable ferroelectric modes and, when appearing, in turn usually compete with and suppress the ferroelectric instability [119] (through a bi-quadratic energy coupling of the form $F \propto g\varnothing^2.P^2$ with $g > 0$ and $\varnothing$ a rotation). Weakening the oxygen rotations should reversibly favour and reintroduce the ferroelectric instability. This physical behaviour was highlighted by Kim *et al.* in $NdNiO_3$, adopting in bulk at low temperatures an AFM and non-polar insulating structure characterized by oxygen cage rotations [120]. In thin films, as a function of the rotation mode amplitude (labelled tilt angle in Figure 10.5(b)), the authors show that the two ferroelectric phonons modes initially stable in the "bulk" (*e.g.* for a tilt angle of approximately 11°) acquire an imaginary frequency upon decreasing the rotations amplitude (Figure 10.5(b)) – we recall that curvature of $E \propto \omega^2$. Consequently, by reducing the amplitude of rotations of O_6 motions, the material can become ferroelectric.

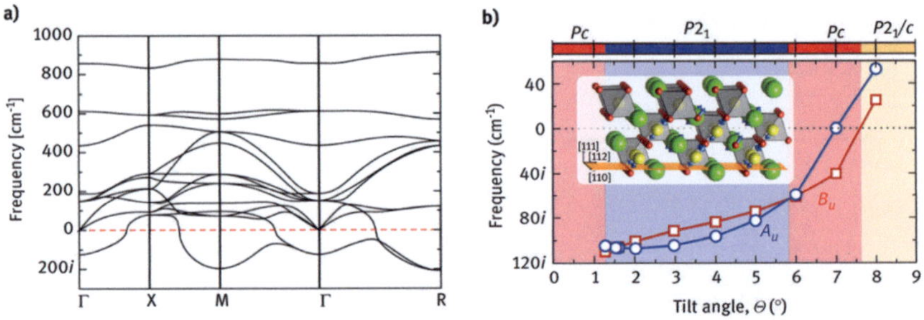

Figure 10.5: (a) Phonon dispersion curve of $CaTiO_3$ in its high symmetry Pm-3 m cubic cell. Figure extracted from Ref. [119]. (b) Evolution of the frequency associated with two ferroelectric phonon modes as a function of the rotation amplitude in $NdNiO_3$. Figure extracted from Ref. [120].

To verify this experimentally, a practical strategy consists in interfacing the material of interest with another perovskite that possesses weaker O_6 groups rotations. At the interface between the two compounds, O positions are locked and the amplitude of antiferrodistortive motions in the material of interest is decreased. In order to maximize the interfacial effect, (111)-oriented films are preferred since three O positions are locked at the interface instead of 1 and 2 for (001) and (110) oriented films, respectively (see Figure 10.6(a)). This is the strategy employed by Kim *et al.*: they interfaced $NdNiO_3$ with $LaAlO_3$, a band insulator with small O_6 rotations. They observe that AFD motions are strongly altered near the interface (see Figure 10.6(b)). Second Harmonic Generation measurements then revealed the existence of polar domains. Combined with its AFM state at low temperature, $NdNiO_3$ is a multiferroic compound. Yet, no report of a magneto-electric effect has been reported nor measured to the best of our knowledge.

Since most ABO_3 perovskites adopt the orthorhombic *Pnma* symmetry – or lower symmetry – characterized by the usual octahedral rotations, growth of oxide interfaces is an appealing strategy to design multiferroic compounds with potentially large magneto-electric effects.

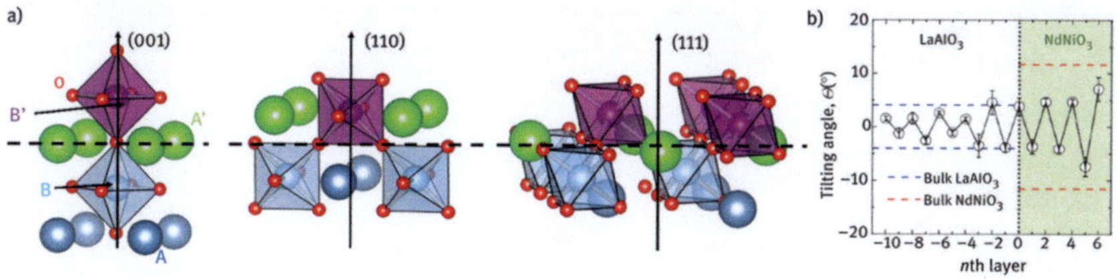

Figure 10.6: (a) Octahedra connectivity at oxide interfaces according to the different growth orientations. (b) Evolution of the rotation angle near the interface between (111) oriented $LaAlO_3$/$NdNiO_3$ interfaces. Bulk angles correspond to the dashed lines. This figure is extracted from Ref. [120].

10.4 Lattice mode couplings

10.4.1 The concept of lattice improper ferroelectricity

So far, we have discussed strategies to induce ferroelectricity by destabilizing an otherwise "stable polar phonon", *i. e.* acting directly on the energy curvature at the origin (from red to blue curve in Figure 10.1). Nevertheless, some materials are ferroelectric despite the fact that the curvature of the ferroelectric mode remains positive at the origin (from red to green curve in Figure 10.1).

$YMnO_3$ is a prototypical example of that type. It is a multiferroic compound that adopts a hexagonal symmetry instead of the usual cubic variants due its very small tolerance factor. At 1258 K, this material undergoes a structural phase transition from a centrosymmetric $P6_3/mmc$ phase to a polar $P6_3cm$ ground state, involving a unit cell tripling. Fennie and Rabe have evidenced that there isn't any polar instability in the centrosymmetric $P6_3/mmc$ phase [17] (see Figure 10.7 upper panel). Instead, they identified that a non-polar K_3 mode at the zone boundary is unstable in the high-symmetry phase, further being responsible of the phase transition and of the unit cell tripling. Interestingly, they identified the following coupling term between the polarization P and the non-polar K_3 mode in the free energy:

$$F \propto c.Q^3_{K_3}.P \tag{10.12}$$

where Q is the amplitude of distortion associated with the K_3 mode and c is a coefficient. Since this term is linear in P, at the structural phase transition, the condensation of the K_3 mode will progressively shift the single well associated with the polar mode to lower energy and finite P amplitude (see Figure 10.7). The polarization is therefore not intrinsically unstable but appears to be a slave of the primary non-polar order parameter. Materials exhibiting such behaviour are coined as "improper ferroelectrics" (green curves in Figure 10.1). Improper ferroelectrics have physical properties distinct from those of proper ferroelectrics: (i) switching the polarization

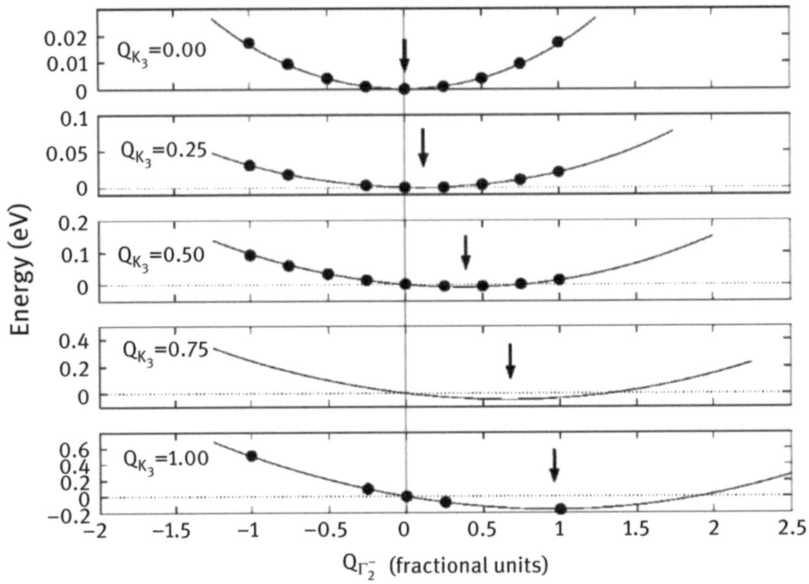

Figure 10.7: Energy potential as a function of the polar distortion at fixed amplitude of the K_3 mode in $YMnO_3$. This figure is extracted from Ref. [17].

will necessarily require the reversal of the primary non-polar distortion (green dashed line in Figure 10.1); (ii) they do not show divergence of their dielectric properties at the phase transition [121] (*i. e.* they don't obey the Curie-Weiss law since the polar mode does not soften) and (iii) they are less sensitive to depolarizing field issues [11, 122, 123].

After this mechanism was revealed in $YMnO_3$, Varignon and Ghosez proposed that $BaMnO_3$ in its 2H hexagonal polymorph also exhibits improper ferroelectricity despite totally different atomic structure and formal oxidation states of cations with respect to $YMnO_3$ [56]. The unit cell tripling in 2H-$BaMnO_3$ was observed experimentally by the group of Kamba but the improper nature of the polarization couldn't be confirmed nor implied [124]. In 2017, Xu *et al.* have proposed general strategies to design new improper ferroelectrics using ABO_3 perovskites as a platform [93]. Most notably, they identified that the otherwise centrosymmetric *R-3c* structure (hexagonal symmetry adopted by $LaAlO_3$ for instance) can exhibit improper ferroelectricity by substituting half of the B site cations using $(ABO_3)_2/(AB'O_3)_2$ superlattices [125]. It results in a polar C_2 symmetry possessing a linear coupling between P and the amplitude Q of a rotation of the form $F \propto P.Q$ in the free energy expansion F starting from a high symmetry *R-3m* phase. This mechanism is identical to that previously established by Young and Rondinelli in 2014 in (111)-oriented $(RAlO_3)_1/((PrAlO_3)_1$ superlattices with R = La or Ce, *i. e.* superlattices based on non-polar materials adopting a *R-3c* structure [126]. Finally, ABO_3 improper ferroelectrics with a coupling between P

and another non-polar motion remains scarce and to the best of our knowledge, these ABO_3 improper ferroelectrics have so far only involved hexagonal systems.

10.4.2 Hybrid improper ferroelectricity

As we already stated in Section 10.3, the oxygen rotations appearing in *Pnma* phases – or lower symmetries – are usually annihilating ferroelectric instabilities identified in many perovskites. To be more precise, Benedek and Fennie have demonstrated that, beyond rotation themselves, the additional presence of A site cation anti-polar displacements minimizing A-O repulsion further suppresses the ferroelectric instability [119]. These A-site anti-polar displacements, associated to the X point of the Brillouin zone and labelled A_X (see Figure 10.8(a)) are usually not intrinsically unstable in the cubic phase but are boosted by the joint presence of out-of-phase $a^-b^0c^-$ (labelled $\varnothing_1$) and in-phase $a^0b^+c^0$ (labelled $\varnothing_2$) rotations in Glazer's notations [88] thanks to the following term in the free energy expansion:

$$F \propto e\varnothing_1\varnothing_2 A_X \tag{10.13}$$

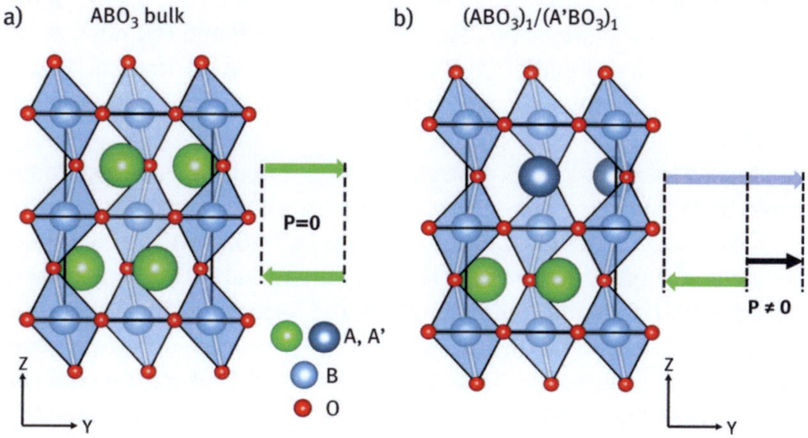

Figure 10.8: Illustration of anti-polar displacements in bulk ABO_3 compounds (a) and ferri-like polar distortions in $(ABO_3)_1/(A'BO_3)_1$ superlattices (b).

These anti-polar motions create locally on each plane a net polarization that is nonetheless cancelled by the very same motion pointing in the opposite direction on the consecutive plane along the z axis, resulting in no net polarization in the material (Figure 10.8(a)). In 2012, Rondinelli and Fennie proposed to substitute half of the A site cations by using $(ABO_3)_1/(A'BO_3)_1$ (001)-oriented superlattices [127]: since A and A' cations do not have similar masses and charges, they produce different local polarizations on consecutive planes along the z axis (see Figure 10.8(b)). Consequently, it results in a net polarization in the material that is coupled to the two rotations by the following unusual trilinear term in the free energy expansion

$$F \propto e' \varnothing_1 \varnothing_2 P \tag{10.14}$$

Since this term is linear in P, the joint condensation of $\varnothing_1$ and $\varnothing_2$ modes will progressively shift the single well associated with the polar mode and produce a spontaneous polarization, similarly to what was previously discussed for improper ferroelectrics (eq. 10.12). This time, however, the joint action of two distinct modes is required to drive the polarization.

The appearance of a spontaneous polarization through such a trilinear term was first discovered by Bousquet *et al.* [128] in 2008 on $SrTiO_3/PbTiO_3$ superlattices – while involving slightly different rotation patterns of octahedra and orientation of the polarization – and is nowadays referred to as "hybrid improper ferroelectricity" (HIF) [129].

HIF is not restricted to 1/1 superlattices and to octahedral rotations and it has since 2008 been extended to (i) 2/2 superlattices involving identical A cations but two different B cations such as $(YFeO_3)_2/(YTiO_3)_2$ superlattices [131]; (ii) thin film forms of oxides such as highly strained $BiFeO_3$, $PbTiO_3$, $SrTiO_3$ or $CaTiO_3$ [97, 98, 131]; (iii) hybrid organic-metal organic perovskites [132, 133]; (iv) double perovskite systems such as $NaLaMnWO_6$ [134] or $RLaMnNiO_6$ [135] (R = Lu-La, Y); (v) Jahn-Teller motions or anti-polar motions [97, 98, 131–133, 136–139] and (vi) other types of layered perovskites such as Ruddlesden-Popper [129, 140, 141], Aurivilius [142] or Dion-Jacobson [143, 144] (we invite the reader to Ref. [145] for a detailed discussions of these materials).

Hybrid improper ferroelectricity is nowadays a well-established strategy to design multiferroic materials since many (magnetic) perovskites adopt the required tilt-pattern of O_6 groups: one can cite layered titanates $(A^{2+}TiO3)_1/(R^{3+}TiO3)_1$ (A and R are an alkaline-earth and a rare-earth element, respectively) that would form ideal multiferroics with coexisting ferromagnetism and ferroelectricity [57] or double layered $(R_2NiMnO_6)_1/(La_2NiMnO_6)_1$ systems that are predicted to be near-room temperature multiferroic compounds [135]. Although *a priori* a theoretical concept, hybrid improper ferroelectricity has been observed experimentally in $(Ca,Sr)_3Ti_2O_7$ and $(Ca_ySr_{1-y})_{1.15}Tb_{1.85}Fe_2O_7$, Ruddlesden-Popper compounds (naturally layered perovskites) [146, 147].

Several theoretical and experimental efforts are still devoted to understand HIF in perovskites and related materials and to rationalize rules maximizing the polarization [57, 100, 125, 127, 140, 141, 145, 148–155]. One key point is to understand the phase transition of these materials: are there single or multiple phases transitions or is it an avalanche effect with all modes appearing at once [148]? Another key aspect is the polarization reversal mechanism: since the trilinear term (eq. 10.14) imposes to switch one of the non-polar distortion together with the polarization to keep the energy invariant, the switching path is not trivial and should be better clarified. Even, the ability to reverse the polarization in hybrid improper ferroelectrics is questionable [129, 156–158].

In addition to producing ferroelectricity in otherwise non-polar materials, HIF is a promising "knob" to tune large magneto-electric responses as suggested by Bousquet *et al.* in their seminal work of 2008 [128]. As highlighted above, in such systems, the reversal of the polarization *P* necessarily requires the switching of a non-polar lattice distortions (either $\varnothing_1$ or $\varnothing_2$) that can be coupled to the magnetism. Fennie and Benedek realized this in $Ca_3Mn_2O_7$ [129]: using *first-principles* calculations, they demonstrated that the two O_6 group rotations not only induce the polarization but that they also produce weak ferromagnetism and a linear magneto-electric effect. Thus, switching the polarization can reverse the magnetization by the reversal of one rotational mode. A similar mechanism has been exploited by Zanolli *et al.* in (001)-oriented $(BiFeO_3)_1/(LaFeO_3)_1$ superlattices [158]. However, we emphasize again that the switching path remains a fundamental aspect for these materials [156].

Since the initial exploitation of rotational modes, other works proposed to harness other lattice distortions intimately coupled to the electronic structure or to magnetism. A key lattice distortion behind many properties of perovskites is the Jahn-Teller motion: it can lift orbital-degeneracies of ions and affect the magnetic-interactions through strongly entangled spin-orbital degrees of freedom [159]. In 2015, Varignon, Bristowe, Bousquet and Ghosez exploited the two competing Jahn-Teller (JT) motions, each forcing a dedicated magnetic order, of rare-earth vanadates to unveil a ferroelectric control of magnetism in (001)-oriented 1/1 superlattices [136] *via* an electrical control of the JT modes. Although no effect on magnetic orderings was discussed, some of these authors also proposed a ferroelectric control of JT distortions in highly strained oxide perovskites [98].

10.4.3 Embedding ferroelectricity and octahedra rotation in a single phonon mode

While the switching pathway is the main drawback of improper and hybrid improper ferroelectrics to have the polarization/magnetization coupling in perovskites, Garcia-Castro *et al.* reported that a solution to this problem is to entangle octahedra rotations and ferroelectric polarization in a single soft phonon mode [160]. They reported such a possibility in $BaCuF_4$, which is polar directly below its melting temperature (1000 K) and is a canted AFM below its Néel temperature ($T_N = 275$ K) with weak ferromagnetism, thus making $BaCuF_4$ multiferroic. In their DFT calculations, Garcia-Castro *et al.* showed that reversing the polarization always reverses the weak ferromagnetic moment in $BaCuF_4$. This systematic flipping of the magnetization with the polarization is attributed to the presence of both polar distortions and octahedra rotations in a single zone centre polar unstable mode of the high symmetry phase *Cmcm*. This means that polarization and octahedra rotations are not coming from the coupling of two different phonon modes but are instead entangled within a single eigenvector, which also means that reversing this single eigenvector necessarily drive the reversing of all the associated distortions (*e. g.* polar displacements plus octahedra rotations). In such situation the coupling between polarization and octahedra rotation is thus ideal and since the weak ferromagnetism is driven by the octahedra rotation, there is also a perfect reversing of the weak ferromagnetism with the polarization. This case thus avoids the drawback of improper ferroelectricity where there is no guaranty to switch the octahedra rotation driving weak ferromagnetism with the polarization. This result thus opens a new direction for seeking new multiferroism in perovskites related materials where the polarization is directly coupled to the weak ferromagnetism.

10.4.4 Triggered-like ferroelectrics

While most perovskites adopt an orthorhombic *Pnma* cell with $a^-b^+a^-$ oxygen cage rotations, some compounds crystallize in a ferroelectric *R3c* structure characterized by $a^-a^-a^-$ oxygen cage rotations due to a tolerance factor largely deviating from 1. This is the case of $ZnSnO_3$ that, surprisingly, exhibits an additional ferroelectric distortion that cannot origin from a lone pair mechanism (Zn^{2+} cations do not have a lone pair, *e. g.* unlike $BiFeO_3$) or a "d^0" rule (Sn^{4+} cations do not possess *d* electrons in conduction band, *e. g.* unlike $BaTiO_3$).

In cubic perovskites, oxygen rotations and polarization can only couple together at even orders yielding therefore the following terms in the energy expansion:

$$F \propto \left(a + b\varnothing^2 + c\varnothing^4 + \ldots\right)P^2 \propto a_{eff}P^2 \tag{10.15}$$

As previously discussed (see Section 10.3), oxygen rotations and polarization typically compete at the bi-quadratic level (*i. e.* $b > 0$ in eq. 10.15) so that appearance of oxygen rotations *a priori* suppresses the ferroelectric distortion. In Ref. [161], Gu *et al.* have shown that, in $ZnSnO_3$, the higher-order coupling term $\varnothing^4P^2$ in eq. 10.15 has, however, a coefficient $c < 0$. This means that, although oxygen rotations and polarization indeed compete at small rotations amplitudes at which the $b\varnothing^2$ contribution dominates, they can nevertheless cooperate at large enough rotation amplitudes for which the negative $c\varnothing^4$ contribution to a_{eff} dominates. They further linked such dual nature of the interaction between oxygen rotations and polarization to steric effects. This highlights that in ABO_3 compounds showing small tolerance factors and large oxygen rotations, the latter can soften the polarization energy well and induce a ferroelectric instability through a so far overlooked cooperative anharmonic coupling $\left(c\varnothing^4P^2\right)$. Since some of these compounds possibly exhibit magnetic properties, this opens a new route to identify promising candidates for multiferroism.

Such a progressive renormalization of the curvature of the polarization energy well by a non-polar distortion is reminiscent of the triggered mechanism proposed by Holakovsky [162]. However, in a "so-called" *triggered* ferroelectric phase transition, the destabilization of P has to be produced by a negative bi-quadratic coupling ($b < 0$) to the primary order parameter while here it is linked to a higher-order term, which will affect the temperature behaviour and related properties. To keep the classification rigorous, we therefore prefer to speak here of a triggered-like transition. We notice that, to date, no example of a real triggered ferroelectric transition has been reported in simple ABO_3 perovskites. To our knowledge the only example of triggered structural phase transitions in simple perovskite is that recently highlighted in rare–earth nickelates $RNiO_3$ (R = Lu-Pr, Y), which concerns the triggering of a breathing distortion of the oxygen cages by oxygen rotations [163, 164].

10.5 Spin, charge and orbital induce ferroelectricity

Up to now, we have addressed systems in which the ionic displacements drive the ferroelectric flavour of the material, irrespective of the proper or improper nature of the polarization. Such phenomenon in magnets produces what one often calls type I multiferroic compounds. However, there are other materials in which the electrons themselves produce the polarization. Although the polarization is often rather small in these compounds (of the order of few $nC.cm^{-2}$) with respect to that of "conventional" ferroelectrics (of the order of few $\mu C.cm^{-2}$), the strong coupling between P and the electronic structure (charge, spin or orbital degrees of freedom) warrants large magneto-electric responses. The electronic rather than ionic origin of P also potentially enables faster switching processes. DFT simulations, aiming at solving the electronic problem, are perfectly suited for studying electronically driven ferroelectrics. We present here the key results established on the basis of DFT simulations and we refer the reader to the reviews of Bousquet and Cano [91] as well as of Barone and Yamauchi [165] for further details.

10.5.1 Magnetically induced ferroelectricity

In some materials, often called type II multiferroics, the magnetic structure itself breaks the inversion symmetry and produces a spontaneous polarization. Such compounds are also called improper ferroelectrics since the polarization is not intrinsically unstable (single well in Figure 10.1) but slave of another primary order parameter, which drive the polarization well to lower energy. Such a mechanism is similar to that explained in Section 10.5.1, except that, there, the primary order parameter was a non-polar lattice mode, while here it is related to the magnetic order.

An example of this type is the ↑↑↓↓ spin chains appearing in the (*ac*)-plane of the *Pbnm* structure of some rare-earth manganites $RMnO_3$ [166, 167] (R = Lu, Tm, Er, Ho, Y). This magnetic order, also called E-AFM order (Figure 10.2(d)), is based on nearest neighbor FM and next-nearest neighbor AFM interactions and it can break the inversion center of the *Pbnm* phase. DFT calculations performed by Picozzi *et al.* have provided a fundamental understanding in the origin of ferroelectricity by demonstrating that the specific spin pattern produces different anion displacements according to the surrounding Mn spins (either ↑↑ or ↑↓) [168]. Assuming that B-O-B bond angles are smaller for AFM interactions than for FM interactions, the O displacements induced by spins are different according to the magnetic moment carried by surrounding B cations. For a FM trilinear chains (Figure 10.9(a)), consecutive anions move with the same magnitude but in opposite direction. Consequently, there is no resultant electric dipole created. For a ↑↑↓ spin chain, anions move differently according to FM or AFM interactions, and their motions do not compensate. A resultant electrical dipole is created (Figure 10.9(b)). Later on, model Hamiltonian based calculations highlighted the importance of the Jahn-Teller motion and orbital-ordering of the Mn^{3+} e_g electron on the ferroelectric properties of $HoMnO_3$ [139]. E-AFM order induced ferroelectricity was further validated experimentally in the orthorhombic polymorph of $YMnO_3$: using X-ray measurements, O displacements, of the order of 10^{-3} Å, are directly observed [170]. DFT studies also clarified the role of strain on the stability of the E-AFM order and on the polarization *P* [171], the effect of the chemical pressure (*i. e.* A site substitution) [172] or revealed a giant spin-driven polarization of the order of $1\,\mu C.cm^{-2}$ in a phase achievable under high pressure [173].

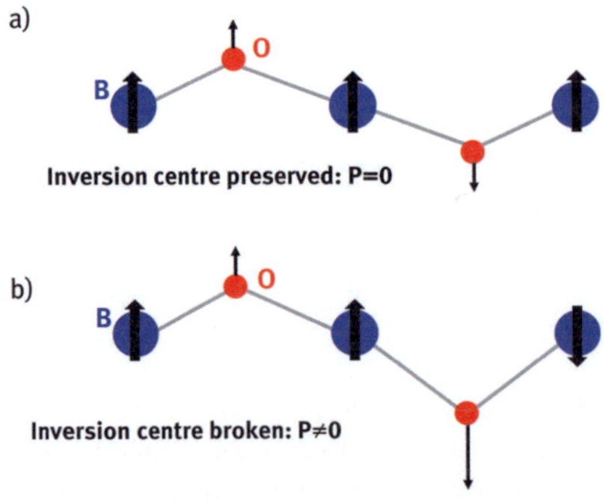

Figure 10.9: Illustration of anion displacements with respect of the neighboring B cation spins.

In 2009, extension of the E-AFM order with different stackings of the ↑↑↓↓ spin chains along the *b* axis of the *Pbnm* structure (S or T-type AFM orders, Figure 10.2(d)) was also predicted on the basis of DFT calculations to produce sizable spontaneous polarizations in $RNiO_3$ compounds [174] (R = Lu-Pr, Y). However, more recent works revealed that the magnetic structure of $RNiO_3$ compounds rather corresponds to ↑ 0 ↓ 0 spin chains due to disproportionation effects, thus not breaking the inversion center [175, 176]. So far, there is no experimental report of ferroelectricity in rare-earth nickelates, but their experimental magnetic structure also remains elusive. We further notice that $YNiO_3$ shows a strong ferroelectric instability in its cubic phase (see Supplementary of Ref. [163]), but it is suppressed by more robust oxygen rotations of the oxygen octahedra bringing the system in the *Pbnm* phase discussed above.

Magnetically induced ferroelectricity can also appear in materials in which spin-orbit interaction (SOI) is playing a key role and produces non-collinear spin structures. This is the case, for instance, in $TbMnO_3$, a multiferroic compound displaying a non-collinear cycloidal spin structure of Mn^{3+} magnetic moments [5]. This peculiar magnetic order breaks the inversion centre and allows a small spontaneous polarization, further unlocking a large magneto-electric response. While the polarization was for a long time thought to originate from purely electronic effects, Malashevich and Vanderbilt as well as Xiang *et al.* demonstrated on the basis of DFT calculations including SOI that the ionic contributions to P is dominant over the pure electronic part [18, 177, 188]. DFT + SOI studies were also able to reproduce and explain the origin of the ferroelectric polarization in cuprates ($LiCu_2O_2$ and $LiCuVO_4$) exhibiting spin-spirals magnetic structures [179]. Apart from of ABO_3 materials, DFT also clarified the origin of the small polarization exhibited by $HoMn_2O_5$ and $TbMn_2O_5$ due to partly cancelling electronic and ionic contributions to P [180, 181] while the polarization in YMn_2O_5 is purely electronic [182].

The polarization in type II multiferroics is not restricted to specific B site magnetic orders but it can be produced by the combination of magnetic orders on A and B sites cations (with or without SOI). For instance, a polarization enhancement is experimentally observed in $DyMnO_3$ once the Dy^{3+} magnetic moments order [183]. This was ascribed to the intimate coupling between *4f* and *3d* electrons *via* exchange-striction by DFT + U and DFT + hybrid calculations performed by Stroppa *et al.* in $DyFeO_3$ [53]. In 2017, Zhao *et al.* performed a systematic study of magnetically induced ferroelectricity in $RCrO_3$ and $RFeO_3$ compounds (where R is a magnetic lanthanide) and observed that simple collinear orders on both A and B site cations, such as A-AFM, C-AFM or G-AFM orders, can give rise to "large" ferroelectric polarizations [184]. Again, a dominant non-relativist exchange-strictive mechanism is at the core of the spontaneous polarization. Finally, Senn and Bristowe used a group theory analysis to enumerate the possible coupling between magnetism and ferroelectricity in ABX_3 (X = O, F, . . .) perovskites [93].

10.5.2 Charge-order induced ferroelectricity

By mixing cations with different valence states, one can achieve charge orderings in materials and create electronic polarization. This was formulated by Efremov, Van den Brink and Khomskii in 2004 [185, 186] and realized in doped manganites [187], with support from DFT to understand underlying mechanisms [188]. Although a non-perovskite, the most famous compound showing charge ordering is Fe_3O_4, also known as magnetite. This is the first magnetic compound discovered and it was known to develop ferroelectricity at low temperature since the eighties [189]. It is naturally based on two iron formal oxidations states, being 2+ and 3+, respectively, and on FeO_4 tetrahedrons [190]. DFT simulations performed by Picozzi *et al.* in 2009 were the first to reveal that the charge ordering between Fe^{3+} and Fe^{2+} breaks the inversion center and produces a spontaneous polarization in the compound [191]. The polarization was since measured and switched experimentally in thin films [192]. Due to its similarity with Fe_3O_4, the spinel $LuFe_2O_4$ in which Fe cations adopt 2+ and 3+ formal oxidation states was long time thought to be the prototypical charge order induced multiferroic compound [193]. However, a debate remains open regarding the ferroelectric properties of this material with a DFT study of 2008 predicting an anti-ferroelectric ground state [194].

Apart from explaining the origin of ferroelectricity in already known compounds, *first-principles* simulations also predicted a vanadium based spinel or other Fe based compounds to be good candidates for multiferroism [195–198]. Finally, Park *et al.* predicted in 2017 that $(La^{3+}VO_3)_1/(Sr^{2+}VO_3)_1$ (001)-oriented superlattices can exhibit a charge-ordered ground state in which electron transfers between V^{3+} and V^{4+} cations produce an additional polarization orthogonal to that coming from the A cation ordering (*i. e.* coming from HIF) [199].

10.5.3 Orbital-order induced ferroelectricity

Along with magnetism and charge ordering, orbitals can also produce a spontaneous polarization even though it is usually linked with or induced by magnetic or charge orderings, and thus the mechanisms remain more elusive. For instance, Varignon and coworkers showed that an electronic instability can produce an orbital ordering irrespective of a Jahn-Teller motion – which is just consequential – and produce a spontaneous electronic polarization of approximately 0.04 $\mu C.cm^{-2}$, reaching 0.34 $\mu C.cm^{-2}$ once the lattice relaxes, in 1/1 (001)-oriented superlattices based on $PrVO_3$ and $LaVO_3$ [136] (*i. e.* this electronic polarization is again orthogonal to that induced by HIF in similar spirit of the work of Park *et al.* [199]). Barone *et al.* proposed that an orbital-ordering is producing the electronic part of the polarization exhibited by undoped rare-earth manganites, although the origin of the orbital-order is likely due the specific E-AFM magnetic order [169]. Ultrathin films of $SrCrO_3$ have been predicted by Gupta *et al.* to become ferroelectric due to the appearance of an orbital ordering driven by lattice distortion [200]. Finally, a joint second-harmonic generation and DFT study revealed that a Jahn-Teller motion and its orbital ordering produce a B cation off-centering in some polar half-doped manganite thin films [201].

10.6 Interfacial systems for efficient magneto-electrics

So far, we have focused on magneto-electric multiferroics combining ferroelectricity and magnetism within the same phase. Nevertheless, there is another route to achieve magneto-electricity: combine two materials exhibiting each only one of the desired properties and expect that the interface will exhibit the coupled functionality. In practice, two main strategies have been considered: (i) exploiting strain mediated coupling at the interface or (ii) harnessing the symmetry breaking occurring at the interface. The former route uses the strain coupling between a piezomagnetic and a piezoelectric material to enable a magneto-electric effect [202]. This strategy has the advantage that the effect can penetrate deep inside the film instead of being confined at the interface. The second strategy benefits from the symmetry breaking occurring at any interface: if one of the materials is a ferromagnet, then a magneto-electric effect is automatically allowed at the interface [203]. We emphasize that the inversion symmetry is also broken locally at the surface of a material that could also allow for magneto-electric effects. Since there are less constraints for realizing “interfacial systems”, they have attracted a large interest both from theoretical and experimental points of view. Nevertheless, in this section, we only report selected discoveries to which *first-principles* simulations have contributed.

Modelling interfacial systems is challenging from the point of view of DFT simulations. Firstly, one must model metal-insulator capacitors under finite electric field. Secondly, DFT is well known for underestimating band gaps and thus band alignment at metal-insulator interfaces becomes a fundamental issue [12, 204]. The first aspect was addressed by Stengel and co-workers during the late 2000’s [204–206]. To better treat the band alignment problem, one could use beyond DFT methods such as DFT + U [207], although the choice of U parameters still has to be handled with care [208], or DFT + hybrid functionals better reproducing band gaps [207, 209, 210]. These problems are nowadays well understood and several key discoveries were enabled by DFT simulations [211].

10.6.1 Charge carrier mediated magneto-electric effect

The most striking advance unveiled by DFT simulations is the carrier mediated magneto-electric effect appearing at the $SrRuO_3/SrTiO_3$ interface [212]. In such a system, spin-polarized carriers of the metal accumulate or deplete at the interfacial region in order to screen the capacitive or bound charges at the interface under electric field (Figure 10.10). The effect is generic to any ferromagnetic (or any magnet) metal-dielectric interface and it should be enhanced at ferroelectric (*e.g.* $BaTiO_3$)-ferromagnetic metal interfaces [213]. Similar effects have been predicted by Tsymbal and co-workers at the surface of various metals such as CrO_2, Fe, Co or Ni under external electric field [214, 215].

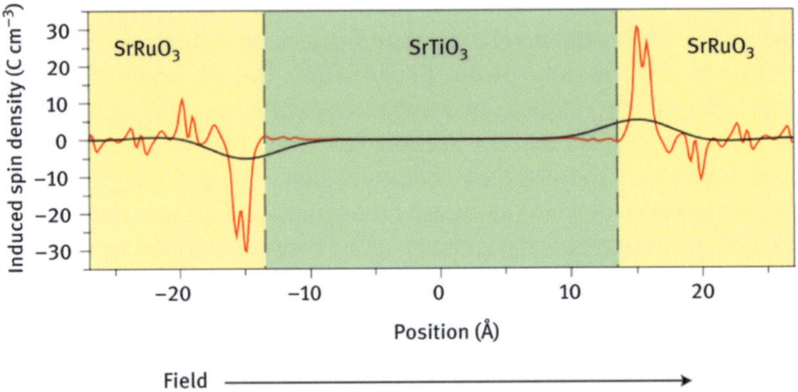

Figure 10.10: Planar (red) and microscopically averaged (black) charge in magnetic density at the interface between $SrTiO_3/SrRuO_3$ capacitor under an applied electric field. This figure is extracted from Ref. [212].

In the aim of realizing the predicted effects at Fe or Co surfaces, one can consider a hybrid system based on a ferroelectric perovskite and a metal. The Fe/$BaTiO_3$ is likely the most studied interfacial system in that respect. At the interface, Fe can induce a magnetic moment on Ti cations, whose magnitude directly depends on the orientation of the polarization in the ferroelectric [216–218] (Figure 10.11(a)). DFT + U calculations from Lee *et al.* ascribed the mechanism underlying the interface magneto-electric effect to hybridization between Fe and Ti (see Figure 10.11) and carrier mediated processes [219]. Nearby the interface, the Ti *d* band peaks around 2 eV above the Fermi level (Figure 10.11(a)) and it can interact with the minority levels of Fe (Figure 10.11(b)). Fe-Ti hybridized *d* levels are created. The polarization then acts on the Fe-Ti distances and *de facto* on the level of hybridization and the resultant interfacial magnetic moment. Other ferroelectric/metal interfaces have been investigated with *first-principles* simulations [207, 217, 220–222], including ferroelectric/AFM metal interfaces [223].

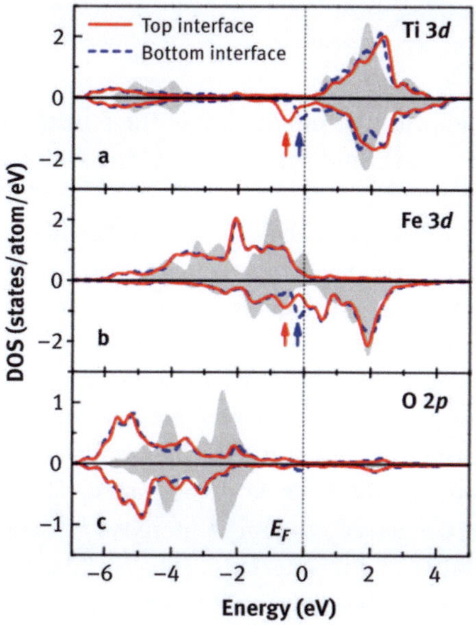

Figure 10.11: Projected density of states (pDOS) on Ti and Fe 3d levels (panels a and b, respectively) and O p levels (c) in the Fe/$BaTiO_3$ interfacial systems. Red and blue curves correspond to the system with a polarization pointing toward and away from the interface, respectively. Grey plots are the pDOS of central atoms. The Fermi level is located to E = 0 eV. This figure is extracted from Ref. [216].

10.6.2 Ferroelectric control of magnetic order

Along with tuning the amplitude of the induced magnetic moment, switching the orientation of the ferroelectric polarization can also change the magnetic order. Some systems such as doped manganites (*e. g.* $La_{1-x}A_xMnO_3$, A = Ca, Sr, Ba) exhibit a rich physical phase diagram as a function of the carrier concentration, and most notably the magnetic order can be altered by continuously changing the doping content. It is straightforward to seek for a ferroelectric control of carrier concentration and magnetic order in such systems. By putting $La_{1-x}A_xMnO_3$ (A = Ca, Sr, Ba) close to a transition, Burton and Tsymbal as well as Bristowe *et al.* demonstrated that reversing the polarization of the ferroelectric can switch the magnetic order of the manganite from ferromagnetic to AFM at $BaTiO_3/La_{1-x}A_xMnO_3$ interfaces [224, 225]. More recently, a ferroelectric control of magnetism was observed at $BaTiO_3$/FeO interfaces covered with Co although the mechanism is rather different to carrier mediated effects [218]. DFT calculations of Plekhanov and Picozzi showed that depending on the orientation of the polarization in $BaTiO_3$, the Fe-O bond length at the FeO-TiO_2 interface as well as the Fe-O-Fe angles are dramatically altered which in turn tune the Fe-Fe magnetic interactions.

10.6.3 Ferroelectric control of magnetic easy axis, orbital occupancies and Curie temperature

Another possibility to tune magnetic properties with the orientation of the ferroelectric polarization is to modify the magnetic easy axis, and *first-principles* simulations have been valuable in predicting this phenomenon [214, 226–233]. In Fe/MgO [230] or other ferromagnetic metals such as (001)-oriented Ni, Fe or Ni [214, 227], the modification of the magnetocrystalline anisotropy is attributed to changes of t_{2g} orbital occupancies due to the electric field. For Fe/ferroelectric oxide interfaces, modification of hybridization between Fe and Ti *d* levels are proposed to be the key factor [226].

Ferroelectricity is also a promising lever to tune orbital occupancies at ferroelectric/manganite interfaces [234]. Other interfacial systems such as short period $(BaTiO_3)_3/(La_{2/3}Sr_{1/3}MnO_3)_3$ superlattices were predicted and experimentally demonstrated to exhibit a strong enhancement of their Curie temperature [210]. Although the role of the polarization is not discussed, the large increase of T_c is assigned to an ordering of Mn orbitals created at the interface, that in turn increases orbital-overlaps and the strength of the double-exchange mechanism.

10.7 Conclusions

In this article, we have highlighted various pathways to achieve magneto-electric compounds, focusing on the role played by DFT methods in boosting many discoveries. At the time of writing this article, advances in understanding and modelling perovskite materials are still on the rise, and most notably, strong dynamical electronic correlations that were believed to be the key aspect of *3d* ABO_3 compounds (to which belong the most famous multiferroic compounds) are demonstrated to a have marginal effects on the perovskite's properties [61]. It constitutes a significant advance in the search of novel multiferroic compounds by alleviating the computational cost for modelling such systems. So, DFT simulations combined with high-throughput methods [235–238], second-principles techniques and "inverse design" strategies [238] appear more than ever very promising to facilitate the discovery of ferroelectric ferromagnets showing large linear magneto-electric effect and operating at room temperature.

Acknowledgements: Ph.G acknowledges support from the F.R.S.-FNRS PDR project HiT4FiT, the ARC project AIMED and the ERA.NET project SIOX.

References

[1] Curie P. Sur la symétrie dans les phénomènes physiques, symétrie d'un champ électrique et d'un champ magnétique. J Phys. 1894;3:393.

[2] Landau LD, Lifshitz EM. Electrodynamics of continuous media, 2nd ed. London: Pergamon Press, 1958.

[3] Fiebig M. Revival of the magnetoelectric effect. J Phys D Appl Phys. 2005;38:R123.

[4] Bibes M, Barthélémy A. Multiferroics: towards a magnetoelectric memory. Nat Mater. 2008;7:425.

[5] Kimura T, Goto T, Shintani H, Ishizaka K, Arima T, Tokura Y. Magnetic control of ferroelectric polarization. Nature. 2003;426:55.

[6] Hur N, Park S, Sharma PA, Ahn JS, Guha S, Cheong SW. Electric polarization reversal and memory in a multiferroic material induced by magnetic fields. Nature. 2004;429:392.

[7] Goto T, Kimura T, Lawes G, Ramirez AP, Tokura Y. Ferroelectricity and giant magnetocapacitance in perovskite rare-earth manganites. Phys Rev Lett. 2004;92:257201.

[8] Brown WF, Hornreich RM, Shtrikman S. Upper bound on the magnetoelectric susceptibility. Phys Rev. 1968;168:574.

[9] Cohen RE. Origin of ferroelectricity in perovskite oxides. Nature. 1992;358:136.

[10] Vanderbilt D. First-principles based modelling of ferroelectrics. Curr Opin Solid State Mater Sci. 1997;2:701.

[11] Junquera J, Ghosez P. Critical thickness for ferroelectricity in perovskite ultrathin films. Nature. 2003;422:506.

[12] Junquera J, Ghosez P. First-Principles study of ferroelectric oxide epitaxial thin films and superlattices: role of the mechanical and electrical boundary conditions. J Comput Theor Nanosci. 2008;5:2071.

[13] Dawber M, Rabe KM, Scott JF. Physics of thin-film ferroelectric oxides. Rev Mod Phys. 2005;77:1083.

[14] Seshadri R, Hill NA. Visualizing the role of Bi 6s "lone pairs" in the off-center distortion in ferromagnetic $BiMnO_3$. Chem Mater. 2001;13:2892.

[15] Krohns S, Lunkenheimer P. Ferroelectric polarization in multiferroics. Phys Sci Rev. 2019.

[16] Van Aken BB, Palstra TT, Filippetti A, Spaldin NA. The origin of ferroelectricity in magnetoelectric $YMnO_3$. Nat Mater. 2004;3:164.

[17] Fennie CJ, Rabe KM. Ferroelectric transition in $YMnO_3$ from first principles. Phys Rev B. 2005;72:100103.

[18] Malashevich A, Vanderbilt D. First-principles theory of magnetically induced ferroelectricity in $TbMnO_3$. Eur Phys J B. 2009;71:345.

[19] Goldschmidt VM. Die gesetze der krystallochemie. Naturwissenschaften. 1926;14:477.

[20] Hill NA. Why are there so few magnetic ferroelectrics? J Phys Chem B. 2000;104:6694.

[21] Baettig P, Spaldin NA. Ab initio prediction of a multiferroic with large polarization and magnetization. Appl Phys Lett. 2005;86:12505.

[22] Nechache R, Harnagea C, Carignan L-P, Gautreau O, Pintilie L, Singh MP, et al. Epitaxial thin films of the multiferroic double perovskite Bi_2FeCrO_6 grown on (100)-oriented $SrTiO_3$ substrates: Growth, characterization, and optimization. J Appl Phys. 2009;105:061621.

[23] Nechache R, Harnagea C, Pignolet A, Normandin F, Veres T, Carignan L-P, et al. Growth, structure, and properties of epitaxial thin films of first-principles predicted multiferroic Bi_2FeCrO_6. Appl Phys Lett. 2006;89:102902

[24] Bhattacharjee S, Bousquet E, Ghosez P. Engineering multiferroism in $CaMnO_3$. Phys Rev Lett. 2009;102:117602.

[25] Hohenberg P, Kohn W. Inhomogeneous electron gas. Phys Rev. 1964;136:B864.
[26] Kohn W, Sham LJ. Self-Consistent equations including exchange and correlation effects. Phys Rev. 1965;140:A1133.
[27] Martin RM. Electronic structure: basic theory and practical methods. Cambridge: Cambridge university press, 2004.
[28] King-Smith RD, Vanderbilt D. Theory of polarization of crystalline solids. Phys Rev B. 1993;47:1651.
[29] Vanderbilt D, King-Smith RD. Electric polarization as a bulk quantity and its relation to surface charge. Phys Rev B. 1993;48:4442.
[30] Resta R. Macroscopic polarization in crystalline dielectrics: the geometric phase approach. Rev Mod Phys. 1994;66:899.
[31] Resta R. Theory of the electric polarization in crystals. Ferroelectrics. 1992;136:51.
[32] Veithen M, Gonze X, Ghosez P. Nonlinear optical susceptibilities, Raman efficiencies, and electro-optic tensors from first-principles density functional perturbation theory. Phys Rev B. 2005;71:125107.
[33] Sai N, Rabe KM, Vanderbilt D. Theory of structural response to macroscopic electric fields in ferroelectric systems. Phys Rev B. 2002;66:104108.
[34] Souza I, Íñiguez J, Vanderbilt D. First-principles approach to insulators in finite electric fields. Phys Rev Lett. 2002;89:117602.
[35] Stengel M, Spaldin NA, Vanderbilt D. Electric displacement as the fundamental variable in electronic-structure calculations. Nat Phys. 2009;5:304.
[36] Thonhauser T, Ceresoli D, Vanderbilt D, Resta R. Orbital magnetization in periodic insulators. Phys Rev Lett. 2005;95:137205.
[37] Xiao D, Shi J, Niu Q. Berry phase correction to electron density of states in solids. Phys Rev Lett. 2005;95:137204.
[38] Resta R. Electrical polarization and orbital magnetization: the modern theories. J Phys Condens Matter. 2010;22:123201.
[39] Malashevich A, Souza I, Coh S, Vanderbilt D. Theory of orbital magnetoelectric response. New J Phys. 2010;12:53032.
[40] Ceresoli D, Gerstmann U, Seitsonen AP, Mauri F. First-principles theory of orbital magnetization. Phys Rev B. 2010;81:60409.
[41] Anisimov VI, Zaanen J, Andersen OK. Band theory and Mott insulators: Hubbard U instead of Stoner I. Phys Rev B. 1991;44:943.
[42] Cococcioni M, De Gironcoli S. Linear response approach to the calculation of the effective interaction parameters in the LDA+U method. Phys Rev B. 2005;71:035105.
[43] Morée JB, Amadon B. First-principles calculation of Coulomb interaction parameters for lanthanides: role of self-consistence and screening processes. Phys Rev B. 2018;98:205101.
[44] Kornev IA, Lisenkov S, Haumont R, Dkhil B, Bellaiche L. Finite-temperature properties of multiferroic BiFeO3. Phys Rev Lett. 2007;99:227602.
[45] Zaanen J, Liechtenstein AI, Anisimov VI. Density-functional theory and strong interactions: Orbital ordering in Mott-Hubbard insulators. Phys Rev B. 1995;52:5467.
[46] Dudarev SL, Botton GA, Savrasov SY, Humphreys CJ, Sutton AP. Electron-energy-loss spectra and the structural stability of nickel oxide: An LSDA+U study. Phys Rev B. 1998;57:1505.
[47] Mellan TA, Cora F, Grau-Crespo R, Ismail-Beigi S. Importance of anisotropic Coulomb interaction in LaMnO3. Phys Rev B. 2015;92:085151.
[48] Bousquet E, Spaldin N. J dependence in the LSDA+U treatment of noncollinear magnets. Phys Rev B. 2010;82:220402.R.
[49] Lee C, Yang W, Parr RG. Development of the Colle-Salvetti correlation-energy formula into a functional of the electron density. Phys Rev B. 1988;37:785.

[50] Heyd J, Scuseria GE, Ernzerhof M. Hybrid functionals based on a screened Coulomb potential. J Chem Phys. 2003;118:8207.
[51] Bilc DI, Orlando R, Shaltaf R, Rignanese GM, Iniguez J, Ghosez P. Hybrid exchange-correlation functional for accurate prediction of the electronic and structural properties of ferroelectric oxides. Phys Rev B. 2008;77:165107.
[52] Stroppa A, Picozzi S. Hybrid functional study of proper and improper multiferroics. Phys Chem Chem Phys. 2010;12:5405.
[53] Stroppa A, Marsman M, Kresse G, Picozzi S. The multiferroic phase of $DyFeO_3$: an ab initio study. New J Phys. 2010;12:093026.
[54] Hong J, Stroppa A, Iñiguez J, Picozzi S, Vanderbilt D, Iñiguez JJ, et al. Spin-phonon coupling effects in transition-metal perovskites: A DFT+U and hybrid-functional study. Phys Rev B. 2012;85:054417.
[55] Prikockyte A, Bilc D, Hermet P, Dubourdieu C, Ghosez P. First-principles calculations of the structural and dynamical properties of ferroelectric $YMnO_3$. Phys Rev B. 2011;84:214301.
[56] Varignon J, Ghosez P. Improper ferroelectricity and multiferroism in 2H-$BaMnO_3$. Phys Rev B. 2013;87:140403.
[57] Varignon J, Fontaine D, Bousquet E, Bristowe NC, Ghosez P. Ferromagnetism induced by entangled charge and orbital orderings in ferroelectric titanate perovskites. Nat Commun. 2015;6:1.
[58] Sun J, Ruzsinszky A, Perdew J. Strongly constrained and appropriately normed semilocal density functional. Phys Rev Lett. 2015;115:036402.
[59] Varignon Julien, Bibes Manuel, Zunger Alex. Mott gapping in 3dABO_3 perovskites without Mott-Hubbard interelectronic repulsion energy U. Physical Review B. 2019;100:035119.
[60] Furness JW, Zhang Y, Lane C, Buda IG, Barbiellini B, Markiewicz RS, et al. An accurate first-principles treatment of doping-dependent electronic structure of high-temperature cuprate superconductors. Commun Phys. 2018;1:11.
[61] Varignon J, Bibes M, Zunger A. Origin of band gaps in 3d perovskite oxides. Nat Commun. 2019;10:1658.
[62] Filippetti A, Spaldin NA. Self-interaction-corrected pseudopotential scheme for magnetic and strongly-correlated systems. Phys Rev B. 2003;67:125109.
[63] Gellé A, Varignon J, Lepetit M-B. Accurate evaluation of magnetic coupling between atoms with numerous open shells: An ab initio method. EPL (Europhysics Lett). 2009;88:37003.
[64] Lepetit MB. How to compute the magneto-electric tensor from ab-initio calculations? Theor Chem Acc. 2016;135:91.
[65] Varignon J, Petit S, Gellé A, Lepetit MB. An ab initio study of magneto-electric coupling of $YMnO_3$. J Phys Condens Matter. 2013;25:496004.
[66] Bousquet E, Spaldin NA, Delaney KT. Unexpectedly large electronic contribution to linear magnetoelectricity. Phys Rev Lett. 2011;106:107202.
[67] Iñiguez J. First-principles approach to lattice-mediated magnetoelectric effects. Phys Rev Lett. 2008;101:117201.
[68] Wojdeł JC, Íñiguez J. Magnetoelectric response of multiferroic $BiFeO_3$ and related materials from first-principles calculations. Phys Rev Lett. 2009;103:267205.
[69] Malashevich A, Coh S, Souza I, Vanderbilt D. Full magnetoelectric response of Cr_2O_3 from first principles. Phys Rev B. 2012;86:094430.
[70] Scaramucci A, Bousquet E, Fechner M, Mostovoy M, Spaldin NA. Linear magnetoelectric effect by orbital magnetism. Phys Rev Lett. 2012;109:197203.
[71] Ricci F, Bousquet E. Unveiling the room-temperature magnetoelectricity of troilite FeS. Phys Rev Lett. 2016;116:227601.

[72] Spaldin NA, Fechner M, Bousquet E, Balatsky A, Nordstr L. Monopole-based formalism for the diagonal magnetoelectric response. Phys Rev B. 2013;88:094429.
[73] Thöle F, Fechner M, Spaldin NA. First-principles calculation of the bulk magnetoelectric monopole density: Berry phase and Wannier function approaches. Phys Rev B. 2016;93:195167.
[74] Tillack N, Yates JR, Radaelli PG. Ab initio cycloidal and chiral magnetoelectric responses in Cr_2O_3. Phys Rev B. 2016;94:100405.R.
[75] Ackermann M, Bohat L, Becker P, Lorenz T, Leo N, Fiebig M. Primary ferrotoroidicity in antiferromagnets. Phys Rev B. 2015;92:094431.
[76] Spaldin NA, Fiebig M, Mostovoy M. The toroidal moment in condensed-matter physics and its relation to the magnetoelectric effect. J Phys Condens Matter. 2008;20:434203.
[77] Gao Y, Vanderbilt D, Xiao D. Microscopic theory of spin toroidization in periodic crystals. Phys Rev B. 2018;97:134423.
[78] Mostovoy M, Scaramucci A, Spaldin NA, Delaney KT. Temperature-dependent magnetoelectric effect from first principles. Phys Rev Lett. 2010;105:087202.
[79] Prosandeev S, Kornev IA, Bellaiche L. Magnetoelectricity in $BiFeO_3$ films: first-principles-based computations and phenomenology. Phys Rev B. 2011;83:020102.R.
[80] Rahmedov D, Wang D, Añiguez J, Bellaiche L. Magnetic cycloid of $BiFeO_3$ from atomistic simulations. Phys Rev Lett. 2012;109:037207.
[81] Zhong W, Rabe K, Vanderbilt D. Phase transitions in $BaTiO_3$ from first principles. Phys Rev Lett. 1994;73:1861.
[82] Zhong W, Rabe M, Vanderbilt D. First-principles theory of ferroelectric phase transitions for perovskites: The case of $BaTiO_3$. Phys Rev B. 1995;52:6301.
[83] Wojdeł JC, Hermet P, Ljungberg MP, Ghosez P, Íñiguez J. First-principles model potentials for lattice-dynamical studies: general methodology and example of application to ferroic perovskite oxides. J Phys Condens Matter. 2013;25:305401.
[84] García-Fernández P, Wojdel JC, Iniguez J, Junquera J. Second-principles method for materials simulations including electron and lattice degrees of freedom. Phys Rev B. 2016;93:195137.
[85] Haeni JH, Irvin P, Chang W, Uecker R, Reiche P, Li YL, et al. Room-temperature ferroelectricity in strained $SrTiO_3$. Nature. 2004;430:758.
[86] Ghosez P, Triscone JM. Multiferroics: Coupling of three lattice instabilities. Nat Mater. 2011;10:269.
[87] Rabe KM, Ahn CH, Triscone J-M. Physics of ferroelectrics, 2007.
[88] Glazer AM. The classification of tilted octahedra in perovskites. Acta crystallogr sect B struct crystallogr cryst chem. 1972;28:3384.
[89] Bousquet E, Spaldin N. Induced magnetoelectric response in pnma perovskites. Phys Rev Lett. 2011;107:197603.
[90] Günter T, Bousquet E, David A, Boullay P, Ghosez P, Prellier W, et al. Incipient ferroelectricity in 2.3% tensile-strained $CaMnO_3$ films. Phys Rev B. 2012;85:214120.
[91] Bousquet E, Cano A. Non-collinear magnetism in multiferroic perovskites. J Phys Condens Matter. 2016;28:123001.
[92] Bellaiche L, Gui Z, Kornev IA. A simple law governing coupled magnetic orders in perovskites. J Phys Condens Matter. 2012;24:312201.
[93] Senn MS, Bristowe NC. A group-theoretical approach to enumerating magnetoelectric and multiferroic couplings in perovskites. Acta Crystallogr Sect A Found Adv. 2018;74:308.
[94] Béa H, Dupé B, Fusil S, Mattana R, Jacquet E, Warot-Fonrose B, et al. Evidence for room-temperature multiferroicity in a compound with a giant axial ratio. Phys Rev Lett. 2009;102:217603.

[95] Diéguez O, González-Vázquez OE, Wojdeł JC, Íñiguez J. First-principles predictions of low-energy phases of multiferroic $BiFeO_3$. Phys Rev B. 2011;83:094105.
[96] Yang Y, Ren W, Stengel M, Yan XH, Bellaiche L. Revisiting properties of ferroelectric and multiferroic thin films under tensile strain from first principles. Phys Rev Lett. 2012;109:057602.
[97] Yang Y, Íñiguez J, Mao AJ, Bellaiche L. Prediction of a novel magnetoelectric switching mechanism in multiferroics. Phys Rev Lett. 2014;112:057202.
[98] Varignon J, Bristowe NC, Ghosez P. Electric field control of Jahn-Teller distortions in bulk perovskites. phys Rev Lett. 2016;116:057602.
[99] Song G, Zhang W. First-principles study on the phase diagram and multiferroic properties of $(SrCoO_3)1/(SrTiO_3)1$ superlattices. Sci Rep. 2014;4:04564.
[100] Aguado-Puente P, García-Fernández P, Junquera J. Interplay of couplings between antiferrodistortive, ferroelectric, and strain degrees of freedom in monodomain $PbTiO_3$/$SrTiO_3$ superlattices. Phys Rev Lett. 2011;107:217601.
[101] Bousquet E, Spaldin NA, Ghosez P. Strain-Induced ferroelectricity in simple rocksalt binary oxides. Phys Rev Lett. 2010;104:037601.
[102] Garcia-Castro AC, Spaldin NA, Romero AH, Bousquet E. Geometric ferroelectricity in fluoroperovskites. Phys Rev B. 2014;89:104107.
[103] Garcia-Castro AC, Romero AH, Bousquet E. Strain-engineered multiferroicity in pnma $NaMnF_3$fluoroperovskite. Phys Rev Lett. 2016;116:117202.
[104] Aschauer U, Pfenninger R, Selbach SM, Grande T, Spaldin NA. Strain-controlled oxygen vacancy formation and ordering in $CaMnO_3$. Phys Rev B. 2013;88:054111.
[105] Chandrasena RU, Yang W, Lei Q, Delgado-Jaime MU, Wijesekara KD, Golalikhani M, et al. Strain-engineered oxygen vacancies in $CaMnO_3$ thin films. Nano Lett. 2017;17:794.
[106] Copie O, Varignon J, Rotella H, Steciuk G, Boullay P, Pautrat A, et al. Chemical strain engineering of magnetism in oxide thin films. Adv Mater. 2017;29:1604112.
[107] Kalinin SV, Spaldin NA. Functional ion defects in transition. Science. 2014;341:858
[108] Xu T, Shimada T, Araki Y, Wang J, Kitamura T. Defect-strain engineering for multiferroic and magnetoelectric properties in epitaxial (110) ferroelectric lead titanate. Phys Rev B. 2015;92:104106.
[109] Wojdeł JC, Íñiguez J. Ab initio indications for giant magnetoelectric effects driven by structural softness. Phys Rev Lett. 2010;105:037208.
[110] Baltensperger W. Influence of magnetic order on conduction electrons and phonons in magnetic semiconductors. J Appl Phys. 1970;41:1052.
[111] Sabiryanov RF, Jaswal SS. Magnons and magnon-phonon interactions in iron. Phys Rev Lett. 1999;83:2062.
[112] Fennie CJ, Rabe KM. Magnetic and electric phase control in epitaxial $EuTiO_3$ from first principles. Phys Rev Lett. 2006;97:267602.
[113] Lee JH, Fang L, Vlahos E, Ke X, Jung YW, Kourkoutis LF, et al. A strong ferroelectric ferromagnet created by means of spin-lattice coupling. Nature. 2010;466:954.
[114] Lee JH, Rabe KM. Epitaxial-strain-induced multiferroicity in $SrMnO_3$ from first principles. Phys Rev Lett. 2010;104:207204.
[115] Srinu Bhadram V, Rajeswaran B, Sundaresan A, Narayana C. Spin-phonon coupling in multiferroic $RCrO_3$ (R-Y, Lu, Gd, Eu, Sm): A Raman study. Eur Phys Lett. 2013;101:17008.
[116] Petit S, Moussa F, Hennion M, Pailhès S, Pinsard-Gaudart L, Ivanov A. Spin phonon coupling in hexagonal multiferroic $YMnO_3$. Phys Rev Lett. 2007;99:266604.
[117] Gupta MK, Mittal R, Zbiri M, Sharma N, Rols S, Schober H, et al. Spin-phonon coupling and high-temperature phase transition in multiferroic material YMnO3. J Mater Chem C. 2015;3:11717.

[118] Schleck R, Nahas Y, Lobo RP, Varignon J, Lepetit MB, Nelson CS, et al. Elastic and magnetic effects on the infrared phonon spectra of MnF2. Phys Rev B. 2010;82:054412.
[119] Benedek NA, Fennie CJ. Why are there so few perovskite ferroelectrics? J Phys Chem C. 2013;117:13339.
[120] Kim TH, Puggioni D, Yuan Y, Xie L, Zhou H, Campbell N, et al. Polar metals by geometric design. Nature. 2016;533:68.
[121] Levanyuk AP, Sannikov DG. Improper ferroelectrics. Sov Phys Uspekhi. 1974;17:199.
[122] Sai N, Fennie CJ, Demkov AA. Absence of critical thickness in an ultrathin improper ferroelectric film. Phys Rev Lett. 2009;102:107601.
[123] Stengel M, Fennie CJ, Ghosez P. Spin-lattice coupling and phonon dispersion of $CdCr_2O_4$from first principles. Phys Rev B. 2012;86:094112.R.
[124] Kamba S, Nuzhnyy D, Savinov M, Tolédano P, Laguta V, Brázda P, et al. Unusual ferroelectric and magnetic phases in multiferroic 2H-$BaMnO_3$ ceramics. Phys Rev B. 2017;95:174103.
[125] Xu K, Lu XZ, Xiang H. Designing new ferroelectrics with a general strategy. Npj Quantum Mater. 2017;2:1.
[126] Young J, Rondinelli JM. Improper ferroelectricity and piezoelectric responses in rhombohedral (A,A′)B2O6 perovskite oxides. Phys Rev B. 2014;89:174110.
[127] Rondinelli JM, Fennie CJ. Octahedral rotation-induced ferroelectricity in cation ordered perovskites. Adv Mater. 2012;24:1961.
[128] Bousquet E, Dawber M, Stucki N, Lichtensteiger C, Hermet P, Gariglio S, et al. Improper ferroelectricity in perovskite oxide artificial superlattices. Nature. 2008;452:732.
[129] Benedek NA, Fennie CJ. Hybrid improper ferroelectricity: A mechanism for controllable polarization-magnetization coupling. Phys Rev Lett. 2011;106:107204.
[130] Zhang H, Weng Y, Yao X, Dong S. Charge transfer and hybrid ferroelectricity in $(YFeO_3)$n/$(YTiO_3)$n magnetic superlattices. Phys Rev B. 2015;91:195145
[131] Zhou Q, Rabe KM. Strain-Induced hybrid improper ferroelectricity in simple perovskites from first principles. ArXiv Prepr. ArXiv1306.1839 2013.
[132] Stroppa A, Barone P, Jain P, Perez-Mato JM, Picozzi S. Hybrid improper ferroelectricity in a multiferroic and magnetoelectric metal-organic framework. Adv Mater. 2013;25:2284.
[133] Stroppa A, Jain P, Barone P, Marsman M, Perez-Mato JM, Cheetham AK, et al. Electric control of magnetization and interplay between orbital ordering and ferroelectricity in a multiferroic metal-organic framework. Angew Chemie - Int Ed. 2011;50:5847.
[134] Young J, Stroppa A, Picozzi S, Rondinelli JM. Tuning the ferroelectric polarization in AA′ MnWO6 double perovskites through A cation substitution. Dalt Trans. 2015;44:10644.
[135] Zhao HJ, Ren W, Yang Y, Íñiguez J, Chen XM, Bellaiche L. Near room-temperature multiferroic materials with tunable ferromagnetic and electrical properties. Nat Commun. 2014;5:4021.
[136] Varignon J, Bristowe NC, Bousquet E, Ghosez P. Coupling and electrical control of structural, orbital and magnetic orders in perovskites. Sci Rep. 2015;5:15364.
[137] Cammarata A, Rondinelli JM. Ferroelectricity from coupled cooperative Jahn-Teller distortions and octahedral rotations in ordered Ruddlesden-Popper manganates. Phys Rev B. 2015;92:014102.
[138] Xu C, Li Y, Xu B, Íñiguez J, Duan W, Bellaiche L. Pressure-Induced multiferroics via pseudo Jahn–Teller effects and novel couplings. Adv Funct Mater. 2017;27:1604513.
[139] Tian Y, Stroppa A, Chai YS, Barone P, Perez-Mato M, Picozzi S, et al. High-temperature ferroelectricity and strong magnetoelectric effects in a hybrid organic–inorganic perovskite framework. Phys Status Solidi - Rapid Res Lett. 2015;9:62.
[140] Liu XQ, Wu JW, Shi XX, Zhao HJ, Zhou HY, Qiu RH, et al. Hybrid improper ferroelectricity in Ruddlesden-Popper $Ca_3(Ti,Mn)_2O_7$ ceramics. Appl Phys Lett. 2015;106:202903.

[141] Senn MS, Bombardi A, Murray CA, Vecchini C, Scherillo A, Luo X, et al. Negative thermal expansion in hybrid improper ferroelectric Ruddlesden-Popper perovskites by symmetry trapping. Phys Rev Lett. 2015;114:035701.
[142] Sim H, Kim BG. Octahedral tilting and ferroelectricity in RbANb2O7 (A = Bi, Nd) from first principles. Phys Rev B. 2014;89:144114.
[143] Benedek NA. Origin of ferroelectricity in a family of polar oxides: the Dion-Jacobson phases. Inorg Chem. 2014;53:3769.
[144] Dixon C, McNulty J, Knight K, Gibbs A, Lightfoot P. Phase transition behavior of the layered perovskite $CsBi0._6La0._4Nb_2O_7$: a hybrid improper ferroelectric. Crystals. 2017;7:135.
[145] Benedek NA, Rondinelli JM, Djani H, Ghosez P, Lightfoot P. Understanding ferroelectricity in layered perovskites: new ideas and insights from theory and experiments. Dalt Trans. 2015;44:10543.
[146] Oh YS, Luo X, Huang FT, Wang Y, Cheong SW. Experimental demonstration of hybrid improper ferroelectricity and the presence of abundant charged walls in (Ca,Sr)3Ti2O7 crystals. Nat Mater. 2015;14:407.
[147] Pitcher MJ, Mandal P, Dyer MS, Alaria J, Borisov P, Niu H, et al. Tilt engineering of spontaneous polarization and magnetization above 300 K in a bulk layered perovskite. Science. 2015;347:420.
[148] Etxebarria I, Perez-Mato JM, Boullay P. The role of trilinear couplings in the phase transitions of aurivillius compounds. Ferroelectrics. 2010;401:17.
[149] Ye F, Wang J, Sheng J, Hoffmann C, Gu T, Xiang HJ, et al. Soft antiphase tilt of oxygen octahedra in the hybrid improper multiferroic $Ca3Mn1._9Ti0._1O_7$. Phys Rev B. 2018;97:041112.
[150] Zhang J, Lin L, Zhang Y, Wu M, Yakobson BI, Dong S. Type-II multiferroic Hf2VC2F2 MXene monolayer with high transition temperature. J Am Chem Soc. 2018;140:9768.
[151] Peng J, Zhang Y, Lin L, Lin L, Liu M. New iron-based multiferroics with improper ferroelectricity. J Phys D Appl Phys. 2018;51:243002.
[152] Dong S, Xiang H, Dagotto E. Magnetoelectricity in multiferroics: a theoretical perspective. Natl Sci Rev. 2019:6:1.
[153] Boström HL, Senn MS, Goodwin AL. Recipes for improper ferroelectricity in molecular perovskites. Nat Commun. 2018;9:2380.
[154] Mulder AT, Benedek NA, Rondinelli JM, Fennie CJ. Turning ABO3 antiferroelectrics into ferroelectrics Design rules for practical rotation-driven ferroelectricity in double perovskites and A3B2O7 Ruddlesden-popper compounds. Adv Funct Mater. 2013;23:4810.
[155] Zhao HJ, Íñiguez J, Ren W, Chen XM, Bellaiche L. Atomistic theory of hybrid improper ferroelectricity in perovskites. Phys Rev B. 2014;89:174101.
[156] Xu B, Wang D, Zhao HJ, Íñiguez J, Chen XM, Bellaiche L. Hybrid improper ferroelectricity in multiferroic superlattices: finite-temperature properties and electric-field-driven switching of polarization and magnetization. Adv Funct Mater. 2015;25:3626.
[157] Nowadnick EA, Fennie CJ. Domains and ferroelectric switching pathways in $Ca_3Ti_2O_7$ from first principles. Phys Rev B. 2016;94:104105.
[158] Zanolli Z, Wojdeł JC, Íñiguez J, Ghosez P. Electric control of the magnetization in $BiFeO_3$/$LaFeO_3$ superlattices. Phys Rev B. 2013;88:060102.R.
[159] Kugel KI, Khomskii DI. Crystal structure and magnetic properties of substances with orbital degeneracy. Sov Phys JETP. 1973;64:1429.
[160] Garcia-Castro AC, Ibarra-Hernandez W, Bousquet E, Romero AH. Direct magnetization-polarization coupling in $BaCuF_4$. Phys Rev Lett. 2018;121:117601.
[161] Gu T, Scarbrough T, Yang Y, Íñiguez J, Bellaiche L, Xiang HJ. Cooperative couplings between octahedral rotations and ferroelectricity in perovskites and related materials. Phys Rev Lett. 2018;120:197602.

[162] Holakovsky J. A new type of the ferroelectric phase transition. Phys Status Solidi. 1973;56:615.

[163] Mercy A, Bieder J, Íñiguez J, Ghosez P. Structurally triggered metal-insulator transition in rare-earth nickelates. Nat Commun. 2017;8:1677.

[164] Liao Z, et al. Metal-insulator-transition engineering by modulation tilt-control in perovskite nickelates for room temperature optical switching. PNAS. 2018;115:9515.

[165] Yamauchi K, Barone P. Electronic ferroelectricity induced by charge and orbital orderings. J Phys Condens Matter. 2014;26:103201.

[166] Muñoz A, Casais MT, Alonso JA, Martínez-Lope MJ, Martinez JL, Fernandez-Diaz MT. Complex magnetism and magnetic structures of the metastable $HoMnO_3$ perovskite. Inorg Chem. 2001;40:1020.

[167] Zhou JS, Goodenough JB. Unusual evolution of the magnetic interactions versus structural distortions in $RMnO_3$ perovskites. Phys Rev Lett. 2006;96:247202.

[168] Picozzi S, Yamauchi K, Sanyal B, Sergienko IA, Dagotto E. Dual nature of improper ferroelectricity in a magnetoelectric multiferroic. Phys Rev Lett. 2007;99:227201.

[169] Barone P, Yamauchi K, Picozzi S. Ferroelectricity due to orbital ordering in E-Type undoped rare-earth manganites. Phys Rev Lett. 2011;106:077201.

[170] Okuyama D, Ishiwata S, Takahashi Y, Yamauchi K, Picozzi S, Sugimoto K, et al. Magnetically driven ferroelectric atomic displacements in orthorhombic $YMnO_3$. Phys Rev B. 2011;84:054440.

[171] Iuşan D, Yamauchi K, Barone P, Sanyal B, Eriksson O, Profeta G, et al. Effects of strain on ferroelectric polarization and magnetism in orthorhombic $HoMnO_3$. Phys Rev B. 2013;87:014403.

[172] Yamauchi K, Freimuth F, Blügel S, Picozzi S. Magnetically induced ferroelectricity in orthorhombic manganites: microscopic origin and chemical trends. Phys Rev B. 2008;78:014403.

[173] Aoyama T, Yamauchi K, Iyama A, Picozzi S, Shimizu K, Kimura T. Giant spin-driven ferroelectric polarization in $TbMnO_3$ under high pressure. Nat Commun. 2014;5:4927.

[174] Giovannetti G, Kumar S, Khomskii D, Picozzi S, Van Den Brink J. Multiferroicity in rare-earth nickelates $RNiO_3$. Phys Rev Lett. 2009;103:156401.

[175] Park H, Millis AJ, Marianetti CA. Site-selective mott transition in rare-earth-element nickelates. Phys Rev Lett. 2012;109:156402.

[176] Varignon J, Grisolia MN, Íñiguez J, Barthélémy A, Bibes M. Complete phase diagram of rare-earth nickelates from first-principles. Npj Quantum Mater. 2017;2:21.

[177] Malashevich A, Vanderbilt D. First principles study of improper ferroelectricity in $TbMnO_3$. Phys Rev Lett. 2008;101:037210.

[178] Xiang HJ, Wei SH, Whangbo MH, Da Silva JLF. Spin-Orbit coupling and ion displacements in multiferroic $TbMnO_3$. Phys Rev Lett. 2008;101:037209.

[179] Xiang HJ, Whangbo MH. Density-functional characterization of the multiferroicity in spin spiral chain cuprates. Phys Rev Lett. 2007;99:257203.

[180] Chang TR, Jeng HT, Ren CY, Hsue CS. Charge-orbital ordering and ferroelectric polarization in multiferroic $TbMn_2O_5$ from first principles. Phys Rev B. 2011;84:024421.

[180] Giovannetti G, Van Den Brink J. Electronic correlations decimate the ferroelectric polarization of Multiferroic $HoMn_2O_5$. Phys Rev Lett. 2008;100:227603.

[181] Unusual evolution of the magnetic interactions versus structural distortions in $RMnO_2O_5$. Phys Rev Lett. 2008;100:227603.

[182] Partzsch S, Wilkins SB, Hill JP, Schierle E, Weschke E, Souptel D, et al. Observation of electronic ferroelectric polarization in multiferroic YMn_2O_5. Phys Rev Lett. 2011;107:057201.

[183] Prokhnenko O, Feyerherm R, Dudzik E, Landsgesell S, Aliouane N, Chapon LC, et al. Enhanced ferroelectric polarization by induced dy spin order in multiferroic $DyMnO_3$. Phys Rev Lett. 2007;98:057206.
[184] Zhao HJ, Bellaiche L, Chen XM, Íñiguez J. Improper electric polarization in simple perovskite oxides with two magnetic sublattices. Nat Commun. 2017;8:14025.
[185] Efremov DV, Van Den Brink J, Khomskii DI. Bond- versus site-centred ordering and possible ferroelectricity in manganites. Nat Mater. 2004;3:853.
[186] Ederer C, Spaldin NA. A new route to magnetic ferroelectrics. Nat Mater. 2004;3:849.
[187] Kadomtseva AM, Popov YF, Vorob'ev GP, Kamilov KI, Ivanov VY, Mukhin AA, et al. Magnetoelectric phenomena in manganites $R0._6Ca0._4MnO_3$ (R = Pr, Nd) with charge ordering suppressed by a magnetic field. J Exp Theor Phys. 2008;106:130.
[188] Giovannetti G, Kumar S, Van Den Brink J, Picozzi S. Magnetically induced electronic ferroelectricity in half-doped manganites. Phys Rev Lett. 2009;103:037601.
[189] Kato K, Iida S, Yanai K, Mizushima K. Ferrimagnetic ferroelectricity of Fe_3O_4. J Magn Magn Mater. 1983;31–34:783.
[190] Senn MS, Wright JP, Attfield JP. Charge order and three-site distortions in the Verwey structure of magnetite. Nature. 2012;481:173.
[191] Yamauchi K, Fukushima T, Picozzi S. Ferroelectricity in multiferroic magnetite Fe_3O_4 driven by noncentrosymmetric Fe_2+/Fe_3+ charge-ordering: First-principles study. Phys Rev B. 2009;79:212404.
[192] Alexe M, Ziese M, Hesse D, Esquinazi P, Yamauchi K, Fukushima T, et al. Ferroelectric switching in multiferroic magnetite (Fe_3O_4) thin films. Adv Mater. 2009;21:4452.
[193] Ikeda N, Ohsumi H, Ohwada K, Ishii K, Inami T, Kakurai K, et al. Ferroelectricity from iron valence ordering in the charge-frustrated system LuFe2O4. Nature. 2005;436:1136.
[194] Angst M, Hermann RP, Christianson AD, Lumsden MD, Lee C, Whangbo MH, et al. Charge order in $LuFe_2O_4$: antiferroelectric ground state and coupling to magnetism. Phys Rev Lett. 2008;101:227601.
[195] Giovannetti G, Stroppa A, Picozzi S, Baldomir D, Pardo V, Blanco-Canosa S, et al. Dielectric properties and magnetostriction of the collinear multiferroic spinel CdV2O4. Phys Rev B. 2011;83:060402.R.
[196] Yamauchi K, Oguchi T, Picozzi S. Ab-initio prediction of magnetoelectricity in infinite-layer $CaFeO_2$ and $MgFeO_2$. J Phys Soc Japan. 2014;83:094712.
[197] Lin L-F, Xu Q-R, Zhang Y, Zhang J-J, Liang Y-P, Dong S. Ferroelectric ferrimagnetic $LiFe_2F_6$: Charge-ordering-mediated magnetoelectricity. Phys Rev Mater. 2017;1:071401.
[198] Yamauchi K, Picozzi S. Interplay between charge order, ferroelectricity, and ferroelasticity: tungsten bronze structures as a playground for multiferroicity. Phys Rev Lett. 2010;105:107202.
[199] Park SY, Kumar A, Rabe KM. Charge-order-induced ferroelectricity in $LaVO_3$/SrV_O3 superlattices. Phys Rev Lett. 2017;118:087602.
[200] Gupta K, Mahadevan P, Mavropoulos P, Ležaić M. Orbital-ordering-induced ferroelectricity in $SrCrO_3$. Phys Rev Lett. 2013;111:077601.
[201] Ogawa N, Ogimoto Y, Ida Y, Nomura Y, Arita R, Miyano K. Polar antiferromagnets produced with orbital order. Phys Rev Lett. 2012;108:157603.
[202] Zheng H, Wang J, Lofland SE, Ma Z, Mohaddes-Ardabili L, Zhao T, et al. Multiferroic $BaTiO_3$-$CoFe_2O_4$ nanostructures. Science. 2004;303:661.
[203] Yamada H, Ogawa Y, Ishii Y, Sato H, Kawasaki M, Akoh H, et al. Engineered interface of magnetic oxides. Science. 2004;305:646.
[204] Stengel M, Aguado-Puente P, Spaldin NA, Junquera J. Band alignment at metal/ferroelectric interfaces: Insights and artifacts from first principles. Phys Rev B. 2011;83:235112.

[205] Stengel M, Spaldin NA. Ab initio theory of metal-insulator interfaces in a finite electric field. Phys Rev B - Condens Matter Mater Phys. 2007;75:205121.

[206] Stengel M, Spaldin NA. Origin of the dielectric dead layer in nanoscale capacitors. Nature. 2006;443:679.

[207] Di Sante D, Yamauchi K, Picozzi S. Beyond standard local density approximation in the study of magnetoelectric effects in Fe/$BaTiO_3$ and Co/$BaTiO_3$ interfaces. J Phys Condens Matter. 2013;25:066001.

[208] Chen H, Ismail-Beigi S. Ferroelectric control of magnetization in La1–xSrxMnO_3 manganites: A first-principles study. Phys Rev B. 2012;86:024433.

[209] Koçak AB, Varignon J, Lemal S, Ghosez P, Lepetit MB. Control of the orbital ordering in manganite superlattices and impact on properties. Phys Rev B. 2017;96:125155.

[210] Sadoc A, Mercey B, Simon C, Grebille D, Prellier W, Lepetit MB. Large increase of the curie temperature by orbital ordering control. Phys Rev Lett. 2010;104:046804.

[211] Ramesh R. Thin films: Theory leads the way to new devices. Nat Nanotechnol. 2008;3:7.

[212] Rondinelli JM, Stengel M, Spaldin NA. Carrier-mediated magnetoelectricity in complex oxide heterostructures. Nat Nanotechnol. 2008;3:46.

[213] Niranjan MK, Burton JD, Velev JP, Jaswal SS, Tsymbal EY. Magnetoelectric effect at the $SrRuO_3$/$BaTiO_3$ (001) interface: An ab initio study. Appl Phys Lett. 2009;95:05250.

[214] Bonfoh N, Tiem S, Lipinski P. Surface magnetoelectric effect in ferromagnetic metal films. Phys Rev Lett. 2007;101:137201.

[215] Duan CG, Nan CW, Jaswal SS, Tsymbal EY. Universality of the surface magnetoelectric effect in half-metals. Phys Rev B. 2009;79:140403.R.

[216] Duan CG, Jaswal SS, Tsymbal EY. Predicted magnetoelectric effect in Fe/$BaTiO_3$ multilayers: ferroelectric control of magnetism. Phys Rev Lett. 2006;97:047201.

[217] Fechner M, Maznichenko IV, Ostanin S, Ernst A, Henk J, Bruno P, et al. Magnetic phase transition in two-phase multiferroics predicted from first principles. Phys Rev B. 2008;78:212406.

[218] Radaelli G, Petti D, Plekhanov E, Fina I, Torelli P, Salles BR, et al. Electric control of magnetism at the Fe/$BaTiO_3$ interface. Nat Commun. 2014;5:3404.

[219] Lee J, Sai N, Cai T, Niu Q, Demkov AA. Interfacial magnetoelectric coupling in tricomponent superlattices. Phys Rev B. 2010;81:144425.

[220] Niranjan MK, Velev JP, Duan C-G, Jaswal SS, Tsymbal EY. Magnetoelectric effect at the Fe_3O_4/$BaTiO_3$ (001) interface: A first-principles study. Phys Rev B. 2008;78:104405.

[221] Yamauchi K, Sanyal B, Picozzi S. Interface effects at a half-metal/ferroelectric junction. Appl Phys Lett. 2007;91:062506.

[222] Hölzer M, Fechner M, Ostanin S, Mertig I. Ab initio study of magnetoelectricity in Fe/$BaTiO_3$: the effects of n-doped perovskite interfaces. J Phys Condens Matter. 2011;23:455902.

[223] Plekhanov E, Stroppa A, Picozzi S. Magneto-electric coupling in antiferromagnet/ferroelectric Mn2Au/$BaTiO_3$ interface. J Appl Phys. 2016;120:074104.

[224] Burton JD, Tsymbal EY. Prediction of electrically induced magnetic reconstruction at the manganite/ferroelectric interface. Phys Rev B. 2009;80:174406.

[225] Bristowe NC, Stengel M, Littlewood PB, Pruneda JM, Artacho E. Electrochemical ferroelectric switching: Origin of polarization reversal in ultrathin films. Phys Rev B. 2012;85:024106.

[226] Duan CG, Velev JP, Sabirianov RF, Mei WN, Jaswal SS, Tsymbal EY. Tailoring magnetic anisotropy at the ferromagnetic/ferroelectric interface. Appl Phys Lett. 2008;92:122905.

[227] Nakamura K, Shimabukuro R, Fujiwara Y, Akiyama T, Ito T, Freeman AJ. Giant modification of the magnetocrystalline anisotropy in transition-metal monolayers by an external electric field. Phys Rev Lett. 2009;102:187201.

[228] Tsujikawa M, Oda T. Finite electric field effects in the large perpendicular magnetic anisotropy surface Pt/Fe/Pt(001): A First-principles study. Phys Rev Lett. 2009;102:247203.
[229] Lukashev PV, Burton JD, Jaswal SS, Tsymbal EY. Ferroelectric control of the magnetocrystalline anisotropy of the Fe/$BaTiO_3$ (001) interface. J Phys Condens Matter. 2012;24:226003.
[230] Niranjan MK, Duan CG, Jaswal SS, Tsymbal EY. Electric field effect on magnetization at the Fe/MgO(001) interface. Appl Phys Lett. 2010;96:222504.
[231] Zhang H, Richter M, Koepernik K, Opahle I, Tasnádi F, Eschrig H. Electric-field control of surface magnetic anisotropy: a density functional approach. New J Phys. 2009;11:043007.
[232] Lee M, Choi H, Chung YC. Ferroelectric control of magnetic anisotropy of FePt/$BaTiO_3$ magnetoelectric heterojunction: A density functional theory study. J Appl Phys. 2013;113:17C729.
[233] Fechner M, Zahn P, Ostanin S, Bibes M, Mertig I. Switching magnetization by 180° with an electric field. Phys Rev Lett. 2112;108:197206.
[234] Chen H, Qiao Q, Marshall MS, Georgescu AB, Gulec A, Phillips PJ, et al. Reversible modulation of orbital occupations via an Interface-induced polar state in metallic manganites. Nano Lett. 2014;14:4965.
[235] Setyawan W, Curtarolo S. High-throughput electronic band structure calculations: Challenges and tools. Comput Mater Sci. 2010;49:299.
[236] Hautier G, Jain A, Ong SP. From the computer to the laboratory: materials discovery and design using first-principles calculations. J Mater Sci. 2012;47:7317.
[237] Jain A, Ong SP, Hautier G, Chen W, Richards WD, Dacek S, et al. The Materials Project: A materials genome approach to accelerating materials innovation. APL Mater. 2013;1:011002.
[238] Bennett JW, Garrity KF, Rabe KM, Vanderbilt D. Hexagonal ABC semiconductors as ferroelectrics. Phys Rev Lett. 2012;109:167602.
[239] Zunger A. Inverse design in search of materials with target functionalities. Nat Rev Chem. 2018;2:121.

Donald M. Evans, Vincent Garcia, Dennis Meier and Manuel Bibes

11 Domains and domain walls in multiferroics

Abstract: Multiferroics are materials combining several ferroic orders, such as ferroelectricity, ferro- (or antiferro-) magnetism, ferroelasticity and ferrotoroidicity. They are of interest both from a fundamental perspective, as they have multiple (coupled) non-linear functional responses providing a veritable myriad of correlated phenomena, and because of the opportunity to apply these functionalities for new device applications. One application is, for instance, in non-volatile memory, which has led to special attention being devoted to ferroelectric and magnetic multiferroics. The vision is to combine the low writing power of ferroelectric information with the easy, non-volatile reading of magnetic information to give a "best of both worlds" computer memory. For this to be realised, the two ferroic orders need to be intimately linked via the magnetoelectric effect. The magnetoelectric coupling – the way polarization and magnetization interact – is manifested by the formation and interactions of domains and domain walls, and so to understand how to engineer future devices one must first understand the interactions of domains and domain walls. In this article, we provide a short introduction to the domain formation in ferroelectrics and ferromagnets, as well as different microscopy techniques that enable the visualization of such domains. We then review the recent research on multiferroic domains and domain walls, including their manipulation and intriguing properties, such as enhanced conductivity and anomalous magnetic order. Finally, we discuss future perspectives concerning the field of multiferroic domain walls and emergent topological structures such as ferroelectric vortices and skyrmions.

Keywords: multiferroic, domains, domain walls, microscopy

11.1 Domain structures in (multi-)ferroics

11.1.1 Introduction to ferroic domains and domain walls

Ferroic materials are defined by the appearance of an order parameter (e.g., elastic, electric or magnetic) at a non-disruptive phase transition. If at least two order parameters coexist in the same phase, the material is called a multiferroic [1]. The order parameter(s) can point in at least two symmetrically equivalent directions (polarities) between which it can be switched by the application of an external

This article has previously been published in the journal Physical Sciences Reviews. Please cite as: Evans, D. M., Garcia, V., Meier, D., Bibes, B. Domains and domain walls in multiferroics *Physical Sciences Reviews* [Online] 2020, 5. DOI: 10.1515/psr-2019-0067

https://doi.org/10.1515/9783110582130-011

field. When cooling through the phase transition in zero-field, the polarities have the same energy and, as a consequence, both polarities appear inside the ferroic material. Regions with the same polarity are called domains and the interfaces that separate them are call domain walls. While this shows, trivially, that multiple domains form naturally in ferroics, the details of how many domains, their size, and where they form, depend on several energy terms, as well as the local defect structure. In the following, we will give a brief outline of how ferroic domains and domain walls form, using a ferroelectric as an illustrative example; although, as we will see, analogous arguments can be made for the formation of domains and domain walls in ferromagnets. For more complete and in-depth descriptions of the physics of ferroelectric and ferromagnetic domains we refer to, for instance, the textbooks by Tagantsev et al. [2] and by Hubert and Schäfer [3].

A proper ferroelectric is a material for which the spontaneous electric polarization plays the role of the primary symmetry breaking order parameter, which can completely describe the phase transition into the ferroic state [4, 5]. Practically, this means energy contributions related to this order parameter are very important in the system, particularly for the domain formation. For instance, in ferroelectrics, surfaces perpendicular to the ferroelectric polarization have bound charges that create a strong depolarizing field (Figure 11.1(a)). This field is a major driving force for domain formation: the depolarizing field can be minimized by either (i) screening the surface charges of the ferroelectric with surface adsorbates or metallic electrodes (Figure 11.1(b)), or (ii) the formation of ferroelectric domains, for example, 180° domains, so that the net polarization at the surface averages to zero (Figure 11.1(c)). The number and size of the ferroelectric domains that form will depend on the details of the boundary conditions (Figure 11.1(d)), such as crystal size, shape, orientation, local defect structure and, critically, the energy costs associated with

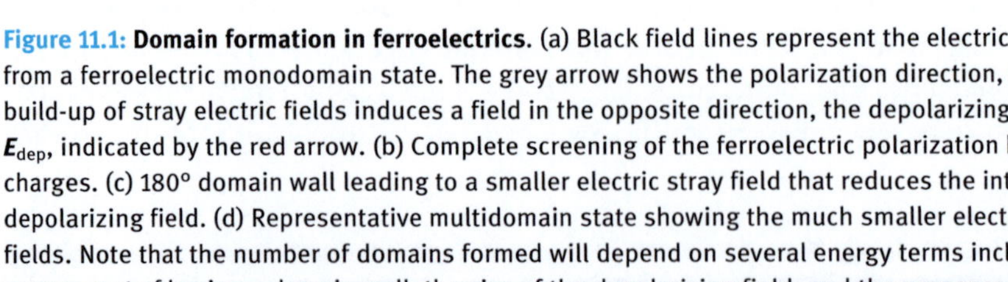

Figure 11.1: Domain formation in ferroelectrics. (a) Black field lines represent the electric stray field from a ferroelectric monodomain state. The grey arrow shows the polarization direction, ***P***. The build-up of stray electric fields induces a field in the opposite direction, the depolarizing field, $\boldsymbol{E}_{dep}$, indicated by the red arrow. (b) Complete screening of the ferroelectric polarization by surface charges. (c) 180° domain wall leading to a smaller electric stray field that reduces the internal depolarizing field. (d) Representative multidomain state showing the much smaller electric stray fields. Note that the number of domains formed will depend on several energy terms including the energy cost of having a domain wall, the size of the depolarizing field, and the presence of uncompensated surface bound charges.

the insertion of domain walls. Thus, depending on the boundary conditions, a thin ferroelectric film can either form a single-domain state or break into multiple nanodomains [2, 6], which has been used to engineer ferroelectric domains [7–10]. This can easily be observed by half covering a large single crystal with electrodes and cooling the sample through its Curie temperature: the area under the electrodes will form larger domains than the uncovered area due to the presence of screening charges from the electrodes, while the area in air will have a finer domain structure due to the depolarizing fields (see, e.g. Gilletta [11]). Importantly, from these simplistic arguments, it can be seen that the formation of domains and domain walls is a natural property of any ferroic crystal. Note that this applies in the case of zero applied electric field. Applying an electric field will affect both the transition temperature and the domain formation, see e.g. Merz [12].

In general, the size of the domains that form can be predicted by what is now known as Kittel's scaling law [13, 14]. Although Kittel attributed the original idea of size effects of domains to Frenkel and Doefman [15], it was Kittel's work on ferromagnets that provided the scaling relationship that now bears his name: the domain width is inversely proportional to one over the square root of their thickness as detailed below. Originally, his mathematical formalism considered only three contributions to the free energy as relevant, i.e., the surface energy of a domain wall, the magnetic field energy of the configuration, and the anisotropy of the spin orientation. Kittel's work on ferromagnets was then expanded to ferroelectrics by Mitsui and Furuichi in 1952 [16], showing that the fundamental scaling remains the same. It has also now been experimentally verified in ferroelectrics and multiferroics down to tens of nanometre sample thicknesses [17, 18]. Some corrections to Kittel's law have been proposed by Scott [19] due to the finite size of domain walls, being particularly pertinent in ferroelectrics, where domain walls are usually about one order of magnitude thinner than in ferromagnets.

To illustrate Kittel's scaling law, we consider the simple case of open boundary conditions without surface screening in a film with 180° striped-domains as illustrated in Figure 11.1(d). Here, the size of domains is determined by the competition between the energy of the domains and that of domain walls. The energy density, E, of the domains depends on their width, ω, as $E = U\omega$, where U is the volume energy density. The energy density of the domain walls depends on their number and so will be inversely proportional to ω as $E = \sigma d/\omega$, where σ is the energy density per unit area of the wall and d the thickness of the film. Minimizing the total energy gives rise to a square-root dependence of the domain size with the film thickness as $\omega = \sqrt{\sigma/U \times d}$, known as Kittel's law [14].

In the original form proposed for ferromagnets, it is the competition between magnetic exchange energy and demagnetizing field contributions that drives the domain formation: the magnetic exchange energy is minimized in single-domain states while the competing demagnetizing field favours the formation of multidomain states. Furthermore, additional factors arise from the size and shape of the magnetic domains and domain walls, which will depend on the magnetocrystalline and magnetostrictive energies of the system. While the magnetic exchange energy favours wide walls, so that neighbouring magnetic moments are almost parallel to each other, magnetocrystalline anisotropy promotes narrow walls so that almost all the magnetization of the system is aligned with the easy magnetic axis. As a result, the thickness of these magnetic domain walls depends strongly on material parameters, but typical values are in the order of a few tens of nanometres and up to hundreds of nanometres [20]. In contrast to ferromagnets, where the exchange energy is much larger than the magnetocrystalline anisotropy, in ferroelectrics the anisotropy and dipole–dipole interactions (the equivalent of exchange in these materials) are of the same order of magnitude [21]. As a consequence, ferroelectrics usually develop much thinner walls than ferromagnets, with typical thicknesses in the order of a few unit cells (see, for instance, Refs [17, 22–26]). The smaller width of ferroelectric domain walls compared to magnetic domain walls implies that the order parameter changes orientation much faster.

At the domain wall itself there are, conceptually, three different ways for the polar order to change, referred to as Ising-, Bloch-, and Néel-type walls. The three domain wall types are schematically illustrated in Figure 11.2. In Ising-type walls (Figure 11.2(a)), the axis along which the order parameter points is fixed. Across the wall, only the magnitude of the polar order changes smoothly from up to down, going through zero at the centre of the wall. These walls are generally the thinnest

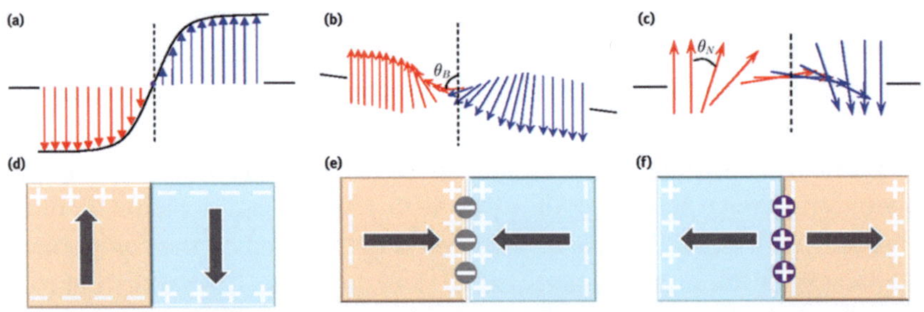

Figure 11.2: **Fundamental types of ferroic domain walls.** Arrows represent the orientation of the order parameter (e.g., polarization or magnetization). (a) Ising-type wall: the polarization/magnetization does not rotate but decreases in magnitude through the wall. (b) Bloch-type wall: the polarization/magnetization does not change magnitude but rotates in the plane of the domain wall. (c) Néel-type wall: the polarization/magnetization does not change magnitude, but rotates perpendicular to the plane of the domain wall. ((a)–(c) are reprinted with permission from [27]. Copyright 2009 by the American Physical Society.) (d)–(f) Charge states at ferroelectric 180° walls. (d) Neutral 180° domain wall. Positive and negative bound charges are denoted by symbols + and –, respectively. (e) Head-to-head domain wall, where the polar discontinuity associated with the positive domain wall bound charges attracts mobile negative charge carriers (grey). (f) Tail-to-tail domain wall, where the polar discontinuity associated with the negative domain wall bound charges attracts mobile positive charge carriers (purple).

of order (sub)nanometre wide. Traditionally, it was assumed that 180° domain walls in ferroelectrics are Ising-type walls, but recent experimental and theoretical work has revealed domain walls with more complex structures (see, e.g., Refs [27–30]). For ferroelectric Ising-type walls, as already mentioned above, both the anisotropy and dipole–dipole interactions are usually strong, and the energy difference between the paraelectric and ferroelectric phases is relatively small. Consequently, an abrupt change of polarization through the wall can readily be accommodated by a reduction of the magnitude of polarization as presented in Figure 11.2(a).

In the Bloch-type wall, the order parameter does not change size but rotates within the plane of the wall (Figure 11.2(b)). The Néel-type wall is characterized by a rotation of the order parameter perpendicular to the domain wall and at the wall centre the respective vector lies orthogonal to the plane of the domain wall (Figure 11.2(c)). Bloch- and Néel-type walls are most common in ferromagnets, where the energy difference between paramagnetic and ferromagnetic phases is large and, hence, the magnitude of magnetic moments does not vary significantly [3]. These magnetic domain walls can be quite wide and extend up to hundreds of nanometres in width [20]. The three extreme cases shown in Figure 11.2(a) to (c) are normally combined to describe domain walls in real systems, leading to mixed-type domain wall states. As already mentioned, it is important to note that Bloch- and Néel-type walls are not restricted to ferromagnets and, recently, there is a growing interest in non-Ising-type domain walls in ferroelectrics and the emergence of walls of Néel- and Bloch-type has been reported for different ferroelectric materials [31], such as $BaTiO_3$ [32] and $Pb(Zr,Ti)O_3$ (PZT) [33–35]. Note that the illustrations in Figure 11.2 are for domain walls at which the order parameter orientation changes by 180°; the details change for other situations, e.g., for 90° walls, but the salient points remain true.

In multiferroics, the situation becomes even more interesting as electric and magnetic domains and domain walls coexist (see, e.g., [36–41]), giving rise to unusual magnetoelectric correlation phenomena as discussed in Section 11.2 and 11.3 and also Chapter 3 and Chapter 4. Because of the coexistence of different types of ferroic order in multiferroics, it becomes important to consider whether the polarization or magnetization plays the role of the symmetry breaking order parameter and, thus, governs the formation of domains and domain walls. In this context, one typically distinguishes between proper and improper systems or, more precisely, between systems in which the ferroic order arises across a proper or improper phase transition [4, 5, 42]. For example, as described above, a ferroelectric phase transition is referred to as proper, if the spontaneous polarization is the primary symmetry breaking order parameter that drives the transition. This is the case in textbook ferroelectrics, such as $BaTiO_3$, PZT, and $LiNbO_3$ [2, 43]. Alternatively, electric order can arise as a secondary effect driven by the coupling to, e.g., a structural [44–47] or magnetic order parameter [48–50] as reviewed, e.g., in Ref [40]. The latter scenario, referred to as improper ferroelectricity (or pseudo-proper; see, e.g., [4, 42] for details), is realized in many multiferroics, such as the hexagonal manganites ($RMnO_3$, with R = Sc, Y, In, Dy to Lu; reviewed in Chapter 3) and orthorhombic $TbMnO_3$ and $DyMnO_3$, with fascinating consequences for the domain walls [39, 51–53]. Here, due to the secondary nature of the spontaneous polarization, exotic *charged* domain walls (Figure 11.2(e) and (f)) arise spontaneously – a situation that is usually avoided due to the high electrostatic energy costs.

Assuming an Ising-type wall, a domain wall is said to be fully charged if the polarization of neighbouring domains comes together head-to-head or tail-to-tail as shown in Figure 11.2(e) and Figure 11.2(f), respectively. The associated discontinuity in polarization $\boldsymbol{P}$ leads to uncompensated domain wall bound charges (div $\boldsymbol{P} \neq 0$), which require screening (see, e.g., [54] and references therein). As a consequence, mobile carriers redistribute, promoting anomalous electronic transport properties at head-to-head and tail-to-tail walls, including highly conducting and insulating states as reviewed in [54–56], as well as the formation of unusual electronic inversion layers [57]. Charged domain walls thus represent a natural type of two-dimensional system with inherent functional properties, which can be injected, moved and erased on demand [55, 58–61]. Additional functionalities arise at domain walls with a strong coupling between electric and magnetic degrees of freedom, enabling magnetic-field control of the local electronic charge state as discussed in Section 11.3 [36, 39]. For a more comprehensive or more technical coverage of charged domain walls in ferroelectrics and multiferroics, we refer to recent reviews [40, 56, 62, 63], and the specific studies highlighted in the following sections.

11.1.2 Visualization of domains

The rapid progress that has been made in the understanding of magnetic and electric domains relies on the recent developments in microscopy techniques with high sensitivity and unprecedented spatial resolution. Nowadays, even atomic level resolution is readily available with transmission electron microscopy (TEM) (see, e.g., Chapter 7 or [64], as well as references therein). In this review, we discuss different microscopy methods that allow for studying the formation and interaction of ferroic domains in spatially resolved measurements on nano- to microscopic length scales. In multiferroics, a specific difficulty arises for domain imaging as one needs to distinguish contributions from at least two coexisting types of ferroic order. Many available techniques, however, are simultaneously sensitive to both magnetism and ferroelectricity, making the distinction between the two signals challenging. For instance, photoemission electron microscopy (PEEM) based on X-ray linear dichroism (XLD) is sensitive to the asymmetry of the electronic charge distribution and, hence, to contributions from both ferroelectric and antiferromagnetic domains. Another example is scanning probe microscopy (SPM); the magnetic tip used for magnetic force magnetometry (MFM), for example, will probe contributions from the stray field of the magnetic domains as well as the electrostatic fields from the ferroelectric domains and surface charges (as in electrostatic force magnetometry (EFM)). In the following, we give examples of different techniques with which magnetic and ferroelectric domains can be observed independently as illustrated based on measurements taken on both intrinsic and artificial multiferroics (Figure 11.3).

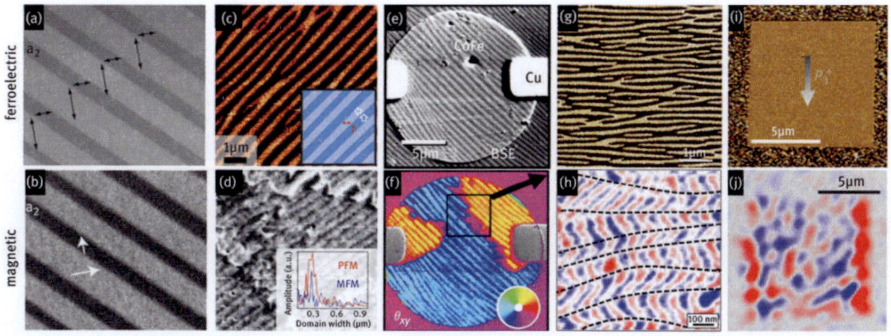

Figure 11.3: Imaging magnetic and ferroelectric domains in artificial and intrinsic multiferroics. (a) Polarized optical microscopy image of (a) the a1-a2 birefringent ferroelectric domains of a $BaTiO_3$ single crystal and (b) the magnetic domains of the uniaxial $Co_{0.6}Fe_{0.4}$ layer (15 nm) grown on top. ((a) and (b) are from [65]). (c) PFM image of a $BiFeO_3$ film grown on a $DyScO_3$(110) substrate and (d) MFM image on the same sample after the growth of a Pt/$Co_{0.9}Fe_{0.1}$ bilayer on top under a magnetic field of 20 mT. In the inset, the Fourier analysis indicates the correlation between the PFM and MFM patterns. ((c) and (d) are reprinted with permission from [66]. Copyright 2013 by the American Physical Society.) (e) Ferroelectric domain structure of $BiFeO_3$ imaged using the back-scattered electron (BSE) intensity with a $Co_{0.9}Fe_{0.1}$ thin-film circular disk and Cu electrode on the side. (f) Simultaneously acquired SEMPA image of the magnetic structure. The magnetization direction is represented by colours as indicated by the colour wheel. ((e) and (f) are reprinted with permission from Springer Nature, taken from [67]. Copyright 2015 by Springer Nature.) (g) In-plane PFM image of the striped-domain structure of a 30-nm-thick $BiFeO_3$ thin film grown on $DyScO_3$(110) and (h) NV-magnetometry image of the stray field produced by the spin cycloid of the same film. ((g) and (h) reprinted with permission from Springer Nature, taken from [68]. Copyright 2017 by Springer Nature.) (i) In-plane PFM image of a single ferroelectric domain written with the trailing field of the SPM tip on a $BiFeO_3$ thin film grown on a $SrTiO_3$(001) substrate. (j) Reconstructed SHG image showing two types of submicron antiferromagnetic domains in the ferroelectric domain showed in (i). ((i) and (j) reprinted with permission from Springer Nature, taken from [69]. Copyright 2017 by Springer Nature.)

11.1.2.1 Optical microscopy

In bulk single crystals, domains may reach sizes in the micron range compatible with optical microscopy. In this case, ferroelectric domains can be imaged with the birefringent contrast [70–72] or selective etching [73], while magnetic domains on this length scale are observed via the magneto-optical Kerr effect. Figure 11.3(a), (b) illustrates this on an artificial multiferroic system based on a thin ferromagnetic film (15 nm of $Co_{0.6}Fe_{0.4}$) deposited on a ferroelectric single crystal ($BaTiO_3$) [65]. The ferroelectric domains of $BaTiO_3$ organize in the form of stripes with 90° ferroelastic domain walls (a1-a2 domains corresponding to in-plane polarization in Figure 11.3(a)). Strain coupling between the $Co_{0.6}Fe_{0.4}$ thin film and the underlying $BaTiO_3$ single crystal induces uniaxial magnetoelastic anisotropy with orthogonal easy axes between neighbouring domains, resulting in an imprinted stripe pattern (Figure 11.3(b)).

11.1.2.2 Scanning probe microscopy

SPM allows nanoscale investigation of both ferroelectric domains, via piezoresponse force microscopy (PFM), and ferromagnetic domains, via MFM. In MFM, the magnetic tip experiences an attracting or repelling force depending on its relative magnetization orientation compared to that of the sample. This force induces a phase lag on the oscillation of the tip that is coverted into a map so that MFM images reveal the spatial distribution of the magnetic stray field coming from the sample.

The probe tips used for MFM are usually coated with ferromagnetic materials and their metallicity enables their subsequent use for PFM. PFM uses the fact that all ferroelectric materials are piezoelectric. By applying an alternating voltage between the SPM tip and the ferroelectric in contact, the ferroelectric will vibrate at the same frequency as the voltage excitation. In the ideal case, the phase shift of the response is directly connected to the out-of-plane polarization orientation of the ferroelectric (see, e.g., [74] for a recent review on PFM). In the same manner, the torsion of the tip contains information on the in-plane components of the polarization. An example is given in Figure 11.3, where the MFM contrast coming from a 2.5-nm-thick $Co_{0.9}Fe_{0.1}$ amorphous layer (Figure 11.3(d)) is correlated to the PFM contrast from the ferroelectric domain pattern (Figure 11.3(c)) in the underlying layer of $BiFeO_3$ [66]. In this configuration, the surface bound charges of $BiFeO_3$ are screened by the magnetic top electrode and the MFM signal is "pure". This correlation attests for the coupling between the ferromagnetic domains in $Co_{0.9}Fe_{0.1}$ and the antiferromagnetic domains in $BiFeO_3$, the latter being magnetoelectrically coupled with the ferroelectric order.

In order to directly image the magnetic stray field coming from the antiferromagnetic domains in multiferroics, such as $BiFeO_3$ (see Chapter 2 for details on $BiFeO_3$), a scanning probe technique with higher sensitivity than MFM is required. Recently, scanning nitrogen-vacancy (NV) magnetometry was developed in which magnetic stray fields down to a few μT can be detected [75]. In bulk [76] and lightly strained films of the antiferromagnetic $BiFeO_3$ [77], the magnetoelectric interaction stabilizes the formation of a spin cycloid – originally probed by Sosnowska et al. using neutron diffraction [78] – whose propagation vector is coupled to the polarization direction. Gross et al. were able to use NV magnetometry to visualize the magnetic stray field emanating from this 70-nm-long spin cycloid (Figure 11.3 (h)) in striped domain patterns of $BiFeO_3$ thin films [68] (Figure 11.3 (g)). The in-plane compressive strain imposed by the $DyScO_3$ substrate lifts the degeneracy between the three possible propagation vectors for the spin cycloid for each ferroelectric domain, resulting in the in-plane 90° rotation of the magnetic signal between alternating ferroelectric domains (Figure 11.3 (h)).

11.1.2.3 Scanning electron microscopy

In a scanning electron microscope with polarization analysis (SEMPA), the low energy secondary electrons are spin-polarized and give information on the magnetization orientation of the ferromagnetic structure, while the high energy elastically back-scattered electrons (BSE) are sensitive to the crystal structure and lattice distortions to reveal the polarization orientation of the ferroelectric structure. Combining SEMPA and BSE, Zhou et al. were able to simultaneously image the local in-plane vector magnetization of the $Co_{0.9}Fe_{0.1}$ layer (Figure 11.3(f)) grown over the striped-domain ferroelectric structure of $BiFeO_3$ (Figure 11.3(e)) [67]. The comparison of the SEMPA and BSE images shows the close correlation between the magnetic structure of the $Co_{0.9}Fe_{0.1}$ film and the striped ferroelectric domain structure of the underlying $BiFeO_3$ layer. Within a stripe, which is about 250 nm wide, the magnetization is aligned parallel or anti-parallel to the in-plane surface projection of the electric polarization.

11.1.2.4 Photoemission electron microscopy

A possible way to discriminate ferroelectric and magnetic domain contrasts is to take advantage of the potentially different Curie and Néel temperatures of the multiferroic. This approach was used by Zhao et al. [79] to distinguish the antiferromagnetic contribution and the ferroelectric one in photoemission electron microscopy (PEEM) images of $BiFeO_3$ thin films. A careful comparison of the PEEM contrast under and above the Néel temperature of $BiFeO_3$ ($T_N = 640$ K) allows for separating the "pure" ferroelectric contrast from the multiferroic one with a resolution down to ≈20 nm. A smart combination of XMLD- (X-ray magnetic linear dichroism) and XMCD- (X-ray magnetic circular dichroism) PEEM imaging (mapping spin and orbital magnetic moments) is a very powerful tool.

11.1.2.5 Second harmonic generation

Second harmonic generation (SHG), that is, the frequency doubling of light in a material, is a powerful technique to sense complex magnetic and electric structures [41, 80, 81]. It is well suited to probe insulators with a spatial resolution in the submicron range limited by the wavelength of the light. In bulk multiferroics, SHG domain imaging was pioneered by Fiebig [41]. More recently, Chauleau et al. used

SHG to probe antiferromagnetic domains across a ferroelectric single-domain of a $BiFeO_3$ thin film (Figure 11.3(i)) [69]. They performed the experiments in transmission while varying the incident polarization of the light and analyzed the full angular dependence of each pixel. The 110-nm-thick $BiFeO_3$ thin film was epitaxially grown on $SrTiO_3$(001) at a compressive strain that destroys the cycloidal spin structure and stabilizes a slightly-canted antiferromagnetic state arising from the Dzyaloshinskii–Moriya interaction [77]. In this G-type antiferromagnetic system, the Fe^{3+} spins are aligned along three possible antiferromagnetic vectors of the (111) plane, perpendicular to the ferroelectric polarization. As depicted in Figure 11.3 (j), only two types of submicron antiferromagnetic domains were observed in this single $BiFeO_3$ ferroelectric domain, suggesting that the magnetoelastic energy lifts the degeneracy between the three types of antiferromagnetic domains.

11.1.2.6 Combined imaging experiments

While the application of just one experimental method that can image both electric and magnetic domains is clearly intriguing, the majority of studies on multiferroics still make use of two or more complementary techniques to image the coexisting domains and domain walls and cover all relevant length scales. The latter is nicely reflected by the research studies on multiferroics involving PFM. PFM by itself provides access to the distribution of ferroelectric domains. The combination of PFM with PEEM and X-ray resonant magnetic scattering led to the discovery of coupled ferroelectric and ferromagnetic domains in $BiFeO_3$ thin films [79] and interfacial multiferroicity in Fe/$BaTiO_3$ and Co/$BaTiO_3$ heterostructures [82]. PFM in combination with NV magnetometry disclosed the coupling between ferroelectric and antiferromagnetic domain in $BiFeO_3$ [68] (see also Section 11.3). In addition, PFM has been combined with many other methods such as TEM, optical microscopy, as well as SHG in order to understand and disentangle the formation of electric and magnetic domains in different multiferroic materials. For a more extended review on PFM, see [74].

11.2 Domain walls in multiferroics

The recent discovery of functional electronic and magnetic properties at multiferroic domain walls triggered world-wide attention and initiated a shift in the research focus away from domains and towards domain walls. The domain wall research is driven by the idea to develop a new generation of agile interfaces/2D systems that remain spatially mobile after a material has been synthesized and implemented into a device structure [55]. Going beyond just conducting domain walls, as observed in ferroelastic and ferroelectric materials, multiferroics offer additional degrees of freedom that arise from the magnetic order and the unusual couplings between their electric, magnetic and structural properties as discussed in the following.

11.2.1 Domain wall types

Across domain walls in multiferroics, one or more order parameters change from one direction to another and this can occur in many different ways depending on the type of involved ferroic orders, the direction of the order parameters in the adjacent domains (e.g., parallel or perpendicular to the wall), and the geometry and dimensions of the sample [83, 84]. The range of structures that arises within a domain wall in the presence of just one order parameter has already been addressed in Section 11.1 (see Figure 11.2). In order to classify the complex domain walls that occur in multiferroics with coexisting electric and magnetic order [24, 62, 85], a first distinction can be made based on the interaction of the respective domain walls as shown in Figure 11.4(a) and Figure 11.4(b) [40].

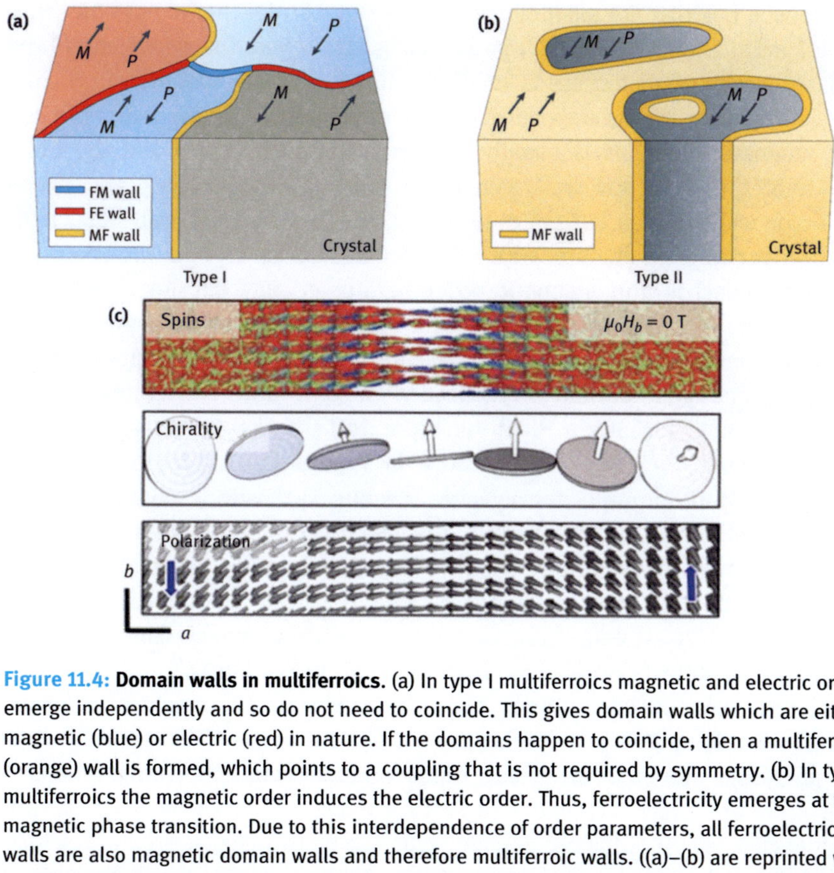

Figure 11.4: Domain walls in multiferroics. (a) In type I multiferroics magnetic and electric order emerge independently and so do not need to coincide. This gives domain walls which are either magnetic (blue) or electric (red) in nature. If the domains happen to coincide, then a multiferroic (orange) wall is formed, which points to a coupling that is not required by symmetry. (b) In type II multiferroics the magnetic order induces the electric order. Thus, ferroelectricity emerges at the magnetic phase transition. Due to this interdependence of order parameters, all ferroelectric domain walls are also magnetic domain walls and therefore multiferroic walls. ((a)–(b) are reprinted with permission from Springer Nature, taken from [40]. Copyright 2016 by Springer Nature.) (c) Landau–Lifshitz–Gilbert simulation showing the complex evolution of spins (top), the spin chirality (middle), and electric polarization (bottom) across a multiferroic domain wall in $Mn_{0.95}Co_{0.05}WO_4$. (Reprinted with permission from Springer Nature, taken from [36]. Copyright 2015 by Springer Nature.)

In multiferroics where electric and magnetic orders occur independently (type I), electric and magnetic domain walls can coincide, but they do not necessarily have to couple due to the different microscopic origin of the polarization (*P*) and magnetization (*M*) (Figure 11.4(a)). This is different in systems where the magnetic order induces the spontaneous polarization (see also Chapter 4), or vice versa, so that both orders occur together (type II, Figure 11.4(b)). In the latter case, every ferroelectric domain wall is also a magnetic domain wall [37, 86], representing a two-dimensional system with multiferroic properties different from the surrounding bulk.

A second, more advanced distinction is related to the conservation (or not) of the amplitude of the order parameter(s) across the wall. As introduced in Section 11.1, we usually consider three fundamental types of ferroic domain walls, namely, Ising-, Néel-, and Bloch-type walls (Figure 11.2). For Ising-type walls, upon crossing the wall the order parameter decreases, becomes zero at the centre and then increases again with the opposite polarity. For example, this is the situation for the ferroelectric walls in the type I multiferroics $BiFeO_3$ [87] and $ErMnO_3$ [22], where the polarization (and the associated structural distortion) is smaller at the core of the wall than in the domains [88]. In contrast, at the coexisting (anti-)ferromagnetic domain walls, the amplitude of magnetic moments is usually conserved (non-Ising-type) and the order parameter rotates through the wall, potentially forming Bloch-, Néel-, and mixed Néel-Bloch-type walls (Figure 11.2), or even more complex vortex-like wall states [89].

Particularly complex domain wall structures arise in type II multiferroics, where both the electric and magnetic order parameter may develop Bloch- and Néel-type like features. In $Mn_{0.95}Co_{0.05}WO_4$, for instance, two antiferromagnetic domain states of opposite chirality occur in the multiferroic state [36]. Across the domain walls, this change in chirality is realized via a continuous 180° rotation of the material's easy-plane, forming a Bloch-type-like domain wall with respect to the spin-chirality vector $C = S_i \times S_j$ as illustrated in Figure 11.4(c). Interestingly, as $Mn_{0.95}Co_{0.05}WO_4$ is an improper ferroelectric (Section 11.1) with a magnetically induced polarization ($P \sim e_{ij} \times (S_i \times S_j)$ [49]), the electric order follows the rotation across the wall, which leads to Néel-type-like ferroelectric walls. Analogous to standard ferroelectric domain walls, such magnetically induced ferroelectric walls can carry a finite domain wall bound charge, which enables magnetic control of electronic domain wall states as we discuss in more detail in Section 11.3.

In summary, pronounced couplings between the electric and magnetic order parameters can occur in multiferroics, giving rise to a huge richness of domain wall structures and properties. Although the electric dipole and magnetic spin configurations at the atomic scale are rarely available – so that the exact wall types often remain unknown – it is clear that multiferroic domain walls can show unexpected and fascinating physical properties beyond the bulk properties. Today, we are only at the verge of discovering these exciting properties. Novel findings and insight into the nano-physics of domain walls are to be expected in the near future, enabled by the continuous and ongoing progress in advanced microscopy techniques. Some of the recent key results in the field are reviewed in the following.

11.2.2 Conduction in domain walls

A little history – The emergence of conducting domain walls was postulated first in the 1960s and 70s based on macroscopic measurements in classic ferroelectrics like TGS (triglycine sulfate), $PbTiO_3$, $BaTiO_3$, $LiNbO_3$, and SbSI [90–94]. Due to experimental limitations of the time, however, it was not possible to measure individual walls directly. Aristov et al. were among the first to spatially resolve domain walls with anomalous electronic properties in ferroelectric $BaTiO_3$ [95] and, later, also in $LiNbO_3$ [96] using scanning electron microscopy [97]. The first direct evidence of conducting domain walls was in pioneering experiments on $BiFeO_3$ by Seidel et al. [87]. Since then, conductance measurements using conductive atomic force microscopy (cAFM) have been performed on a diverse range of prototypical (proper) ferroelectrics, including $PbZr_{0.2}Ti_{0.8}O_3$ [98–100], $LiNbO_3$ [101–103], and $BaTiO_3$ [104], as well as improper ferroelectrics, such as hexagonal $RMnO_3$ [105–109], $Cu_3B_7O_{13}Cl$ [110], and $(Ca,Sr)_3Ti_2O_7$ [111], demonstrating that domain wall conductivity is a quite general phenomenon. Although conducting domain walls have already been analyzed and discussed more than half a century ago, only the recent in-depth studies have revealed their full technological potential and triggered world-wide interest. During the last decade, domain-wall-based multi-configurational devices, atomic-scale electronic components and memory technology have been proposed [58, 112, 113]. Another idea is to use domain walls in order to achieve reconfigurable doping: While semiconductor technology enables the precise control of charged dopants during the fabrication process, their location remains fixed. Reconfigurable channels of charge carriers are, in principle, achievable using polarization charges as quasi-dopants [114]. Then, doping may be achieved in ferroelectrics within the domain walls. For more insight into the physics and properties of ferroelectric domain walls, we refer to, e.g., the recent review articles by Catalan et al. [55], Meier [54], Jiang et al. [113], and Bednyakov et al. [56].

The first direct observation of conductivity at a domain wall was also the first in a multiferroic: Seidel et al. observed room-temperature conduction at 180° and 109° domain walls in multiferroic $BiFeO_3$ thin films (Figure 11.5(a), (b)), while no conduction could be detected at 71° domain walls [87]. This seminal paper triggered a plethora of experimental and theoretical works on conductive domain walls partly fuelled by the controversy regarding the origin of these effects. The initial interpretation from Seidel et al., supported by density functional theory (DFT), suggested that electrostatic potential steps at the domain walls were responsible for the enhanced conduction, as well as a reduced bandgap induced by structural transitions in the wall. In addition, DFT calculations showed that these effects were minimized for 71° domain walls in agreement with experiments. However, following reports concluded that the conduction through these neutral walls was mostly related to extrinsic contributions [83, 115]. Farokhipoor & Noheda found that 71° domain walls were also conducting in $BiFeO_3$ thin films (Figure 11.5(d), (e)) [115]. Through a temperature analysis of the conduction, they demonstrated that it is governed by thermally activated transport from defects (oxygen vacancies) at low voltage (Figure 11.5(f)) and Schottky emission at higher voltage, just as for the domains. The different conduction for domains and domain walls is explained by an increased density of oxygen vacancies at the walls, giving rise to in-gap states and lower electron hopping energy. In the same vein, Seidel et al. reported that the oxygen pressure during the cool-down process after the growth of La-doped $BiFeO_3$ had a large influence on the 109° domain wall conduction, with a current increase by two orders of magnitude under low oxygen pressure (Figure 11.5(c)) [83].

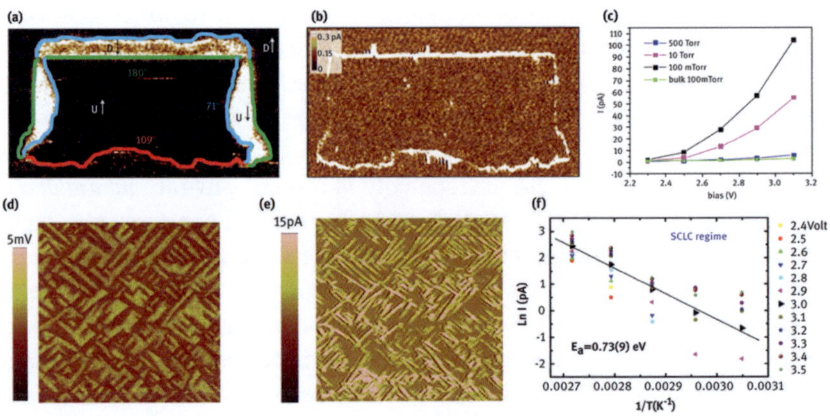

Figure 11.5: Conduction at charge-neutral domain walls in multiferroic $BiFeO_3$ thin films. (a) 109°, 180° and 71° domain walls in a (110)-oriented $BiFeO_3$ thin film observed by PFM and the corresponding (b) cAFM map showing enhanced conduction for 180° and 109° domain walls. ((a)–(b) reprinted with permission from Springer Nature, taken from [87]. Copyright 2009 by Springer Nature.) (c) Influence of the oxygen pressure after growth on the conduction of 109° domain walls in La-doped $BiFeO_3$ thin films. ((c) is reprinted with permission from [83]. Copyright 2010 by the American Physical Society.) (d) PFM amplitude of a (001)-oriented $BiFeO_3$ film with a majority of 71° domain walls and (e) corresponding cAFM image showing enhanced conduction at the domain walls. (f) Arrhenius plot of the current vs. temperature showing a thermally activated behaviour of the conduction through 71° domain walls. ((d)–(f) are reprinted with permission from [115]. Copyright 2011 by the American Physical Society.)

Somewhat easier and more straightforward to understand are electrostatics-driven contributions to the domain wall conductivity. At domain walls where the polarization meets either fully or partially in head-to-head or tail-to-tail configuration (see Figure 11.2(e), (f)), positive or negative domain wall bound charges exist, respectively, creating diverging electrostatic potentials. The compensation of the bound charge can be achieved by the redistribution of mobile charge carriers as we addressed in Section 11.1. Though charged domain walls are not favoured in proper ferroelectrics, they naturally arise in various multiferroic materials. This abundance is due to the fact that multiferroics are often improper ferroelectrics (see Section 11.1 and 11.2), where the formation of domains is governed by, e.g., a structural or magnetic order parameter and not the electric polarization [40]. Multiferroic hexagonal manganites, $RMnO_3$, exhibit an intriguing ferroelectric domain pattern [106, 116] and are an interesting example for the natural emergence of both neutral [106, 117] and charged domain walls [57, 105, 118–121]: Here, a trimerizing lattice distortion leads to stable 180° charged domain walls with anomalous electronic transport properties (Figure 11.6 (a), (b)) [105, 119]. The extraordinary stability of these charged domain walls is reflected by recent electrostatic force microscopy measurements [122], which showed that partially unscreened walls arise at low temperature, representing a rare example of a stable, electrically uncompensated oxide interface (Figure 11.6(c)).

Meier et al. observed that in $ErMnO_3$, the conduction of the domain walls is a continuous function of the wall orientation (Figure 11.6(a)) [105]. Enhanced conductance was found at tail-to-tail walls while head-to-head walls showed a lower conduction than the domains (Figure 11.6(b)). This can be explained by the p-type semiconducting nature of $ErMnO_3$; here, mobile holes are available to screen the negative bound charges at tail-to-tail walls. The electrostatic potential at the walls shifts the Fermi level into the broad valence band where the effective mass is low, leading to an enhanced conduction at the walls [105]. In order to control and optimize the electronic domain wall properties in hexagonal manganites, effects related to off-stoichiometry [118, 123–128] and aliovalent cation substitution on the A- and B-site were studied systematically [129, 130]. Interestingly, recent studies suggest that the emergence of anomalous conductance is not restricted to isolated head-to-head and tail-to-tail walls in $RMnO_3$: Wherever the walls intersect in the characteristic cloverleaf-like arrangement [106, 116], ferroelectric vortex cores with emergent U(1) symmetry form [131–133]. These vortex cores are quasi-1D objects and exhibit quite unusual bound-charge distributions and electrostatics (Figure 11.6(d)), which – similar to the charged walls – is likely to change the electronic transport properties locally [22].

The results highlighted here, however, represent only a fraction of the research devoted to the domain and domain wall physics in hexagonal manganites. Due to the stability and abundance of charged domain walls, the material has become an important model system for the theoretical and experimental study of the complex nano-physics of functional domain walls and different application opportunities as nanoscale digital switches and half-wave rectifiers have been proposed [57, 108].

The majority of envisioned domain wall applications in nanoelectronics, such as domain-wall-based memories and multi-configurational devices, however, require highly mobile domain walls that can readily be injected and deleted at will. The intentional creation of charged domain walls was first demonstrated in 2013 in ferroelectric $BaTiO_3$ single crystals [104]. The stabilization of 90° charged domain

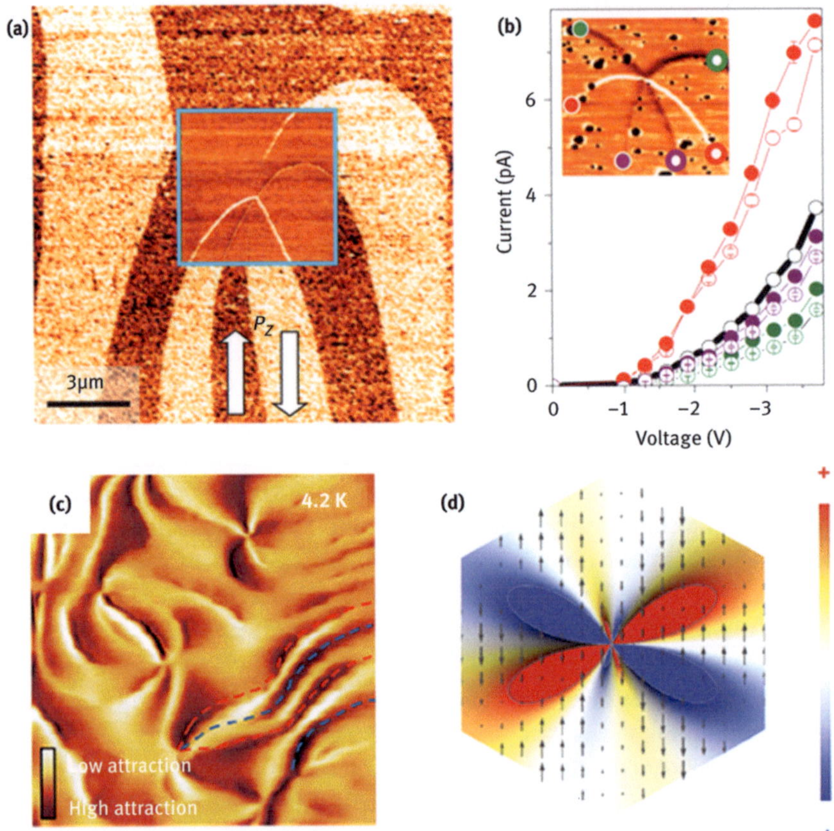

Figure 11.6: Charged domain walls in hexagonal manganites. (a) PFM image in the yz-plane of an $ErMnO_3$ single crystal. The inset shows a cAFM map recorded in the same region. The bright and dark lines show that domain walls have enhanced and reduced conduction relative to the domains, respectively. (b) Local current-voltage characteristics obtained at the locations shown in the inset. Tail-to-tail walls (red) show an enhanced conduction while head-to-head walls (purple and green) are less conducting than the domains (black). (reprinted with permission from Springer Nature, taken from [105]. Copyright 2012 by Springer Nature.) (c) EFM image of $Er_{0.99}Ca_{0.01}MnO_3$ taken at 4.2 K, showing the different fields (imaged as high/low attraction) arising from partially uncompensated charges at domain walls. (Reprinted with permission from [122]. Copyright 2019 American Chemical Society.) (d) Density plot of the bound-charge distribution emerging from the centre of a vortex core in the hexagonal manganites; the arrows indicate the direction of ferroelectric polarization. ((d) is taken from [22] and reused with permission from ACS. All further permissions must be directed to ACS.)

walls was realized by cooling down the single crystal under a strong electric field from above the Curie temperature. Then, the authors compared the conduction of individual head-to-head and tail-to-tail domain walls with the domains by contacting them with 200-micron-wide metallic electrodes (Figure 11.7(a)). While the conduction at tail-to-tail walls was comparable to that of domains, a conduction 10^4 to 10^6 times higher than in the domains was observed at head-to-head domain walls (Figure 11.7(b)) with a metallic character (Figure 11.7(c)). However, this frustrated poling process is not technology transferrable for the realization of reconfigurable devices based on ferroelectric and multiferroic thin films. In 2015, Crassous et al. succeeded in creating charged walls, using the trailing field of the SPM probe tip (similar to Balke et al. [134]) to design charged ferroelastic domain walls in a multiferroic thin film, this time using La-doped $BiFeO_3$ grown on $DyScO_3(110)$ orthorhombic substrates (Figure 11.7(d)) [114]. CAFM maps of the artificial domains revealed a remarkable enhancement of the conductivity at head-to-head domain walls only (Figure 11.7(e)) with unprecedently high currents (two orders of magnitude higher than previous reports with neutral walls). Moreover, temperature-dependent measurements indicated a metallic behaviour for these charged walls as opposed to the thermally activated behaviour at neutral 71° domain walls (Figure 11.7(f)). On the other hand, ferroelectric tail-to-tail walls do not show a significant conduction as this configuration cannot be compensated by mobile holes, which gives rise to a roughening of the wall as well. In addition, the authors demonstrated

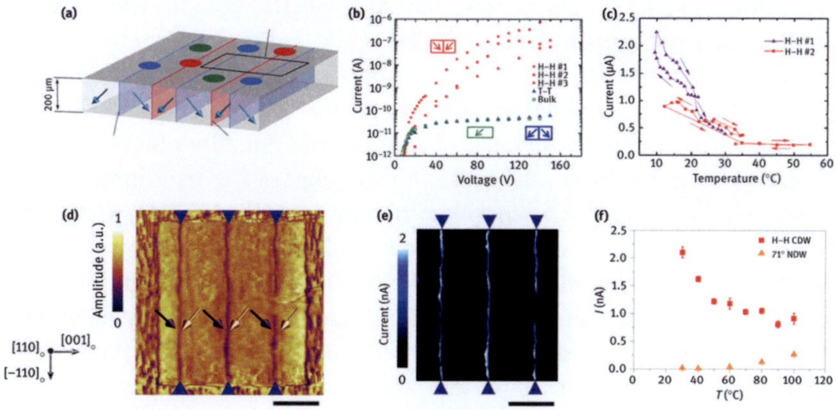

Figure 11.7: Conduction at created charged domain walls in ferroelectrics and multiferroics. (a) Sketch of the (110) $BaTiO_3$ crystal where the surface is covered with 200 μm diameter electrodes. The domain walls are irregularly distributed with periods from 100 to 300 μm. (b) Current-voltage characteristics of the tail-to-tail and head-to-head walls, as well as the domains. (c) Temperature dependence of the current through the head-to-head domain walls indicating a metallic behavior. ((a)–(c) are reprinted with permission from Springer Nature, taken from [104]. Copyright 2013 by Springer Nature.) (d) PFM amplitude of the domain pattern with alternating head-to-head and tail-to-tail domain walls defined by the trailing field of the SPM tip in La-doped $BiFeO_3$ thin films. (e) CAFM maps on the same domain pattern show an enhanced conduction at the head-to-head domain walls. The scale bars in (d) and (e) are 1 μm. (f) Temperature dependence of the current through the head-to-head walls compared to that of native 71° domain walls. ((d)–(f) are reprinted with permission from Springer Nature, taken from [114]. Copyright 2015 by Springer Nature.)

the reconfigurability of the writing process as well as the nanoscale manipulation of conductive head-to-head walls.

11.2.3 Magnetism at domain walls

In $BiFeO_3$, the coupling between ferroelectric and antiferromagnetic domain walls has been extensively investigated [135–137]. In $BiFeO_3$ thin films grown under high compressive strain, the rhombohedral-like phase and the tetragonal-like phase co-exist in the form of nanoscale structures. Using XMCD-PEEM, Zhang et al. observed an enhancement of magnetism at the nanoscale boundaries of these mixed-phase thin films [138]. The magnetotransport properties of these walls have also been investigated [139].

Domain wall magnetism and its correlation to the ferroelectric order was first investigated in hexagonal manganites. These materials order antiferromagnetically below a Néel temperature, $T_N \approx 100$ K, whereas ferroelectricity arises at much higher temperature, $T_C \approx 1000$ K (see, e.g., Chapter 3). Using SHG imaging at 6 K, Fiebig et al. were able to discriminate antiferromagnetic domains from ferroelectric domains in a $YMnO_3$ single crystal [41]. They found that ferroelectric domain walls always coincide with antiferromagnetic domain walls, while antiferromagnetic domain walls can also exist independently (Figure 11.8(a)). Although the resolution of the applied SHG experiment ($\gtrsim 1$ µm) did not allow to resolve the domain walls directly, this was the first experimental evidence of multiferroic domain walls. While no magnetoelectric coupling exists in the bulk, the coupling of the two ferroic orders suggests a possible magnetoelectric coupling at domain walls. In 2012, Geng et al. studied the magnetic properties of multiferroic domain walls in an $ErMnO_3$ single crystal using low-temperature MFM [140]. They first used PFM to image the ferroelectric domain structure at room temperature (Figure 11.8(b)) and the MFM data showed that each multiferroic domain wall carried a net magnetization (Figure 11.8 (c)–(d)), which they attributed to the presence of uncompensated Er^{3+} spins at the domain walls [140, 141].

Multiferroic domains and domain walls with explicitly strong coupling between the electric and magnetic order were discovered in the spin-spiral multiferroic $MnWO_4$ (type II) [37, 86]. In $MnWO_4$, hybrid ferroelectric/antiferromagnetic domains arise below 15 K (Figure 11.8(e), (f)) driven by an elliptical spin spiral that breaks the inversion symmetry and, thereby, induces improper ferroelectricity [142]. Because of the magnetic origin of the ferroelectric order, and in strong contrast to the hexagonal manganites, electric and magnetic domain walls always coexist in $MnWO_4$. The inner structure of these walls remains to be measured, but theory predicts that they are antiferromagnetic in nature, similar to the case illustrated in Figure 11.4(c) [36].

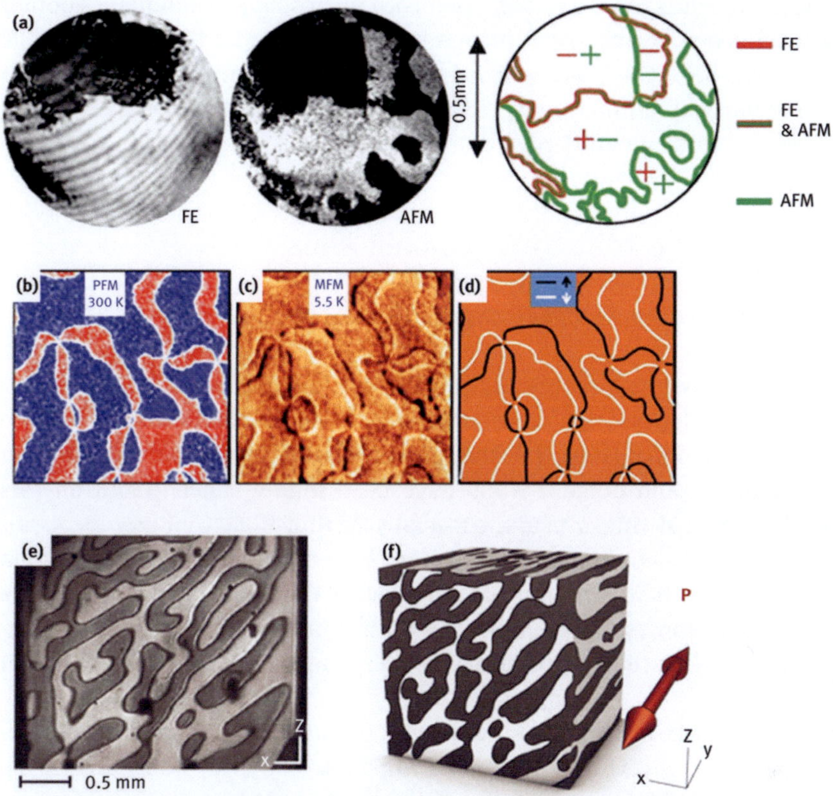

Figure 11.8: Electric and magnetic domains in multiferroics. (a) Coexistence of ferroelectric and antiferromagnetic domains in hexagonal $YMnO_3$ visualized by SHG. Bright and dark areas correspond to regions with different orientations of the order parameter (*P* on the left image – ferroelectric (FE), *L* on the right image – antiferromagnetic (AFM)). The sketch on the right shows that some of the domain walls are both antiferromagnetic and ferroelectric. ((a) is reprinted with permission from Springer Nature, taken from [41]. Copyright 2002 by Springer Nature.) (b) PFM image of hexagonal $ErMnO_3$ showing up- and down-polarized ferroelectric domains. (c) MFM image at low temperature at the same location as in (b) showing an induced magnetic moment at ferroelectric domain walls. (d) Sketch from image in (c), emphasizing the magnetic moments at domain walls of the 16x16 μm^2 image. ((b)–(d) reprinted (adapted) with permission from [140]. Copyright 2012 American Chemical Society.) (e) Multiferroic domain walls in the spin-spiral system $MnWO_4$ visualized by SHG. (f) Illustration showing the 3D distribution of the multiferroic domains in (e). ((e)–(f) reprinted with permission from [37]. Copyright 2009 by the American Physical Society.)

Recently, Farokhipoor et al. reported enhanced magnetism at domain walls in epitaxial thin films of the orthorhombic spin-spiral multiferroic $TbMnO_3$ grown on $SrTiO_3$ [143]. They found out that the macroscopic magnetic signal of the samples is correlated to the density of domain walls of $TbMnO_3$ with different thicknesses. While the results are supported by DFT calculations, direct evidence of magnetism at domain walls remains to be demonstrated with microscopy techniques. Another idea is the emergence of dynamical magnetic fields at moving ferroelectric walls proposed by Juraschek et al. [144], which extends magnetism in domain walls into the realm of dynamical multiferroicity, foreshadowing exciting perspectives for future experimental work on multiferroic domain walls.

11.3 Manipulating domains and domain walls

11.3.1 Electric control of antiferromagnetic domains

$BiFeO_3$ has been involved in a number of scientific breakthroughs, such as the observation of conducting ferroelectric domain walls [87] and room-temperature electric field control of magnetism [145]. It is thus not surprising that research efforts on multiferroic domains and domain walls have been highly concentrated on this material during the last fifteen years, especially in thin films (Chapter 2). A primary goal of magnetoelectric multiferroics was to control magnetic domains via the low power electric-field switching of ferroelectric domains in view of possible applications in spintronics. $BiFeO_3$ being an antiferromagnetic ferroelectric, concepts based on the electric control of exchange bias of an adjacent ferromagnet were proposed [146].

In 2014, Heron et al. investigated the kinetics of the polarization switching process in 100-nm-thick $BiFeO_3$ thin films grown on $DyScO_3(110)$ substrates [147]. Using time-dependent PFM, they found out that the out-of-plane electric field was giving rise to a 180° reversal of polarization in a two-step switching process combining 71° and 109° switching (Figure 11.9(a)). Ab initio calculations supported these observations with a direct 180° switching having a too high energy barrier compared to the two-step process. In addition, it was predicted that the two-step switching enables an in-plane switching of the canted moment by 180° through the reversal of the rotation of the oxygen octahedra (Figure 11.9(b)). This was experimentally confirmed by visualizing the influence of the ferroelectric switching onto the magnetization of a $Co_{0.9}Fe_{0.1}$ ferromagnet deposited on the film of $BiFeO_3$. Using XMCD-PEEM, the authors observed that, after the application of a 6 V voltage pulse to switch the ferroelectric polarization, the net magnetization of the ferromagnet was reversed (Figure 11.9(c)).

However, these experiments consider that $BiFeO_3$ is in a canted G-type antiferromagnetic phase [148] in contradiction with the measurements of a spin cycloid in low-strained $BiFeO_3$ films grown on substrates such as $DyScO_3(110)$ using Mossbauer and Raman spectroscopies [83]. Using NV magnetometry, Gross et al. recently confirmed that this state was favoured by imaging in real-space the stray field coming from the spin-cycloid of $BiFeO_3$, with an imprint of the ferroelectric stripe-domains onto the magnetic texture (Figure 11.3(h)) [68]. In addition, by defining single ferroelectric

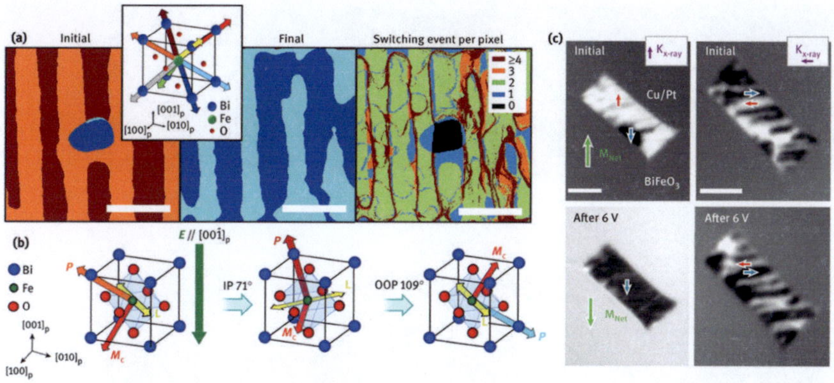

Figure 11.9: **Deterministic switching of magnetization with an electric field.** (a) Polarization vector images determined from PFM measurements before (initial) and after applying an out-of-plane field (final) to the 100-nm-thick $BiFeO_3$ grown on $SrRuO_3/DyScO_3(110)$. The image on the right (number of switching events per pixel) shows that in average a two-step switching is favored. Scale bars are 500 nm. (b) Schematic of the two-step 180° switching of polarization (*P*) with a 71° and a subsequent 109° switching. The antiferromagnetic vector (*L*) and canted moment (M_C) are also represented. Consequently, the two-step switching is accompanied by a 180° rotation of the canted moment of $BiFeO_3$. (c) XMCD-PEEM images of the in-plane moment of a $Co_{0.9}Fe_{0.1}$ layer deposited on $BiFeO_3$ with components viewed perpendicular (vertical $K_{X\text{-ray}}$, where $K_{X\text{-ray}}$ defines the in-plane component of the incident X-ray beam) and parallel to the stripe domains (horizontal $K_{X\text{-ray}}$). The directions of the magnetization in each domain are highlighted with blue and red arrows, which correspond to the local moment direction being perpendicular or parallel to $K_{X\text{-ray}}$. The net $Co_{0.9}Fe_{0.1}$ magnetization (green arrows) reverses after the voltage is applied. Scale bars are 2 μm. (Figure is reprinted with permission from Springer Nature, all panels taken from [147]. Copyright 2014 by Springer Nature.)

domains using the trailing field of the SPM tip, they observed a single spin cycloid within the plane of the film (Figure 11.10(a)). The spin cycloid period is about 70 nm and quantitative NV magnetometry indicates that both the magnetoelectric and the Dzyaloshinskii–Moriya interactions play a role, resulting in a wiggling cycloid (Figure 11.10(b)). Hence, thanks to the magnetoelectric exchange coupling in $BiFeO_3$, the propagation vector of the spin cycloid can be controlled deterministically, envisioning potential applications in the field of magnonics [149] or antiferromagnetic spintronics [150]. In general, the exchange coupling between a strong ferromagnet and a non-collinear antiferromagnetic multiferroic is an intriguing topic that is worth being investigated more thoroughly both theoretically and experimentally.

For large epitaxial strain on $BiFeO_3$, the spin cycloid order is destabilized and a canted G-type antiferromagnetic order emerges [77]. Using SHG, Chauleau et al. observed that two types of submicron antiferromagnetic domains coexist within a single $10 \times 10\ \mu m^2$ ferroelectric domain of $BiFeO_3$ grown on $SrTiO_3(001)$ (Figure 11.3(j)). In addition, they were able to manipulate these antiferromagnetic domains with multiple

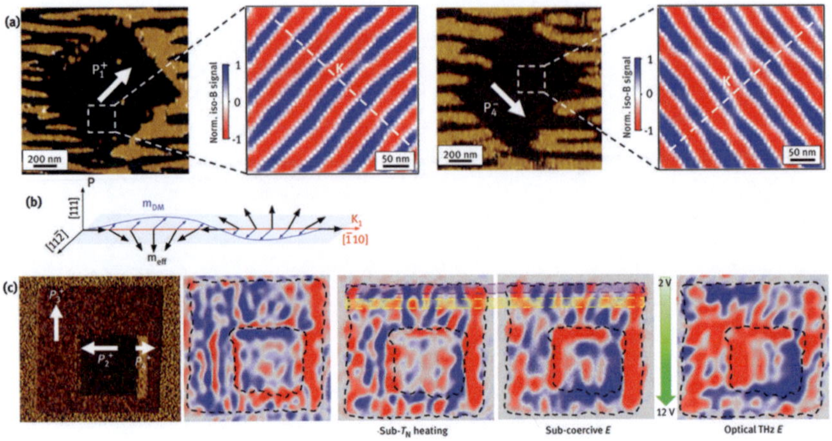

Figure 11.10: Manipulation of the antiferromagnetic order in $BiFeO_3$ thin films. (a) Micron square single domain in $BiFeO_3$ grown on $DyScO_3$(110) read by in-plane PFM and corresponding NV magnetometry image in the center of this square showing the periodic magnetic signal of a spin cycloid. On the right, another micron square single domain with the corresponding magnetic image showing that the propagation vector of the spin cycloid can be rotated by 90° via the magnetoelectric coupling. (b) Sketch of the wiggling cycloid in $BiFeO_3$ with two corresponding canted moments, m_{eff} and m_{DM}, corresponding to magnetoelectric and Dzyaloshinskii–Moriya interactions, respectively. ((a)–(b) are reprinted with permission from Springer Nature, taken from [68]. Copyright 2017 by Springer Nature.) (c) In-plane PFM of a $BiFeO_3$ thin film grown on $SrTiO_3$(001) in which three domains were defined in a 10 × 10 μm^2 square. On the right, corresponding SHG images and manipulation with temperature (570 K), electric field (2 to 10 V) and optical THz. ((c) is reprinted with permission from Springer Nature, taken from [69]. Copyright 2017 by Springer Nature.)

stimuli (Figure 11.10(c)). First, manipulating the ferroelectric domains gave rise to a new pattern of antiferromagnetic domains. This is because, due to the magnetoelectric coupling, switching the polarization variants toggles the direction of magnetic anisotropy, which can induce a rotation of the antiferromagnetic vector. Then, they heated the $BiFeO_3$ sample close to the Néel temperature, which led to a reinitialization of the antiferromagnetic domains. They showed that sub-coercive electric fields were also manipulating these antiferromagnetic domains without changing the ferroelectric domain pattern. Finally, 100-fs laser pulses were used to generate a terahertz electrical pulse in the sample. A profound modification of the antiferromagnetic domain pattern resulted, as this matches the range of the antiferromagnetic resonance in $BiFeO_3$. Thus, one can envisage controlling the antiferromagnetic order in a contactless manner using ultrafast light pulses for an all-optical information technology approach.

Another system which is less established than $BiFeO_3$, but which shows significant potential for electric field control of magnetism, is lead zirconium titanate (PZT)-lead iron tantalate (PFT) [151–154]. Using lamellae prepared out of PZT-PFT ceramics, it has been demonstrated that the ferroelectric order switches under applied magnetic fields at room temperature. An intriguing interplay of electric and magnetic degrees of freedom in PZT-PFT was observed in spatially resolved and integrated measurements, and large coupling constants have been reported [155–158]. The latter is likely to be due to the low structural energy costs of switching between ferroelectric states, evidenced by the relaxor-like domain pattern sometimes observed in thin TEM samples [155] and the small variation in lattice parameters [159]. To evaluate the full potential of PZT-PFT, however, additional comprehensive investigations at the level of domains and domain walls is desirable.

11.3.2 Magnetic control of domain wall charge states in improper ferroelectrics

Spin-driven multiferroics develop domains and domain walls with inseparably entangled electric and magnetic degrees of freedom as introduced in Sections 11.1 and 11.2 (see Chapter 4 for an extended discussion on spin-driven multiferroics). The strong magnetoelectric coupling represents a unique feature of this material class, offering interesting opportunities for the manipulation of domain walls and their electronic properties. The outstanding potential is reflected by proof-of-concept work on iron garnet, where the electrical polarity of domain walls was switched with a magnetic field [160]. In more recent work, a magnetic field was applied to the spin-driven multiferroic $Mn_{0.95}Co_{0.05}WO_4$ to continuously rotate the magnetic easy-plane by 90° and with it the orientation of the ferroelectric polarization (Figure 11.11(a)) [36]. Leo et al. exploited this phenomenon to gain control of the domain wall charge state. They used SHG to image the ferroelectric spin-spiral domains in a $Mn_{0.95}Co_{0.05}WO_4$ single crystal and created neutral 180° ferroelectric domain walls by applying an electric field through macroscopic electrodes (Figure 11.11(b), top). As the primary magnetic order prevents the motion of the wall, a magnetic field of 6 T was used to switch the polarization by 90° without changing the location of the domain wall (Figure 11.11(b), bottom). This way, a magnetic-field-driven change of the charge state was achieved switching from electrically nominally neutral to positively/negatively charged walls [36]. Landau–Lifshitz–Gilbert simulations support the observations of the magnetic-field-induced change of the ferroelectric domain wall charge and give additional insight into the complex Néel-type structure of the domain walls (Figure 11.4(c)).

A similar but discontinuous effect was observed for domain walls in $TbMnO_3$ [39]. In orthorhombic $TbMnO_3$, a spin cycloid appears below $T_C = 27$ K, which breaks inversion symmetry and leads to a spontaneous polarization along the c-axis [50]. Applying a magnetic field along the b-axis alters the orientation of the magnetic easy-plane, giving rise to a polarization flop to the a-axis through a first-order phase transition (Figure 11.11(c)) [50]. Using SHG, Matsubara et al. [39] observed the evolution of multiferroic domains in a $TbMnO_3$ single crystal under the application of an electric field applied along the c-axis. The crystal is naturally forming neutral

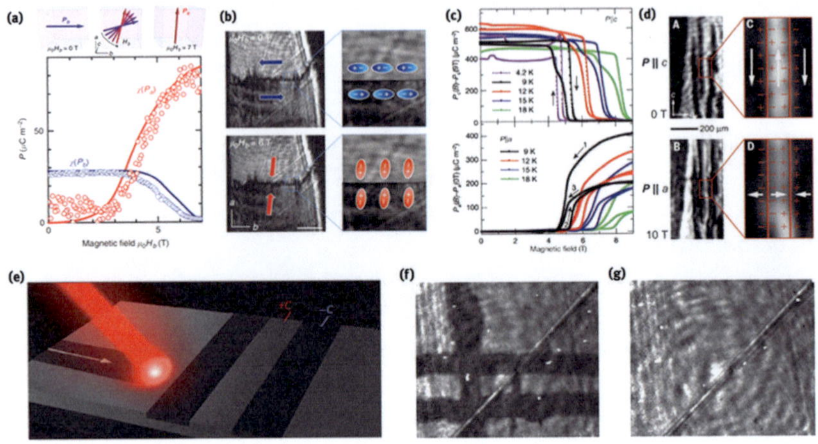

Figure 11.11: **Magnetic control of the domain wall charge state in type II multiferroics.** (a) Top, sketch illustrating the magnetic-field-induced polarization re-orientation from the b- to the a-axis in $Mn_{0.95}Co_{0.05}WO_4$. Bottom, pyroelectric current (solid lines) and integrated SHG measurements (open symbols) at 5 K, showing the magnetic-field induced rotation of polarization. (b) Top, spatially resolved SHG data of a side-by-side 180° domain wall prepared by applying an electric field along the b-axis at zero magnetic field. Bottom, SHG image of the same area under a magnetic field of 6 T, showing that the wall did not move and becomes a tail-to-tail wall with polarization along the a-axis. The scale bar is 250 μm. ((a) – (b) are reprinted with permission from Springer Nature, taken from [36]. Copyright 2015 by Springer Nature.) (c) Electric polarization flop from the c- to the a-axis in $TbMnO_3$ with a magnetic field applied along the b-axis. ((c) is reprinted with permission from Springer Nature, taken from [50]. Copyright 2003 by Springer Nature.) (d) Domain structure of $TbMnO_3$ across the first-order polarization flop observed with SHG. Top, ferroelectric domain structure in the ground state. Bottom, the application of a magnetic field of 10 T flops the polarization along the a-axis but the domain structure does not change, implying that the walls become charged. ((d) is taken from [39]. Reprinted with permission from AAAS.) (e) Schematic image of the optical poling procedure used by Manz et al. [161] to reversibly write antiferromagnetic domains on $TbMnO_3$. (f) SHG image showing laser-written antiferromagnetic domains with positive (+C, bright) and negative (−C, dark) spin chirality. (g) SHG image of the same region shown in (f) after erasing the laser-written −C domains. ((e) – (g) are reprinted with permission from Springer Nature, taken from [161]. Copyright 2016 by Springer Nature.)

ferroelectric 180° domain walls elongated along the c-axis (Figure 11.9(d), top). As the magnetic exchange interaction is also stronger along the c-axis, both magnetic and electric energies are lowered by the formation of these walls. The team then used SHG at 9 K to track the evolution of domains with polarization parallel to the a- and c-axes under a magnetic field of 10 T applied along the b-axis. Within the resolution of the technique (a few microns), they did not detect any change in the domain pattern from 0 to 10 T while the polarization flopped from the c- to the a-axis (Figure 11.11(d), bottom). This implies that the walls changed from a neutral side-by-side configuration to either tail-to-tail or head-to-head charged states. LLG simulations concluded that the magnetic field along the b-axis exerts an effective torque on the magnetization around this axis and leads to an identical flop for both up and down polarization previously aligned along the c-axis. The deterministic nature of this multiferroic phase transition allows to convert nominally neutral walls into charged domain walls.

The opportunity to magnetically convert neutral domain walls into charged domain walls is intriguing as it allows, in principle, reversible control of the density of screening charges and, hence, conductivity. This would enable the design of nanosized electrical gates, where ON and OFF states are set by the magnetic field. Going beyond the application of external magnetic (or electric) fields for controlling the domain wall charge state, intense light fields have been used to selectively create nominally neutral and charged walls in spin-driven multiferroics. Manz et al. showed that both nominally neutral and charged walls can be written and erased in $TbMnO_3$ using pulsed- or continuous-wave lasers (Figure 11.11(e)–(g)) [161]. The findings demonstrate the possibility to optically engineer individual domain walls or patterns of higher complexity as required, e.g., for domain-wall-based circuitry. At present, however, any technological merit is clearly suspended by the cryogenic range of the multiferroic phase in spin-spiral multiferroics such as $TbMnO_3$ and $Mn_{0.95}Co_{0.05}WO_4$, and a one-to-one correlation between domain wall charge state and local conductivity is yet to be demonstrated in these materials. Still, the conceptual results reveal completely new pathways for domain wall engineering and highlight the additional functionalities that arise from the strong coupling between electric and magnetic order available in multiferroics.

11.3.3 Tailoring topological states in multiferroics

The ability to electrically write ferroelectric domain patterns with a conductive tip offers exciting possibilities to realize and harness configurations beyond domain walls that will show topological properties (i.e., possessing an integer or semi–integer topological charge Q). The most classical topological textures are vortices ($Q = 1/2$) and skyrmions ($Q = 1$) (antivortices have $Q = -1/2$ and antiskyrmions $Q = -1$) (see also Chapter 9). So far, research has concentrated mostly on topological magnetic structures, but more and more examples of electric textures with non-trivial topology are reported.

After early theoretical predictions by Naumov et al. [162], ferroelectric vortices were first observed in microscale capacitors of PZT by Gruverman et al. in 2008 [163]. Since then, there have been numerous reports in various types of systems (mostly $Pb(Zr,Ti)O_3$, $BaTiO_3$ and $BiFeO_3$). Two main strategies have been pursued, namely: (i) nanostructuration by, e.g., self-organized growth (Figure 12(a)–(c)) [164, 165] or deposition through sacrificial templates [166], and (ii) electric-field manipulation by scanning probe tips [167]. Only in a few of these studies, the topological character of the vortices has been explicitly quantified, for instance, by computing the

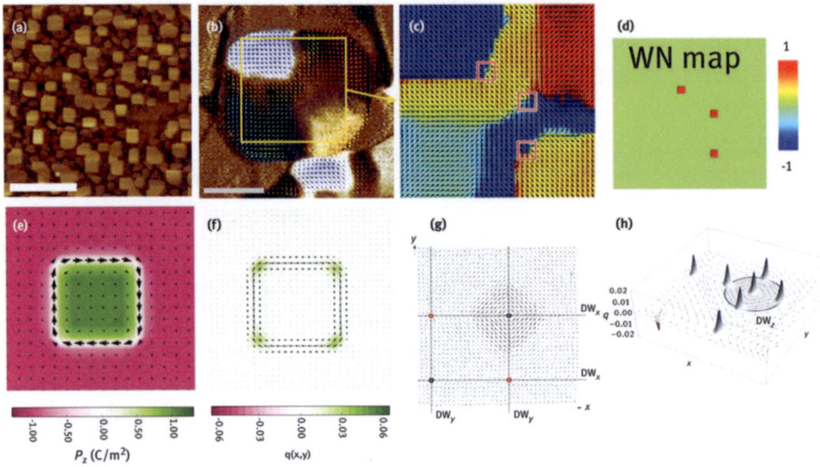

Figure 11.12: Topological structures in multiferroics. (a) Atomic force microscopy image of a $BiFeO_3$ film on $LaAlO_3$ with self-assembled nanoplates. (b) Vector maps of in-plane piezoresponse overlaid on the corresponding out-of-plane PFM contrast of one such nanoplate. (c) In-plane piezoresponse angle maps extracted from the yellow box in (b). (d) Winding number (WN) maps determined from (c). ((a)–(d) are reprinted with permission from Springer Nature, taken from [165]. Copyright 2018 by Springer Nature.) Calculated (e) polarization and (f) Pontryagin density of a nanodomain in $PbTiO_3$ ((e)-(f) reprinted with permission of AAAS from [169] © The Authors, some rights reserved; exclusive licensee American Association for the Advancement of Science. Distributed under a Creative Commons Attribution NonCommercial License 4.0 (CC BY-NC) http://creativecommons.org/licenses/by-nc/4.0/). (g) Vortex structure of a $BaTiO_3$ nanowire embedded in $SrTiO_3$. (h) Associated Pontryagin density map. ((g)–(h) are reprinted with permission from Springer Nature, taken from [168]. Copyright 2015 by Springer Nature.)

vorticity from vector-PFM maps (Figure 11.12(d)) or phase-field simulations mimicking the observed polarization textures. Some works report the electrical manipulation of these vortices, paving the way towards their use in nanoelectronics. Ferroelectric skyrmions were predicted to be stable in nanocomposites (Figure 11.12(g)–(h)) [168] and nanodomains (Figure 11.12(e)–(f)) [169] and, most recently, their existence has been confirmed experimentally in ferroelectric superlattices of $(PbTiO_3)_n/(SrTiO_3)_n$ [170]. To date, whether these ferroelectric topological structures translate into magnetic ones through the local magnetoelectric coupling remains elusive, but new high-sensitivity magnetic microscopy techniques, such as scanning NV microscopy, may shed light on this issue.

In metallic systems with magnetic skyrmion structures, the topological charges can influence the electron transport to produce a topological Hall effect. Many questions arise regarding the consequence of the presence of electric and/or magnetic topological textures in multiferroics in view of their very low conductivity, practically hampering Hall measurements, and more generally on the influence of an electrical topological charge on the macroscopic physical properties. Several papers have predicted other phenomena arising from the presence of magnetic topological charges, such as the appearance of a topological orbital moment [171, 172] or a spin-chirality-driven inverse Faraday effect [173]. The mounting interest for topological structures in ferroelectrics and multiferroics now calls for similar predictions of exotic effects, driven by electrical topology.

11.4 Conclusions

In this article, we reviewed the recent developments in the field of domains and domain walls in multiferroics. More complete review articles dedicated to domain wall nanoelectronics [55, 56, 174] or functional domain and domain walls in multiferroics [40, 54, 62] can be found elsewhere. Charged domain walls have emerged within the last few years as nanoscale objects with enhanced conductivity or distinct magnetism. They are found naturally in improper ferroelectrics [105] or designed in multiferroics by SPM [114], which opens the way towards reconfigurable nanocircuitry [30, 55] and on-demand spintronic nanocomponents.

New tools are now available to image magnetic textures in antiferromagnets at the local scale. These techniques could help understanding the interplay between ferroelectricity and magnetism in multiferroic domain walls. In addition, the field of antiferromagnetic spintronics [150] could be extended to multifunctional materials, such as multiferroics, in which the propagation of spin waves [175] could be controlled by the electric control of ferroelectric domains.

Finally, the possibility to design topological states [176] with electric-field input is particularly exciting for multiferroics. Theoretical works suggest the possibility to fabricate the electric counterpart of magnetic skyrmions. In addition, recent observations reveal the presence of non-Ising walls in 180° domain walls of regular ferroelectrics ($Pb(Zr,Ti)O_3$ and $LiTaO_3$) [35]. This indicates that complex interconnected chiral electric and magnetic structures could be realized in multiferroics.

References

[1] Schmid H. Multi-ferroic magnetoelectrics. Ferroelectrics. 1994;162:317–38.
[2] Tagantsev AK, Cross LE, Fousek J. Domains in ferroic crystals and thin films. New York: Springer-Verlag, 2010.
[3] Hubert A, Schäfer R. Magnetic domains the analysis of magnetic microstructures. Berlin Heidelberg: Springer-Verlag, 1998.
[4] Tolédano JC, Tolédano P. The landau theory of phase transitions. Singapore: World Scientific, 1987.
[5] Wadhawan VK. Introduction to ferroic materials. Singapore: Gordon and Breach Science Publisher 2, 2000.

[6] Fong DD, Kolpak AM, Eastman JA, Streiffer SK, Fuoss PH, Stephenson GB, et al. Stabilization of monodomain polarization in ultrathin $PbTiO_3$ Films. Phys Rev Lett. 2006;96:127601.

[7] Schilling A, Prosandeev S, McQuaid RG, Bellaiche L, Scott JF, Gregg JM. Shape-induced phase transition of domain patterns in ferroelectric platelets. Phys Rev B Condens Matter Mater Phys. 2011;84:64110.

[8] Schilling A, Byrne D, Catalan G, Webber KG, Genenko YA, Wu GS, et al. Domains in ferroelectric nanodots. Nano Lett. 2009;9:3359–64.

[9] Kalinin S, Gruverman A. Scanning probe microscopy, vol. 2. New York: Springer US, 2007.

[10] Eliseev EA, Fomichv YM, Kalinin SV, Vysochanskii YM, Maksymovich P, Morozovska AN. Labyrinthine domains in ferroelectric nanoparticles: manifestation of a gradient-induced morphological transition. Phys Rev B. 2018;98:54101.

[11] Gilletta F. Evolution of ferroelectric domains in TGS single crystals. Phys Status Solidi. 1972;11:721–7.

[12] Merz WJ. Double hysteresis loop of BaTiO3 at the Curie point. Phys Rev. 1953;91:513–17.

[13] Kittel C. Physical theory of ferromagnetic domains. Rev Mod Phys. 1949;21:541–83.

[14] Kittel C. Theory of the structure of ferromagnetic domains in films and small particles. Phys Rev. 1946;70:965–71.

[15] Frenkel J, Doefman J. Spontaneous and induced magnetisation in ferromagnetic bodies. Nature. 1930;126:274–5.

[16] Mitsui T, Furuichi J. Domain structure of rochelle salt and KH2PO4. Phys Rev. 1953;90:193–202.

[17] Catalan G, Scott JF, Schilling A, Gregg JM. Wall thickness dependence of the scaling law for ferroic stripe domains. J Phys Condens Matter. 2007;19:022201.

[18] Catalan G, Béa H, Fusil S, Bibes M, Paruch P, Barthélémy A, et al. Fractal dimension and size scaling of domains in thin films of multiferroic BiFeO3. Phys Rev Lett. 2008;100:027602.

[19] Scott JF. Nanoferroelectrics: statics and dynamics. J Phys Condens Matter. 2006;18:R361–86.

[20] Kläui M. Head-to-head domain walls in magnetic nanostructures. J Phys Condens Matter. 2008;20:313001.

[21] Eyring EM. Topics in applied physics. Nucl Technol. 1978;39:327–8.

[22] Holtz ME, Shapovalov K, Mundy JA, Chang CS, Yan Z, Bourret E, et al. Topological defects in hexagonal manganites: inner structure and emergent electrostatics. Nano Lett. 2017;17:5883–90.

[23] Choudhury S, Li Y, Odagawa N, Vasudevarao A, Tian L, Capek P, et al. The influence of 180° ferroelectric domain wall width on the threshold field for wall motion. J Appl Phys. 2008;104:84107.

[24] Scott JF, Evans DM, Gregg JM, Gruverman A. Hydrodynamics of domain walls in ferroelectrics and multiferroics: impact on memory devices. Appl Phys Lett. 2016;109:42901.

[25] Tsai F, Khiznichenko V, Cowley JM. High-resolution electron microscopy of 90° ferroelectric domain boundaries in BaTiO3 and Pb(Zr0.52Ti0.48)O3, Ultramicroscopy. 1992;45:55–63.

[26] Dennis MD, Bradt RC. Thickness of 90° ferroelectric domain walls in (Ba,Pb)TiO3 single crystals. J Appl Phys. 1974;45:1931–3.

[27] Lee D, Behera RK, Wu P, Xu H, Sinnott SB, Phillpot SR, et al. Mixed Bloch-Néel-Ising character of 180° ferroelectric domain walls. Phys Rev B. 2009;80:2–5.

[28] Goncalves-Ferreira L, Redfern SA, Artacho E, Salje EK. Ferrielectric twin walls in CaTiO3. Phys Rev Lett. 2008;101:97602.

[29] Tagantsev AK, Courtens E, Arzel L. Prediction of a low-temperature ferroelectric instability in antiphase domain boundaries of strontium titanate. Phys Rev B Condens Matter Mater Phys. 2001;64:2241071–4.

[30] Gregg JM. Exotic domain states in ferroelectrics: searching for vortices and skyrmions. Ferroelectrics. 2012;433:74–87.

[31] Morozovska AN. Domain wall conduction in ferroelectrics. Ferroelectrics. 2012;438:3–19.

[32] Stepkova V, Marton P, Hlinka J. Stress-induced phase transition in ferroelectric domain walls of $BaTiO_3$. J Phys Condens Matter. 2012;24:212201.

[33] De Luca G, Rossell MD, Schaab J, Viart N, Fiebig M, Trassin M. Domain wall architecture in tetragonal ferroelectric thin films. Adv Mater. 2017;29:1605145.

[34] Wei X-K, Jia C-L, Sluka T, Wang B-X, Ye Z-G, Setter N. Néel-like domain walls in ferroelectric $Pb(Zr,Ti)O_3$ single crystals. Nat Commun. 2016;7:12385.

[35] Cherifi-Hertel S, Bulou H, Hertel R, Taupier G, Dorkenoo KD(H), Andreas C, et al. Non-Ising and chiral ferroelectric domain walls revealed by nonlinear optical microscopy. Nat Commun. 2017;8:15768.

[36] Leo N, Bergman A, Cano A, Poudel N, Lorenz B, Fiebig M, et al. Polarization control at spin-driven ferroelectric domain walls. Nat Commun. 2015;6:4–9.

[37] Meier D, Leo N, Maringer D, Lottermoser Th, Fiebig M, Becker P, et al. Topology and manipulation of multiferroic hybrid domains in MnWO4. Phys Rev B. 2009;80:224420.

[38] Domingo N, Farokhipoor S, Santiso J, Noheda B, Catalan G. Domain wall magnetoresistance in BiFeO3 thin films measured by scanning probe microscopy. J Phys Condens Matter. 2017;29:334003.

[39] Matsubara M, Manz S, Mochizuki M, Kubacka T, Iyama A, Aliouane N, et al. Magnetoelectric domain control in multiferroic TbMnO3. Science. 2015;348:1112–15.

[40] Fiebig M, Lottermoser T, Meier D, Trassin M. The evolution of multiferroics. Nat Rev Mater. 2016;1:16046.

[41] Fiebig M, Lottermoser T, Fröhlich D, Goltsev AV, Pisarev RV. Observation of coupled magnetic and electric domains. Nature. 2002;419:818–20.

[42] Redfern SA, Carpenter MA. Transformation processes in mineral. Berlin, Boston: De Gruyter, 2018.

[43] Lines ME, Glass AM. Principles and applications of ferroelectrics and related materials. Oxford: Oxford University Press, 2010.

[44] Salje EK, Carpenter MA. Linear - Quadratic order parameter coupling and multiferroic phase transitions. J Phys Condens Matter. 2011;23:462202.

[45] Salje EK. Phase transitions in ferroelastic and co-elastic crystals. Cambridge University Press, 1991.

[46] Carpenter MA, Salje EK. Elastic anomalies in minerals due to structural phase transitions. Eur J Mineral. 1998;10:693–812.

[47] Fiebig M. Revival of the magnetoelectric effect. J Phys D Appl Phys. 2005;38:R123–R152.

[48] Kimura T. Spiral magnets as magnetoelectrics. Annu Rev Mater Res. 2007;37:387–413.

[49] Mostovoy M. Ferroelectricity in spiral magnets. Phys Rev Lett. 2006;96:67601.

[50] Kimura T, Goto T, Shintani H, Ishizaka K, Arima T, Tokura Y. Magnetic control of ferroelectric polarization. Nature. 2003;426:55–8.

[51] Kagawa F, Mochizuki M, Onose Y, Murakawa H, Kaneko Y, Furukawa N, et al. Dynamics of multiferroic domain wall in spin-cycloidal ferroelectric DyMnO3. Phys Rev Lett. 2009;102:57604.

[52] Lorenz B. Hexagonal manganites—(RMnO 3): class (I) multiferroics with strong coupling of magnetism and ferroelectricity. ISRN Condens Matter Phys. 2013;2013:1–43.

[53] Wu W, Geng Y, Fennie C, Mostovoy M, Cheong S. Visualizing domain wall magnetism and domain magnetoelectricity in multiferroic hexagonal manganites. In: 2015 IEEE International Magnetics Conference (INTERMAG), 2015:1–1.

[54] Meier D. Functional domain walls in multiferroics. J Phys Condens Matter. 2015;27.

[55] Catalan G, Seidel J, Ramesh R, Scott JF. Domain wall nanoelectronics. Rev Mod Phys. 2012;84:119–56.

[56] Bednyakov PS, Sturman BI, Sluka T, Tagantsev AK, Yudin PV. Physics and applications of charged domain walls. npj Comput Mater. 2018;4:65.

[57] Mundy JA, Schaab J, Kumagai Y, Cano A, Stengel M, Krug IP, et al. Functional electronic inversion layers at ferroelectric domain walls. Nat Mater. 2017;16:622.

[58] Sharma P, Zhang Q, Sando D, Lei CH, Liu Y, Li J, et al. Nonvolatile ferroelectric domain wall memory. Sci Adv. 2017;3.

[59] Whyte JR, McQuaid RG, Sharma P, Canalias C, Scott JF, Gruverman A, et al. Ferroelectric domain wall injection. Adv Mater. 2014;26:293–8.

[60] Whyte JR, Gregg JM. A diode for ferroelectric domain-wall motion. Nat Commun. 2015;6:7361.

[61] Bednyakov PS, Sluka T, Tagantsev AK, Damjanovic D, Setter N. Formation of charged ferroelectric domain walls with controlled periodicity. Sci Rep. 2015;5:15819.

[62] Spaldin NA, Ramesh R. Advances in magnetoelectric multiferroics. Nat Mater. 2019;18:203–12.

[63] Dong S, Xiang H, Dagotto E. Magnetoelectricity in multiferroics: a theoretical perspective. Natl Sci Rev. 2019;6:629–41.

[64] Campanini M, Erni R, Rossell MD. Probing local order in multiferroics by transmission electron microscopy. Phys Sci Rev. 2019;0:1–36.

[65] Lahtinen THE, Franke KJA, van Dijken S. Electric-field control of magnetic domain wall motion and local magnetization reversal. Sci Rep. 2012;2:258.

[66] Trassin M, Clarkson JD, Bowden SR, Liu J, Heron JT, Paull RJ, et al. Interfacial coupling in multiferroic/ferromagnet heterostructures. Phys Rev B. 2013;87:134426.

[67] Zhou Z, Trassin M, Gao Y, Gao Y, Qiu D, Ashraf K, et al. Probing electric field control of magnetism using ferromagnetic resonance. Nat Commun. 2015;6:6082.

[68] Gross I, Akhtar W, Garcia V, MartÎnez LJ, Chouaieb S, Garcia K, et al. Real-space imaging of non-collinear antiferromagnetic order with a single-spin magnetometer. Nature. 2017;549:252–6.

[69] Chauleau JY, Haltz E, Carrétéro C, Fusil S, Viret M. Multi-stimuli manipulation of antiferromagnetic domains assessed by second-harmonic imaging. Nat Mater. 2017;16:803–7.

[70] Merz WJ. The electric and optical behavior of BaTiO3 single-domain crystals. Phys Rev. 1949;76:1221–5.

[71] Forsbergh PW. Domain structures and phase transitions in barium titanate. Phys Rev. 1949;76:1187–201.

[72] Matthias B, von Hippel A. Domain structure and dielectric response of barium titanate single crystals. Phys Rev. 1948;73:1378–84.

[73] Hooton JA, Merz WJ. Etch patterns and ferroelectric domains in BaTiO3 single crystals. Phys Rev. 1955;98:409–13.

[74] Gruverman A, Alexe M, Meier D. Piezoresponse force microscopy and nanoferroic phenomena. Nat Commun. 2019;10:1661.

[75] Rondin L, Tetienne JP, Hingant T, Roch JF, Maletinsky P, Jacques V. Magnetometry with nitrogen-vacancy defects in diamond. Reports Prog Phys. 2014;77:056503.

[76] Lebeugle D, Colson D, Forget A, Viret M, Bataille AM, Gukasov A. Electric-field-induced spin flop in $BiFeO_3$ single crystals at room temperature. Phys Rev Lett. 2008;100:227602.

[77] Sando D, Agbelele A, Rahmedov D, Liu J, Rovillain P, Toulouse C, et al. Crafting the magnonic and spintronic response of BiFeO3 films by epitaxial strain. Nat Mater. 2013;12:641–6.

[78] Sosnowska I, Neumaier TP, Steichele E. Spiral magnetic ordering in bismuth ferrite. J Phys C Solid State Phys. 1982;15:4835–46.

[79] Zhao T, Scholl A, Zavaliche F, Lee K, Barry M, Doran A, et al. Electrical control of antiferromagnetic domains in multiferroic BiFeO3 films at room temperature. Nat Mater. 2006;5:823–9.

[80] Fiebig M, Fröhlich D, Kohn K, Leute St, Lottermoser Th, Pavlov VV, et al. Determination of the magnetic symmetry of hexagonal manganites by second harmonic generation. Phys Rev Lett. 2000;84:5620–3.
[81] Nordlander J, De Luca G, Strkalj N, Fiebig M, Trassin M. Probing ferroic states in oxide thin films using optical second harmonic generation. Appl Sci. 2018;8.
[82] Valencia S, Crassous A, Bocher L, Garcia, V, Moya X, Cherifi RO, et al. Interface-induced room-temperature multiferroicity in BaTiO3. Nat Mater. 2011;10:753.
[83] Seidel J, Maksymovych P, Batra Y, Katan A, Yang SY, He Q, et al. Domain wall conductivity in La-doped BiFeO3. Phys Rev Lett. 2010;105:197603.
[84] Salje EK, Lashley JC. Domain boundary engineering in ferroic and multiferroic materials: a simple introduction. In: Kakeshita T, Fukuda T, Saxena A, Planes A editors, Springer series in materials science. vol. 148 no. 1. Berlin, Heidelberg: Springer Berlin Heidelberg, 2011:1–18.
[85] Scott JF. Room-temperature multiferroic magnetoelectrics. Npg Asia Mater. 2013;5:e72.
[86] Meier D, Maringer M, Lottermoser T, Becker P, Bohatý L, Fiebig M. Observation and coupling of domains in a spin-spiral multiferroic. Phys Rev Lett. 2009;102:107202.
[87] Seidel J, Martin LW, He Q, Zhan Q, Chu YH, Rother A, et al. Conduction at domain walls in oxide multiferroics. Nat Mater. 2009;8:229–34.
[88] Padilla J, Zhong W, Vanderbilt D. First-principles investigation of 180° domain walls in BaTO3. Phys Rev B Condens Matter Mater Phys. 1996;53:R5969–R5973.
[89] McMichael RD, Donahue MJ. Head to head domain wall structures in thin magnetic strips. IEEE Trans Magn. 1997;33, PART 2:4167–9.
[90] Chynoweth AG, Feldmann WL. Ferroelectric domain delineation in triglycine sulphate and domain arrays produced by thermal shocks. J Phys Chem Solids. 1960;15:225–33.
[91] Grekov AA, Adonin AA, Protsenko NP. Encountering domains in sbsi. Ferroelectrics. 1976;13:483–5.
[92] Fesenko EG, Gavrilyatchenko VG, Martinenko MA, Semenchov AF, Lapin IP. Domain structure peculiarities of lead-titanate crystals. Ferroelectrics. 1973;6:61–5.
[93] Surowiak Z, Dec J, Skulski R, Fesenko EG, Gavrilyatchenko VG, Semenchev AF. The domain structure formation at phase transitions. Ferroelectrics. 1978;20:277–9.
[94] Jona F, Shirane G. Ferroelectric crystals, vol, 43. John Wiley & Sons, Ltd, 1963:10–11.
[95] Aristov VV, Kokhanchik LS, Meyer K-P, Blumtritt H. Scanning electron microscopic investigations of peculiarities of the BaTiO3 ferroelectric domain contrast. Phys Status Solidi. 1983;78:229–36.
[96] Aristov VV, Kokhanchik LS, Voronovskii YI. Voltage contrast of ferroelectric domains of lithium niobate in SEM. Phys Status Solidi. 1984;86:133–41.
[97] Sogr AA. Domain structure of ferroelectrics observed in the scanning electron microscope. Ferroelectrics. 1989;97:47–57.
[98] Guyonnet J, Gaponenko I, Gariglio S, Paruch P. Conduction at domain walls in insulating Pb (Zr0.2Ti0.8)O3 thin films. Adv Mater. 2011;23:5377–82.
[99] Tselev A, Yu P, Cao Y, Dedon LR, Martin LW, Kalinin SV, et al. Microwave a.c. conductivity of domain walls in ferroelectric thin films. Nat Commun. 2016;7:11630.
[100] Stolichnov I, Feigl L, McGilly LJ, Sluka T, Wei XK, Colla E, et al. Bent ferroelectric domain walls as reconfigurable metallic–like channels. Nano Lett. 2015;15:8049–55.
[101] Schröder M, Haußmann A, Thiessen A, Soergel E, Woike T, Eng LM. Conducting domain walls in lithium niobate single crystals. Adv Funct Mater. 2012;22:3936–44.
[102] Godau C, Kämpfe T, Thiessen A, Eng LM, Haußmann A. Enhancing the domain wall conductivity in lithium niobate single crystals. ACS Nano. 2017;11:4816–24.

[103] Werner CS, Herr SJ, Buse K, Sturman B, Soergel E, Razzaghi C, et al. Large and accessible conductivity of charged domain walls in lithium niobate. Sci Rep. 2017;7:9862.
[104] Sluka T, Tagantsev AK, Bednyakov P, Setter N. Free-electron gas at charged domain walls in insulating BaTiO3. Nat Commun. 2013;4:1808.
[105] Meier D, Seidel J, Cano A, Delaney K, Kumagai Y, Mostovoy M, et al. Anisotropic conductance at improper ferroelectric domain walls. Nat Mater. 2012;11:284–8.
[106] Choi T, Horibe Y, Yi HT, Choi YJ, Wu W, Cheong SW. Insulating interlocked ferroelectric and structural antiphase domain walls in multiferroic YMnO3. Nat Mater. 2010;9:253–8.
[107] Campbell MP, McConville JP, McQuaid RG, Prabhakaran D, Kumar A, Gregg JM. Hall effect in charged conducting ferroelectric domain walls. Nat Commun. 2016;7:13764.
[108] Schaab J, Skærvø SH, Krohns S, et al. Electrical half-wave rectification at ferroelectric domain walls. Nat Nanotechnol. 2018;13:1028–34.
[109] Schaab J, Krug IP, Nickel F, Dai X, Holtz ME, Cano A, et al. Imaging and characterization of conducting ferroelectric domain walls by photoemission electron microscopy. Appl Phys Lett. 2014;104:232904.
[110] McQuaid RGP, Campbell MP, Whatmore RW, Kumar A, Gregg JM. Injection and controlled motion of conducting domain walls in improper ferroelectric Cu-Cl boracite. Nat Commun. 2017;8:15105.
[111] Oh YS, Luo X, Huang FT, Wang Y, Cheong SW. Experimental demonstration of hybrid improper ferroelectricity and the presence of abundant charged walls in (Ca,Sr)3Ti2O7 crystals. Nat Mater. 2015;14:407–13.
[112] Jiang J, Bai ZL, Chen ZH, He L, Zhang DW, Zhang QH, et al. Temporary formation of highly conducting domain walls for non-destructive read-out of ferroelectric domain-wall resistance switching memories. Nat Mater. 2018;17:49–55.
[113] Jiang AQ, Zhang Y. Next-generation ferroelectric domain-wall memories: principle and architecture. NPG Asia Mater. 2019;11:2.
[114] Crassous A, Sluka T, Tagantsev AK, Setter N. Polarization charge as a reconfigurable quasi-dopant in ferroelectric thin films. Nat Nanotechnol. 2015;10:614–18.
[115] Farokhipoor S, Noheda B. Conduction through 71° domain walls in BiFeO3 thin films. Phys Rev Lett. 2011;107:1–4.
[116] Jungk T, Hoffmann Á, Fiebig M, Soergel E. Electrostatic topology of ferroelectric domains in $YMnO_3$. Appl Phys Lett. 2010;97:012904.
[117] Kumagai Y, Spaldin NA. Structural domain walls in polar hexagonal manganites. Nat Commun. 2013;4:1540.
[118] Småbråten DR, Meier QN, Skjærvø SH, Inzani K, Meier D, Selbach SM. Charged domain walls in improper ferroelectric hexagonal manganites and gallates. Phys Rev Mater. 2018;2:114405.
[119] Wu W, Horibe Y, Lee N, Cheong S-W, Guest JR. Conduction of topologically protected charged ferroelectric domain walls. Phys Rev Lett. 2012;108:77203.
[120] Mosberg AB, Roede ED, Evans DM, Holstad TS, Bourret E, Yan Z, et al. FIB lift-out of conducting ferroelectric domain walls in hexagonal manganites. Appl Phys Lett. 2019;115:122901.
[121] Turner PW, McConville JP, McCartan SJ, Campbell MH, Schaab J, McQuaid RG, et al. Large carrier mobilities in $ErMnO_3$ conducting domain walls revealed by quantitative hall effect measurements. Nano Lett. 2018;18:6381–6.
[122] Schoenherr P, Shapovalov K, Schaab J, Yan Z, Bourret ED, Hentschel M, et al. Observation of uncompensated bound charges at improper ferroelectric Domain Walls. Nano Lett. 2019;19:1659–64.

[123] Skjærvø SH, Småbråten DR, Spaldin NA, Tybell T, Selbach SM. Oxygen vacancies in the bulk and at neutral domain walls in hexagonal YMnO3. Phys Rev B. 2018;98:184102.

[124] Skjærvø SH, Wefring ET, Nesdal SK, Gaukås NH, Olsen GH, Glaum J, et al. Interstitial oxygen as a source of p-type conductivity in hexagonal manganites. Nat Commun. 2016;7:13745.

[125] Wang X, Huang FT, Hu R, Fan F, Cheong SW. Self-poling with oxygen off-stoichiometry in ferroelectric hexagonal manganites. APL Mater. 2015;3:41505.

[126] Selbach SM, Nordli Løvik A, Bergum K, Tolchard JR, Einarsrud M-A, Grande T. Crystal structure, chemical expansion and phase stability of $HoMnO_3$ at high temperature. J Solid State Chem. 2012;196:528–35.

[127] McCarthy GJ, Gallagher PV, Sipe C. Crystal chemistry of catalyst materials. I. Composition and unit cell parameters of $REMnO_3$ phases prepared in air. Mater Res Bull. 1973;8:1277–84.

[128] Griffin SM, Reidulff M, Selbach SM, Spaldin NA. Defect chemistry as a crystal structure design parameter: intrinsic point defects and Ga substitution in $InMnO_3$. Chem Mater. 2017;29:2425–34.

[129] Holstad TS, Evans DM, Ruff A, Småbråten DR, Schaab J, Tzschaschel Ch, et al. Electronic bulk and domain wall properties in B -site doped hexagonal ErMnO3. Phys Rev B. 2018;97:85143.

[130] Hassanpour E, Wegmayr V, Schaab J, Yan Z, Bourret E, Lottermoser Th, et al. Robustness of magnetic and electric domains against charge carrier doping in multiferroic hexagonal $ErMnO_3$. New J Phys. 2016;18:43015.

[131] Meier QN, Lilienblum M, Griffin SM, Conder K, Pomjakushina E, Yan Z, et al. Global formation of topological defects in the multiferroic hexagonal manganites. Phys Rev X. 2017;7:41014.

[132] Griffin SM, Lilienblum M, Delaney KT, Kumagai Y, Fiebig M, Spaldin NA. Scaling behavior and beyond equilibrium in the hexagonal manganites. Phys Rev X. 2012;2:41022.

[133] Griffin SM, Spaldin NA. On the relationship between topological and geometric defects. J Phys Condens Matter. 2017;29:343001.

[134] Balke N, Choudhury S, Jesse S, Huijben M, Chu YH, Baddorf AP, et al. Deterministic control of ferroelastic switching in multiferroic materials. Nat Nanotechnol. 2009;4:868–75.

[135] Lubk A, Gemming S, Spaldin NA. First-principles study of ferroelectric domain walls in multiferroic bismuth ferrite. Phys Rev B. 2009;80:104110.

[136] Daraktchiev M, Catalan G, Scott JF. Landau theory of domain wall magnetoelectricity. Phys Rev B. 2010;81:224118.

[137] Scheinfein MR, Unguris J, Kelley MH, Pierce DT, Celotta RJ. Scanning electron microscopy with polarization analysis (SEMPA). Rev Sci Instrum. 1990;61:2501–27.

[138] Zhang JX, Zeches RJ, He Q, Chu YH, Ramesh R. Nanoscale phase boundaries: a new twist to novel functionalities. Nanoscale. 2012;4:6196–204.

[139] He Q, Yeh CH, Yang JC, Singh-Bhalla G, Liang CW, Chiu PW, et al. Magnetotransport at domain walls in BiFeO3. Phys Rev Lett. 2012;108:1–5.

[140] Geng Y, Lee N, Choi YJ, Cheong SW, Wu W. Collective magnetism at multiferroic vortex domain walls. Nano Lett. 2012;12:6055–9.

[141] Meier D, Ryll H, Kiefer K, Klemke B, Hoffmann JU, Ramesh R, et al. Mutual induction of magnetic 3d and 4f order in multiferroic hexagonal ErMnO3. Phys Rev B. 2012;86:184415.

[142] Taniguchi K, Abe N, Takenobu T, Iwasa Y, Arima T. Ferroelectric polarization flop in a frustrated magnet MnWO4 induced by a magnetic field. Phys Rev Lett. 2006;97:097203.

[143] Farokhipoor S, Magén C, Venkatesan S, Íñiguez J, Daumont CJ, Rubi D, et al. Artificial chemical and magnetic structure at the domain walls of an epitaxial oxide. Nature. 2014;515:379–83.

[144] Juraschek DM, Fechner M, Balatsky AV, Spaldin NA. Dynamical multiferroicity. Phys Rev Mater. 2017;1:14401.

[145] Heron JT, Trassin M, Ashraf K, Gajek M, He Q, Yang SY, et al. Electric-field-induced magnetization reversal in a ferromagnet-multiferroic heterostructure. Phys Rev Lett. 2011;107:217202.
[146] Bibes M, Barthélémy A. Multiferroics: towards a magnetoelectric memory. Nat Mater. 2008;7:425–6.
[147] Heron JT, Bosse JL, He Q, Gao Y, Trassin M, Ye L, et al. Deterministic switching of ferromagnetism at room temperature using an electric field. Nature. 2014;516:370–3.
[148] Ederer C, Spaldin NA. Influence of strain and oxygen vacancies on the magnetoelectric properties of multiferroic bismuth ferrite. Phy Rev B. 2005;71:224103.
[149] Rovillain P, de Sousa R, Gallais Y, Sacuto A, Méasson MA, Colson D, et al. Electric-field control of spin waves at room temperature in multiferroic BiFeO3. Nat Mater. 2010;9:975–9.
[150] Jungwirth T, Marti X, Wadley P, Wunderlich J. Antiferromagnetic spintronics. Nat Nanotechnol. 2016;11:231–41.
[151] Scott JF, Evans DM, Katiyar RS, McQuaid RG, Gregg JM. Nonequilibrium ferroelectric-ferroelastic 10 nm nanodomains: wrinkles, period-doubling, and power-law relaxation. J Phys Condens Matter. 2017;29:304001.
[152] Glinchuk MD, Eliseev EA, Morozovska AN. Novel room temperature multiferroics on the base of single-phase nanostructured perovskites. J Appl Phys. 2014;116:54101.
[153] Glinchuk MD, Eliseev EA, Morozovska AN. Landau-Ginzburg description of anomalous properties of novel room temperature multiferroics Pb(Fe1/2Ta1/2)x(Zr0.53Ti0.47)1-xO3 and Pb(Fe1/2Nb1/2)x(Zr0.53Ti0.47)1-xO3. J Appl Phys. 2016;119:24102.
[154] Wu Y-C, Ho SZ, Liu YC, Liou Y-D, Liu W-Y, Huang S-W, et al. Room temperature multiferroic PZTFT thin films. ACS Appl Electron Mater, 2019.
[155] Evans DM, Schilling A, Kumar A, Sanchez D, Ortega N, Arredondo M, et al. Magnetic switching of ferroelectric domains at room temperature in multiferroic PZTFT. Nat Commun. 2013;4:1534.
[156] Evans DM, Schilling A, Kumar A, Sanchez D, Ortega N, Katiyar RS, et al. Switching ferroelectric domain configurations using both electric and magnetic fields in Pb(Zr,Ti)O_3-Pb(Fe,Ta)O_3 single-crystal lamellae. Philos Trans R Soc A. 2014;372:20120450.
[157] Schiemer J, Carpenter MA, Evans DM, Gregg, JM, Schilling A, Arredondo M, et al. Studies of the room-temperature multiferroic Pb(Fe0.5Ta0.5)0.4(Zr0.53Ti0.47)0.6O3: resonant ultrasound spectroscopy, dielectric, and magnetic phenomena. Adv Funct Mater. 2014;24:2993–3002.
[158] Evans DM, Alexe M, Schilling A, Kumar A, Sanchez D, Ortega N, et al. The Nature of magnetoelectric coupling in Pb(Zr,Ti)O3-Pb(Fe,Ta)O3. Adv Mater. 2015;27:6068–73.
[159] Sanchez DA, Ortega N, Kumar A, Roque-Malherbe R, Polanco R, Scott JF, et al. Symmetries and multiferroic properties of novel room-temperature magnetoelectrics: lead iron tantalate – lead zirconate titanate (PFT/PZT). AIP Adv. 2011;1:42169.
[160] Pyatakov AP, Sechin DA, Sergeev AS, Nikolaev AV, Nikolaeva EP, Logginov AS, et al. Magnetically switched electric polarity of domain walls in iron garnet films. Epl. 2011;93:17001.
[161] Manz S, Matsubara M, Lottermoser T, Büchi J, Iyama A, Kimura T, et al. Reversible optical switching of antiferromagnetism in TbMnO3. Nat Photonics. 2016;10:653–6.
[162] Naumov II, Bellaiche L, Fu H. Unusual phase transitions in ferroelectric nanodisks and nanorods. Nature. 2004;432:737–40.
[163] Gruverman A, Wu D, Fan HJ, Vrejoiu I, Alexe M, Harrison RJ, et al. Vortex ferroelectric domains. J Phys Condens Matter. 2008;20:342201.

[164] Ma J, Ma J, Zhang Q, Peng R, Wang J, Liu C, et al. Controllable conductive readout in self-assembled, topologically confined ferroelectric domain walls. Nat Nanotechnol. 2018;13:947–52.
[165] Kim KE, Jeong S, Chu K, Lee JH, Kim GY, Xue F, et al. Configurable topological textures in strain graded ferroelectric nanoplates. Nat Commun. 2018;9:1–12.
[166] Li Z, Wang Y, Tain G. High-density array of ferroelectric nanodots with robust and reversibly switchable topological domain states. Sci Adv. 2017;3:e1700919.
[167] Balke N, Winchester B, Ren W, Chu, YH, Morozovska AN, Eliseev EA, et al. Enhanced electric conductivity at ferroelectric vortex cores in BiFeO3. Nat Phys. 2012;8:81–8.
[168] Nahas Y, Prokhorenko S, Louis L, Gui Z, Kornev I, Bellaiche L. Discovery of stable skyrmionic state in ferroelectric nanocomposites. Nat Commun. 2015;6:1–6.
[169] Gonçalves MA, Escorihuela-Sayalero C, Garca-Fernández P, Junquera J, Íñiguez J. Theoretical guidelines to create and tune electric skyrmion bubbles. Sci Adv. 2019;5:eaau7023.
[170] Das S, Tang YL, Hong Z, GonÇalves MA, McCarter MR, Klewe C, et al. Observation of room-temperature polar skyrmions. Nature. 2019;568:368–72.
[171] Hoffmann M, Weischenberg J, Dupé B, Freimuth F, Ferriani P, Mokrousov Y, et al. Topological orbital magnetization and emergent Hall effect of an atomic-scale spin lattice at a surface. Phys Rev B. 2015;92:020401.
[172] Hanke J, Freimuth F, Blügel S, Mokrousov Y. Prototypical topological orbital ferromagnet γ-FeMn. Sci Rep. 2017;7:41078.
[173] Taguchi K, Tatara G. Inverse faraday effect driven by spin chirality in weak ferromagnets. J Korean Phys Soc. 2013;62:1759–62.
[174] Vasudevan RK, Wu W, Guest JR, Baddorf AP, Morozovska AN, Eliseev EA, et al. Domain wall conduction and polarization-mediated transport in ferroelectrics. Adv Funct Mater. 2013;23:2592–616.
[175] Lebrun R, Ross A, Bender SA, Qaiumzadeh A, Baldrati L, Cramer J, et al. Tunable long-distance spin transport in a crystalline antiferromagnetic iron oxide. Nature. 2018;561:222–5.
[176] Huang F-T, Cheong S-W. Aperiodic topological order in the domain configurations of functional materials. Nat Rev Mater. 2017;2:17004.

Elzbieta Gradauskaite, Peter Meisenheimer, Marvin Müller, John Heron and Morgan Trassin

12 Multiferroic heterostructures for spintronics

Abstract: For next-generation technology, magnetic systems are of interest due to the natural ability to store information and, through spin transport, propagate this information for logic functions. Controlling the magnetization state through currents has proven energy inefficient. Multiferroic thin-film heterostructures, combining ferroelectric and ferromagnetic orders, hold promise for energy efficient electronics. The electric field control of magnetic order is expected to reduce energy dissipation by 2–3 orders of magnitude relative to the current state-of-the-art. The coupling between electrical and magnetic orders in multiferroic and magnetoelectric thin-film heterostructures relies on interfacial coupling though magnetic exchange or mechanical strain and the correlation between domains in adjacent functional ferroic layers. We review the recent developments in electrical control of magnetism through artificial magnetoelectric heterostructures, domain imprint, emergent physics and device paradigms for magnetoelectric logic, neuromorphic devices, and hybrid magnetoelectric/spin-current-based applications. Finally, we conclude with a discussion of experiments that probe the crucial dynamics of the magnetoelectric switching and optical tuning of ferroelectric states towards all-optical control of magnetoelectric switching events.

Keywords: multiferroics, $BiFeO_3$, Cr_2O_3, magnetoelectrics, artificial multiferroics, spintronics, ferroic domain imprint, exchange bias, magnetostriction, memory and logic

12.1 Introduction

Our review is composed of five sections. We begin with a brief introduction dealing with thin-film multilayers combining ferroelectric and ferromagnetic order parameters, so-called multiferroic magnetoelectric heterostructures. The ferroelectricity of the first layer, together with accompanying electrically tunable strain state and surface bound charges, influence the magnetization of the adjacent layer. We present multiferroic magnetoelectric heterostructures as an energy-efficient alternative to existing magnetic-based data storage and logic devices. We continue, in Section 12.2, with an overview of the recent developments concerning the electric field control of magnetism using multiferroic heterostructures. We then, in Section 12.3, discuss different

This article has previously been published in the journal Physical Sciences Reviews. Please cite as: Gradauskaite, E., Meisenheimer, P., Müller, M, Heron, J., Trassin, M. Multiferroic heterostructures for spintronics *Physical Sciences Reviews* [Online] 2021, 6. DOI: 10.1515/psr-2019-0072

https://doi.org/10.1515/9783110582130-012

ways to engineer ferroelectric domains and domain walls. Their role in the magnetoelectric switching mechanism will be presented. We then have a look at the current device concepts, involving multiferroic heterostructures in Section 12.4. Finally, Section 12.5 highlights the prospects for determination of magnetoelectric switching dynamics and for ultrafast all-optical magnetoelectric switch.

12.1.1 Magnetoelectric multiferroic heterostructures

Multiferroic materials, exhibiting multiple order parameters in a single-phase, were first discussed in the 50's [1, 2] and in the recent years have been at the center of interest with the development of experimental techniques to probe and design such complex materials [3–5]. Here, we will consider materials exhibiting ferroelectricity and ferromagnetism for spintronic applications [6, 7]. When the two orders coexist but are not coupled, so-called multiferroic memory elements can be envisioned [8]. In such memories, the information can be stored independently in each order parameter state, increasing the storage density in a given active space. The focus of the scientific community, however, is shifting towards magnetoelectric multiferroic materials, where the magnetic and electric orders are coupled, meaning that a magnetic state can be electrically altered. Magnetoelectric-based device paradigms [9] could solve the current limitations accompanying the magnetic data storage development and the energy costs associated with magnetization reversal, as state-of-the-art magnetization switching processes involve high-density spin polarized currents, which result in extreme heat dissipation and limited energy efficiency. Using an electric field to reverse a magnetic bit could lead to current-less and extremely localized switching events through voltage application to a capacitor.

Technologically relevant magnetoelectric multiferroic materials, i.e. those possessing coexisting and robustly coupled magnetization and ferroelectric polarization at elevated temperature, are scarce. The conditions for the establishment of a polarization and magnetization in a material are often mutually exclusive [10], yet, several mechanisms allow for a polarization to coexist with magnetic order [1, 2]. For example, improper ferroelectric materials, in which the polarization is not the primary order parameter, can in some cases exhibit magnetic order. Sometimes the magnetism itself can trigger a polarization via the inverse Dzyaloshinskii-Moriya interaction (DMI) [11], causing an induced ferroelectricity to emerge at low temperatures. In this case, however, the polarization values are too small for device integration [1]. Most single-phase multiferroic magnetoelectric materials exhibit a ferroelectric polarization in an antiferromagnetic matrix, the prototypical materials being $YMnO_3$ and $BiFeO_3$. We emphasize here that Cr_2O_3, discussed in the following sections is not multiferroic, however, its established magnetoelectric effect at room temperature [12, 13] brings this material in the focus regarding magnetoelectric-based device research. Hexagonal $YMnO_3$ belongs to the improper ferroelectric category, in which the ferroelectricity emerges as a consequence of steric effects [1, 14]; it is also antiferromagnetic below ~ 130 K [1, 15–17]. The integration of $YMnO_3$ into thin-film heterostructures and the accompanying impact of epitaxial interface on the exotic ferroelectric state have only been investigated recently [18]. Most of the multiferroic and application relevant advances in hexagonal manganites are dealing with bulk measurements [19–21].

In contrast, $BiFeO_3$ is the most studied multiferroic thin-film material [1, 22]. Its room temperature ferroelectricity with extremely high polarization values, reaching up to 150 μC cm^{-2} [23], and G-type antiferromagnetic order [24, 25] offer a unique platform to study the potential applications of multiferroic magnetoelectric materials. We recommend recent reviews dedicated to the material for further reading [26–29].

With the progress in growth techniques, new multiferroic magnetoelectric candidates have emerged as alternatives to $BiFeO_3$-based heterostructures. For instance, $GaFeO_3$ [30–32], exhibiting a higher magnetoelectric coefficient than the reference Cr_2O_3 material, can now be deposited as high-quality thin films [33]. Similarly, layered materials such as Aurivillius phases [34, 35], only recently grown in a single-crystalline epitaxial form [36], show robust ferroelectric properties in the ultrathin regime [36–39] in a structure that is compatible with long-range magnetic order [40, 41].

In order to compensate for the current lack of materials exhibiting strong and coupled ferroelectric and magnetic orders, artificial multiferroic heterostructures were introduced to the mix and are now playing a major role [7]. Here, the ferroelectric and magnetically ordered systems are coupled in multilayers, which is triggered via interface-based interactions, as depicted in Figure 12.1. The geometries of the magnetoelectric heterostructures, i.e. thin film interface or nano-sized pillar self-

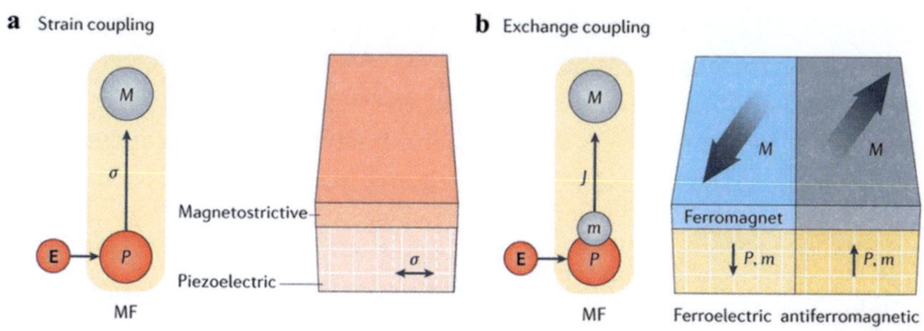

Figure 12.1: In artificial magnetoelectric multiferroic heterostructures, electric-field (E) control of magnetization (M) is based on strain coupling (a) or on interfacial magnetic exchange (b). In (a) the electric field induces a strain σ variation in the piezoelectric constituent which is transferred to the magnetostrictive magnetic layer, while in (b) the ferroelectric order, P, is coupled to the antiferromagnetic order, m, within the multiferroic layer. The antiferromagnetic order couples to the ferromagnetic order of the adjacent layer, the magnetization M, via magnetic exchange, J. Reprinted with permission from [1]. Copyright © 2016, Springer Nature.

assembled architectures, have been reviewed in detail previously [4], thus here we restrict ourselves to the most recent development in the design and integration of these multilayers into emerging technologies discussed in the following sections.

12.1.2 Energy-efficient spintronics

The demand for smaller and energy-efficient devices is driving the integration of complex oxides into new device architectures with reduced energy consumption. With the technological burst that followed the discovery of giant magnetoresistance [42, 43], magnet-based data storage devices are at the forefront of current research. In state-of-the-art device concepts, spin currents replace heat- or magnetic field- assisted switching of the magnetic bit. Spin transfer torque (STT) technology is robust and offers ultrafast dynamics for various applications [44, 45]. Spin currents directly generated via the spin Hall effect in heavy metals electrodes, so-called spin-orbit torques (SOT), are also emerging as technologically relevant magnetic bit manipulation process [46–48]. Both STT- and SOT-based switching, however, dissipate significant energy through Joule heating, which, ultimately limits their integration into ever-smaller device architectures.

Ferroelectric-based memory devices offer the advantage of a non-volatile polarization state that can be controlled using a low-energy-consuming electric field. Electroresistance then allows the direct reading of the polarization state by monitoring the tunneling current across the ferroelectric [49, 50]. Here however, interface related effects can drastically alter the polarization state in the ultrathin regime and hinder the implementation of ferroelectric materials in the required metal-ferroelectric-metal capacitor design. Uncompensated bound charges at ferroelectric epitaxial interfaces trigger a depolarizing field [51], which in return can suppress the ferroelectric behavior in the application relevant ultrathin films [52, 53].

The combined ferroelectric and magnetic states in artificial multiferroic heterostructures offer the possibility to control with an electric field a robust and easy to read magnetic state in the ultrathin regime. Here, we review the advances in the field of electrical control of magnetization in such architectures.

12.2 Acting on magnetic order with electric fields in multiferroic heterostructures

The development focus for the field of multiferroic heterostructures primarily lies in systems coupled through strain and interlayer magnetic exchange [4]. This is because the largest converse magnetoelectric coefficients and most reasonable switching times reported to date are observed in these systems [54, 55]. Single phase multiferroics that are operable at room temperature, such as $BiFeO_3$, tend to have low converse magnetoelectric coefficients [1, 56, 57] that need to be enhanced to facilitate real applications.

12.2.1 Magnetostriction and electric field control of magnetic anisotropy

In this section, we focus on the use of ferroelectric and piezoelectric crystals to modify or switch magnetocrystalline anisotropy in an adjacent magnetically ordered coupled layer, most commonly through the application of strain by the piezoelectric layer and utilizing the magnetostriction of the ferromagnet. A magnetostrictive ferromagnetic layer strained by a piezoelectric/ferroelectric crystal [58, 59] can produce changes in magnetic anisotropy [60–63] and control the associated energy landscape, as represented in Figure 12.2(a).

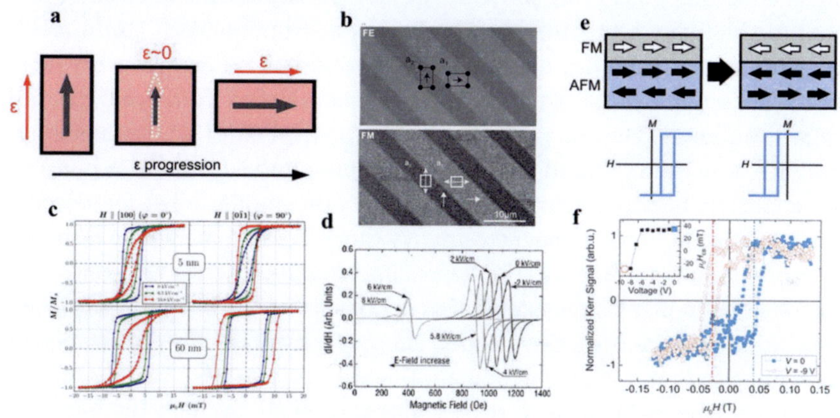

Figure 12.2: Strain- and exchange-coupled multiferroic composites. (a) Schematic representation showing how strain-coupled composites operate, where strain (ε) is generated by a piezoelectric layer and transduced into a magnetization state (black arrow). The figure shows both enhancement of uniaxial anisotropy due to strain and a potential switching pathway for a composite structure. (b) Birefringent (top) and magneto-optical Kerr effect (MOKE, bottom) images showing domain mapping from ferroelectric $BaTiO_3$ to ferromagnetic CoFe domains based on the in-plane strain state. Reprinted with permission from [60]. Copyright © 2012, Springer Nature. (c) Hysteresis of a uniaxial $Fe_{0.81}Ga_{0.19}$/PMN-PT strain-coupled composite, where the magnetic anisotropy is modulated by the application of an electric field. Reprinted with permission from [64]. Copyright 2020 by the American Physical Society. (d) Shift in FMR resonance as a function of applied electric field in FeGaB/PZN-PT ($Pb(Zn_{1/3}Nb_{2/3})O_3$-$PbTiO_3$). Reproduced with permission from [65] (copyright 2009 Wiley-VCH Verlag GmbH & Co. KGaA). (e) A schematic of switching of the exchange bias field, induced when a magnetoelectric antiferromagnet (AFM) is interfaced with a ferromagnetic layer (FM). The magnetoelectric effect drives the direction of the bias field. (f) MOKE hysteresis loop of exchange biased $(CoPd)_n/Cr_2O_3$, where the bias field is switched with a voltage and magnetic field pulse. Reprinted with permission from [66]. Copyright © 2010, Springer Nature.

The magnetostriction coefficient, usually denoted as λ_{ijkl}, describes how a material transduces magnetization into strain according to the relation $\varepsilon_{ij} = \lambda_{ijkl} m_k m_l = -C^{-1}_{ijmn} B_{mnkl} m_k m_l$, where C_{ijmn} is the stiffness tensor, B_{mnkl} is the magnetoelastic energy, and m_k, m_l are the directions of magnetization [67, 68]. Physically, this originates from changes in the band structure of the material due to the mechanical distortion. Effectively, as a crystal is strained and the atoms move, the band structure will change which can deplete or add to the spin-dependent density of states of a ferromagnet [69–71]. Macroscopically, this is observed as the change in magnetocrystalline anisotropy per change in strain, $B_{mnkl} = \frac{\partial K_{mn}}{\partial \varepsilon_{kl}}$, where K_{mn} is the magnetic anisotropy, and ε_{kl} is the strain [72]. As this is an effect of the magnetic anisotropy energy, it is proportional to the spin-orbit coupling of the material [71] and ferromagnets with large spin-orbit coupling, like rare-earth-containing metal alloys, tend to also exhibit large magnetostriction coefficients. Additionally, materials that are mechanically isotropic (i.e. the ratio of components of C_{ijmn} approach 1 or the shear modulus goes to 0) show larger magnetostriction values since λ is proportional to the inverse of stiffness, $C^{-1}_{ijmn} B_{mnkl}$. Experiments with strain-mediated composite magnetoelectrics usually employ conductive magnets with large magnetostrictive coefficients (λ) such as rare-earth containing Terfenol-D [58, 73] ($\lambda \sim 1200$ ppm), Galfenol ($Fe_{1-x}Ga_x$) [74, 75] ($\lambda \sim 250$ ppm), FeRh [76] (ferromagnet-antiferromagnet transition), CoFeB [73, 77] ($\lambda \sim 50$ ppm), and Ni [78] (λ~34 ppm).

One static demonstration of this concept has been the imprinting of ferroelectric domains from a substrate onto a magnetostrictive film via strain-mediation (see section 3.2 for more on domain imprint). By depositing magnetostrictive $Co_{0.9}Fe_{0.1}$ on a $BaTiO_3$ substrate, the in-plane strain states of a_1-a_2 domains in the $BaTiO_3$ create localized easy axes in the magnetostrictive magnet, effectively mapping the polarization of the substrate directly into a magnetic domain (Figure 12.2(b)) [79]. These imprinted magnetic domains have been seen to persist unambiguously through the temperature-driven phase transitions of $BaTiO_3$ [80] and the effect has been noted with other magnetic films, such as $La_{1-x}Sr_xMnO_3$ (LSMO) (a) and $CoFe_2O_4$ [81]. Beyond this, magnetic domain imprinting has been used to control domain walls in the pre-switching regime, as the magnetic domain will move with the ferroelectric domain.

Magnetoelectric switching of magnetic anisotropy with the use of magnetostrictive ferromagnets has resulted in energy dissipations per area per switch on the order of 1–100 μJ cm^{-2} [78], making these devices competitive for post-Si memory and logic applications [82–84]. While much of the early work in strain-coupled multiferroics was done using bonded piezoelectric transducers [85, 86], scale devices necessitate thin-film magnets deposited on high-strain piezoelectric crystals, typically relaxor-type $PbTiO_3$ (PTO) derivatives such as $Pb(Zr,Ti)O_3$ (PZT) [78, 87, 88] and $Pb(MgNb)O_3$-$PbTiO_3$ (PMN-PT) [77, 89, 90]. Strain mediation has widely been used to manipulate the energy landscape in composite materials [91], changing coercive fields and the depth of the potential well along the easy axis (an example of which is shown in Figure 12.2(c)), however, full switching of the anisotropy axis is much rarer [55, 92] and will be discussed further below.

Strain-mediated magnetoelectric coupling is widely used for the production of extremely sensitive magnetic field detectors, particularly for biomedical applications [93–95]. In addition to the directional dependence of magnetization in strain-mediated composites, small changes in linear anisotropy energies can be detected via ferromagnetic resonance (FMR), shown in Figure 12.2(d), and acoustic resonance of a cantilever structure [65, 93, 95, 96]. These multiferroic antennae have demonstrated frequency tunability in the GHz regime on the order of $\Delta f \cong 6$ GHz [65]. The use of this technology for biomagnetic sensors is very promising, with significant work in the field having progressed to devices testable *in vivo* [97, 98].

12.2.2 Electric field control of exchange bias

Due to the mutually exclusive conditions for the onset of simultaneous ferromagnetic and ferroelectric order in a material (refer to Section 12.1 and [10]), most single-phase multiferroic materials exhibit a ferroelectric order coexisting with an antiferromagnetic state. Therefore, single-phase multiferroic materials, such as $BiFeO_3$, tend to have low converse magnetoelectric coefficients due to their small ferromagnetic moment that results from canted antiferromagnetic order [1, 56, 57, 99, 100]. Conductive ferromagnetic layers, such as Co [101], CoFe [102, 103] and Ni [78] are employed and exchange-coupled to the multiferroic magnetic order, effectively amplifying the weak magnetic moment of the multiferroic. In so-called exchange-coupled multiferroic heterostructures, a single-phase magnetoelectric antiferromagnetic multiferroic is coupled to a ferromagnetic layer through interfacial coupling. Exchange bias, see Figure 12.2(e), is a phenomenon that occurs when a thin ferromagnet is in contact with the surface of an antiferromagnetic crystal. Below the ordering temperature of the antiferromagnetic state, the Néel temperature, the relatively hard spins of the antiferromagnetic surface interact with the soft spins in the ferromagnet, pinning the magnetization direction to that of the surface spins in the antiferromagnet [104]. This results in a significant broadening of the ferromagnetic hysteresis under magnetic field application – an enhancement in the coercive field, and a shift of the loop along the field direction called the exchange bias field. At zero magnetic field, the ferromagnet is robustly pinned into one state by the bias field. Here, a magnetoelectric switching event can trigger a change in the antiferromagnetic state and lead to an exchange-coupled net magnetization switch. The efficiency of this switching mechanism has been demonstrated in heterostructures based on prototypical antiferromagnetic magnetoelectric materials Cr_2O_3 [105, 106] and $BiFeO_3$ [102, 107–109]. An example of this switching using Cr_2O_3 is shown in Figure 12.2(f).

Cr_2O_3 is a non-ferroelectric antiferromagnetic material which shows a linear magnetoelectric effect above room temperature [12, 13]. It is a uniaxial antiferromagnet pointing along the *c*-axis of the hexagonal unit cell, and the magnetic cations are arranged in such a way that a *c*-plane terminated crystal can be engineered to produce a single, global, perpendicular surface magnetization [110] . This out-of-plane antiferromagnet would exchange couple to a ferromagnetic film or multilayer with perpendicular magnetic anisotropy, in contrast to most in-plane AFM/FM systems, creating an out-of-plane magnetized multiferroic heterostructure. Electric-field-driven switching of this configuration was first demonstrated in bulk crystals with out-of-plane ferromagnet $(Co/Pt)_n$, however it required a static magnetic field to achieve full switching of the surface magnetization [106, 110]. Linear magnetoelectric switching is defined by a critical energy product of electric field and magnetic field [13], meaning magnetoelectric switching of Cr_2O_3 requires either substantial static magnetic fields [111, 112] or extraordinarily large voltages. This presents a significant engineering challenge for scaling of devices, as a Cr_2O_3 thin film must be able to withstand voltages large enough for switching, and significant work is currently focused on increasing the breakdown voltage or decreasing the switching threshold for thin films [113, 114]. Thus far, magnetoelectric switching in thin-film heterostructures has required substantial magnetic fields on the order of ~ 0.6–1 T [66, 111, 112].

In room temperature ferroelectric antiferromagnetic $BiFeO_3$, the antiferromagnetic structure can cant and result in a weak moment that forms a long-range spin cycloid [115]. With enough epitaxial strain, a uniform canted moment emerges [116]. The antiferromagnetic and canted moment structure have been observed to track the ferroelectric polarization, thus, the moments of an exchange-coupled ferromagnet not only amplify the effective moment but also correlate to the ferroelectric domain structure of the multiferroic [25, 27, 101, 103, 107, 117, 118]. This coupling between the magnetic and ferroelectric domain structure then permits the electric field control of the magnetization through the control of the ferroelectric polarization. This has been leveraged in $BiFeO_3$/LSMO heterostructures and the sign of the bias field could be reversed after electrical switching [108, 109]. In this system, however, the exchange bias field is not large enough to fully switch the magnetization of the ferromagnet, requiring the use of other magnetic layers to realize deterministic switching.

12.2.3 Electric-field-induced magnetization reversal

In strain-coupled multiferroics, the use of epitaxy to maximize strain transfer to the magnetostrictive layer has enabled electric field-controlled, non-volatile 90° switching of the ferromagnetic magnetization in a Galfenol (FeGa) film [74] (Figure 12.3(a,b)). This, however, may be the limit for purely strain-based devices, as symmetry considerations state that deterministic 90° switching should only be possible from a single

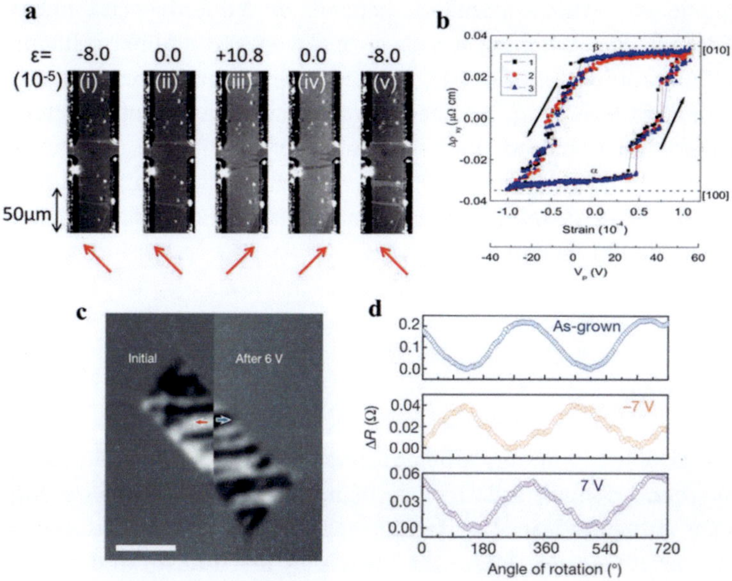

Figure 12.3: Magnetoelectric switching of magnetization. a) Magneto-optical Kerr effect (MOKE) micrographs of a strain-coupled composite multiferroic consisting of magnetostrictive $Fe_{0.81}Ga_{0.19}$ epitaxially grown on a GaAs substrate and bonded to a piezoelectric transducer. We see that, as the device is cycled through a full ferroelectric loop, the magnetization (red arrow) is deterministically switched between the two strain states. b) Transverse anisotropic magnetoresistance (AMR) showing repeatable resistive switching of the device. Reprinted with permission form [74]. Copyright 2012, AIP Publishing LLC. c) X-ray magnetic circular dichroism photoemission electron microscopy (XMCD-PEEM) images of a $Co_{0.9}Fe_{0.1}/BiFeO_3$ exchange-coupled multiferroic showing deterministic switching of the magnetization that follows the ferroelectric domain pattern of the $BiFeO_3$. The scale bar is 2 μm. d) Low-field AMR versus applied voltage, showing 180° switching of magnetization under voltage. The upper, middle and lower panels show AMR from the as-grown (as-deposited) state, after −7 V and after + 7 V. Reprinted with permission from [102]. Copyright © 2014, Springer Nature.

elastic switching event. Experimental reports have demonstrated that enough magnetoelastic energy can be generated to drive 90° reorientation of anisotropy [60, 74, 90], but 180° switching, which is desired for maximum magnetic readout, may not be achievable unless sequential voltages are applied [119, 120]. As an engineering challenge, this problem can be circumvented if material symmetries are broken by external stimuli or clever architectures utilizing the switching kinetics of the ferroelectric [119]. One way this can be accomplished is through external magnetic fields [121–123] or hybrid magnetoelectric-spintronic devices [124, 125], which are discussed in Section 12.4.

Full switching of magnetization induced by an electric field has also been observed in exchange coupled $BiFeO_3$-ferromagnetic heterostructures, shown in Figure 12.3(c,d). The magnetization of the ferromagnetic layer is strongly coupled to the antiferromagnetic axis, whose orientation depends on the ferroelectric polarization direction. Deterministic switching has been observed and studied primarily with $Co_{0.9}Fe_{0.1}$ metallic ferromagnets [102, 107, 126, 127], using both microscopy [102] and transport measurements [107, 127]. This deterministic switching event, which requires no external fields, has spurred the development of multiferroic computing technologies based on the material system [83, 84], as well as significant efforts to increase its efficiency [127, 128]. Its implementation into energy efficient device heterostructures will be described in Section 12.4.

12.3 Ferroic domains and domain imprint in thin-film heterostructures

In multiferroic heterostructures, energy efficient functionalities rely on the ability to alter the ferromagnetic domain architecture using an electric field in a controlled fashion once integrated in the ferroelectric/ferromagnetic multilayer. Ferroelectric domain architectures define the switching mechanism and can be transferred into an adjacent ferromagnetic layer and therefore drive the magnetoelectric behavior of the multiferroic multilayer. Thus, engineering of ferroelectric domains and corresponding domain walls is the key towards reliable magnetoelectric switching events.

12.3.1 Ferroelectric domain engineering

Epitaxy – Strong lattice-polarization coupling in ferroelectrics enables the design of the ferroelectric domain configuration in epitaxial thin films through the surface symmetry, surface morphology, and epitaxial strain imposed by the selected single-crystal substrate. Substrate step edges are often preferential nucleation sites for ferroelectric domains [129], therefore substrate orientation and miscut angle become tuning parameters to control the degeneracy of domains [130, 131]. In-plane strain anisotropy induced by the substrate often results in anisotropic stripe domain patterns [132, 133]. Epitaxial strain, in extreme cases, can even induce polarity in otherwise non-polar structures, such as $SrTiO_3$ [134], $EuTiO_3$ [135], $SrMnO_3$ [136, 137] or $BaSnO_3$, $SrSnO_3$ [138]. Domain engineering in ferroelectric thin films using epitaxial strain has been reviewed in great detail, see [51] for references. Here, we highlight some recent developments involving electrostatic boundary conditions, surface symmetry, and atomic terminations to achieve a predefined polarization state.

In the particular case of multiferroic magnetoelectric $BiFeO_3$ thin films, the configuration of ferroelectric domains can be fully controlled by the choice of the crystallographic orientation of the substrate: thin films have one, two, or four allowed polarization variants when grown on (111), (110) and (001) orientations of the lattice matching $SrTiO_3$ substrate, respectively [130]. A high interfacial strain gradient can produce mosaic ferroelectric domains in $BiFeO_3$ that can be turned into stripes upon substrate annealing [139]. It is worth noting that the antiferromagnetic order in multiferroic $BiFeO_3$ is strongly correlated with the ferroelectric domain state and also depends on epitaxial strain: spin cycloid can be suppressed by epitaxial strain [140, 141]. $BiFeO_3$ epitaxial films thus provide a unique platform towards the antiferromagnetic order parameter engineering using epitaxial strain in thin films. Its antiferromagnetic-axis orientation can be tuned [142] from in-plane to out-of-plane by compressive and tensile strain, respectively. Furthermore, once stabilized in a single ferroelectric and magnetic domain state, the magnetic polarity in $BiFeO_3$ domains can be readily inverted following 180° ferroelectric switch on the macroscopic scale [143, 144].

Electrostatics – A depolarizing field emerges with the accumulation of the ferroelectric bound charges at the surfaces of ferroelectric thin films [5]. Oppositely oriented to the polarization direction, it usually has a detrimental impact on the polar properties of the films in the technology-relevant ultrathin regime. The depolarizing field can cause the polarization suppression [53, 145, 146] and limit the integration of ultrathin ferroelectric materials into devices and is therefore often viewed in a negative light. As most ferroelectric systems grow epitaxially in the ferroelectric phase [146,147,148], the depolarizing field already impacts the domain state during the design of functional heterostructures based on ferroelectric films [52]. Nevertheless, the progress in epitaxial design and nanoscale ferroelectric domain probe capacity recently allowed the integration of depolarizing field effects into the domain engineering process to tailor ferroelectric domains in a controlled manner with desired electrostatic boundary conditions [149, 150].

In addition to the impact of the bound charge screening on the ferroelectric polarization state in the ultrathin regime, the electrode material selection and its corresponding work function drastically impact the bulk ferroelectric properties. Any asymmetry in the electronic and chemical environments of a ferroelectric layer (usually at the interfaces) results in an internal built in field [151, 152], which affects the polarization orientation and induces a preferred polarization direction and retention problems [153, 154]. The insertion of a dielectric material into the ferroelectric capacitor architecture results in reduced bound charge screening efficiency and an onset of the depolarizing field, which induces a domain nucleation and leads to a multi-domain state in the ferroelectric layer. It attenuates the built-in field influence on the ferroelectric switching properties, i.e. hysteresis loop. Thickness control of the dielectric layer allows for tuning of the depolarizing field strength and the corresponding domain formation [150]. Furthermore, the existence of domains in an as-grown state and diminished dielectric leakage further improves the switching performance [149, 150, 153]. Such an approach previously restricted to prototypical tetragonal ferroelectric ($BaTiO_3$, PZT and PTO) has been used to engineer multiferroic domains in $BiFeO_3$

films. Chen and co-workers [155] showed that an ultrathin dielectric layer of La-doped $BiFeO_3$ radically alters the domain state in $BiFeO_3$. On $DyScO_3$ substrates, when a conducting buffer is used, a uniform out-of-plane polarization state is stable and the $BiFeO_3$ systems exhibits a 71° stripe domain architecture. However, the insertion of the dielectric buffer results in the conversion towards a fully 109°-type stripe domain pattern, without any net out-of-plane polarization component and technology-relevant conducting 109° domain walls [156, 157] (see Figure 12.4(a–d)).

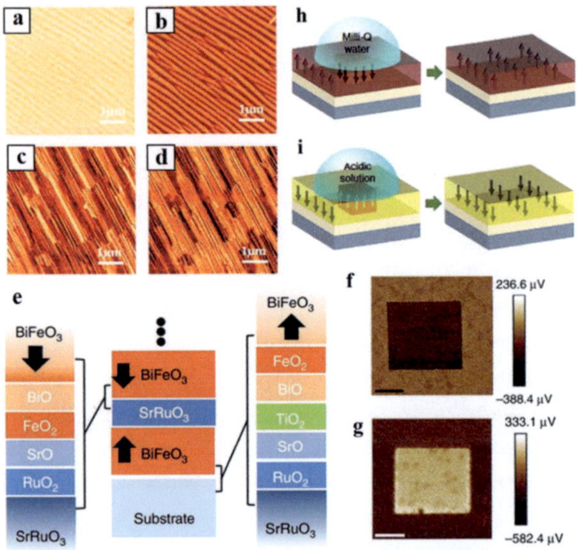

Figure 12.4: (a–d) $BiFeO_3$ domain engineering by insertion of a dielectric La-doped $BiFeO_3$ buffer layer. Out-of-plane (a) and in-plane (b) piezoresponse force microscopy (PFM) images of the $BiFeO_3$ 71° stripe domain structure in $BiFeO_3/SrRuO_3/DyScO_3$. 109° stripe domains appear upon the insertion of La-doped $BiFeO_3$ buffer in vertical (c) and lateral (d) PFM images. Reprinted with permission from [155]. Copyright © 2017, American Chemical Society. (e–g) Engineering ferroelectric multilayers with user-defined up/down polarization sequence (e): A-site growth and upwards polarization (black arrow) of $BiFeO_3$ is achieved by insertion of a TiO_2 monolayer. B-site growth and downwards polarization of $BiFeO_3$ is achieved by SrO self-termination of SRO buffer, confirmed by PFM scans (f) and (g), respectively. Scale bars correspond to 1 µm. Reprinted with permission from [146]. Copyright © 2017, Springer Nature (h-i) Water poling of ultrathin ferroelectric $BiFeO_3$: polarization is reversed to the upward (h) and downward (i) direction after exposing the film to water and acidic solution, respectively. Reprinted with permission from [158]. Copyright © 2018, Springer Nature.

Atomic surface termination and interface chemistry, symmetry – In the ultrathin regime, the polarization direction (up or down in a single domain state) and overall domain architecture can also be controlled by selecting an appropriate atomic surface termination of the substrate or buffer layer. Using chemical surface treatments or thermal annealing, a single substrate atomic termination can be selected. Considering the perovskite ABO_3 model system, the thin film (001)-oriented growth can take place from either the AO or BO_2 plane. In the case of ABO_3 type ferroelectrics containing charged planes, there is an electrostatic potential step across the interface with the buffer. This results in a deterministic control on the direction of polarization of the film. For instance, because the $BiFeO_3$ unit cell consists of charged planes, i.e. $(BiO)^+$ and $(FeO_2)^-$, the polarization can be predefined by selecting the buffer surface termination for the growth. On negatively charged $(MnO_2)^{-0.7}$ planes of an LSMO buffer, the growth starts with positively charged $(BiO)^+$ planes. The resulting potential sets the polarization in the upwards direction. The deposition on $(La_{0.7}Sr_{0.3}O)^{+0.7}$ plane results in downward oriented films [159]. Because the stacking sequence in atomic plane can propagate during the growth of several successive perovskite materials, the design of multilayers with predefined polarization configuration [146], i.e. with parallel or antiparallel relative polarization state, is possible as shown in Figure 12.4(e–g). Moreover, different interfacial atomic sequences and the corresponding electrostatic potential also result in modulations of magnetoelectric effect and exchange bias in oxide multilayers [160]. The atomic surface termination and symmetry provide a handle on the polarization state from the bottom interface during the multilayer design process itself, adding a degree of freedom for the design of domain architecture and the surface topography of the film [161].

In order to stabilize ferroelectricity in the ultrathin regime, polar metals crystallizing in non-centrosymmetric point groups compatible with ferroelectric symmetries are emerging as a promising alternative to standard metals. Their intrinsically broken parity can even help to surpass critical thicknesses [162], which would benefit the integration of ultrathin ferroelectrics in technology-relevant capacitors. The scarcity of polar metals that are structurally compatible with application-relevant ferroelectrics, however, limits their integration into ferroelectric-based devices. Alternatively, the chemistry at the top interface of the ferroelectric material can be tuned in order to achieve a deterministic polarization state [158]. The importance of tunable electrostatics at ferroelectric $BiFeO_3$ films surfaces was explored by a reversible aqueous switching. Depending on the H^+/OH^- concentration of the solution in direct contact with the film, the formation of polarization-selective chemical bonds at the interface is controlled, as shown in Figure 12.4(h–i).

12.3.2 Domain imprint

The interface between ferroelectric and ferromagnetic thin films can enable magnetoelectric coupling. In particular, electrically tunable ferroelectric domain architectures can be imprinted into the magnetic state in an adjacent layer. There are three distinct ways to achieve artificial magnetoelectric coupling [7]: via (i) strain, (ii) direct (spin) exchange, and (iii) charge coupling, see Section 12.2. In strain-coupled artificial multiferroics, a piezoelectric crystal or thin film and magnetostrictive layer are elastically-coupled leading to controllable magnetoelastic anisotropy due to propagation of electrostrain [79, 163–170] with a key requirement of strong elastic pinning of magnetic domain walls onto ferroelectric domain walls [171]. In exchange-biased multiferroic composites, the interaction occurs between a ferromagnet and intrinsic multiferroic with uncompensated antiferromagnetic order, such as $BiFeO_3$ [25, 103, 172, 173], hexagonal $YMnO_3$ [174] and $LuMnO_3$ [123]. In these systems, an applied electric field changes both the polarization and the direction of antiferromagnetic spin ordering in a multiferroic. This allows a direct exchange effect between antiferromagnetic and ferromagnetic spin ordering. Although the physics of exchange coupling is different from strain coupling, both produce lateral modulations of magnetic anisotropy leading to domain transfer. Note that because strain is uniaxial, the electrical directional control of a magnetic state, i.e. 180° electric-field-induced magnetization reversal, has been restricted to exchange-coupled systems. Charge-coupled artificial multiferroics make use of the ferroelectric field effect: non-volatile bound charges at the ferroelectric interface are screened by the ferromagnetic layer, leading to either accumulation (hole doped) or depletion (electron doped) states in ferromagnet when the polarization is pointing away or towards the ferromagnet, respectively. If the ferromagnet has multiple accessible ground states, this might result in drastic alterations of magnetization and domain imprint [87, 175–181]. This mechanism is different from the two mentioned previously, because changes in magnetization are limited to interfaces, up to a screening length [181, 182]. However, in the ultrathin regime (especially when superlattices are considered), full domain transfer could be achieved.

Most recent studies on domain imprint have focused on the dynamics investigation aiming to understand the origin of the domain imprint and the domain pattern evolution upon the application of electric fields. De Luca and coworkers [183] investigated magnetoelectric coupling dynamics between exchange-coupled $BiFeO_3$ and ferromagnetic $Co_{0.9}Fe_{0.1}$ using optical second harmonic generation and magnetic force microscopy (MFM) techniques *operando*; see Figure 12.5(a–f) . An electric field in the order of the coercive field of $BiFeO_3$ is required to activate the domain pattern transfer. In another work, magneto-optical Kerr effect (MOKE) microscopy was employed to show that ferroelectric and ferromagnetic domain walls move in unison upon the application of out-of-plane electric field pulses [186–188]. These experiments confirmed that the process is reversible and neither magnetic fields, nor currents are required. The domain wall velocity is hence fully determined by the electric field strength and can reach several 100 m s^{-1} as predicted by numerical simulations [189]. Recent studies on strain-mediated magnetoelectric coupling revealed that rotations of the local magnetization upon electric field pulses are normally smaller than expected 90° due to electrically driven shear strain [184], see Figure 12.5(g–l). This guarantees a deterministic return to the initial magnetization direction with no possibility of magnetization reversal and offers the prospect of writing data both magnetically and electrically [184].

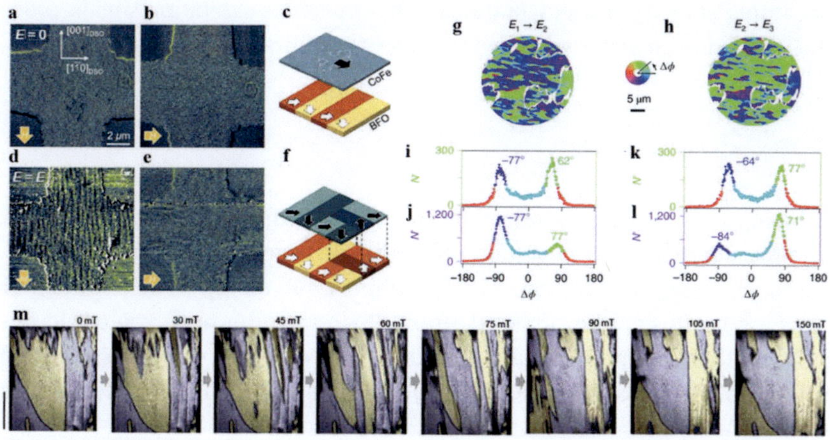

Figure 12.5: (a–f) Domain transfer from $BiFeO_3$ to $Co_{0.9}Fe_{0.1}$. The spatially MFM scans show ferromagnetic domains (black arrows) in $Co_{0.9}Fe_{0.1}$ with the magnetic field (yellow arrow) applied along $[001]_{DyScO3}$ (a,d) and $[1–10]_{DyScO3}$. (b,e) The pristine $Co_{0.9}Fe_{0.1}$film has a uniform in-plane magnetization indicating the suppression of the magnetoelectric $BiFeO_3$/ $Co_{0.9}Fe_{0.1}$ coupling as shown schematically in (c). The application of an electric field $E \approx E_c$ activates the ferroelectric domain pattern (white arrows) transfer (f). Reprinted with permission from [183]. Copyright 2018 by the American Physical Society. (g-l) Electrically driven shear strain effects in magnetoelectric switching of PMN–PT/Ni heterostructure. 50-μm-diameter maps (g,h) showing changes of magnetization direction $-180° \le \Delta\varphi \le 180°$ of each pixel upon electric field steps (E1→E2) and (E2→E3) anticipated to interconvert the hard and easy magnetization directions in Ni. Due to shear strain magnetization rotates by angles smaller than expected 90°, as shown by the number of pixels N′ that undergo a change of magnetization direction Δφ, for pixels that are green (i,k) and purple (j,l) with modal angles specified. The data color represents a modal angle on the color wheel. Reprinted with permission form [184]. Copyright © 2019, Springer Nature. (m) Inversion of the ferroelectric domain pattern in multiferroic Mn_2GeO_4 evidenced by sequentially taken SHG images of out-of-plane P_z domains at the given magnetic fields H_z. Scale bar, 500 μm. Reprinted with permission from [185]. Copyright © 2018, Springer Nature.

A novel route towards exact matching between ferroelectric and ferromagnetic domain patterns was proposed by Leo et al. [185]. Complete magnetoelectric domain inversion was observed in multiferroic Mn_2GeO_4 and magnetoelectric Co_3TeO_6 bulk crystals: the ferroic order parameter in these materials is fully reversed in in each domain, but the initial domain pattern is perfectly reproduced. Figure 12.5 (m) shows how ferroelectric domains in Mn_2GeO_4 are inverted upon application of magnetic field. The observation is rationalized by the coupling between a complex set of order parameters where one of them is independently switched with an external field, while another retains the memory of the domain pattern. A related effect was observed in $Dy_{0.7}Tb_{0.3}FeO_3$ [190]: here multiferroic domains defined by multiple order parameters can be reversibly converted to multiferroic domain walls and back at a specified location within a non-multiferroic matrix. Such interconversion can be a key to tailoring elusive domain architectures in antiferromagnets. Although both effects have only been demonstrated at cryogenic temperatures, other materials exhibiting similar properties are anticipated to be discovered, owing to the generality of the two effects.

12.3.3 Domain wall engineering

Domain architecture is also characterized by its domain walls configuration. Domain walls, separating different regions of uniformly oriented order parameter (here, ferroelectricity and ferromagnetism), can play an important role in functionalities of thin-film heterostructures. The physical properties emerging at ferroelectric domain walls and the corresponding applications have been the subject of several reviews; refer to [191–194]. Here, we place the emphasis on the magnetoelectric systems and applications.

Tremendous progress has been achieved in the field of ferroelectric domain walls over the last couple of years. Ways of injecting ferroelectric domain walls [195–197] and controlling their motion [196, 198, 199] as well as their speed [200] are now established. Sharma and coworkers [197] have demonstrated a prototype domain wall switch in $BiFeO_3$ with three distinct states: no domain wall, neutral domain wall and charged domain wall as shown in Figure 12.6(a–f). Domain wall detection is now possible not only via conventional scanning probe techniques [192], but also with optical second harmonic generation [201, 202], Raman spectroscopy [203] as well as synchrotron X-ray diffraction (XRD) [204]. Moreover, PZT domain walls were recently identified as optically active objects when interfaced with MoS_2 flakes offering novel tunability of interfacial optical response [205]. All these advances point towards the future development of magnetoelectric devices based on domain walls.

Magnetic domain walls in a ferroelectric matrix – Magnetoelectric coupling in heterostructures can alternatively be achieved through the domain walls as opposed to the direct domain imprint discussed previously. The importance of the domain wall role first emerged upon nanoscale investigation of $BiFeO_3$ interfaced with $Co_{0.9}Fe_{0.1}$, which showed that exchange-bias directly depends on the type and crystallography of the domain walls in $BiFeO_3$ [117]. An enhancement of the coercive field is observed regardless of the domain wall type, while exchange bias itself, manifesting by shifted magnetization hysteresis, is ascribed to uncompensated spins at 109° domain walls exclusively [117], as seen in Figure 12.6 (g–l). The domain wall dynamics itself influences the domain imprint or domain coupling in artificial multiferroic heterostructures. The application of an electric field can induce a ferroelectric domain wall motion, releasing the imprint or magnetoelectric coupling locally until the domain coupling is recovered at a new ferroelectric domain wall location. This was used to control the magnetic exchange bias and magnetotransport properties in the model system hexagonal $YMnO_3$/Py and $LuMnO_3$/Py bilayers [174]. Antiferromagnetic and ferroelectric domain walls in multiferroic manganites can be unclamped and change the sign of exchange bias after each successive electric pulse [123]. Such clamped

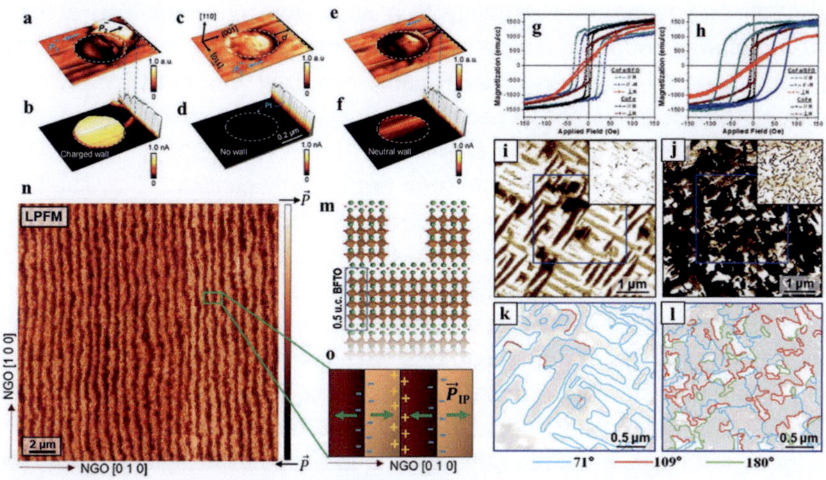

Figure 12.6: (a–f) Tristate conformational domain wall switch in $BiFeO_3$. Lateral PFM amplitude (a,c,e) and the corresponding c-AFM images (b,d,f): in the absence of domain walls (a,b) and in the presence of neutral (c,d), and charged (e,f) domain wall achieved with the device. Reproduced with permission from [197]. (g–l) Exchange bias dependence on domain wall type in $BiFeO_3$. Magnetic measurements on $BiFeO_3$/ $Co_{0.9}Fe_{0.1}$ heterostructures exhibiting enhanced coercivity (g) and exchange bias (h). (i,j) In-plane and out-of-plane (inset) PFM contrast for $BiFeO_3$ films corresponding to (g) and (h), respectively. Detailed domain wall analysis for (k) stripe-like and (l) mosaic-like $BiFeO_3$ films. 109° domain walls are directly related to exchange bias, while other domain walls still enhance coercivity. Reprinted with permission from [117]. Copyright © 2008, American Chemical Society. Copyright 2019 Wiley-VCH Verlag GmbH & Co. KGaA. (m–o) Nominally charged domain walls in ultrathin $Bi_5FeTi_3O_{15}$ Aurivillius films (film structure depicted in m). Lateral PFM contrast (n) recorded for a half-unit-cell film revealing nominally charged head-to-head and tail-to-tail domain walls, sketched in a (o), where green arrows represent in-plane polarization directions. Reproduced with permission from [36]. Copyright 2020 Wiley-VCH Verlag GmbH & Co. KGaA.

domain walls, possessing multiple ferroic order parameters, can in turn be referred to as multiferroic domain walls [1]. They sometimes are the sole reason for magnetoelectric domain coupling, e. g. as in rare-earth (R) hexagonal manganites h-$RMnO_3$ [15]. Multiferroic domain walls have also been reported in multiferroics such as $MnWO_4$ [206, 207], $RFeO_3$ [208, 209] or $TbMnO_3$ [210]. A finite magnetic moment at the domain wall can sometimes accompany polar and antiferromagnetic domains as measured in h-$ErMnO_3$ [211], oxygen deficient $PbTiO_3$ [212] or ferroelastic $TbMnO_3$ [213]. We note that polar domain walls in non-ferroelectric structures, e. g. Néel-like domain walls in iron garnet films [214], ferroelastic domain boundaries [215–217] have also been reported. All of this holds potential for magnetoelectric coupling in heterostructures right at the wall location in a non-magnetoelectric host compound.

Charged domain walls in the vicinity of a magnetic state – Bound charge accumulation at so called charged ferroelectric domain walls [19, 156] or ferroelastic boundaries [218] creates another pathway to a magnetoelectric effect at domain walls: localized charge coupling can be induced at interfaces of metallic ferromagnets [176]. This could trigger electrically tunable localized interfacial reconstructions, leading to the ferromagnetic to antiferromagnetic phase transition in colossal magnetoresistive ferromagnets such as LSMO [177, 219]. A challenge remains in the stabilization of charged domain walls and their unfavorable electrostatics. The family of improper ferroelectrics, i.e. geometric ferroelectrics h-$RMnO_3$ [19] or electronically-driven improper ferroelectric $CsNbW_2O_9$ [220] are prominent candidates for the design of functional and stable charged domain walls.

In proper ferroelectrics, fully in-plane polarized systems may open new avenues towards stable charged domain wall design. In in-plane polarized thin films, the depolarizing field influence on the domain pattern does not depend on the layer thickness and may be linked to the domain width [36, 37]. Stable periodic arrays of nominally charged domain walls have been recently reported in ultrathin Aurivillius films (see Figure 12.6(m–o) [36], putting this class of layered ferroelectrics [35, 221] in the forefront of domain wall engineering research. These compounds can further host magnetic ions driving multiferroicity [40, 222–224]. However, the impact of the layered anisotropic structure [225, 226] and B-site doping on the domain wall electronic configuration remains to be explored.

12.4 Devices based on magnetoelectric thin films

The coupling between electric and magnetic orders in magnetoelectric and multiferroic heterostructures opens pathways to energy efficient spintronic devices and other novel devices (such as sensors, motors, voltage-controlled RF devices, and reconfigurable electronics) that leverage the electric to magnetic transduction. Over the past 20 years, the embodiment of devices that employ magnetoelectrics or multiferroics has grown from a magnetic memory element to reconfigurable memory-in-logic devices and circuit architectures. Here we will focus on the modern spin-based device concepts that integrate magnetoelectric multiferroic heterostructures and briefly discuss the evolution of the concepts.

12.4.1 Magnetoelectric/multiferroic memory

The magnetic random-access memory (MRAM) was one of the earliest devices identified where integration of a magnetoelectric or multiferroic could improve performance. This would occur through a reduced write energy via magnetoelectric switching or a multi-state output through independent control of the electric and magnetic polarizations. Here we will focus on the former as it has had the most impact on devices to date.

Spin-based MRAM technology uses spin-transfer or spin-orbit torques (STT and SOT, respectively) to convert an electric charge current into a spin current that transfers angular momentum to the magnetization of the free layer in a memory element. From an energy perspective, the conversion of charge to spin current is energy inefficient as the current densities to drive switching are large (~ 10^6–10^7 A cm^{-2}) and losses from resistive heating are significant. Magnetoelectric switching mitigates the large write currents and dissipation energy as an electric field is applied across a dielectric insulator. The energy dissipation is then dominated by charging of the capacitor or ferroelectric switching and the switching of the magnetization.

For example, the state-of-the-art SOT switching of a ferromagnetic layer employed low-resistivity AuPt alloys and demonstrated switching at a current density of 1.2×10^7 A cm^{-2}. The low resistivity (80 μΩ cm) and high spin Hall conductivity ($4.4 \times 10^5\ \Omega^{-1}\,m^{-1}$) leads to a power dissipation per write operation that is ~ 10 times lower than that using tungsten [227]. Follow-up work has reported sub-50 fJ dissipation (~ 580 μJ cm^{-2}) at 1 ns current pulses and less than 10^{-5} error rate in the $Au_{0.25}Pt_{0.75}$ system [228]. In contrast, the electric field control of magnetism in rather unoptimized multiferroic heterostructures have shown energy dissipations in the range of 1–500 μJ cm^{-2} [55]. The primary energy dissipation comes from the switching of the ferroelectric polarization, or the charging of and leakage in the capacitor in the case of non-ferroelectric magnetoelectrics. Thus, the most immediate embodiment of a magnetoelectric memory element uses the magnetoelectric switching of a layer in a giant magnetoresistance (GMR) or magnetic tunnel junction (MTJ) element, whereby a small current can be used to readout the state of the device (Figure 12.7).

One of the first materials proposed for such applications was the room temperature magnetoelectric multiferroic $BiFeO_3$. While the idea was put forward in the 2000's [9, 231], it was not until the mid- and late 2010's that reversible magnetoelectric switching of a GMR element was demonstrated at room temperature (shown in Figure 12.7) [102, 230]. By exchange coupling the free layer of the GMR stack to the $BiFeO_3$ layer, the free layer then becomes controlled by the applied electric field to the $BiFeO_3$. This is observed in current in-plane resistance measurements of the GMR structure versus applied voltage (Figure 12.7). Notably, the hysteresis and coercive voltages in the ferroelectric response of the $BiFeO_3$ agree well with the hysteresis observed in the GMR device, demonstrating a clear coupling between electrical and magnetic parameters. In the devices that involve switching the ferroelectric order parameter, the energy dissipation is dominated by that contribution. The energy dissipation (per unit area) can then be approximated by $2P_SV_C$ where P_S is the saturation polarization of the ferroelectric and V_C is the coercive voltage. Thus, the energy dissipation for these pioneering structures is ~ 450 μJ cm^{-2} [55, 102].

While the pioneering devices have energy dissipation per unit area that is comparable to the leading SOT devices, there are clear routes to tune the magnetoelectric energy landscape for improving performance. La substitution in $BiFeO_3$ systematically destabilizes the ferroelectric order thereby reducing saturation polarization P_S and

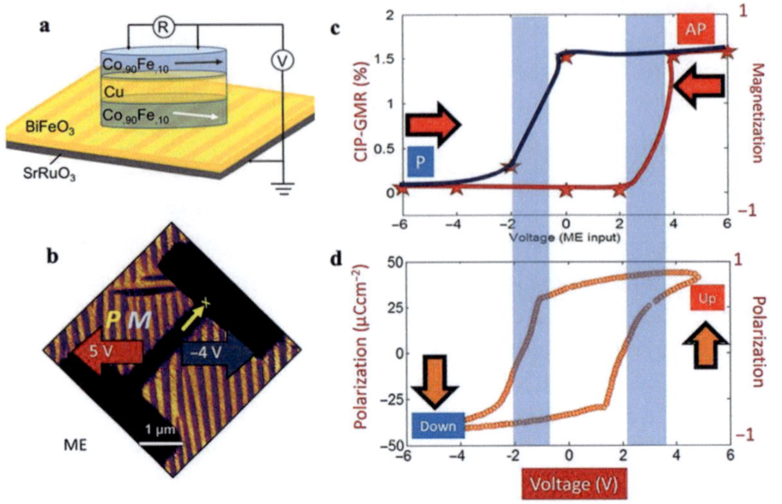

Figure 12.7: (a–d) Electric field control of a GMR device using $BiFeO_3$. (a) Schematic of the magnetoelectric memory element. A voltage is applied across the multiferroic $BiFeO_3$ which controls the magnetization state of the exchange-coupled layer. Current is applied in the plane of the GMR stack for readout. Arrows represent magnetization directions. Reprinted with permission from [229]. (b) PFM image of an actual device with the current bar oriented at ~ 45° from the $BiFeO_3$ domains. (c,d) Measured GMR (c) and ferroelectric polarization (d) versus voltage applied across the $BiFeO_3$ layer. The resistance of the GMR device (c) is hysteretic, non-volatile, and with coercive voltages in agreement with that from the ferroelectric switching (d) revealing the magnetoelectric switching of the $BiFeO_3$ and the reversal of the exchange-coupled magnetic layer in the GMR stack. Arrows represent magnetization and polarization directions. Reprinted with permission from [230].

coercive voltage V_C. Yet recent work has found that the substitution progressively increases the degeneracy of the potential energy landscape of the system and rotates the polar axis [118, 128]. While the magnetoelectric coupling seems preserved, the increased complexity in ferroelectric and antiferromagnetic domain states compared to the undoped $BiFeO_3$ films demands further investigation.

Recently, the field of antiferromagnetic spintronics has garnered attention due to the new physical insights, probe development of the otherwise inert state, and the new avenues for device implementation. For electrical insulators, the longitudinal and Hall resistance of a spin Hall metal in contact with the antiferromagnetic insulator will depend on the orientation of the Néel vector. This occurs either through spin Hall magnetoresistance or a proximity-induced Hall effect. Work by Kosub and colleagues [112] demonstrated a purely antiferromagnetic magnetoelectric memory in a Cr_2O_3/Pt heterostructure (Figure 12.8). Here the Pt layer couples to the surface magnetization state of Cr_2O_3, which can be readout through a Hall measurement. The surface magnetization state is determined by the antiferromagnetic domain state which can be switched by simultaneous application of an electric and magnetic field with a critical electric and magnetic field product [110]. A few features of this purely antiferromagnetic magnetoelectric memory are particularly noteworthy. First, contrary to the $BiFeO_3$-based devices discussed previously, the absence of an exchange-coupled ferromagnetic layer is expected to decrease the device switching times due to the faster spin dynamics in antiferromagnetic systems. Second, Cr_2O_3 is not ferroelectric, therefore, energy dissipation is capacitive and ultra-low (~ 6×10^{-2} μJ cm^{-2} – assuming ideal dielectric behavior and neglecting the dissipation associated with the applied 6.3 kOe magnetic field [55]). Finally, extending the idea of coupling to the surface magnetization to quantum materials such as topological insulators, topological ferromagnetic insulators, or spin-valley Hall materials opens doors to electrical control of spin-momentum locked phenomena [232, 233].

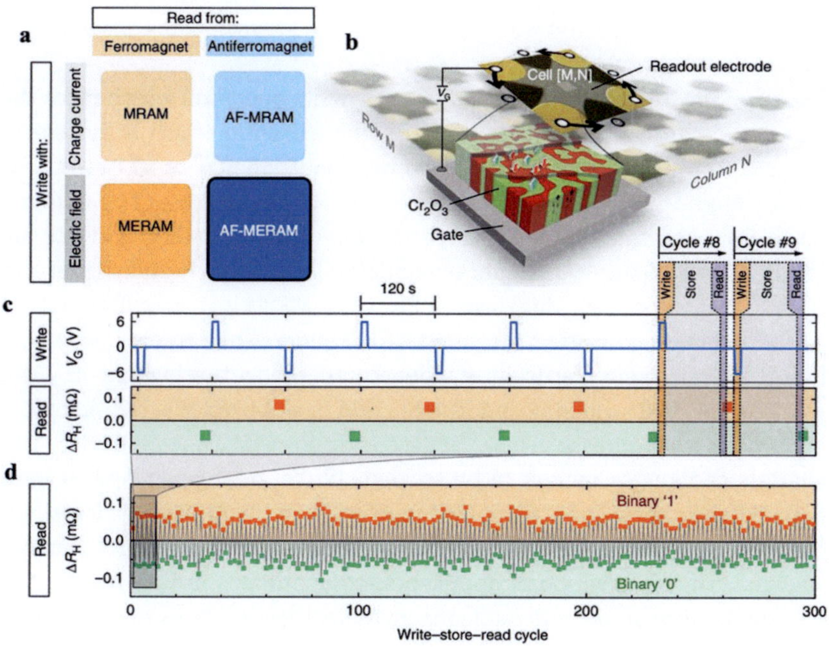

Figure 12.8: (a–d) Cr_2O_3-based antiferromagnetic magnetoelectric memory. (a) Schematic of spin-based memory elements categorized by the writing stimulus and order parameter for readout. (b) The surface magnetization of antiferromagnetic magnetoelectric Cr_2O_3 enables readout of the antiferromagnetic state of a bit through a Hall effect in an adjacent Pt layer. (c,d) Write voltage and Hall signal showing the fundamental write-read procedure (c) and robustness of the device with cycle number (d). Reprinted with permission from [112]. Copyright © 2017, Springer Nature.

12.4.2 Magnetoelectric/multiferroic Logic

Computation based on complimentary metal-oxide-semiconductor (CMOS) transistors is inherently inefficient due to the volatility of the device and the fundamental losses as the devices are scaled down. In the search for a new materials platform for the next generation of computation, non-volatile technologies have emerged as promising avenues as the total energy dissipation, write energy, and scalability can be shown to outperform CMOS. Work led by Intel and researchers at UC Berkeley has proposed a Magneto-Electric-Spin-Orbit (MESO) logic as one such technology (Figure 12.9) [84]. The fundamental device operates with a magnetoelectric write operation that determines the magnetization orientation of a magnetic bit. A supply current injects electrons into the magnetic bit and spin-polarizes them. This spin polarized current is then injected into a spin-orbit layer that uses the interface Rashba-Edelstein effect or the inverse spin Hall effect to create a transverse current that is used for readout or cascading logic devices.

At the heart of the device is the magnetoelectric write operation which takes the voltage input and controls the orientation of the magnetization which is hysteretic, non-volatile, and 10 to 1000 times more energy efficient than SOT or STT switching [55]. Critical for applications is the efficiency of the device at scale. The authors, using accepted bulk and thin-film material parameters, find that MESO will operate at 1–10 aJ dissipation and sub-100 mV per operation at $20 \times 30\ nm^2$ nanomagnet and $10 \times 10 \times 10\ nm^3$ magnetoelectric dimensions. Proposed MESO logic circuits compare favorably to CMOS fanout-4 inverter, two-input NAND adder, 32-bit ripple-carry adder and 32-bit ALU logic circuits. Particularly when compared to the inverter in 2018, MESO shows a 10–100 fold switching energy reduction in a non-volatile technology.

The estimations are impressive, and the race is on to realize such a device. Yet, several materials challenges remain to be tackled. While great progress has been made to realize the electrical switching of magnetism, demonstrating this at technologically relevant scale and solving defect pinning at metal-oxide interfaces remain largely untackled. The observation of application-compatible spin Hall resistivities in material systems compatible with oxide magnetoelectrics has been unexplored.

12.4.3 Neuromorphic devices

Biologically inspired computing mimics the neural structure of the brain and is highly efficient at difficult tasks. Of current interest is the development of neuromorphic circuits that are able to analyze videos and images for recognition and filtering applications. While such neural networks can be realized using CMOS technology, the inherent inefficiencies of CMOS devices leave room for non-volatile beyond-CMOS technologies to enhance the performance and parallelization. Spin-based neurons have been proposed and simulated. While showing promise, many rely on magnetic

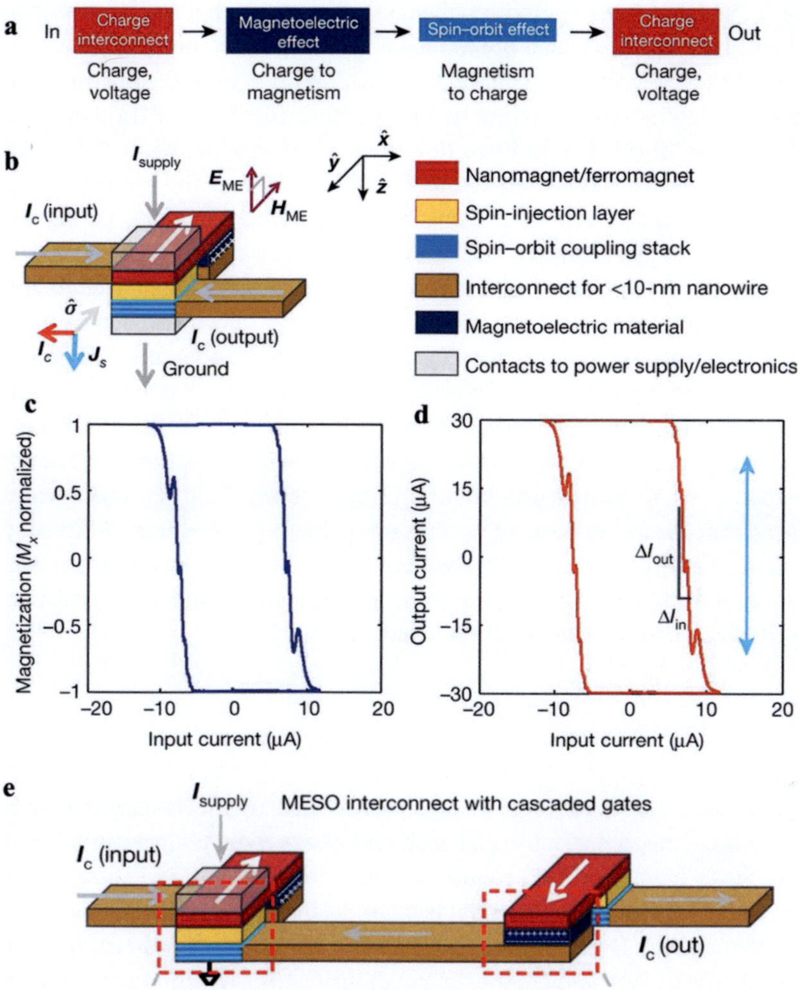

Figure 12.9: (a–e) The MESO device and connectivity. (a) Shows the fundamental mechanisms employed in the MESO devices from an input signal to output. (b) Schematic of the MESO device. Input, output, supply and spin currents are represented by I_c (input), I_c (output), I_{supply}, and J_s (with polarization direction σ), respectively. The electric field (E_{ME}) applied to the magnetoelectric at the input and the effective magnetic field (H_{ME}) generated that couples to the ferromagnetic bit are shown. (c,d) Normalized magnetization direction (c) and output current (d) versus input current. The output is hysteretic and non-volatile. The output current is proportional to the orientation of the magnetization which is controlled by the input to the magnetoelectric layer. (e). Schematic showing the connectivity to form logic structures. Reprinted with permission from [84]. Copyright © 2018, Springer Nature.

tunnel junctions, which makes readout challenging due to the nature of resistance and spin Hall and spin torques which are not ideal due to resistive heating [234].

Building on the MESO device, simulations of a cellular neural network with inverse Rashba-Edelstein magnetoelectric (IRME) neurons and CMOS synapses performing a low-pass filtering have been reported to be fast and efficient [235]. In this case, the IRME neuron state is written with a voltage across the antiferromagnetic magnetoelectric layer, which controls the magnetization state of an exchange-coupled ferromagnetic layer. Readout is done with a supply current that is spin polarized by the ferromagnet and injected in the inverse Rashba-Edelstein heterostructure, which converts the injected spin current into a transverse voltage. Assuming materials parameters for $Co_{0.9}Fe_{0.1}$ as the ferromagnet, magnetoelectric coefficients comparable to $BiFeO_3$, and an inverse Rashba-Edelstein parameter for that found in $LaAlO_3/SrTiO_3$ polar interface [236], the IRME performs state transitions in ~ 100 ps, network operations with a per-neuron energy cost on the order of ~ 1 fJ, and is non-volatile [235]. The IRME device has several notable features that lead to enhanced performance. Beyond the energy efficiency of the IRME neuron and its non-volatility which are enabled by the materials, the IRME network is parallelized as each neuron operates simultaneously and performs both write and read operations at the same time.

12.4.4 Hybrid magnetoelectric-spin-orbit torque heterostructures

Composite magnetoelectric multiferroic heterostructures are relatively easy to realize and demonstrate large magnetoelectric effects at and above room temperature. While such heterostructures have displayed robust coupling and even 90° switching of the magnetization, a full magnetization reversal is often desired to maximize readout signal. Thus, hybrid magnetoelectric-spin-orbit torque heterostructures have emerged to mitigate the weakness of SOT and composite magnetoelectric switching and leveraging their respective strengths. One embodiment of this concept is the hybrid magnetoelectric-spin-orbit torque (ME-SOT) heterostructure, which consists of a composite magnetoelectric with a spin-orbit torque active layer.

The main steps to switching in such a heterostructure first involve a voltage applied to the piezoelectric layer to drive the 90° switching of the magnetization. Then a current is applied to the SOT layer to drive the remaining 90° switch. Simulations of this two-step hybrid switching process indicate that the current pulse width for reliable spin-orbit torque switching may decrease by a factor of ~ 10–1000, depending on the anisotropy and damping parameter of the magnetostrictive layer [124]. The faster switching times is the result of the reduced processional motion of the magnetization as the magnetoelectric has already initialized it to the peak of the magnetic anisotropy barrier (assuming a uniaxial anisotropy). The device demonstrates some key advantages, in particular high temperature operation, simple materials synthesis, deterministic 180° magnetization switching at an ~10–1000 times improved energy-delay performance.

Other manifestations of hybrid ME-SOT heterostructures have been explored to break critical symmetries and lift degeneracies in SOT switching [237–240]. For instance, SOT switching of a perpendicular magnetic anisotropy (PMA) magnet by the spin Hall effect of isotropic heavy metals is complicated by non-deterministic switching. The symmetry of the system can be changed to realize deterministic switching by applying an in-plane magnetic field, switching in the presence of an in-plane exchange bias [240], but also by magnetoelectrically induced changes to the magnetic anisotropy and electric field gradients [241, 242].

12.5 Future perspectives and concluding remarks

The current achievements dealing with the electric field control of magnetism have been reviewed. Let us now address one of the main challenges limiting the insertion of multiferroics into spintronic devices: the determination of the magnetoelectric switching dynamics. Switching dynamics are typically estimated from Landau-Lifshitz-Gilbert-Slonczewski (LLGS) simulations and have seen little experimental work. Therefore, despite all the progress, the competitiveness of magnetoelectric-based devices with respect to established spin-current-based technology is still under debate. We review emergent probes of the dynamics of multiferroic antiferromagnetic states in thin-film heterostructures. We finally present some new avenues in the field, highlighting the potential of light to probe and even assist multiferroic switching.

12.5.1 Magnetoelectric switching dynamics

Magnetoelectric random-access memories or magnetoelectric spin-orbit logic devices with strong charge-to-spin coupling promise attojoule switching energies. If those concepts shall become reality, the scientific community needs to intensify efforts towards a deeper understanding of multiferroic and interlayer exchange dynamics. The lack of existing data on multiferroic switching dynamics partly originates from the difficulty to simultaneously and *operando* probe two order parameters, i.e. electric polarization and net magnetization. The direct observation of the ferroelectric-ferromagnetic domain correlation during a magnetoelectric switching event is a prerequisite to the understanding of the switching mechanism and its dynamics.

In the model system, $BiFeO_3$/ferromagnetic heterostructures, the electric-field-induced magnetization reversal involves a two-step switching process. Within the unit cell, the polarization projection onto the interface plane rotates twice by 90° [102]. The resulting 180° reversal of the polarization projection and the corresponding antiferromagnetic configuration in $BiFeO_3$ in the final state thus leads to the 180° switch of the exchange-coupled magnetization in the ferromagnetic layer [102, 107]. While the ferroelectric switching has been unambiguously demonstrated, the dynamics of the magnetic order in $BiFeO_3$ and in the magnetically coupled ferromagnetic layer have remained unexplored.

On the one hand, several experimental techniques allow probing of the ferroelectric switching and domain dynamics, even in a buried state. Beyond the investigation of the remnant hysteresis using positive-up negative-down (PUND) method, hard XRD [243, 244], hard X-ray photoemission spectroscopy [245], optical second harmonic generation [183, 246], X-ray photon correlation spectroscopy [247], and piezoresponse force spectroscopy [131, 248] provide direct access to the polarization with time resolution and are compatible with *operando* analysis. On the other hand, antiferromagnetic states, dominating the magnetoelectric switching mechanism, lack a net magnetization and have therefore remained difficult to address. Over the last few years, the following techniques have emerged as efficient probe of antiferromagnetic states in thin-film heterostructures.

The single-spin magnetometer scanning probe technique – Based on a point-like impurity nitrogen-vacancy (NV) defect in diamond mounted on a scanning tip [249–251], this technique provides an outstanding magnetic sensitivity (down to a few femtotesla) with a spatial resolution comparable to the one achieved with atomic force microscopy. It has been used to spatially resolve the local magnetic order in multiferroic antiferromagnetic $BiFeO_3$ thin films [5, 252]. The periodic modulation in the magnetic response corresponds to the spin cycloid in the films. Such a measurement allows the correlation of the antiferromagnetic domain state with the ferroelectric domain architecture obtained by PFM.

Magnetoelectric force microscopy (MeFM) – The MeMFM is another scanning-probe-based technique. A magnetically tip is used to probe the magnetoelectric response of the material exposed to an AC voltage. It is based on the lock-in detection of the magnetic force microscopy signal from the electric-field-induced magnetization in magnetoelectric materials. First developed in order to probe magnetoelectric domains in hexagonal $ErMnO_3$ crystals at low temperature [253], its efficiency to monitor the magnetoelectric response of the Cr_2O_3 crystal model system has been recently demonstrated [254], see Figure 12.10(a–b). The MeFM requires the magnetoelectric material to be integrated in a capacitor to trigger the magnetoelectric response. On one hand, such capping may prevent simultaneous piezoresponse-based ferroelectric domain imaging. On the other hand, the use of heavy metals for the electrode might provide a platform for the investigation of hybrid magnetoelectric STT, SOT type of heterostructures, see Section 12.4.4.

Optical second harmonic generation (SHG) – SHG can be sensitive to the reduction of symmetry through the magnetic ordering of spins. We refer the reader to several reviews dealing with the potential of SHG to probe ferroic states in bulk and thin-film materials [246, 257, 258]. Here we highlight the seminal demonstration in antiferromagnetic magnetoelectric Cr_2O_3 crystal, in which SHG was employed to image

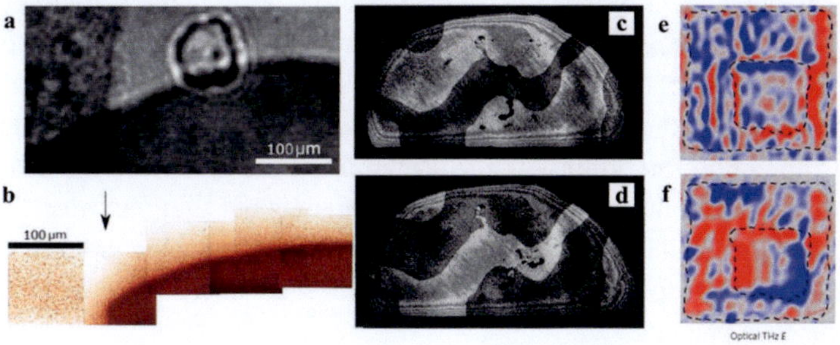

Figure 12.10: (a–b) Cr_2O_3 crystal imaged using optical second harmonic (a) and magnetoelectric force microscopy (b). Reprinted with permission from [254]. (c–d) Optical second harmonic generation images of antiferromagnetic domains in Cr_2O_3 obtained with right (c) and (d) left circularly polarized light. The sample is 6 mm long and 3 mm wide. Reprinted with permission from [255]. Copyright 1995, AIP Publishing LLC. (e–f) Optical second-harmonic-generation-based reconstruction of the antiferromagnetic configurations of the patterns in $BiFeO_3$ thin films in virgin state (e) and after sub-coercive picosecond electric pulses delivered by rectification of intense laser pulses (f). The image size is $10 \times 10\ \mu m^2$. Reprinted with permission from [256]. Copyright © 2017, Springer Nature.

antiferromagnetic domains [255], see Figure 12.10(c–d). Such an optical technique further offers time resolution matching the ultrafast dynamics of the ferroic switching speed. The potential of time-resolved SHG to determine antiferromagnetic ultrafast dynamics has recently been revealed [259]. In bulk $YMnO_3$, the spin ordering modulation upon coherent magnon excitation could be tracked using a combination of Faraday rotation and SHG. In multiferroic thin-film systems, SHG has been used to identify the antiferromagnetic domains in epitaxially strained $BiFeO_3$ [256]. Furthermore, taking advantage of the optical rectification process in intense light fields, 100 fs light pulses were used to generate electric pulses in the antiferromagnetic resonance regime of $BiFeO_3$. The corresponding change in the antiferromagnetic domains were imaged by SHG after the optically induced electric pulses [256], see Figure 12.10(e–f). This work suggests new avenues towards the optical control of multiferroic and magnetoelectric switching in thin films. We address the non-invasive optical control of such switching events in the next section.

12.5.2 Towards all-optical magnetoelectric switching

Light, which has been mainly used to probe materials [257, 258], is now used to act on ferroic states. Its potential to induce magnetization reversal in thin films was recently demonstrated [260, 261]. The long working distance and ultrafast timescale of optical processes render it further compatible with device integration. The light-matter interaction in ferroelectrics however drastically differs from the ferromagnetic analogs. The light-induced processes in ferroelectric materials span from local heating [262] to photo-induced flexoelectric effect [263], photovoltaic effect [264] or photostriction [265]. In particular, photo-induced charge carriers may migrate towards domain walls or to the film surface and act on the bound charge screening at the ferroelectric surfaces. Considering the depolarizing impact on domain state in epitaxial thin ferroelectric films, see Section 12.3 and [52, 150], light irradiation on thin films may therefore emerge as an efficient approach towards the domain modulation in ferroelectric and multiferroic magnetoelectric systems.

A pioneering work dealing with light-induced polarization changes focused mainly on bulk materials and led to local domain wall motions or phase transitions [264, 266–268]. The direct monitoring of an optically induced domain wall motion and domain configuration change in ferroelectric material opened new avenues for tuning macroscopic polarization states by means of a non-contact external control. Using *in-situ* XRD measurements, the $BaTiO_3$ in-plane oriented a-domains reorient in an orthogonal axis point either in-plane or out-of-plane upon irradiation under polarized light as shown in Figure 12.11(a–c). The process is fully reversible and the domains return to their original configuration in the dark.

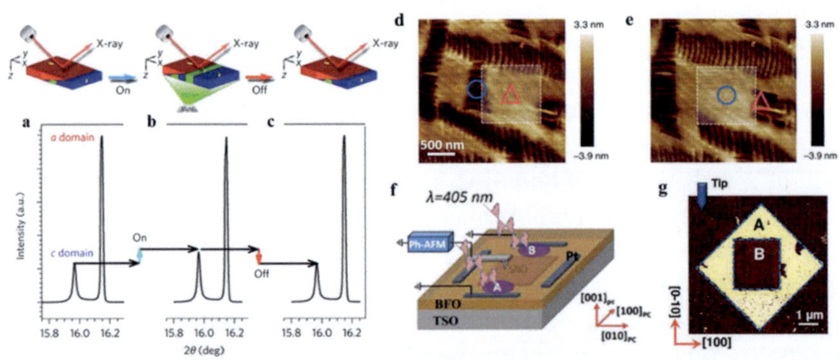

Figure 12.11: (a–c) Reversible optical control of macroscopic polarization in $BaTiO_3$. Sequence of synchrotron radiation high-resolution XRD patterns of (002)/(200) reflections corresponding to the off (a)–on (b)–off (c) light succession. The c-domain signal increases when the light is on due to the domain wall motion and is reversible. Reprinted with permission from [267]. Copyright © 2017, Springer Nature. (d–e) Deterministic optical control of phase distribution in mixed-phase $BiFeO_3$. Topography images of mixed-phase $BiFeO_3$ after illumination of areas marked with a triangle (d) and circle (e). Reprinted with permission from [263]. Copyright© 2019, Springer Nature. (f–g) Light-induced reversible switching of ferroelectric $BiFeO_3$ polarization. (f) Schematic showing photoelectric atomic force microscopy (Ph-AFM) setup and illumination areas resulting in photocurrent flow in different directions and hence reversed electric-field polarity under the tip leading to a reversible switch shown in PFM image (g). Reproduced with permission from [264]. Copyright 2018 Wiley-VCH Verlag GmbH & Co. KGaA).

In highly strained multiferroic $BiFeO_3$ films, the flexophotovoltaic effect [269], combining strain gradient and light sensitivity, appears as a potential degree of freedom for all-optical multiferroic magnetoelectric switch. The strain-induced morphotropic boundary in $BiFeO_3$ films grown on highly compressive epitaxial strain leads to a rhombohedral and tetragonal phase coexistence. Upon illumination with a coherent light source, the balance between the two phases can be controlled, as illustrated in Figure 12.11(d–e). The mechanism involves a combination of flexoelectricity and thermal heating. Most importantly, the first experimental data further indicate that the light-induced changes in the film domain state are accompanied by an alteration of the antiferromagnetic state.

Finally, photocurrents generated via the bulk photovoltaic may be engineered in order to achieve local ferroelectric switching events. Using a tip enhanced electric field created by increased photocurrent density at the tip, the light irradiation induces an electric field surpassing the coercive electric field, as shown in Figure 12.11(f–g). The bulk photovoltaic effect depends on the crystallographic direction and on the light polarization such that an all-optical light polarization dependent switching can be achieved in multiferroic $BiFeO_3$ films by simply changing the light polarization. Because strain strongly impacts the photocurrent generation [270], the recent development in epitaxial design might further support the development of optically induced functionality in magnetoelectric multilayers.

12.6 Conclusion

In this article, we reviewed the most recent developments in the field of magnetoelectric heterostructures. More in-depth review articles dedicated to artificial multiferroic heterostructures [6, 7] or domain manipulation [51, 271] can be found elsewhere. Magnetoelectric heterostructures have first emerged as an alternative to the scarcity of single-phase multiferroic magnetoelectric materials [10]. They now play a major role in the pursuit of low energy consuming devices paradigms.

New probing tools are pushing the understanding of ferroic domain coupling and imprint in the multilayers [3]. In particular non-invasive, *operando* investigations [183, 246] are the key towards determination of magnetoelectric switching dynamics. Finally, a heavy metal capping of the ferromagnetic multiferroic heterostructures enables the realization of hybrid devices combining magnetoelectric and SOT effects (discussed in Section 12.4.4). This should lead to ultra-low energy consumption magnetic logic operation. We further foresee the development of all-optical switching schemes to multiferroic thin films for ultra-fast dynamics.

Acknowledgements: E.G. and M.T. acknowledge the financial support by the Swiss National Science Foundation under project No. 200021_188414. P.B.M. and J.T.H. acknowledge that this work was supported in part by the Semiconductor Research Corporation (SRC) as the NEWLIMITS Center and NIST through award number 70NANB17H041.

References

[1] Fiebig M, Lottermoser T, Meier D, Trassin M. The evolution of multiferroics. Nat Rev Mater. 2016;1:16046.
[2] DOI: 10.1515/PSR.2020.0032.
[3] Spaldin NA, Ramesh R. Advances in magnetoelectric multiferroics. Nat Mater. 2019;18:203–12.
[4] Ramesh R, Spaldin NA. Multiferroics: progress and prospects in thin films. Nat Mater. 2007;6:21–9.
[5] DOI: 10.1515/PSR.2019.0067.
[6] Trassin M. Low energy consumption spintronics using multiferroic heterostructures. J Phys Condens Matter. 2016;28:033001.
[7] Fernandes Vaz CA, Staub U. Artificial multiferroic heterostructures. J Mater Chem C. 2013;1:6731–42.
[8] Scott JF. Multiferroic memories. Nat Mater. 2007;6:256–7.
[9] Bibes M. Towards a magnetoelectric memory. Nat Mater. 2008;7:7–8.
[10] Hill NA. Why are there so few magnetic ferroelectrics? J Phys Chem B. 2000;104:6694.
[11] DOI: 10.1515/PSR.2019.0016.
[12] Wiegelmann H, Jansen AG, Wyder P, Rivera JP, Schmid H. Magnetoelectric effect of Cr_2O_3 in strong static magnetic fields. Ferroelectrics. 1994;162:141–6.
[13] Fiebig M, Fröhlich D, Krichevtsov BB, Pisarev RV. Second harmonic generation and magnetic-dipole-electric-dipole interference in antiferromagnetic Cr_2O_3. Phys Rev Lett. 1994;73:2127–30.
[14] Van Aken BB, Palstra TT, Filippetti A, Spaldin NA. The origin of ferroelectricity in magnetoelectric $YMnO_3$. Nat Mater. 2004;3:164–70.
[15] Fiebig M, Lottermoser T, Fröhlich D, Goltsev AV, Pisarev RV. Observation of coupled magnetic and electric domains. Nature. 2002;419:818–20.
[16] Bertaut EF, Mercier M. Structure magnetique de $MnYO_3$. Phys Lett. 1963;5:27–9.
[17] DOI: 10.1515/PSR.2019.0014.
[18] Nordlander J, Campanini M, Rossell MD, Erni R, Meier QN, Cano A, et al. The ultrathin limit of improper ferroelectricity. Nat Commun. 2019;10:5591.
[19] Meier D, Seidel J, Cano A, Delaney K, Kumagai Y, Mostovoy M, et al. Anisotropic conductance at improper ferroelectric domain walls. Nat Mater. 2012;11:284–8.
[20] Schaab J, Skjærvø SH, Krohns S, Dai X, Holtz ME, Cano A, et al. Electrical half-wave rectification at ferroelectric domain walls. Nat Nanotechnol. 2018;13:1028–34.
[21] Evans DM, Holstad TS, Mosberg AB, Småbråten DR, Vullum PE, Dadlani AL, et al. Conductivity control via minimally invasive anti-Frenkel defects in a functional oxide. Nat Mater. 2020;19:1195–1200.
[22] Wang J, Neaton JB, Zheng H, Nagarajan V, Ogale SB, Liu B, et al. Epitaxial $BiFeO_3$ multiferroic thin film heterostructures. Science. 2003;299:1719–22.
[23] Zhang JX, He Q, Trassin M, Luo W, Yi D, Rossell MD, et al. Microscopic origin of the giant ferroelectric polarization in tetragonal-like $BiFeO_3$. Phys Rev Lett. 2011;107:1–5.

[24] Lebeugle D, Colson D, Forget A, Viret M, Bataille AM, Gukasov A. Electric-field-induced spin flop in $BiFeO_3$ single crystals at room temperature. Phys Rev Lett. 2008;100:1–4.

[25] Lebeugle D, Mougin A, Viret M, Colson D, Ranno L. Electric field switching of the magnetic anisotropy of a ferromagnetic layer exchange coupled to the multiferroic compound $BiFeO_3$. Phys Rev Lett. 2009;103:257601.

[26] Sando D, Barthélémy A, Bibes M. $BiFeO_3$ epitaxial thin films and devices: Past, present and future. J Phys Condens Matter. 2014;26:473201.

[27] Heron JT, Schlom DG, Ramesh R. Electric field control of magnetism using $BiFeO_3$-based heterostructures. Appl Phys Rev. 2014;1:21303.

[28] DOI: 10.1515/PSR.2019.0070.

[29] Burns SR, Paull O, Juraszek J, Nagarajan V, Sando D. The experimentalist's guide to the cycloid, or noncollinear antiferromagnetism in Epitaxial $BiFeO_3$. Adv Mater. 2020;2003711:1–51.

[30] Trassin M, Viart N, Versini G, Barre S, Pourroy G, Lee J, et al. Room temperature ferrimagnetic thin films of the magnetoelectric $Ga_{2-x}Fe_xO_3$ *J*. Mater Chem. 2009;19:8876–80.

[31] Trassin M, Viart N, Versini G, Loison JL, Vola JP, Schmerber G, et al. Epitaxial thin films of multiferroic $GaFeO_3$ on conducting indium tin oxide (001) buffered yttrium-stabilized zirconia (001) by pulsed laser deposition. Appl Phys Lett. 2007;91:2–5.

[32] Thomasson A, Cherifi S, Lefevre C, Roulland F, Gautier B, Albertini D, et al. Room temperature multiferroicity in $Ga_{0.6}Fe_{1.4}O_3$:Mg thin films. J Appl Phys. 2013;113:3–7.

[33] Homkar S, Preziosi D, Devaux X, Bouillet C, Nordlander J, Trassin M, et al. Ultrathin regime growth of atomically flat multiferroic gallium ferrite films with perpendicular magnetic anisotropy. Phys Rev Mater. 2019;3:1–10.

[34] Aurivillius B. Mixed bismuth oxides with layer lattices. 1. the structure type of $CaNb_2Bi_2O_9$. Ark Kemi. 1949;1:463–80.

[35] Benedek NA, Rondinelli JM, Djani H, Ghosez P, Lightfoot P. Understanding ferroelectricity in layered perovskites: new ideas and insights from theory and experiments. Dalt Trans. 2015;44:10543–58.

[36] Gradauskaite E, Campanini M, Biswas B, Schneider CW, Fiebig M, Rossell MD, et al. Robust in-plane ferroelectricity in ultrathin epitaxial Aurivillius films. Adv Mater Interfaces. 2020;2000202:1–8.

[37] Keeney L, Smith RJ, Palizdar M, Schmidt M, Bell AJ, Coleman JN, et al. Ferroelectric behavior in exfoliated 2D Aurivillius oxide flakes of sub-unit cell thickness. Adv Electron Mater. 2020;6:1901264.

[38] Wang C, Ke X, Wang J, Liang R, Luo Z, Tian Y, et al. Ferroelastic switching in a layered-perovskite thin film. Nat Commun. 2016;7:10636.

[39] Lee HN, Hesse D, Zakharov N, Gösele U. Ferroelectric $Bi_{3.25}La_{0.75}Ti_3o_{12}$ films of uniform a-axis orientation on silicon substrates. Science. 2002;296:2006–9.

[40] Faraz A, Maity T, Schmidt M, Deepak N, Roy S, Pemble ME, et al. Direct visualization of magnetic-field-induced magnetoelectric switching in multiferroic aurivillius phase thin films. J Am Ceram Soc. 2017;100:975–87.

[41] Keeney L, Downing C, Schmidt M, Pemble ME, Nicolosi V, Whatmore RW. Direct atomic scale determination of magnetic ion partition in a room temperature multiferroic material. Sci Rep. 2017;7:1–11.

[42] Baibich MN, Broto JM, Fert A, Van Dau FN, Petroff F, Eitenne P, et al. Giant magnetoresistance of (001)Fe/(001)Cr magnetic superlattices. Phys Rev Lett. 1988;61:2472–5.

[43] Binasch G, Grünberg P, Saurenbach F, Zinn W. Enhanced magnetoresistance in layered magnetic structures with antiferromagnetic interlayer exchange. Phys Rev B. 1989;39:4828–30.

[44] Myers EB, Ralph DC, Katine JA, Louie RN, Buhrman RA. Current-induced switching of domains in magnetic multilayer devices. Science. 1999;285:867–70.

[45] Jian-Gang ZH. Magnetoresistive random access memory: the path to competitiveness and scalability. Proc IEEE. 2008;96:1786–98.

[46] Miron IM, Gaudin G, Auffret S, Rodmacq B, Schuhl A, Pizzini S, et al. Current-driven spin torque induced by the Rashba effect in a ferromagnetic metal layer. Nat Mater. 2010;9:230–4.

[47] Miron IM, Garello K, Gaudin G, Zermatten PJ, Costache MV, Auffret S, et al. Perpendicular switching of a single ferromagnetic layer induced by in-plane current injection. Nature. 2011;476:189–93.

[48] Vélez S, Schaab J, Wörnle MS, Müller M, Gradauskaite E, Welter P, et al. High-speed domain wall racetracks in a magnetic insulator. Nat Commun. 2019;10:4750.

[49] Zhuravlev MY, Sabirianov RF, Jaswal SS, Tsymbal EY. Giant electroresistance in ferroelectric tunnel junctions. Phys Rev Lett. 2005;94:1–4.

[50] Garcia V, Fusil S, Bouzehouane K, Enouz-Vedrenne S, Mathur ND, Barthélémy A, et al. Giant tunnel electroresistance for non-destructive readout of ferroelectric states. Nature. 2009;460:81–4.

[51] Strkalj N, Gradauskaite E, Nordlander J, Trassin M. Design and manipulation of ferroic domains in complex oxide heterostructures. Materials. 2019;12:3108.

[52] Strkalj N, De Luca G, Campanini M, Pal S, Schaab J, Gattinoni C, et al. Depolarizing-field effects in epitaxial capacitor heterostructures. Phys Rev Lett. 2019;123:147601.

[53] Junquera J, Ghosez P. Critical thickness for ferroelectricity in perovskite ultrathin films. Nature. 2003;422:506–9.

[54] Hu J-M, Duan C-G, Nan C-W, Chen L-Q. Understanding and designing magnetoelectric heterostructures guided by computation: progresses, remaining questions, and perspectives. npj Comput Mater. 2017;3:18.

[55] Meisenheimer PB, Novakov S, Vu NM, Heron JT. Perspective: magnetoelectric switching in thin film multiferroic heterostructures. J Appl Phys. 2018;123:240901.

[56] Bibes M, Villegas JE, Barthélémy A. Ultrathin oxide films and interfaces for electronics and spintronics. Adv Phys. 2011;60:5–84.

[57] Zhou J, Trassin M, He Q, Tamura N, Kunz M, Cheng C, et al. Directed assembly of nano-scale phase variants in highly strained $BiFeO_3$ thin films. J Appl Phys. 2012;112:64102.

[58] Liu M, Li S, Zhou Z, Beguhn S, Lou J, Xu F, et al. Electrically induced enormous magnetic anisotropy in Terfenol-D/lead zinc niobate-lead titanate multiferroic heterostructures. J Appl Phys. 2012;112:63917.

[59] Liu M, Sun NX. Voltage control of magnetism in multiferroic heterostructures. Phil Trans R Soc A. 2014;372:20120439.

[60] Lahtinen TH, Franke KJ, Van Dijken S. Electric-field control of magnetic domain wall motion and local magnetization reversal. Sci Rep. 2012;2:258.

[61] Franke KJ, López González D, Hämäläinen SJ, van Dijken S. Size dependence of domain pattern transfer in multiferroic heterostructures. Phys Rev Lett. 2014;112:17201.

[62] Buzzi M, Chopdekar RV, Hockel JL, Bur A, Wu T, Pilet N, et al. Single domain spin manipulation by electric fields in strain coupled artificial multiferroic nanostructures. Phys Rev Lett. 2013;111:27204.

[63] Lou J, Reed D, Pettiford C, Liu M, Han P, Dong S, et al. Giant microwave tunability in FeGaB/lead magnesium niobate-lead titanate multiferroic composites. Appl Phys Lett. 2008;92:262502.

[64] Jahjah W, Jay J-P, Le Grand Y, Fessant A, Prinsloo AR, Sheppard CJ, et al. Electrical manipulation of magnetic anisotropy in a $Fe_{81}Ga_{19}/Pb(Mg_{1/3}Nb_{2/3})O_3$-$Pb(Zr_xTi_{1-x})O_3$ magnetoelectric multiferroic composite. Phys Rev Appl. 2020;13:34015.

[65] Lou J, Liu M, Reed D, Ren Y, Sun NX. Giant electric field tuning of magnetism in novel multiferroic FeGaB/Lead Zinc Niobate–Lead Titanate (PZN-PT) heterostructures. Adv Mater. 2009;21:4711–5.
[66] Wang J-L, Echtenkamp W, Mahmood A, Binek C. Voltage controlled magnetism in Cr_2O_3 based all-thin-film systems. J Magn Magn Mater. 2019;486:165262.
[67] Callen ER, Callen HB. Static magnetoelastic coupling in cubic crystals. Phys Rev. 1963;129:578–93.
[68] Newnham RE. Properties of materials: anisotropy, symmetry, structure. Oxford: OUP, 2004.
[69] Wu R. Origin of large magnetostriction in FeGa alloys. J Appl Phys. 2002;91:7358–60.
[70] Du Y, Huang M, Chang S, Schlagel DL, Lograsso TA, McQueeney RJ. Relation between Ga ordering and magnetostriction of Fe-Ga alloys studied by X-ray diffuse scattering. Phys Rev B. 2010;81:54432.
[71] Wang H, Zhang YN, Wu RQ, Sun LZ, Xu DS, Zhang ZD. Understanding strong magnetostriction in $Fe_{100-x}Ga_x$ alloys. Sci Rep. 2013;3:1–5.
[72] Furthmuller J, Fahnle M, Herzer G. Theory of magnetostriction in amorphous and polycrystalline ferromagnets. J Phys F Met Phys. 1986;16:L255–L258.
[73] Wang Q, Li X, Liang C-Y, Barra A, Domann J, Lynch C, et al. Strain-mediated 180° switching in CoFeB and Terfenol-D nanodots with perpendicular magnetic anisotropy. Appl Phys Lett. 2017;110:102903.
[74] Parkes DE, Cavill SA, Hindmarch AT, Wadley P, McGee F, Staddon CR, et al. Non-volatile voltage control of magnetization and magnetic domain walls in magnetostrictive epitaxial thin films. Appl Phys Lett. 2012;101:72402.
[75] Ahmad H, Atulasimha J, Bandyopadhyay S. Reversible strain-induced magnetization switching in FeGa nanomagnets: pathway to a rewritable, non-volatile, non-toggle, extremely low energy straintronic memory. Sci Rep. 2015;5:srep18264.
[76] Lee Y, Liu ZQ, Heron JT, Clarkson JD, Hong J, Ko C, et al. Large resistivity modulation in mixed-phase metallic systems. Nat Commun. 2015;6:ncomms6959.
[77] Zhang S, Zhao Y, Xiao X, Wu Y, Rizwan S, Yang L, et al. Giant electrical modulation of magnetization in $Co_{40}Fe_{40}B_{20}/Pb(Mg_{1/3}Nb_{2/3})_{0.7}Ti_{0.3}O_3(011)$ heterostructure. Sci Rep. 2014;4:3727.
[78] Feng M, Wang J, Hu J-M, Wang J, Ma J, Li H-B, et al. Optimizing direct magnetoelectric coupling in $Pb(Zr,Ti)O_3$/Ni multiferroic film heterostructures. Appl Phys Lett. 2015;106:72901.
[79] Lahtinen TH, Tuomi JO, Van Dijken S. Pattern transfer and electric-field-induced magnetic domain formation in multiferroic heterostructures. Adv Mater. 2011;23:3187–91.
[80] Lahtinen TH, van Dijken S. Temperature control of local magnetic anisotropy in multiferroic CoFe/$BaTiO_3$. Appl Phys Lett. 2013;102:112406.
[81] Kim JY, Yao L, Van Dijken S. Coherent piezoelectric strain transfer to thick epitaxial ferromagnetic films with large lattice mismatch. J Phys Condens Matter. 2013;25:082205.
[82] Nikonov DE, Young IA. Benchmarking of beyond-CMOS exploratory devices for logic integrated circuits. IEEE J Explor Solid-State Comput Devices Circuits. 2015;1:3–11.
[83] Jaiswal A, Roy K. MESL: proposal for a non-volatile cascadable magneto-electric spin logic. Sci Rep. 2017;7:39793.
[84] Manipatruni S, Nikonov DE, Lin -C-C, Gosavi TA, Liu H, Prasad B, et al. Scalable energy-efficient magnetoelectric spin–orbit logic. Nature. 2019;565:35–42.
[85] Ma J, Shi Z, Nan C-W. Magnetoelectric properties of composites of single $Pb(Zr,Ti)O_3$ rods and Terfenol-D/Epoxy with a single-period of 1-3-type structure. Adv Mater. 2007;19:2571–3.
[86] Nan C-W, Liu G, Lin Y. Influence of interfacial bonding on giant magnetoelectric response of multiferroic laminated composites of $Tb_{1-x}Dy_xFe_2$ and $PbZr_xTi_{1-x}O_3$. Appl Phys Lett. 2003;83:4366–8.

[87] Vaz CA, Hoffman J, Segal Y, Reiner JW, Grober RD, Zhang Z, et al. Origin of the magnetoelectric coupling effect in $Pb(Zr_{0.2}Ti_{0.8})O_3$ / $La_{0.8}Sr_{0.2}MnO_3$ multiferroic heterostructures. Phys Rev Lett. 2010;104:127202.
[88] Evans DM, Schilling A, Kumar A, Sanchez D, Ortega N, Arredondo M, et al. Magnetic switching of ferroelectric domains at room temperature in multiferroic PZTFT. Nat Commun. 2013;4:1534.
[89] Zhang S, Zhao YG, Li PS, Yang JJ, Rizwan S, Zhang JX, et al. Electric-field control of nonvolatile magnetization in $Co_{40}Fe_{40}B_{20}/Pb(Mg_{1/3}Nb_{2/3})_{0.7}Ti_{0.3}O_3$ structure at room temperature. Phys Rev Lett. 2012;108:137203.
[90] Wu T, Bur A, Wong K, Leon Hockel J, Hsu C-J, Kim HK, et al. Electric-poling-induced magnetic anisotropy and electric-field-inducedmagnetization reorientation in magnetoelectric Ni/(011) $[Pb(Mg_{1/3}Nb_{2/3})O_3]_{(1-x)}$-$[PbTiO_3]_x$ heterostructure. J Appl Phys. 2011;109:07D732.
[91] Mardana A, Ducharme S, Adenwalla S. Ferroelectric Control of Magnetic Anisotropy. Nano Lett. 2011;11:3862–7.
[92] Ghidini M, Pellicelli R, Prieto JL, Moya X, Soussi J, Briscoe J, et al. Non-volatile electrically-driven repeatable magnetization reversal with no applied magnetic field. Nat Commun. 2013;4:1453.
[93] Kittmann A, Durdaut P, Zabel S, Reermann J, Schmalz J, Spetzler B, et al. Wide band low noise love wave magnetic field sensor system. Sci Rep. 2018;8:278.
[94] Lin H, Page MR, McConney M, Jones J, Howe B, Sun NX. Integrated magnetoelectric devices: filters, pico-Tesla magnetometers, and ultracompact acoustic antennas. MRS Bull. 2018;43:841–7.
[95] Nan T, Lin H, Gao Y, Matyushov A, Yu G, Chen H, et al. Acoustically actuated ultra-compact NEMS magnetoelectric antennas. Nat Commun. 2017;8:296.
[96] Liu M, Obi O, Lou J, Chen Y, Cai Z, Stoute S, et al. Giant electric field tuning of magnetism in novel multiferroic FeGaB/Lead Zinc Niobate–Lead Titanate (PZN-PT) heterostructures. Adv Funct Mater. 2009;19:1826–31.
[97] Zaeimbashi M, Lin H, Dong C, Liang X, Nasrollahpour M, Chen H, et al. NanoNeuroRFID: a wireless implantable device based on magnetoelectric antennas. IEEE J Electromagn RF Microwaves Med Biol. 2019;3:206–15.
[98] Emam S, Sun N-X, Nasrollahpour M. P3-226: electrochemical gas sensor arrays for detecting volatile organic compound biomarkers in the exhaled breath. Alzheimer's Dement. 2019;15: P1018–P1018.
[99] Béa H, Bibes M, Ott F, Dupé B, Zhu X-H, Petit S, et al. Mechanisms of exchange bias with multiferroic $BiFeO_3$ epitaxial thin films. Phys Rev Lett. 2008;100:17204.
[100] Trassin M, Luca GD, Manz S, Fiebig M. Probing Ferroelectric Domain Engineering in $BiFeO_3$ Thin Films by Second Harmonic Generation. 2015:4871–6.
[101] Gao T, Zhang X, Ratcliff W, Maruyama S, Murakami M, Varatharajan A, et al. Electric-field induced reversible switching of the magnetic easy axis in Co/$BiFeO_3$ on $SrTiO_3$. Nano Lett. 2017;17:2825–32.
[102] Heron JT, Bosse JL, He Q, Gao Y, Trassin M, Ye L, et al. Deterministic switching of ferromagnetism at room temperature using an electric field. Nature. 2014;516:370–3.
[103] Trassin M, Clarkson JD, Bowden SR, Liu J, Heron JT, Paull RJ, et al. Interfacial coupling in multiferroic/ferromagnet heterostructures. Phys Rev B Condens Matter Mater Phys. 2013;87:134426.
[104] Radu F, Zabel H. Exchange bias effect of ferro-/antiferromagnetic heterostructures. Springer Tracts Mod Phys. 2007;227:97–184.

[105] Toyoki K, Shiratsuchi Y, Kobane A, Mitsumata C, Kotani Y, Nakamura T, et al. Magnetoelectric switching of perpendicular exchange bias in Pt/Co/α-Cr_2O_3/Pt stacked films. Appl Phys Lett. 2015;106:162404.

[106] Borisov P, Hochstrat A, Chen X, Kleemann W, Binek C. Magnetoelectric switching of exchange bias. Phys Rev Lett. 2005;94:117203.

[107] Heron JT, Trassin M, Ashraf K, Gajek M, He Q, Yang SY, et al. Electric-field-induced magnetization reversal in a ferromagnet-multiferroic heterostructure. Phys Rev Lett. 2011;107:1–5.

[108] Wu SM, Cybart SA, Yu P, Rossell MD, Zhang JX, Ramesh R, et al. Reversible electric control of exchange bias in a multiferroic field-effect device. Nat Mater. 2010;9:756–61.

[109] Wu SM, Cybart SA, Yi D, Parker JM, Ramesh R, Dynes RC. Full electric control of exchange bias. Phys Rev Lett. 2013;110:67202.

[110] He X, Wang Y, Wu N, Caruso AN, Vescovo E, Belashchenko KD, et al. Robust isothermal electric control of exchange bias at room temperature. Nat Mater. 2010;9:579–85.

[111] Nozaki T, Al-Mahdawi M, Pati SP, Ye S, Shiokawa Y, Sahashi M. Magnetoelectric switching energy in Cr_2O_3/Pt/Co perpendicular exchange coupled thin film system with small Cr_2O_3 magnetization. Jpn J Appl Phys. 2017;56:70302.

[112] Kosub T, Kopte M, Hühne R, Appel P, Shields B, Maletinsky P, et al. Purely antiferromagnetic magnetoelectric random access memory. Nat Commun. 2017;8:13985.

[113] Borisov P, Ashida T, Nozaki T, Sahashi M, Lederman D. Magnetoelectric properties of 500-nm Cr_2O_3 films. Phys Rev B. 2016;93:174415.

[114] Vu NM, Luo X, Novakov S, Jin W, Nordlander J, Meisenheimer PB, et al. Bulk-like dielectric and magnetic properties of sub 100 nm thick single crystal Cr_2O_3 films on an epitaxial oxide electrode. Sci Rep. 2020;10:14721.

[115] Gross I, Akhtar W, Garcia V, Martínez LJ, Chouaieb S, Garcia K, et al. Real-space imaging of non-collinear antiferromagnetic order with a single-spin magnetometer. Nature. 2017;549:252–6.

[116] Sando D, Agbelele A, Rahmedov D, Liu J, Rovillain P, Toulouse C, et al. Crafting the magnonic and spintronic response of $BiFeO_3$ films by epitaxial strain. Nat Mater. 2013;12:641–6.

[117] Martin LW, Chu YH, Holcomb MB, Huijben M, Yu P, Han SJ, et al. Nanoscale control of exchange bias with $BiFeO_3$ thin films. Nano Lett. 2008;8:2050–5.

[118] Huang Y-L, Nikonov D, Addiego C, Chopdekar RV, Prasad B, Zhang L, et al. Manipulating magnetoelectric energy landscape in multiferroics. Nat Commun. 2020;11:2836.

[119] Wang JJ, Hu JM, Ma J, Zhang JX, Chen LQ, Nan CW. Full 180° magnetization reversal with electric fields. Sci Rep. 2014;4:7507.

[120] Hu J-M, Yang T, Wang J, Huang H, Zhang J, Chen L-Q, et al. Purely electric-field-driven perpendicular magnetization reversal. Nano Lett. 2015;15:616–22.

[121] Imamura H, Nozaki T, Yuasa S, Suzuki Y. Deterministic magnetization switching by voltage control of magnetic anisotropy and dzyaloshinskii-moriya interaction under an in-plane magnetic field. Phys Rev Appl. 2018;10:54039.

[122] Zavaliche F, Zhao T, Zheng H, Straub F, Cruz MP, Yang P-L, et al. Electrically assisted magnetic recording in multiferroic nanostructures. Nano Lett. 2007;7:1586–90.

[123] Skumryev V, Laukhin V, Fina I, Martí X, Sánchez F, Gospodinov M, et al. Magnetization reversal by electric-field decoupling of magnetic and ferroelectric domain walls in multiferroic-based heterostructures. Phys Rev Lett. 2011;106:057206.

[124] Kani N, Heron JT, Naeemi A. Strain-mediated magnetization reversal through spin-transfer torque. IEEE Trans Magn. 2017;53:4300808.

[125] Roy K, Bandyopadhyay S, Atulasimha J. Hybrid spintronics and straintronics: A magnetic technology for ultra low energy computing and signal processing. Appl Phys Lett. 2011;99:63108.
[126] Morelli A, Johann F, Burns SR, Douglas A, Gregg JM. Deterministic switching in bismuth ferrite nanoislands. Nano Lett. 2016;16:5228–34.
[127] Prasad B, Huang Y-L, Chopdekar RV, Chen Z, Steffes J, Das S, et al. Ultralow voltage manipulation of ferromagnetism. Adv Mater. 2020;32:2001943.
[128] Parsonnet E, Huang Y-L, Gosavi T, Qualls A, Nikonov D, Lin -C-C, et al. Toward intrinsic ferroelectric switching in multiferroic $BiFeO_3$. Phys Rev Lett. 2020;125:67601.
[129] Nagarajan V, Ganpule CS, Li H, Salamanca-Riba L, Roytburd AL, Williams ED, et al. Control of domain structure of epitaxial $PbZr_{0.2}Ti_{0.8}O_3$ thin films grown on vicinal (001) $SrTiO_3$ substrates. Appl Phys Lett. 2001;79:2805–7.
[130] Chu YH, Cruz MP, Yang CH, Martin LW, Yang PL, Zhang JX, et al. Domain control in multiferroic $BiFeO_3$ through substrate vicinality. Adv Mater. 2007;19:2662–6.
[131] Xu R, Liu S, Grinberg I, Karthik J, Damodaran AR, Rappe AM, et al. Ferroelectric polarization reversal via successive ferroelastic transitions. Nat Mater. 2015;14:79–86.
[132] Chu YH, He Q, Yang CH, Yu P, Martin LW, Shafer P, et al. Nanoscale control of domain architectures in $BiFeO_3$ thin films. Nano Lett. 2009;9:1726–30.
[133] Feigl L, Yudin P, Stolichnov I, Sluka T, Shapovalov K, Mtebwa M, et al. Controlled stripes of ultrafine ferroelectric domains. Nat Commun. 2014;5:4677.
[134] Haeni JH, Irvin P, Chang W, Uecker R, Reiche P, Li YL, et al. Room-temperature ferroelectricity in strained $SrTiO_3$. Nature. 2004;430:758–61.
[135] J H L, Fang L, Vlahos E, Ke X, Jung YW, Kourkoutis LF, et al. A strong ferroelectric ferromagnet created by means of spin-lattice coupling. Nature. 2010;466:954–8.
[136] Becher C, Maurel L, Aschauer U, Lilienblum M, Magén C, Meier D, et al. Strain-induced coupling of electrical polarization and structural defects in $SrMnO_3$ films. Nat Nanotechnol. 2015;10:661–5.
[137] Guo JW, Wang PS, Yuan Y, He Q, Lu JL, Chen TZ, et al. Strain-induced ferroelectricity and spin-lattice coupling in $SrMnO_3$ thin films. Phys Rev B. 2018;97:235135.
[138] Zhang Y, Wang J, Sahoo MPK, Shimada T, Kitamura T. Strain-induced ferroelectricity and lattice coupling in $BaSnO_3$ and $SrSnO_3$. Phys Chem Chem Phys. 2017;19:26047–55.
[139] Sando D, Han MJ, Govinden V, Paull O, Appert F, Carrétéro C, et al. Interfacial strain gradients control nanoscale domain morphology in epitaxial $BiFeO_3$ multiferroic films. Adv Func Mat. 2020;30:2000343.
[140] Agbelele A, Sando D, Toulouse C, Paillard C, Johnson RD, Rüffer R, et al. Strain and magnetic field induced spin-structure transitions in multiferroic $BiFeO_3$. Adv Mater. 2017;29:1602327.
[141] Sando D, Appert F, Xu B, Paull O, Burns SR, Carrétéro C, et al. A magnetic phase diagram for nanoscale epitaxial $BiFeO_3$ films. Appl Phys Rev. 2019;6:041404.
[142] Chen Z, Chen Z, Kuo CY, Tang Y, Dedon LR, Li Q, et al. Complex strain evolution of polar and magnetic order in multiferroic $BiFeO_3$ thin films. Nat Commun. 2018;9:3764.
[143] Waterfield Price N, Vibhakar AM, Johnson RD, Schad J, Saenrang W, Bombardi A, et al. Strain Engineering a Multiferroic Monodomain in Thin-Film $BiFeO_3$. Phys Rev Appl. 2019;11:024035.
[144] Saenrang W, Davidson BA, Maccherozzi F, Podkaminer JP, Irwin J, Johnson RD, et al. Deterministic and robust room-temperature exchange coupling in monodomain multiferroic $BiFeO_3$ heterostructures. Nat Commun. 2017;8:1583.
[145] Fong DD, Stephenson GB, Streiffer SK, Eastman JA, Auciello O, Fuoss PH, et al. Ferroelectricity in ultrathin perovskite films. Science. 2004;304:1650–3.

[146] De Luca G, Strkalj N, Manz S, Bouillet C, Fiebig M, Trassin M. Nanoscale design of polarization in ultrathin ferroelectric heterostructures. Nat Commun. 2017;8:1419.
[147] Choi KJ, Biegalski M, Li YL, Sharan A, Schubert J, Uecker R, et al. Enhancement of ferroelectricity in strained $BaTiO_3$ thin films. Science. 2004;306:1005–9.
[148] Sarott MF, Fiebig M, Trassin M. Tracking ferroelectric domain formation during epitaxial growth of $PbTiO_3$ films. Appl Phys Lett. 2020;117:132901.
[149] Liu G, Chen J, Lichtensteiger C, Triscone JM, Aguado-Puente P, Junquera J, et al. Positive effect of an internal depolarization field in Ultrathin Epitaxial ferroelectric films. Adv Electron Mater. 2016;2:1500288.
[150] Lichtensteiger C, Fernandez-Pena S, Weymann C, Zubko P, Triscone JM. Tuning of the depolarization field and nanodomain structure in ferroelectric thin films. Nano Lett. 2014;14:4205–11.
[151] Liu Y, Lou X, Bibes M, Dkhil B. Effect of a built-in electric field in asymmetric ferroelectric tunnel junctions. Phys Rev B Condens Matter Mater Phys. 2013;88:1–8.
[152] Simmons JG. Intrinsic fields in thin insulating films between dissimilar electrodes. Phys Rev Lett. 1963;10:10–12.
[153] Lichtensteiger C, Weymann C, Fernandez-Pena S, Paruch P, Triscone JM. Built-in voltage in thin ferroelectric $PbTiO_3$ films: the effect of electrostatic boundary conditions. New J Phys. 2016;18:043030.
[154] Balke N, Ramesh R, Yu P. Manipulating ferroelectrics through changes in surface and interface properties. ACS Appl Mater Interfaces. 2017;9:39736–46.
[155] Chen D, Chen Z, He Q, Clarkson JD, Serrao CR, Yadav AK, et al. Interface engineering of domain structures in $BiFeO_3$ thin films. Nano Lett. 2017;17:486–93.
[156] Seidel J, Martin LW, He Q, Zhan Q, Chu YH, Rother A, et al. Conduction at domain walls in oxide multiferroics. Nat Mater. 2009;8:229–34.
[157] Campanini M, Gradauskaite E, Trassin M, Yi D, Yu P, Ramesh R, et al. Imaging and quantification of charged domain walls in $BiFeO_3$. Nanoscale. 2020;12:9186–93.
[158] Tian Y, Wei L, Zhang Q, Huang H, Zhang Y, Zhou H, et al. Water printing of ferroelectric polarization. Nat Commun. 2018;9:3809.
[159] Yu P, Luo W, Yi D, Zhang JX, Rossell MD, Yang C-H, et al. Interface control of bulk ferroelectric polarization. Proc Natl Acad Sci. 2012;109:9710–5.
[160] Yi D, Yu P, Chen YC, Lee HH, He Q, Chu YH, et al. Tailoring Magnetoelectric Coupling in $BiFeO_3/La_{0.7}Sr_{0.3}MnO_3$ Heterostructure through the Interface Engineering. Adv Mater. 2019;30:1806335.
[161] Li L, Jokisaari JR, Zhang Y, Cheng X, Yan X, Heikes C, et al. Control of Domain Structures in Multiferroic Thin Films through Defect Engineering. Adv Mater. 2018;30:1802737.
[162] Puggioni D, Giovannetti G, Rondinelli JM. Polar metals as electrodes to suppress the critical-thickness limit in ferroelectric nanocapacitors. J Appl Phys. 2018;124:174102.
[163] Chopdekar RV, Malik VK, Fraile Rodríguez A, Le Guyader L, Takamura Y, Scholl A, et al. Spatially resolved strain-imprinted magnetic states in an artificial multiferroic. Phys Rev B Condens Matter Mater Phys. 2012;86:014408.
[164] Lahtinen TH, Franke KJ, Dijken SV. Electric-field control of magnetic domain wall motion and local magnetization reversal. Sci Rep. 2012;2:258.
[165] You L, Wang B, Zou X, Lim ZS, Zhou Y, Ding H, et al. Origin of the uniaxial magnetic anisotropy in $La_{0.7}Sr_{0.3}MnO_3$ on stripe-domain $BiFeO_3$. Phys Rev B Condens Matter Mater Phys. 2013;88:1–9.
[166] Streubel R, Köhler D, Schäfer R, Eng LM. Strain-mediated elastic coupling in magnetoelectric nickel/barium-titanate heterostructures. Phys Rev B Condens Matter Mater Phys. 2013;87:054410.

[167] Brandl F, Franke KJ, Lahtinen TH, Van Dijken S, Grundler D. Spin waves in CoFeB on ferroelectric domains combining spin mechanics and magnonics. Solid State Commun. 2014;198:13–17.
[168] Fackler SW, Donahue MJ, Gao T, Nero PN, Cheong SW, Cumings J, et al. Local control of magnetic anisotropy in transcritical permalloy thin films using ferroelectric $BaTiO_3$ domains. Appl Phys Lett. 2014;105:212905.
[169] Wang J, Pesquera D, Mansell R, Van Dijken S, Cowburn RP, Ghidini M, et al. Giant non-volatile magnetoelectric effects via growth anisotropy in $Co_{40}Fe_{40}B_{20}$ films on PMN-PT substrates. Appl Phys Lett. 2019;114:092401.
[170] Phillips LC, Cherifi RO, Ivanovskaya V, Zobelli A, Infante IC, Jacquet E, et al. Local electrical control of magnetic order and orientation by ferroelastic domain arrangements just above room temperature. Sci Rep. 2015;5:10026.
[171] Franke KJ, Lahtinen TH, Van Dijken S. Field tuning of ferromagnetic domain walls on elastically coupled ferroelectric domain boundaries. Phys Rev B Condens Matter Mater Phys. 2012;85:094423.
[172] Kim SH, Choi H, No K, Shin SC. Electric-field induced ferromagnetic domain changes in exchange biased Co-$BiFeO_3$ composites. J Phys D Appl Phys. 2010;43:165001.
[173] Cybart SA, Rossell MD, Wu SM, Zhang JX, Yu P, Ramesh R, et al. Reversible electric control of exchange bias in a multiferroic field-effect device. Nat Mater. 2010;9:756–61.
[174] Laukhin V, Skumryev V, Martí X, Hrabovsky D, Sánchez F, García-Cuenca MV, et al. Electric-field control of exchange bias in multiferroic epitaxial heterostructures. Phys Rev Lett. 2006;97:227201.
[175] Duan CG, Jaswal SS, Tsymbal EY. Predicted magnetoelectric effect in Fe/$BaTiO_3$ multilayers: ferroelectric control of magnetism. Phys Rev Lett. 2006;97:047201.
[176] Rondinelli JM, Stengel M, Spaldin NA. Carrier-mediated magnetoelectricity in complex oxide heterostructures. Nat Nanotechnol. 2008;3:46–50.
[177] Molegraaf HJ, Hoffman J, Vaz CA, Gariglio S, van der Marel D, Ahn CH, et al. Magnetoelectric effects in complex oxides with competing ground states. Adv Mater. 2009;21:3470–4.
[178] Cai T, Ju S, Lee J, Sai N, Demkov AA, Niu Q, et al. Magnetoelectric coupling and electric control of magnetization in ferromagnet/ferroelectric/normal-metal superlattices. Phys Rev B Condens Matter Mater Phys. 2009;80:140415(R).
[179] Brivio S, Cantoni M, Petti D, Bertacco R. Near-room-temperature control of magnetization in field effect devices based on $La_{0.67}Sr_{0.33}MnO_3$ thin films. J Appl Phys. 2010;108:113906.
[180] Spurgeon SR, Balachandran PV, Kepaptsoglou DM, Damodaran AR, Karthik J, Nejati S, et al. Polarization screening-induced magnetic phase gradients at complex oxide interfaces. Nat Commun. 2015;6:6735.
[181] Avula SR, Heidler J, Dreiser J, Vijayakumar J, Howald L, Nolting F, et al. Study of magneto-electric coupling between ultra-thin Fe films and PMN-PT using X-ray magnetic circular dichroism. J Appl Phys. 2018;123:064103.
[182] Hong X, Posadas A, Ahn CH. Examining the screening limit of field effect devices via the metal-insulator transition. Appl Phys Lett. 2005;86:142501.
[183] De Luca G, Schoenherr P, Mendil J, Meier D, Fiebig M, Trassin M. Domain-pattern transfer across an artificial magnetoelectric interface. Phys Rev Appl. 2018;10:054030.
[184] Ghidini M, Mansell R, Maccherozzi F, Moya X, Phillips LC, Yan W, et al. Shear-strain-mediated magnetoelectric effects revealed by imaging. Nat Mater. 2019;18:840–5.
[185] Leo N, Carolus V, White JS, Kenzelmann M, Hudl M, Tolédano P, et al. Magnetoelectric inversion of domain patterns. Nature. 2018;560:466–70.
[186] Franke KJ, Van de Wiele B, Shirahata Y, Hämälainen SJ, Taniyama T, van Dijken S. Reversible electric-field-driven magnetic domain-wall motion. Phys Rev X. 2015;**5**:011010.

[187] Gao Y, Hu JM, Wu L, Nan CW. Dynamic in situ visualization of voltage-driven magnetic domain evolution in multiferroic heterostructures. J Phys Condens Matter. 2015;27:504005.
[188] López González D, Shirahata Y, Van de Wiele B, Franke KJ, Casiraghi A, Taniyama T, et al. Electric-field-driven domain wall dynamics in perpendicularly magnetized multilayers. AIP Adv. 2017;7:035119.
[189] Van De Wiele B, Leliaert J, Franke KJ, Van Dijken S. Electric-field-driven dynamics of magnetic domain walls in magnetic nanowires patterned on ferroelectric domains. New J Phys. 2016;18:033027.
[190] Hassanpour E, Weber MC, Bortis A, Tokunaga Y, Taguchi Y, Tokura Y, et al. Interconversion of multiferroic domains and domain walls. 2019:arXiv: 1908.06876.
[191] Meier D. Functional domain walls in multiferroics. J Phys Condens Matter. 2015;27:463003.
[192] Gruverman A, Alexe M, Meier D. Piezoresponse force microscopy and nanoferroic phenomena. Nat Commun. 2019;10:1661.
[193] Evans DM, Garcia V, Meier D, Bibes M. Domains and domain walls in multiferroics. Phys Sci Rev. 2020;5:20190067.
[194] Nataf GF, Guennou M, Gregg JM, Meier D, Hlinka J, Salje EK, et al. Domain-wall engineering and topological defects in ferroelectric and ferroelastic materials. Nat Rev Phys. 2020;2:634–648.
[195] Whyte JR, McQuaid RG, Sharma P, Canalias C, Scott JF, Gruverman A, et al. Ferroelectric domain wall injection. Adv Mater. 2014;26:293–8.
[196] McQuaid RG, Campbell MP, Whatmore RW, Kumar A, Marty Gregg J. Injection and controlled motion of conducting domain walls in improper ferroelectric Cu-Cl boracite. Nat Commun. 2017;8:15105.
[197] Sharma P, Sando D, Zhang Q, Cheng X, Prosandeev S, Bulanadi R, et al. Conformational domain wall switch. Adv Funct Mater. 2019;29:1807523.
[198] Whyte JR, Gregg JM. A diode for ferroelectric domain-wall motion. Nat Commun. 2015;6:7361.
[199] McGilly LJ, Yudin P, Feigl L, Tagantsev AK, Setter N. Controlling domain wall motion in ferroelectric thin films. Nat Nanotechnol. 2015;10:145–50.
[200] Guo EJ, Roth R, Herklotz A, Hesse D, Dörr K. Ferroelectric 180° domain wall motion controlled by biaxial strain. Adv Mater. 2015;27:1615–18.
[201] De Luca G, Rossell MD, Schaab J, Viart N, Fiebig M, Trassin M. Domain wall architecture in tetragonal ferroelectric thin films. Adv Mater. 2017;29:1605145.
[202] Cherifi-Hertel S, Bulou H, Hertel R, Taupier G, Dorkenoo KDH, Andreas C, et al. Non-Ising and chiral ferroelectric domain walls revealed by nonlinear optical microscopy. Nat Commun. 2017;8:15768.
[203] Nataf GF, Barrett N, Kreisel J, Guennou M. Raman signatures of ferroic domain walls captured by principal component analysis. J Phys Condens Matter. 2018;30:035902.
[204] Hadjimichael M, Zatterin E, Fernandez-Peña S, Leake SJ, Zubko P. Domain wall orientations in ferroelectric superlattices probed with synchrotron X-ray diffraction. Phys Rev Lett. 2018;120:037602.
[205] Li D, Huang X, Xiao Z, Chen H, Zhang L, Hao Y, et al. Polar coupling enabled nonlinear optical filtering at MoS_2/ferroelectric heterointerfaces. Nat Commun. 2020;11:1422.
[206] Meier D, Maringer M, Lottermoser T, Becker P, Bohatý L, Fiebig M. Observation and coupling of domains in a spin-spiral multiferroic. Phys Rev Lett. 2009;102:107202.
[207] Leo N, Bergman A, Cano A, Poudel N, Lorenz B, Fiebig M, et al. Polarization control at spin-driven ferroelectric domain walls. Nat Commun. 2015;6:6661.
[208] Tokunaga Y, Furukawa N, Sakai H, Taguchi Y, Arima TH, Tokura Y. Composite domain walls in a multiferroic perovskite ferrite. Nat Mater. 2009;8:558–62.
[209] Tokunaga Y, Taguchi Y, Arima TH, Tokura Y. Electric-field-induced generation and reversal of ferromagnetic moment in ferrites. Nat Phys. 2012;8:838–44.

[210] Matsubara M, Manz S, Mochizuki M, Kubacka T, Iyama A, Aliouane N, et al. Magnetoelectric domain control in multiferroic $TbMnO_3$. Science. 2015;348:1112–15.
[211] Geng Y, Lee N, Choi YJ, Cheong SW, Wu W. Collective magnetism at multiferroic vortex domain walls. Nano Lett. 2012;12:6055–9.
[212] Xu T, Shimada T, Araki Y, Wang J, Kitamura T. Multiferroic domain walls in ferroelectric $PbTiO_3$ with oxygen deficiency. Nano Lett. 2016;16:454–8.
[213] Farokhipoor S, Magén C, Venkatesan S, Íñiguez J, Daumont CJ, Rubi D, et al. Artificial chemical and magnetic structure at the domain walls of an epitaxial oxide . Nature. 2014;515:379–83.
[214] Logginov AS, Meshkov GA, Nikolaev AV, Nikolaeva EP, Pyatakov AP, Zvezdin AK. Room temperature magnetoelectric control of micromagnetic structure in iron garnet films. Appl Phys Lett. 2008;93:182510.
[215] Fontcuberta J, Skumryev V, Laukhin V, Granados X, Salje EK. Polar domain walls trigger magnetoelectric coupling. Sci Rep. 2015;5:13784.
[216] Salje EK, Ding X. Ferroelastic domain boundary-based multiferroicity. Crystals. 2016;6:163.
[217] Nataf GF, Guennou M, Kreisel J, Hicher P, Haumont R, Aktas O, et al. Control of surface potential at polar domain walls in a nonpolar oxide. Phys Rev Mater. 2017;1:074410.
[218] Wei XK, Sluka T, Fraygola B, Feigl L, Du H, Jin L, et al. Controlled charging of Ferroelastic domain walls in oxide ferroelectrics. ACS Appl Mater Interfaces. 2017;9:6539–46.
[219] Burton JD, Tsymbal EY. Prediction of electrically induced magnetic reconstruction at the manganite/ferroelectric interface. Phys Rev B Condens Matter Mater Phys. 2009;80:174406.
[220] McNulty JA, Tran TT, Halasyamani PS, McCartan SJ, MacLaren I, Gibbs AS, et al. An electronically driven improper Ferroelectric: tungsten bronzes as microstructural analogs for the hexagonal manganites. Adv Mater. 2019;31:1903620.
[221] Moure A. Review and perspectives of Aurivillius structures as a lead-free piezoelectric system. Appl Sci. 2018;8:62.
[222] Birenbaum AY, Ederer C. Potentially multiferroic Aurivillius phase $Bi_5FeTi_3O_{15}$: Cation site preference, electric polarization, and magnetic coupling from first principles. Phys Rev B. 2014;90:214109.
[223] Zhai X, Grutter AJ, Yun Y, Cui Z, Lu Y. Weak magnetism of Aurivillius-type multiferroic thin films probed by polarized neutron reflectivity. Phys Rev Mater. 2018;2:044405.
[224] Yang J, Zhu X, Wang F, Song D, Chen L, Yang B. Evolution of structure and ferroelectricity in Aurivillius $Bi_4Bi_{n-3}Fe_{n-3}Ti_3O_{3n+3}$ thin films. J Mater Chem C. 2018;6:8618–27.
[225] Djani H, Garcia-Castro AC, Tong WY, Barone P, Bousquet E, Picozzi S, et al. Rationalizing and engineering Rashba spin-splitting in ferroelectric oxides. Npj Quantum Mater. 2019;4:51.
[226] Campanini M, Trassin M, Ederer C, Erni R, Rossell MD. Buried in-plane ferroelectric domains in Fe-Doped single-crystalline aurivillius thin films. ACS Appl Electron Mater. 2019;1:1019–28.
[227] Zhu L, Ralph DC, Buhrman RA. Highly efficient spin-current generation by the spin hall effect in $Au_{1-x}Pt_x$. Phys Rev Appl. 2018;10:31001.
[228] Zhu L, Zhu L, Shi S, Ralph DC, Buhrman RA. Energy-efficient ultrafast SOT-MRAMs based on low-resistivity spin hall metal $Au_{0.25}Pt_{0.75}$. Adv Electron Mater. 2020;6:1–7.
[229] Heron JT 2013 Electric field control of ferromagnetism and magnetic devices using multiferroics *PhD Thesis, Univ. California, Berkeley*
[230] Manipatruni S, Nikonov DE, Lin -C-C, Prasad B, Huang Y-L, Damodaran AR, et al. Voltage control of unidirectional anisotropy in ferromagnet-multiferroic system. Sci Adv. 2018;4: eaat4229.
[231] Binek C, Doudin B. Magnetoelectronics with magnetoelectrics. J Phys Condens Matter. 2005;17:L39–44.

[232] Dowben PA, Binek C, Zhang K, Wang L, Mei WN, Bird JP, et al. Towards a strong spin-orbit coupling magnetoelectric transistor. IEEE J Explor Solid-State Comput Devices Circuits. 2018;4:1–9.
[233] Chang S.-C. Inversion Charge Boost and Transient Steep-Slope Induced by Free-Charge-Polarization Mismatch in a Ferroelectric-Metal–Oxide–Semiconductor Capacitor. IEEE J Explor Solid-State Comput Devices Circuits. 2018;4:44–49.
[234] Pan C, Lou Q, Niemier M, Hu S, Naeemi A. Energy-efficient convolutional neural network based on cellular neural network using beyond-CMOS technologies. IEEE J Explor Solid-State Comput Devices Circuits. 2019;5:85–93.
[235] Stephan AW, Hu J, Koester SJ. Performance estimate of inverse Rashba-Edelstein magnetoelectric devices for neuromorphic computing. IEEE J Explor Solid-State Comput Devices Circuits. 2019;5:25–33.
[236] Lesne E, Fu Y, Oyarzun S, Rojas-Sánchez JC, Vaz DC, Naganuma H, et al. Highly efficient and tunable spin-to-charge conversion through Rashba coupling at oxide interfaces. Nat Mater. 2016;15:1261–6.
[237] Yu G, Upadhyaya P, Fan Y, Alzate JG, Jiang W, Wong KL, et al. Switching of perpendicular magnetization by spin–orbit torques in the absence of external magnetic fields. Nat Nanotechnol. 2014;9:548–54.
[238] Wang Z, Li Z, Wang M, Wu B, Zhu D, Zhao W. Field-free spin–orbit-torque switching of perpendicular magnetization aided by uniaxial shape anisotropy. Nanotechnology. 2019;30:375202.
[239] Lau Y-C, Betto D, Rode K, Coey JMD, Stamenov P. Spin–orbit torque switching without an external field using interlayer exchange coupling. Nat Nanotechnol. 2016;11:758–62.
[240] Van Den Brink A, Vermijs G, Solignac A, Koo J, Kohlhepp JT, Swagten HJ, et al. Field-free magnetization reversal by spin-hall effect and exchange bias. Nat Commun. 2016;7:1–6.
[241] Cai K, Yang M, Ju H, Wang S, Ji Y, Li B, et al. Electric field control of deterministic current-induced magnetization switching in a hybrid ferromagnetic/ferroelectric structure. Nat Mater. 2017;16:712–6.
[242] Nan T, Hu J-M, Dai M, Emori S, Wang X, Hu Z, et al. A strain-mediated magnetoelectric-spin-torque hybrid structure. Adv Funct Mater. 2019;29:1806371.
[243] Laanait N, Saenrang W, Zhou H, Eom CB, Zhang Z. Dynamic X-ray diffraction imaging of the ferroelectric response in bismuth ferrite. Adv Struct Chem Imaging. 2017;3:11.
[244] Liu L, Rojac T, Damjanovic D, Di Michiel M, Daniels J. Frequency-dependent decoupling of domain-wall motion and lattice strain in bismuth ferrite. Nat Commun. 2018;9:1–10.
[245] Matveyev Y, Negrov D, Chernikova A, Lebedinskii Y, Kirtaev R, Zarubin S, et al. Effect of polarization reversal in ferroelectric $TiN/Hf_{0.5}Zr_{0.5}O_2/TiN$ devices on electronic conditions at interfaces studied in operando by hard X-ray photoemission spectroscopy. ACS Appl Mater Interfaces. 2017;9:43370–6.
[246] Nordlander J, De Luca G, Strkalj N, Fiebig M, Trassin M. Probing ferroic states in oxide thin films using optical second harmonic generation. Appl Sci. 2018;8:570.
[247] Gorfman S, Bokov AA, Davtyan A, Reiser M, Xie Y, Ye ZG, et al. Ferroelectric domain wall dynamics characterized with X-ray photon correlation spectroscopy. Proc Natl Acad Sci U S A. 2018;115:E6680–9.
[248] Agar JC, Naul B, Pandya S, van der Walt S, Maher J, Ren Y, et al. Revealing ferroelectric switching character using deep recurrent neural networks. Nat Commun. 2019;10:1–11.
[249] Degen CL. Scanning magnetic field microscope with a diamond single-spin sensor. Appl Phys Lett. 2008;92:24311.
[250] Maze JR, Stanwix PL, Hodges JS, Hong S, Taylor JM, Cappellaro P, et al. Nanoscale magnetic sensing with an individual electronic spin in diamond. Nature. 2008;455:644–7.

[251] Rondin L, Tetienne JP, Hingant T, Roch JF, Maletinsky P, Jacques V. Magnetometry with nitrogen-vacancy defects in diamond. Reports Prog Phys. 2014;77:056503.
[252] Gross I, Akhtar W, Garcia V, Martínez LJ, Chouaieb S, Garcia K, et al. Real-space imaging of non-collinear antiferromagnetic order with a single-spin magnetometer. Nature. 2017;549:252.
[253] Geng Y, Das H, Wysocki AL, Wang X, Cheong S-W, Mostovoy M, et al. Direct visualization of magnetoelectric domains. Nat Mater. 2014;13:163–7.
[254] Schoenherr P, Giraldo L, Lilienblum M, Trassin M, Meier D, Fiebig M. Magnetoelectric force microscopy on antiferromagnetic 180° domains in Cr_2O_3. Materials. 2017;10:1051.
[255] Fiebig M, Fröhlich D, Sluyterman G, Pisarev RLV. Domain topography of antiferromagnetic Cr_2O_3 by second-harmonic generation. Appl Phys Lett. 1995;66:2906.
[256] Chauleau JY, Haltz E, Carrétéro C, Fusil S, Viret M. Multi-stimuli manipulation of antiferromagnetic domains assessed by second-harmonic imaging. Nat Mater. 2017;16:803–7.
[257] Denev SA, Lummen TT, Barnes E, Kumar A, Gopalan V. Probing ferroelectrics using optical second harmonic generation. J Am Ceram Soc. 2011;94:2699–727.
[258] Fiebig M, Pavlov VV, Pisarev RV. Second-harmonic generation as a tool for studying electronic and magnetic structures of crystals: review. J Opt Soc Am B. 2005;22:96.
[259] Tzschaschel C, Satoh T, Fiebig M. Tracking the ultrafast motion of an antiferromagnetic order parameter. Nat Commun. 2019;10:1–6.
[260] Stanciu CD, Hansteen F, Kimel AV, Kirilyuk A, Tsukamoto A, Itoh A, et al. All-optical magnetic recording with circularly polarized light. Phys Rev Lett. 2007;99:1–4.
[261] Lambert CH, Mangin S, Varaprasad BS, Takahashi YK, Hehn M, Cinchetti M, et al. All-optical control of ferromagnetic thin films and nanostructures. Science. 2014;345:1337–40.
[262] Manz S, Matsubara M, Lottermoser T, Büchi J, Iyama A, Kimura T, et al. Reversible optical switching of antiferromagnetism in $TbMnO_3$. Nat Photonics. 2016;10:653–6.
[263] Liou YD, Chiu YY, Hart RT, Kuo CY, Huang YL, Wu YC, et al. Deterministic optical control of room temperature multiferroicity in $BiFeO_3$ thin films. Nat Mater. 2019;18:580–7.
[264] Yang MM, Alexe M. Light-induced reversible control of ferroelectric polarization in $BiFeO_3$. Adv Mater. 2018;30:1704908.
[265] Kundys B, Viret M, Colson D, Kundys DO. Light-induced size changes in $BiFeO_3$ crystals. Nat Mater. 2010;9:803–5.
[266] Vats G, Bai Y, Zhang D, Juuti J, Seidel J. Optical control of ferroelectric domains: nanoscale insight into macroscopic observations. Adv Opt Mater. 2019;7:1800858.
[267] Rubio-Marcos F, Ochoa DA, Del Campo A, García MA, Castro GR, Fernández JF, et al. Reversible optical control of macroscopic polarization in ferroelectrics. Nat Photonics. 2018;12:29–32.
[268] Stoica VA, Laanait N, Dai C, Hong Z, Yuan Y, Zhang Z, et al. Optical creation of a supercrystal with three-dimensional nanoscale periodicity. Nat Mater. 2019;18:377–83.
[269] Yang M, Kim DJ, Alexe M. Flexo-photovoltaic effect. Science. 2018;907:904–7.
[270] Chu K, Jang BK, Sung JH, Shin YA, Lee ES, Song K, et al. Enhancement of the anisotropic photocurrent in ferroelectric oxides by strain gradients. Nat Nanotechnol. 2015;10:972–9.
[271] Matzen S, Fusil S. Domains and domain walls in multiferroics. Comptes Rendus Phys. 2015;16:227–40.

Index

https://doi.org/10.1515/9783110582130-013

Bei Fragen zur Produktsicherheit wenden Sie sich bitte an:
If you have any questions regarding product safety,
please contact:

Walter de Gruyter GmbH
Genthiner Straße 13
10785 Berlin
productsafety@degruyterbrill.com